Graduate Texts in Mathematics 131

Springer
New York
Berlin
Heidelberg
Barcelona
Hong Kong
London
Milan
Paris
Singapore
Tokyo

Graduate Texts in Mathematics

1 TAKEUTI/ZARING. Introduction to Axiomatic Set Theory. 2nd ed.
2 OXTOBY. Measure and Category. 2nd ed.
3 SCHAEFER. Topological Vector Spaces. 2nd ed.
4 HILTON/STAMMBACH. A Course in Homological Algebra. 2nd ed.
5 MAC LANE. Categories for the Working Mathematician. 2nd ed.
6 HUGHES/PIPER. Projective Planes.
7 SERRE. A Course in Arithmetic.
8 TAKEUTI/ZARING. Axiomatic Set Theory.
9 HUMPHREYS. Introduction to Lie Algebras and Representation Theory.
10 COHEN. A Course in Simple Homotopy Theory.
11 CONWAY. Functions of One Complex Variable I. 2nd ed.
12 BEALS. Advanced Mathematical Analysis.
13 ANDERSON/FULLER. Rings and Categories of Modules. 2nd ed.
14 GOLUBITSKY/GUILLEMIN. Stable Mappings and Their Singularities.
15 BERBERIAN. Lectures in Functional Analysis and Operator Theory.
16 WINTER. The Structure of Fields.
17 ROSENBLATT. Random Processes. 2nd ed.
18 HALMOS. Measure Theory.
19 HALMOS. A Hilbert Space Problem Book. 2nd ed.
20 HUSEMOLLER. Fibre Bundles. 3rd ed.
21 HUMPHREYS. Linear Algebraic Groups.
22 BARNES/MACK. An Algebraic Introduction to Mathematical Logic.
23 GREUB. Linear Algebra. 4th ed.
24 HOLMES. Geometric Functional Analysis and Its Applications.
25 HEWITT/STROMBERG. Real and Abstract Analysis.
26 MANES. Algebraic Theories.
27 KELLEY. General Topology.
28 ZARISKI/SAMUEL. Commutative Algebra. Vol.I.
29 ZARISKI/SAMUEL. Commutative Algebra. Vol.II.
30 JACOBSON. Lectures in Abstract Algebra I. Basic Concepts.
31 JACOBSON. Lectures in Abstract Algebra II. Linear Algebra.
32 JACOBSON. Lectures in Abstract Algebra III. Theory of Fields and Galois Theory.
33 HIRSCH. Differential Topology.
34 SPITZER. Principles of Random Walk. 2nd ed.
35 ALEXANDER/WERMER. Several Complex Variables and Banach Algebras. 3rd ed.
36 KELLEY/NAMIOKA et al. Linear Topological Spaces.
37 MONK. Mathematical Logic.
38 GRAUERT/FRITZSCHE. Several Complex Variables.
39 ARVESON. An Invitation to C^*-Algebras.
40 KEMENY/SNELL/KNAPP. Denumerable Markov Chains. 2nd ed.
41 APOSTOL. Modular Functions and Dirichlet Series in Number Theory. 2nd ed.
42 SERRE. Linear Representations of Finite Groups.
43 GILLMAN/JERISON. Rings of Continuous Functions.
44 KENDIG. Elementary Algebraic Geometry.
45 LOÈVE. Probability Theory I. 4th ed.
46 LOÈVE. Probability Theory II. 4th ed.
47 MOISE. Geometric Topology in Dimensions 2 and 3.
48 SACHS/WU. General Relativity for Mathematicians.
49 GRUENBERG/WEIR. Linear Geometry. 2nd ed.
50 EDWARDS. Fermat's Last Theorem.
51 KLINGENBERG. A Course in Differential Geometry.
52 HARTSHORNE. Algebraic Geometry.
53 MANIN. A Course in Mathematical Logic.
54 GRAVER/WATKINS. Combinatorics with Emphasis on the Theory of Graphs.
55 BROWN/PEARCY. Introduction to Operator Theory I: Elements of Functional Analysis.
56 MASSEY. Algebraic Topology: An Introduction.
57 CROWELL/FOX. Introduction to Knot Theory.
58 KOBLITZ. p-adic Numbers, p-adic Analysis, and Zeta-Functions. 2nd ed.
59 LANG. Cyclotomic Fields.
60 ARNOLD. Mathematical Methods in Classical Mechanics. 2nd ed.
61 WHITEHEAD. Elements of Homotopy Theory.
62 KARGAPOLOV/MERLZJAKOV. Fundamentals of the Theory of Groups.
63 BOLLOBAS. Graph Theory.
64 EDWARDS. Fourier Series. Vol. I. 2nd ed.
65 WELLS. Differential Analysis on Complex Manifolds. 2nd ed.

(continued after index)

T.Y. Lam

A First Course in Noncommutative Rings

Second Edition

Springer

T.Y. Lam
Department of Mathematics
University of California, Berkeley
Berkeley, CA 94720-0001

Mathematics Subject Classification (2000): 16-01, 16D10, 16D30, 16D60

Library of Congress Cataloging-in-Publication Data
Lam, T.Y. (Tsit-Yuen), 1942–
A first course in noncommutative rings / T.Y. Lam. — 2nd ed.
p. cm. — (Graduate texts in mathematics ; 131)
Includes bibliographical references and index.
ISBN 0-387-95183-0 (hc; alk. paper)
ISBN 0-387-95325-6 (sc; alk. paper)
1. Noncommutative rings. I. Title. II. Series.
QA251.4 .L36 2001
512′.4—dc21 00-052277

Printed on acid-free paper.

Production managed by Terry Kornak; manufacturing supervised by Jerome Basma.
Typeset by Asco Typesetters, North Point, Hong Kong.
Printed and bound by Maple-Vail Book Manufacturing Group, York, PA.
Printed in the United States of America.

9 8 7 6 5 4 3 2 1

ISBN 0-387-95183-0 SPIN 10789436 (hardcover)
ISBN 0-387-95325-6 SPIN 10842006 (softcover)

Springer-Verlag New York Berlin Heidelberg
A member of BertelsmannSpringer Science+Business Media GmbH

To Juwen, Fumei, Juleen, and Dee-Dee

my most delightful ring

Preface to the Second Edition

The wonderful reception given to the first edition of this book by the mathematical community was encouraging. It gives me much pleasure to bring out now a new edition, exactly ten years after the book first appeared.

In the 1990s, two related projects have been completed. The first is the problem book for "*First Course*" (Lam [95]), which contains the solutions of (and commentaries on) the original 329 exercises and 71 additional ones. The second is the intended "sequel" to this book (once called "*Second Course*"), which has now appeared under the different title "*Lectures on Modules and Rings*" (Lam [98]). These two other books will be useful companion volumes for this one. In the present book, occasional references are made to "*Lectures*", but the former has no logical dependence on the latter. In fact, all three books can be used essentially independently.

In this new edition of "*First Course*", the entire text has been retyped, some proofs were rewritten, and numerous improvements in the exposition have been included. The original chapters and sections have remained unchanged, with the exception of the addition of an Appendix (on uniserial modules) to §20. All known typographical errors were corrected (although no doubt a few new ones have been introduced in the process!). The original exercises in the first edition have been replaced by the 400 exercises in the problem book (Lam [95]), and I have added at least 30 more in this edition for the convenience of the reader. As before, the book should be suitable as a text for a one-semester or a full-year graduate course in noncommutative ring theory.

I take this opportunity to thank heartily all of my students, colleagues, and other users of "*First Course*" all over the world for sending in corrections on the first edition, and for communicating to me their thoughts on possible improvements in the text. Most of their suggestions have been followed in this new edition. Needless to say, I will continue to welcome such feedback from my readers, which can be sent to me by email at the address "lam@math.berkeley.edu".

T.Y.L.

Berkeley, California
01/01/01

Preface to the First Edition

One of my favorite graduate courses at Berkeley is Math 251, a one-semester course in ring theory offered to second-year level graduate students. I taught this course in the Fall of 1983, and more recently in the Spring of 1990, both times focusing on the theory of noncommutative rings. This book is an outgrowth of my lectures in these two courses, and is intended for use by instructors and graduate students in a similar one-semester course in basic ring theory.

Ring theory is a subject of central importance in algebra. Historically, some of the major discoveries in ring theory have helped shape the course of development of modern abstract algebra. Today, ring theory is a fertile meeting ground for group theory (group rings), representation theory (modules), functional analysis (operator algebras), Lie theory (enveloping algebras), algebraic geometry (finitely generated algebras, differential operators, invariant theory), arithmetic (orders, Brauer groups), universal algebra (varieties of rings), and homological algebra (cohomology of rings, projective modules, Grothendieck and higher K-groups). In view of these basic connections between ring theory and other branches of mathematics, it is perhaps no exaggeration to say that a course in ring theory is an indispensable part of the education for any fledgling algebraist.

The purpose of my lectures was to give a general introduction to the theory of rings, building on what the students have learned from a standard first-year graduate course in abstract algebra. We assume that, from such a course, the students would have been exposed to tensor products, chain conditions, some module theory, and a certain amount of commutative algebra. Starting with these prerequisites, I designed a course dealing almost exclusively with the theory of noncommutative rings. In accordance with the historical development of the subject, the course begins with the Wedderburn–Artin theory of semisimple rings, then goes on to Jacobson's general theory of the radical for rings possibly not satisfying any chain conditions. After an excursion into representation theory in the style of Emmy Noether, the course continues with the study of prime and semiprime rings, primitive and semiprimitive rings, division rings, ordered rings, local and semilocal rings, and finally, perfect and semiperfect rings. This material, which was as much as I managed to cover in a one-semester course, appears here in a somewhat expanded form as the eight chapters of this book.

Of course, the topics described above correspond only to part of the foundations of ring theory. After my course in Fall, 1983, a self-selected group of students from this course went on to take with me a second course (Math 274, Topics in Algebra), in which I taught some further basic topics in the subject. The notes for this second course, at present only partly written, will hopefully also appear in the future, as a sequel to the present work. This intended second volume will cover, among other things, the theory of modules, rings of quotients and Goldie's Theorem, noetherian rings, rings with polynomial identities, Brauer groups and the structure theory of finite-dimensional central simple algebras. The reasons for publishing the present volume first are two-fold: first it will give me the opportunity to class-test the second volume some more before it goes to press, and secondly, since the present volume is entirely self-contained and technically indepedent of what comes after, I believe it is of sufficient interest and merit to stand on its own.

Every author of a textbook in mathematics is faced with the inevitable challenge to do things differently from other authors who have written earlier on the same subject. But no doubt the number of available proofs for any given theorem is finite, and by definition the best approach to any specific body of mathematical knowledge is unique. Thus, no matter how hard an author strives to appear original, it is difficult for him to avoid a certain degree of "plagiarism" in the writing of a text. In the present case I am all the more painfully aware of this since the path to basic ring theory is so well-trodden, and so many good books have been written on the subject. If, of necessity, I have to borrow so heavily from these earlier books, what are the new features of this one to justify its existence?

In answer to this, I might offer the following comments. Although a good number of books have been written on ring theory, many of them are monographs devoted to specialized topics (e.g., group rings, division rings, noetherian rings, von Neumann regular rings, or module theory, PI-theory, radical theory, loalization theory). A few others offer general surveys of the subject, and are encyclopedic in nature. If an instructor tries to look for an introductory graduate text for a one-semester (or two-semester) course in ring theory, the choices are still surprisingly few. It is hoped, therefore, that the present book (and its sequel) will add to this choice. By aiming the level of writing at the novice rather than the connoisseur, we have sought to produce a text which is suitable not only for use in a graduate course, but also for self-study in the subject by interested graduate students.

Since this book is a by-product of my lectures, it certainly reflects much more on my teaching style and my personal taste in ring theory than on ring theory itself. In a graduate course one has only a limited number of lectures at one's disposal, so there is the need to "get to the point" as quickly as possible in the presentation of any material. This perhaps explains the often business-like style in the resulting lecture notes appearing here. Nevertheless, we are fully cognizant of the importance of motivation and examples, and we have tried hard to ensure that they don't play second fiddle to theorems and proofs. As far as the choice of the material is concerned, we have perhaps given more than the usual emphasis to a few of the famous open problems in ring theory, for instance, the Köthe Conjecture for rings with zero upper nilradical (§10), the semiprimitivity problem and the zero-divisor problem for group rings (§6), etc. The fact that these natural and very easily stated problems have remained unsolved for so long seemed to have captured the students' imagination. A few other possibly "unusual" topics are included in the text: for instance, noncommutative ordered rings are treated in §17, and a detailed exposition of the Mal'cev–Neumann construction of general Laurent series rings is given in §14. Such material is not easily available in standard textbooks on ring theory, so we hope its inclusion here will be a useful addition to the literature.

There are altogether twenty five sections in this book, which are consecutively numbered independently of the chapters. Results in Section x will be labeled in the form (x.y). Each section is equipped with a collection of exercises at the end. In almost all cases, the exercises are perfectly "doable" problems which build on the text material in the same section. Some exercises are accompanied by copious hints; however, the more self-reliant readers should not feel obliged to use these.

As I have mentioned before, in writing up these lecture notes I have consulted extensively the existing books on ring theory, and drawn material from them freely. Thus I owe a great literary debt to many earlier authors in the field. My graduate classes in Fall 1983 and Spring 1990 at Berkeley were attended by many excellent students; their enthusiasm for ring theory made the class a joy to teach, and their vigilance has helped save me from many slips. I take this opportunity to express my appreciation for the role they played in making these notes possible. A number of friends and colleagues have given their time generously to help me with the manuscript. It is my great pleasure to thank especially Detlev Hoffmann, André Leroy, Ka-Hin Leung, Mike May, Dan Shapiro, Tara Smith and Jean-Pierre Tignol for their valuable comments, suggestions, and corrections. Of course, the responsibility for any flaws or inaccuracies in the exposition remains my own. As mathematics editor at Springer-Verlag, Ulrike Schmickler-Hirzebruch has been most understanding of an author's plight, and deserves a word of special thanks for bringing this long overdue project to fruition. Keyboarder Kate MacDougall did an excellent job in transforming my handwritten manuscript into LaTex, and the Production Department's efficient handling of the entire project has been exemplary.

Last, first, and always, I owe the greatest debt to members of my family. My wife Chee-King graciously endured yet another book project, and our four children bring cheers and joy into my life. Whatever inner strength I can muster in my various endeavors is in large measure a result of their love, devotion, and unstinting support.

T.Y.L.

Berkeley, California
November, 1990

Contents

CHAPTER 8
Perfect and Semiperfect Rings 335

Notes to the Reader

As we have explained in the Preface, the twenty five sections in this book are numbered independently of the eight chapters. A cross-reference such as (12.7) refers to the result so labeled in §12. On the other hand, Exercise 12.7 will refer to Exercise 7 appearing at the end of §12. In referring to an exercise appearing (or to appear) in the same section, we shall sometimes drop the section number from the reference. Thus, when we refer to "Exercise 7" anywhere *within* §12, we shall mean Exercise 12.7.

Since this is an exposition and not a treatise, the writing is by no means encyclopedic. In particular, in most places, no systematic attempt is made to give attributions, or to trace the results discussed to their original sources. References to a book or a paper are given only sporadically where they seem more essential to the material under consideration. A reference in brackets such as Amitsur [56] (or [Amitsur: 56]) shall refer to the 1956 paper of Amitsur listed in the reference section at the end of the book.

Occasionally, references will be made to the intended sequel of this book, which will be briefly called *Lectures*. Such references will always be peripheral in nature; their only purpose is to point to material which lies ahead. In particular, no result in this book will depend logically on any result to appear later in *Lectures*.

Throughout the text, we use the standard notations of modern mathematics. For the reader's convenience, a partial list of the notations commonly used in basic algebra and ring theory is given on the following pages.

Some Frequently Used Notations

$\mathbb{Z}$	ring of integers
$\mathbb{Q}$	field of rational numbers
$\mathbb{R}$	field of real numbers
$\mathbb{C}$	field of complex numbers
$\mathbf{F}_q$	finite field with q elements
$\mathbf{M}_n(S)$	set of $n \times n$ matrices with entries from S
$\subset, \subseteq$	used interchangeably for inclusion
$\subsetneq$	strict inclusion
$\vert A\vert$, $Card\ A$	used interchangeably for the cardinality of the set A
$A \backslash B$	set-theoretic difference
$A \twoheadrightarrow B$	surjective mapping from A onto B
δ_{ij}	Kronecker deltas
E_{ij}	matrix units
tr	trace (of a matrix or a field element)
$\langle x \rangle$	cyclic group generated by x
$Z(G)$	center of the group (or the ring) G
$C_G(A)$	centralizer of A in G
$[G : H]$	index of subgroup H in a group G
$[K : F]$	field extension degree
$[K : D]_\ell$, $[K : D]_r$	left, right dimensions of $K \supseteq D$ as D-vector space
K^G	G-fixed points on K
M_R, ${}_RN$	right R-module M, left R-module N
$M \otimes_R N$	tensor product of M_R and ${}_RN$
$Hom_R(M, N)$	group of R-homomorphisms from M to N
$End_R(M)$	ring of R-endomorphisms of M
nM (or M^n)	$M \oplus \cdots \oplus M$ (n times)
$\prod_i R_i$	direct product of the rings $\{R_i\}$
$char\ R$	characteristic of a ring R
$U(R), R^*$	group of units of the ring R
$U(D)$, D^*, $\dot{D}$	multiplicative group of the division ring D
$GL_n(R)$	group of invertible $n \times n$ matrices over R
$GL(V)$	group of linear automorphisms of a vector space V
$rad\ R$	Jacobson radical of R
$Nil^*(R)$	upper nilradical of R
$Nil_*(R)$	lower nilradical (or prime radical) of R
$Nil\ R$	ideal of nilpotent elements in a commutative ring R
$ann_\ell(S)$, $ann_r(S)$	left, right annihilators of the set S
kG, $k[G]$	(semi)group ring of the (semi)group G over the ring k
$k[x_i : i \in I]$	polynomial ring over k with (commuting) variables $\{x_i : i \in I\}$
$k\langle x_i : i \in I \rangle$	free ring over k generated by $\{x_i : i \in I\}$

ACC	ascending chain condition
DCC	descending chain condition
LHS	left-hand side
RHS	right-hand side

CHAPTER 1

Wedderburn–Artin Theory

Modern ring theory began when J.H.M. Wedderburn proved his celebrated classification theorem for finite dimensional semisimple algebras over fields. Twenty years later, E. Noether and E. Artin introduced the Ascending Chain Condition (ACC) and the Descending Chain Condition (DCC) as substitutes for finite dimensionality, and Artin proved the analogue of Wedderburn's Theorem for general semisimple rings. The Wedderburn–Artin theory has since become the cornerstone of noncommutative ring theory, so in this first chapter of our book, it is only fitting that we devote ourselves to an exposition of this basic theory.

In a (possibly noncommutative) ring, we can add, subtract, and multiply elements, but we may not be able to "divide" one element by another. In a very natural sense, the most "perfect" objects in noncommutative ring theory are the *division rings*, i.e. (nonzero) rings in which each nonzero element has an inverse. From division rings, we can build up matrix rings, and form finite direct products of such matrix rings. According to the Wedderburn–Artin Theorem, the rings obtained in this way comprise exactly the all-important class of semisimple rings. This is one of the earliest (and still one of the nicest) complete classification theorems in abstract algebra, and has served for decades as a model for many similar results in the structure theory of rings.

There are several different ways to define semisimplicity. Wedderburn, being interested mainly in finite-dimensional algebras over fields, defined the radical of such an algebra R to be the largest nilpotent ideal of R, and defined R to be *semisimple* if this radical is zero, i.e., if there is no nonzero nilpotent ideal in R. Since we are interested in rings in general, and not just finite-dimensional algebras, we shall follow a somewhat different approach. In this chapter, we define a semisimple ring to be a ring all of whose modules are semisimple, i.e., are sums of simple modules. This module-theoretic definition of semisimple rings is not only easy to work with, but also leads quickly and naturally to the Wedderburn–Artin Theorem on their complete classification. The consideration of the radical is postponed to the next chapter, where the Wedderburn radical for finite-dimensional algebras is generalized to the Jacobson radical for arbitrary rings. With this more general notion of the radical, it will be seen that semisimple rings are exactly the (left or right) artinian rings with a zero (Jacobson) radical.

Before beginning our study of semisimple rings, it is convenient to have a quick review of basic facts and terminology in ring theory, and to look at some illustrative examples. The first section is therefore devoted to this end. The development of the Wedderburn–Artin theory will occupy the rest of the chapter.

§1. Basic Terminology and Examples

In this beginning section, we shall review some of the basic terminology in ring theory and give a good supply of examples of rings. We assume the reader is already familiar with most of the terminology discussed here through a good course in graduate algebra, so we shall move along at a fairly brisk pace.

Throughout the text, the word "ring" means a ring with an identity element 1 which is not necessarily commutative. The study of commutative rings constitutes the subject of commutative algebra, for which the reader can find already excellent treatments in the standard textbooks of Zariski–Samuel, Atiyah–Macdonald, and Kaplansky. In this book, instead, we shall focus on the *noncommutative* aspects of ring theory. Of course, we shall not exclude commutative rings from our study. In most cases, the theorems proved in this book remain meaningful for commutative rings, but in general these theorems become much easier in the commutative category. The main point, therefore, is to find good notions and good tools to work with in the possible absence of commutativity, in order to develop a general theory of possibly noncommutative rings. Most of the discussions in the text will be self-contained, so technically speaking we need not require much prior knowledge of commutative algebra. However, since much of our work is an attempt to extend results from the commutative setting to the general setting, it will pay handsomely if the reader already has a good idea of what goes on in the commutative case. To be more specific, it would be helpful if the reader has already acquired from a graduate course in algebra some acquaintance with the basic notions and foundational results of commutative algebra, for this will often supply the motivation needed for the general treatment of noncommutative phenomena in the text.

Generally, rings shall be denoted by letters such as R, R', or A. By a *subring* of a ring R, we shall always mean a subring containing the identity element 1 of R. If R is commutative, it is important to consider ideals in R. In the general case, we have to differentiate carefully between *left ideals* and *right ideals* in R. By an *ideal* I in R, we shall always mean a 2-sided ideal in R; i.e., I is both a left ideal and a right ideal. For such an ideal I in R, we can form the quotient ring $\bar{R} := R/I$, and we have a natural surjective ring homomorphism from R to $\bar{R}$ sending $a \in R$ to $\bar{a} = a + I \in \bar{R}$. The kernel of this ring homomorphism is, of course, the ideal I, and the quotient ring $\bar{R}$ has the universal property that any ring homomorphism φ from R to another ring R' with $\varphi(I) = 0$ "factors uniquely" through the natural homomorphism $R \to \bar{R}$.

A nonzero ring R is said to be a *simple ring* if (0) and R are the only ideals in R. This requires that, for any nonzero element $a \in R$, the ideal generated by a is R. Thus, a nonzero ring R is simple iff, for any $a \neq 0$ in R, there exists an equation $\sum b_i a c_i = 1$ for suitable $b_i, c_i \in R$. Using this, it follows easily that, if R is commutative, then R is simple iff R is a field. The class of noncommutative simple rings is, however, considerably larger, and much more difficult to describe.

In general, rings may have lots of zero-divisors. A nonzero element $a \in R$ is said to be a *left* 0-*divisor* if there exists a nonzero element $b \in R$ such that $ab = 0$ in R. Right 0-divisors are defined similarly. In the commutative setting, of course, we can drop the adjectives "left" and "right" and just speak of 0-divisors, but for noncommutative rings, a left 0-divisor need not be a right 0-divisor. For instance, let R be the ring $\begin{pmatrix} \mathbb{Z} & \mathbb{Z}/2\mathbb{Z} \\ 0 & \mathbb{Z} \end{pmatrix}$, by which we mean the ring of matrices of the form $\begin{pmatrix} x & y \\ 0 & z \end{pmatrix}$, where $x, z \in \mathbb{Z}$ and $y \in \mathbb{Z}/2\mathbb{Z}$, with formal matrix multiplication. (For more details, see Example 1.14 below.) If we let

$$a = \begin{pmatrix} 2 & 0 \\ 0 & 1 \end{pmatrix} \quad \text{and} \quad b = \begin{pmatrix} 0 & \bar{1} \\ 0 & 0 \end{pmatrix},$$

then $ab = 0 \in R$, so a is a left 0-divisor, but a is not a right 0-divisor since

$$0 = \begin{pmatrix} x & y \\ 0 & z \end{pmatrix} \begin{pmatrix} 2 & 0 \\ 0 & 1 \end{pmatrix} = \begin{pmatrix} 2x & y \\ 0 & z \end{pmatrix}$$

clearly implies that $x, z = 0$ in $\mathbb{Z}$ and $y = 0$ in $\mathbb{Z}/2\mathbb{Z}$. On the other hand, $b^2 = 0$, so b is both a left 0-divisor and a right 0-divisor.

A ring R is called a *domain* if $R \neq 0$, and $ab = 0$ implies $a = 0$ or $b = 0$ in R. In such a ring, we have no left (or right) 0-divisors. The reader no doubt knows many examples of commutative domains (= integral domains); some examples of noncommutative domains will be given later in this section.

A ring R is said to be *reduced* if R has no nonzero nilpotent elements, or, equivalently, if $a^2 = 0 \Rightarrow a = 0$ in R. For instance, the direct product of any family of domains is reduced.

An element a in a ring R is said to be *right-invertible* if there exists $b \in R$ such that $ab = 1$. Such an element b is called a *right inverse* of a. Left-invertible elements and their left inverses are defined analogously. If a has *both* a right inverse b and a left inverse b', then

$$b' = b'(ab) = (b'a)b = b.$$

In this case, we shall say that a is *invertible* (or a *unit*) in R, and call $b = b'$ the *inverse* of a. (The definite article is justified here since in this case b is easily seen to be unique.) We shall write $U(R)$ (or sometimes R^*) for the set of units in R; this is a group under the multiplication of R (with identity 1).

If $a \in R$ has a right inverse b, then $a \in U(R)$ iff we also have $ba = 1$. In the literature, a ring R is said to be *Dedekind-finite* (or *von Neumann-finite*) if $ab = 1 \Rightarrow ba = 1$, so these are the rings in which right-invertibility of elements implies left-invertibility. Many rings satisfying some form of "finiteness conditions" can be shown to be Dedekind-finite, but there do exist non-Dedekind-finite rings. For instance, let V be the k-vector space $ke_1 \oplus ke_2 \oplus \cdots$ with a countably infinite basis $\{e_i \colon i \geq 1\}$ over a field k, and let $R = End_k(V)$ be the k-algebra of all vector space endomorphisms of V. If $a, b \in R$ are defined on the basis by

$$b(e_i) = e_{i+1} \quad \text{for all } i \geq 1, \quad \text{and}$$

$$a(e_1) = 0, \quad a(e_i) = e_{i-1} \quad \text{for all } i \geq 2,$$

then clearly $ab = 1 \neq ba$, so a is right-invertible without being left-invertible, and R gives an example of a non-Dedekind-finite ring. On the other hand, if V_0 is a *finite-dimensional* k-vector space, then $R_0 = End_k(V_0)$ is Dedekind-finite: this is a well-known fact in linear algebra.

In some sense, the most "perfect" objects in noncommutative ring theory are the division rings: we say that a ring R is a *division ring* if $R \neq 0$ and $U(R) = R \backslash \{0\}$. (Note that commutative division rings are just fields.) To check that a nonzero ring R is a division ring, it is sufficient to show that every element $a \neq 0$ is right-invertible (this is an elementary exercise in group theory). From this, it is easy to see that $R \neq 0$ is a division ring iff the only right ideals in R are $\{0\}$ and R. Of course, the analogous statements also hold if we replace the word "right" by the word "left" in the above. In general, in the sequel, if we have proved certain results for rings "on the right," then we shall use such results freely also "on the left," provided that these results can indeed be proved by the same arguments applied "to the other side."

In connection with the remark just made, it is useful to recall the formation of the opposite ring R^{op} to a given ring R. By definition, R^{op} consists of elements of the form a^{op} in 1-1 correspondence with the elements a of R, with multiplication defined by

$$a^{\text{op}} \cdot b^{\text{op}} = (ba)^{\text{op}} \quad (\text{for } a, b \in R).$$

Generally speaking, if we have a result for rings "on the right," then we can obtain analogous results "on the left" by applying the known results to opposite rings. Of course, this has to be done carefully in order to avoid unpleasant mistakes.

We shall now record our list of basic examples of rings. (We have to warn our readers in advance that a few of these are somewhat sketchy in details.) Since the first noncommutative system was discovered by Sir William Rowan Hamilton, it seems most appropriate to begin this list with Hamilton's real quaternions.

(1.1) Example. Let $\mathbb{H} = \mathbb{R}1 \oplus \mathbb{R}i \oplus \mathbb{R}j \oplus \mathbb{R}k$, with multiplication defined by $i^2 = -1$, $j^2 = -1$, and $ij = -ji = k$. This is a 4-dimensional $\mathbb{R}$-algebra with center $\mathbb{R}$. If $\alpha = a + bi + cj + dk$ where $a, b, c, d \in \mathbb{R}$, we define $\bar{\alpha} = a - bi - cj - dk$, and check easily that

$$\alpha\bar{\alpha} = \bar{\alpha}\alpha = a^2 + b^2 + c^2 + d^2 \in \mathbb{R}.$$

Thus, if $\alpha \neq 0$, then $\alpha \in U(\mathbb{H})$ with

$$\alpha^{-1} = (a^2 + b^2 + c^2 + d^2)^{-1}\bar{\alpha}.$$

In particular, $\mathbb{H}$ is a division ring (we say that $\mathbb{H}$ is a division algebra over $\mathbb{R}$). Note that everything we said so far remains valid if we replace $\mathbb{R}$ by any field in which

$$(a, b, c, d) \neq (0, 0, 0, 0) \Rightarrow a^2 + b^2 + c^2 + d^2 \neq 0$$

(or, equivalently, -1 is not a sum of two squares). For instance, the "rational quaternions" $a + bi + cj + dk$ with $a, b, c, d \in \mathbb{Q}$ form a 4-dimensional division $\mathbb{Q}$-algebra R_1. In R_1, we have the subring R_2 consisting of

$$\{a + bi + cj + dk\colon a, b, c, d \in \mathbb{Z}\}.$$

This is not a division ring any more. In fact, its group of units is very small: we see easily that

$$U(R_2) = \{\pm 1, \pm i, \pm j, \pm k\} \quad \text{(the quaternion group)}.$$

There is a somewhat bigger subring R_3 of R_1 containing R_2, called *Hurwitz' ring of integral quaternions*. By definition, R_3 is the set of quaternions of the form $(a + bi + cj + dk)/2$, where $a, b, c, d \in \mathbb{Z}$ are either all even, or all odd. This is easily checked to be a subring of R_1. As an abelian group, R_3 is free on the basis

$$\{(1 + i + j + k)/2, i, j, k\},$$

so the (additive) index $[R_3 : R_2]$ is 2. The unit group of R_3 can be checked to be

$$U(R_3) = \{\pm 1, \pm i, \pm j, \pm k, (\pm 1 \pm i \pm j \pm k)/2\},$$

where the signs "± 1" are arbitrarily chosen. This group of 24 elements is the binary tetrahedral group—a nontrivial 2-fold covering of the tetrahedral group A_4. In fact, $U(R_3)/\{\pm 1\} \cong A_4$. The reader can also check easily that $U(R_3)$ contains the quaternion group $U(R_2)$ as a normal subgroup, so

$U(R_3)$ is a split extension of the quaternion group of order 8 by a cyclic group of order 3.

(1.2) Example (Free k-Rings). Let k be any ring, and $\{x_i: i \in I\}$ be a system of independent, noncommuting indeterminates over k. Then we can form the "free k-ring" generated by $\{x_i: i \in I\}$, which we denote by

$$R = k\langle x_i: i \in I\rangle.$$

The elements of R are polynomials in the noncommuting variables $\{x_i\}$ with coefficients from k. Here, the coefficients are supposed to commute with each x_i. The "freeness" of R refers to the following universal property: if $\varphi_0: k \to k'$ is any ring homomorphism, and $\{a_i: i \in I\}$ is any subset of k' such that each a_i commutes with each element of $\varphi_0(k)$, then there exists a unique ring homomorphism $\varphi: R \to k'$ such that $\varphi|k = \varphi_0$, and $\varphi(x_i) = a_i$ for every $i \in I$. The free k-ring $k\langle x_i: i \in I\rangle$ behaves rather differently from the polynomial ring $k[x_i: i \in I]$ (in which the x_i's commute). For instance, in the free k-ring $k\langle x, y\rangle$ in two variables, the subring generated over k by

$$z_i = xy^i \quad (0 \le i \le n)$$

is a free k-ring on $(n+1)$-generators. This is easily verified by showing that different monomials in $\{z_0, \ldots, z_n\}$ convert into different monomials in $\{x, y\}$. Therefore $k\langle x, y\rangle$ contains copies of $k\langle x_0, \ldots, x_n\rangle$ for every n. In fact, by the same reasoning, the subring of $k\langle x, y\rangle$ generated over k by $\{z_i: i \ge 0\}$ is seen to be isomorphic to $k\langle x_0, x_1, \ldots\rangle$, so $k\langle x, y\rangle$ even contains a copy of the free k-ring generated by countably many (noncommuting) indeterminates. This kind of phenomenon does not occur for polynomial rings in commuting indeterminates.

(1.3) Examples (Rings with Generators and Relations). Let k and R be as above, and let $F = \{f_j: j \in J\} \subseteq R$. Writing (F) for the ideal generated by F in R, we can form the quotient ring $\bar{R} = R/(F)$. We refer to $\bar{R}$ as the ring "generated over k by $\{x_i\}$ with relations F" (the latter term reflects the fact $f_j(\{\bar{x}_i: i \in I\}) = 0 \in \bar{R}$ for all j). The following are some specific examples.

(a) If we use the relations $\bar{x}_i\bar{x}_{i'} - \bar{x}_{i'}\bar{x}_i = 0$ for all $i, i' \in I$, the quotient ring $\bar{R}$ is the "usual" polynomial ring $k[\bar{x}_i: i \in I]$ in the *commuting* variables $\{\bar{x}_i\}$.

(b) If $R = \mathbb{R}\langle x, y\rangle$ and $F = \{x^2+1, y^2+1, xy+yx\}$, then $R/(F)$ is the $\mathbb{R}$-algebra of quaternions.

(c) If $R = k\langle x, y\rangle$ and $F = \{xy - yx - 1\}$, then $\bar{R} = R/(F)$ is the (first) *Weyl algebra*[1] over k, which we shall denote by $A_1(k)$. The relation

$$\bar{x}\bar{y} - \bar{y}\bar{x} = 1$$

---

[1] Since k need not be commutative, it is actually not quite right to use the term "algebra" in this context. But the nomenclature of Weyl algebras is so well established in the literature that we have to make an exception here.

in $A_1(k)$ arose naturally in the work on the mathematical foundations of quantum mechanics by Dirac, Weyl, Jordan–Wigner, D.E. Littlewood and others. (Indeed, $A_1(k)$ has been referred to by some as the "algebra of quantum mechanics.") In the case when k is a field of characteristic 0, $A_1(k)$ can also be viewed as a ring of differential operators on the polynomial ring $P = k[y]$. Indeed, if D denotes the operator d/dy on P and L denotes left multiplication on P by y, then for any $f(y) \in P$, Newton's law for the differentiation of a product yields

$$(DL)(f) = \frac{d}{dy}(yf) = y\frac{df}{dy} + f = (LD + I)f,$$

where I denotes the identity operator on P. Thus we have a k-algebra homomorphism φ of $A_1(k)$ into the endomorphism algebra $End_k\, P$ sending $\bar{x}$ to D and $\bar{y}$ to L. It is not difficult to see that the image of φ is exactly the ring S of differential operators of the form

$$\sum_i a_i(L)D^i,$$

where the a_i's are polynomials in y. From this one can check that φ is an isomorphism from $A_1(k)$ onto S. In a later example, we shall see that $A_1(k)$ may also be thought of as a ring of twisted polynomials in the variable x over the ring $P = k[y]$. Once $A_1(k)$ is defined, we can define the higher Weyl algebras inductively by

$$A_n(k) = A_1(A_{n-1}(k)),$$

or, equivalently, $A_n(k)$ is generated by a set of elements $\{x_1, y_1, \ldots, x_n, y_n\}$, each commuting with elements of k, with the relations:

$$x_iy_i - y_ix_i = 1 \quad (1 \le i \le n), \qquad x_iy_j - y_jx_i = 0 \quad (i \ne j),$$
$$x_ix_j - x_jx_i = 0 \quad (i \ne j), \qquad y_iy_j - y_jy_i = 0 \quad (i \ne j).$$

For some more details on these algebras, see (3.17).

(d) Let $R = \mathbb{Z}\langle x, y\rangle$ and $F = \{xy\}$. The ring $\bar{R} = R/(F)$ is then generated by $\bar{x}, \bar{y}$, with a "generic" relation $\bar{x}\bar{y} = 0$. In this ring, $\bar{x}$ is a left 0-divisor, but it can be shown that it is not a right 0-divisor. Similarly, if $R = \mathbb{Z}\langle x, y\rangle$ and $F = \{xy - 1\}$, then $\bar{R} = R/(F)$ is generated by $\bar{x}, \bar{y}$, with a "generic" relation $\bar{x}\bar{y} = 1$. It is not hard to show (e.g. by specialization) that $\bar{y}\bar{x} \ne 1$ in $\bar{R}$. Thus, $\bar{x}$ has a right inverse in $\bar{R}$, but is not a unit.

(1.4) Example. Let k be any ring, and G be a group or a semigroup (with identity), written multiplicatively. Then we can form the (*semi*)*group ring*

$$A = kG = \bigoplus_{\sigma \in G} k\sigma.$$

Elements of A are finite formal sums of the shape $\sum_{\sigma \in G} a_\sigma \sigma$, and are multiplied by using the multiplication in G. Thus,

$$\left(\sum_{\sigma} a_\sigma \sigma\right)\left(\sum_{\tau} b_\tau \tau\right) = \sum c_\mu \mu,$$

where $c_\mu = \sum a_\sigma b_\tau$, with summation over all $(\sigma, \tau) \in G \times G$ such that $\sigma\tau = \mu$. Note that under this multiplication in A, elements of k $(= k \cdot 1)$ commute with elements of G $(= 1 \cdot G)$. Clearly, A is commutative iff both k and G are commutative. This enables us to construct lots of examples of noncommutative rings. Note that if G is the free semigroup generated by $\{x_i: i \in I\}$, then kG is just the free k-ring $k\langle x_i: i \in I\rangle$ discussed in (1.2). Assuming further that k is a domain, it is easy to see that $U(kG) = U(k)$. If, however, G is a group (instead of just a semigroup), then clearly G is a subgroup of $U(kG)$. In general, $U(kG)$ may be much larger than $U(k) \cdot G$. For instance, when G is a cyclic group of order 5 generated by x, then in the integral group ring $\mathbb{Z}G$, we have $ab = 1$ for

$$a = 1 - x^2 - x^3 \quad \text{and} \quad b = 1 - x - x^4,$$

so a, b are units of $\mathbb{Z}G$ not belonging to $U(\mathbb{Z})G = \pm G$. In general, the problem of determining the group of units for a group ring kG is quite difficult, and has been solved only in certain special cases.

(1.5) Example. Let k be a ring and $\{x_i: i \in I\}$ be independent variables over k. In this example, the variables may be taken to be either pairwise commuting or otherwise, but we shall assume that they all commute with elements of k. With this convention, we can form the ring of formal power series $R = k[[x_i: i \in I]]$. The elements of R have the form $f_0 + f_1 + f_2 + \cdots$, where each f_n is a homogeneous polynomial in $\{x_i: i \in I\}$ over k with degree n, and we multiply these power series formally. It is not difficult to calculate the units of R; indeed,

$$F = f_0 + f_1 + f_2 + \cdots$$

is a unit in R iff the constant term f_0 is a unit in k. It suffices to do the "if" part, so let us assume that $f_0 \in U(k)$. To find a power series

$$G = g_0 + g_1 + g_2 + \cdots$$

such that $FG = 1$, we have to solve the equations:

$$1 = f_0 g_0, \quad 0 = f_0 g_1 + f_1 g_0, \quad 0 = f_0 g_2 + f_1 g_1 + f_2 g_0, \quad \ldots, \quad \text{etc.}$$

Since $f_0 \in U(k)$, we can solve for $g_0, g_1, g_2, \ldots$ inductively. This shows that F is right-invertible in R, and by symmetry we see that F is also left-invertible in R.

(1.6) Example. For any ring k, we can define the ring $k((x))$ of Laurent series in one variable x over k to be the set of formal Laurent series $F = \sum_{-\infty}^{\infty} f_i x^i$, where, among the coefficients $f_i \in k$ with $i < 0$, only finitely many can be nonzero. Again, these Laurent series are multiplied formally, with the elements of k commuting with the variable x. One particularly good feature of $R = k((x))$ is that, *if k is a division ring, then so is R*. To see this, let F above be a nonzero element of R. Choose a suitable power x^i ($i \in \mathbb{Z}$) such that

$$F \cdot x^i = g_0 + g_1 x + g_2 x^2 + \cdots$$

with $g_0 \neq 0$. If k is a division ring, then

$$g_0 + g_1 x + g_2 x^2 + \cdots \in U(k[[x]])$$

by an earlier remark in (1.5). Since x^i is obviously in $U(R)$, it follows that $F \in U(R)$. Therefore, the Laurent series construction enables us to produce new division rings from old division rings, and, of course, this construction can be repeated to give division rings of iterated Laurent series over a given division ring.

(1.7) Example (Hilbert's Twist). Let k be a ring and σ be a ring endomorphism of k. We can construct "twisted" (or skew) versions of the polynomial ring and the power series ring over k in one variable x by relaxing our earlier assumption that elements of k commute with x. Instead of $xb = bx$ for $b \in k$, we shall now stipulate that $xb = \sigma(b)x$. Thus, elements of the skew polynomial ring $k[x;\sigma]$ are "left polynomials" of the form $\sum_{i=0}^{n} a_i x^i$, with multiplication defined by:

$$\left(\sum a_i x^i\right)\left(\sum b_j x^j\right) = \sum a_i \sigma^i(b_j) x^{i+j}.$$

It is easy to check that $k[x;\sigma]$ is indeed a ring (and the skew power series ring $k[[x;\sigma]]$ is defined similarly). Note that if σ is not the identity, then $k[x;\sigma]$ (and $k[[x;\sigma]]$) will be noncommutative rings even though k may be commutative. In $k[x;\sigma]$, we can talk about the right polynomials (with the coefficients appearing on the right): $c_0 + xc_1 + \cdots + x^n c_n$, but these are left polynomials of the special form

$$c_0 + \sigma(c_1)x + \cdots + \sigma^n(c_n)x^n,$$

so not every member of $k[x;\sigma]$ can be written as a right polynomial. Of course, if σ is onto, then every left polynomial will be a right polynomial. If σ is not injective, say $\sigma(b) = 0$ for some $b \in k\backslash\{0\}$, then $xb = \sigma(b)x = 0$, although $f(x)x \neq 0$ for any $f \neq 0$ in R. This provides another example of a left 0-divisor in a ring which is not a right 0-divisor. On the other hand, if σ is injective and k is a domain, then a simple consideration of lowest-term coefficients shows that $k[x;\sigma]$ and $k[[x;\sigma]]$ are also domains. The unit groups of $k[x;\sigma]$ and $k[[x;\sigma]]$ are easy to determine: we have

$$U(k[x;\sigma]) = U(k), \quad \text{and}$$

$$U(k[[x;\sigma]]) = \{a_0 + a_1 x + \cdots : \ a_0 \in U(k)\},$$

without any assumptions on the endomorphism σ. The necessary arguments are easy generalizations of the ones used earlier, combined with the additional observation that

$$a_0 \in U(k) \Rightarrow \sigma^i(a_0) \in U(k) \quad \text{for all } i \geq 0.$$

(1.8) Example. Continuing in the spirit of (1.7), we can form a twisted (or skew) Laurent series ring $k((x;\sigma))$. For this, however, it is necessary to assume that σ is an *automorphism* of k. Under this assumption, we can "commute $b \in k$ past powers of x" by the rule $x^i b = \sigma^i(b)x^i$ for all $i \in \mathbb{Z}$, including negative integers. Again, it is easy to see that this leads to an associative multiplication on left Laurent series of the form $\sum_{-\infty}^{\infty} a_i x^i$ (with finitely many terms involving negative exponents). This gives the ring $k((x;\sigma))$ of skew Laurent series, in which we have in particular

$$x^{-1}\sigma(b) = \sigma^{-1}(\sigma b)x^{-1} = bx^{-1}.$$

Thus, $\sigma(b) = xbx^{-1}$ for every $b \in k$, so the automorphism σ may now be viewed as the conjugation by x on $k((x;\sigma))$ restricted to the subring k. Just as before, we can show that *if k is a division ring, then so is $k((x;\sigma))$*, as long as σ is an automorphism of k. For instance, if $k = \mathbb{Q}(t)$ and σ is the $\mathbb{Q}$-automorphism of k sending t to $2t$, then in $k((x;\sigma))$, we have the relation $xt = 2tx$. Hilbert was the first one to use the skew Laurent series construction to produce examples of noncommutative ordered division rings. Indeed, once the notion of an ordering on a division ring is defined, it is not difficult to see that the noncommutative division ring $k((x;\sigma))$ constructed above can be ordered. An introduction to the theory of orderings on rings will be given in Chapter 6.

In the ring $k((x;\sigma))$ of skew Laurent series, there is also the interesting subring consisting of $\sum_{-\infty}^{\infty} a_i x^i$ with only finitely many nonzero terms. (These are called the (skew) Laurent polynomials.) Since this ring is generated over k by x and x^{-1}, we shall denote it by $k[x, x^{-1}; \sigma]$.

(1.9) Example (Differential Polynomial Rings). In multiplying left polynomials, there is another thing we can do if we want to relax the assumption that elements of k commute with the variable x. To commute $a \in k$ past x, we can try to use the new rule: $xa = ax + \delta(a)$, where $\delta(a) \in k$ depends on a. If this is to lead to an associative multiplication among left polynomials, we must have $x(ab) = (xa)b$, so

$$(ab)x + \delta(ab) = (ax + \delta(a))b = a(bx + \delta(b)) + \delta(a)b.$$

Canceling $(ab)x = a(bx)$, we get

$$\delta(ab) = a\delta(b) + \delta(a)b,$$

and, of course, to guarantee the distributive law, we also need

$$\delta(a+b) = \delta(a) + \delta(b).$$

A map $\delta: k \to k$ satisfying these two properties (for all $a, b \in k$) is called a *derivation* on k. Given such a derivation, we can introduce a multiplication on left polynomials in x by repeatedly using the rule $xa = ax + \delta(a)$. The task of checking that this indeed leads to an associative multiplication is nontrivial, but we shall not dwell on the details here. (The interested reader should carry out this check as a supplementary exercise.) With the multiplication described above, the left polynomials $\sum a_i x^i$ form a ring, denoted by $k[x;\delta]$. In the literature, this is known as a *differential polynomial ring*. Note that if k is a domain, then so is $k[x;\delta]$. In the special case when δ is an inner derivation, $k[x;\delta]$ turns out to be isomorphic to the usual polynomial ring $k[t]$. By definition, δ is an *inner derivation* on k if there exists $c \in k$ such that $\delta(a) = ca - ac$ for every $a \in k$. (It is easy to check that such a δ is indeed a derivation.) For such a δ, we have

$$(x - c)a = ax + \delta(a) - ca = a(x - c)$$

for all $a \in k$, so $t = x - c$ commutes with k and we can show easily from this that $k[x;\delta] \cong k[t]$. In general, however, a derivation δ need not be inner. For instance, let $k = k_0[y]$ where k_0 is some (nonzero) ring, and let δ be the derivation on k defined by formal differentiation with respect to y (treating elements of k_0 as constants):

$$\delta\left(\sum b_i y^i\right) = \sum i b_i y^{i-1},$$

then δ is not inner since y is in the center of k but $\delta(y) = 1$. In the differential polynomial ring $k[x;\delta] = k_0[y][x;\delta]$, elements have the form $\sum a_i(y)x^i$ ($a_i(y) \in k_0[y]$), and we have the relation

$$xy = yx + \delta(y) = yx + 1.$$

From this, one can check without much difficulty that $k_0[y][x;\delta]$ is isomorphic to the Weyl algebra

$$A_1(k_0) = k_0\langle x, y\rangle/(xy - yx - 1)$$

defined in (1.3)(c). In particular, one sees that a k_0-basis for $A_1(k_0)$ is given by

$$\{\bar{x}^i \bar{y}^j : \ i \geq 0, j \geq 0\} \quad \text{as well as by} \quad \{\bar{y}^j \bar{x}^i : \ j \geq 0, i \geq 0\}.$$

It also follows by induction on n that, if k_0 is a domain, then the higher Weyl algebras $A_n(k_0)$ are all domains.

(1.10) Examples. Let V be an n-dimensional vector space over a field k, with $n < \infty$. Then we can form the *tensor algebra* $T(V)$ over k. If $\{e_1, \ldots, e_n\}$ is a k-basis on V, $T(V)$ is essentially the free k-algebra $R = k\langle e_1, \ldots, e_n\rangle$. Various quotient algebras of R are of interest. First, the *symmetric algebra*

$S(V)$ obtained from R by quotienting out the ideal generated by all

$$u \otimes v - v \otimes u \quad (u, v \in V)$$

is just the ordinary polynomial algebra $k[e_1, \ldots, e_n]$ (with commuting e_i's). Secondly, we have the *exterior algebra* $\bigwedge(V)$ obtained from R by quotienting out the ideal generated by $v \otimes v$ for all $v \in V$. This is a finite-dimensional k-algebra, with $dim_k \bigwedge(V) = 2^n$. The ideal J of $\bigwedge(V)$ generated by $\bar{e}_1, \ldots, \bar{e}_n$ has the property that $J^n \neq 0$ and $J^{n+1} = 0$. In the terminology to be introduced in §19, $\bigwedge(V)$ is a (generally noncommutative) local ring, with residue field $\bigwedge(V)/J \cong k$. If V has some further algebraic structure, we can define other quotients of $T(V)$, as follows.

(a) If k has characteristic $\neq 2$ and V is equipped with a quadratic form q: $V \rightarrow k$, then we can form the Clifford algebra $C(V, q)$ by quotienting out the ideal of $T(V)$ generated by $v \otimes v - q(v)$ for all $v \in V$. Again, it can be shown that $dim_k\ C(V, q) = 2^n$. In the special case when the quadratic form q is the zero form, we get back the exterior algebra: $C(V, 0) \cong \bigwedge(V)$.

(b) If V has a given structure as a Lie algebra over k with a bracket operation

$$[\ ,\ ]:\ V \times V \rightarrow V,$$

we can form the *universal enveloping algebra* U of $(V, [\ ,\ ])$ by quotienting out the ideal of $T(V)$ generated by

$$u \otimes v - v \otimes u - [u, v] \quad \text{for all } u, v \in V.$$

If we fix a k-basis $\{e_1, \ldots, e_n\}$ on V, and let $\{a_{ij\ell}\}$ be the structure constants of the Lie bracket operation defined by

$$[e_i, e_j] = \sum_{\ell} a_{ij\ell} e_{\ell},$$

then U is just the k-algebra generated with $\bar{e}_1, \ldots, \bar{e}_n$ with relations

$$\bar{e}_i \bar{e}_j - \bar{e}_j \bar{e}_i = \sum_{\ell} a_{ij\ell} \bar{e}_{\ell}.$$

(According to a famous theorem of Poincaré–Birkhoff–Witt, a k-basis of U is given by the "monomials"

$$\{\bar{e}_i^{i_1} \bar{e}_2^{i_2} \cdots \bar{e}_n^{i_n} :\ i_1, \ldots, i_n \geq 0\}.$$

However, we shall not make use of this result here.) In the special case when V is an abelian Lie algebra (that is, $[u, v] = 0$ for all u, v), we get back the symmetric algebra: $U \cong S(V)$. On the other hand, if V is the binary space $ke_1 \oplus ke_2$ with a Lie algebra structure given by the Lie product $[e_1, e_2] = e_2$, it can be checked that all relations in U boil down to a single one: $\bar{e}_1\bar{e}_2 - \bar{e}_2\bar{e}_1 - \bar{e}_2 = 0$, so

$$U \cong k\langle x, y\rangle/(xy - yx - y).$$

The latter algebra U' is isomorphic to the skew polynomial ring $k[x][y;\sigma]$, where σ is the k-automorphism of $k[x]$ sending x to $x-1$. (In this ring, $yx = \sigma(x)y = (x-1)y$, so we have $xy - yx = y$.) Another description of U' is $U' \cong k[y][x,\delta]$, where δ is the derivation on $k[y]$ given by

$$\delta(f) = y\frac{df}{dy}.$$

(In $k[y][x,\delta]$, we have again $xy = yx + \delta(y) = yx + y$.) Yet another description of U is given by identifying U with a certain subalgebra of the Weyl algebra $A_1(k) = k\langle t,s\rangle/(ts - st - 1)$. To do this, just note that, by left multiplication of $ts - st - 1$ by s, we get $(st)s - s(st) - s$, so we can define a k-algebra homomorphism

$$\varphi\colon\ k\langle x,y\rangle/(xy - yx - y) \longrightarrow A_1(k)$$

by taking $\varphi(x) = \bar{s}\bar{t}$ and $\varphi(y) = \bar{s}$. It follows easily that U is isomorphic to the subalgebra of $A_1(k)$ generated by $\bar{s}$ and $\bar{s}\bar{t}$.

As another example, consider the $(2n+1)$-dimensional *Heisenberg Lie algebra* V with basis $\{x_1, \ldots, x_n, y_1, \ldots, y_n, z\}$ and Lie products:

$$[x_i, y_i] = z = -[y_i, x_i] \quad (1 \le i \le n),$$

with all other Lie products equal to 0. If we "identify" z with 1 in the universal enveloping algebra U of V, we have the relations

$$x_i y_i - y_i x_i = 1 \quad (1 \le i \le n),$$

$$x_i y_j - y_j x_i = 0, \quad x_i x_j - x_j x_i = 0, \quad y_i y_j - y_j y_i = 0 \quad (\forall i \ne j).$$

These are exactly the relations defining the nth Weyl algebra $A_n(k)$. Thus, we have an isomorphism $U/(z-1) \cong A_n(k)$. The examples given in this and the last paragraph suggest that, generally speaking, universal enveloping algebras of Lie algebras are somewhat related to higher Weyl algebras and iterated differential polynomial rings.

(1.11) Example (Skew Group Rings). Let k be a ring and let G be a group acting on k as a group of automorphisms. Then we can form a skew group ring $R = k * G$ by taking its elements to be finite formal combinations $\sum_{\sigma \in G} a_\sigma \sigma$, with multiplication induced by:

$$(a_\sigma \sigma)(b_\tau \tau) = a_\sigma \sigma(b_\tau)(\sigma\tau).$$

For instance, if G is an infinite cyclic group $\langle\sigma\rangle$ where σ acts on k, then $k * G$ is isomorphic to the skew Laurent polynomial ring $k[x, x^{-1}; \sigma]$. To show how naturally skew group rings arise in practice, let us consider a group G which is a semidirect product of a normal subgroup T with a complement H. Here, H acts on T by conjugation, and this action can be extended uniquely to an action on the (usual) group ring kT. We express this action by writing

$$^h\left(\sum_{\tau \in T} a_\tau \tau\right) = \sum_{\tau \in T} a_\tau {}^h\tau,$$

where $^h\tau = h\tau h^{-1}$ for $h \in H$ and $\tau \in T$. Then in the (usual) group ring kG, we have

$$h \cdot \sum_{\tau \in T} a_\tau \tau = \left(\sum_{\tau \in T} a_\tau h\tau h^{-1}\right)h = {}^h\left(\sum_{\tau \in T} a_\tau \tau\right)h.$$

This shows that $kG \cong (kT) * H$, where the skew group ring on the *RHS* is formed with respect to the action of H on kT as described above. From this example, we see that the formation of skew group rings is helpful already in understanding the structure of the ordinary group rings kG.

(1.12) Example. If A is any object in an additive category $\mathscr{C}$, then $End_{\mathscr{C}}\, A$ (consisting of $\mathscr{C}$-endomorphisms of A) is a ring. For instance, if $\mathscr{C}$ is the category of right modules over a ring R, then we have the ring of endomorphisms $End_{\mathscr{C}}\, A = End_R(A)$ associated to any right R-module A. In the special case when $A = R$ (viewed as a right module over itself), we can define a mapping $L\colon R \to End_R(R)$ by sending $r \in R$ to the left multiplication map $L(r)$ on R defined by $L(r)(a) = ra$ for any $a \in R$. Since

$$L(r)(ab) = r(ab) = (ra)b = (L(r)(a))b,$$

we have indeed $L(r) \in End_R(R)$. A similar calculation shows that L is a ring homomorphism. If $L(r) = 0$, then $0 = L(r)(1) = r$, so L is one-one. Finally L is also onto, for, if $\varphi \in End_R(R)$, then for $r := \varphi(1)$, we have

$$L(r)(a) = ra = \varphi(1)a = \varphi(a).$$

Since this holds for all $a \in R$, we have $L(r) = \varphi$. Thus, we have a ring isomorphism $R \cong End_R(R)$.

(1.13) Examples. Let V be an n-dimensional right vector space over a division ring k. Then, using a fixed basis $\{e_1, \ldots, e_n\}$ on V, we can identify $End_k\, V$ as usual with the ring $R = \mathbb{M}_n(k)$ of $n \times n$ matrices over k. This matrix ring R has many interesting subrings, some of which are described below.

(a) The subring T of R consisting of all upper triangular matrices. The set I of matrices of T with a zero diagonal is easily seen to be an ideal of T, with $T/I \cong k \times \cdots \times k$ (direct product of n copies of k). Moreover, using linear algebra considerations, one sees that $I^{n-1} \neq 0$ but $I^n = 0$.

(b) The set of all matrices (a_{ij}) in T with $a_{2n} = a_{3n} = \cdots = a_{n-1,n} = 0$ can be checked to be a subring of T.

(c) The set of all matrices (a_{ij}) in T with $a_{11} = a_{22}$ and all off-diagonal elements zero except perhaps a_{1n} is another subring of T.

(d) Let $k = \mathbb{C}$ and $n = 2$. Then the set of matrices of the form $\begin{pmatrix} \alpha & -\bar{\beta} \\ \beta & \bar{\alpha} \end{pmatrix}$ (where $\alpha, \beta \in \mathbb{C}$ and "bar" denotes taking complex conjugates) is $\mathbb{R}$-isomorphic to the division ring $\mathbb{H}$ of real quaternions. An explicit isomorphism is given by mapping a quaternion

$$a + bi + cj + dk \quad (a, b, c, d \in \mathbb{R})$$

to the matrix $\begin{pmatrix} a+bi & -c-di \\ c-di & a-bi \end{pmatrix}$. Under this isomorphism, we have

$$1 \leftrightarrow \begin{pmatrix} 1 & 0 \\ 0 & 1 \end{pmatrix}, \quad i \leftrightarrow \begin{pmatrix} i & 0 \\ 0 & -i \end{pmatrix}, \quad j \leftrightarrow \begin{pmatrix} 0 & -1 \\ 1 & 0 \end{pmatrix}, \quad \text{and} \quad k \leftrightarrow \begin{pmatrix} 0 & -i \\ -i & 0 \end{pmatrix}.$$

(e) Continuing the notations in (d), consider the isomorphism

$$\varphi \colon \mathbb{H} \to End_{\mathbb{H}}(\mathbb{H})$$

obtained in (1.12), where the last $\mathbb{H}$ is viewed as a right $\mathbb{H}$-module. Since

$$End_{\mathbb{H}}(\mathbb{H}) \subseteq End_{\mathbb{R}}(\mathbb{H}) \cong \mathbb{M}_4(\mathbb{R})$$

(using the basis $\{1, i, j, k\}$ on $\mathbb{H}$), $\varphi(\mathbb{H})$ is the set of all 4×4 real matrices of the form

$$\begin{pmatrix} a & -b & -c & -d \\ b & a & -d & c \\ c & d & a & -b \\ d & -c & b & a \end{pmatrix}.$$

Therefore, these real matrices form an $\mathbb{R}$-subalgebra of $\mathbb{M}_4(\mathbb{R})$ isomorphic to the algebra $\mathbb{H}$ of all real quaternions.

(f) For $k = \mathbb{Q}$, let

$$S = \left\{ \begin{pmatrix} x+y & 4y \\ -y & x-y \end{pmatrix} : x, y \in \mathbb{Q} \right\}$$

and

$$S' = \left\{ \begin{pmatrix} x & y \\ -y & x+y \end{pmatrix} : x, y \in \mathbb{Q} \right\}.$$

Then S, S' are both subfields of $\mathbb{M}_2(\mathbb{Q})$ isomorphic to the field $\mathbb{Q}(\sqrt{-3})$. In fact, for $\alpha = \begin{pmatrix} 1 & 4 \\ -1 & -1 \end{pmatrix}$ and $\alpha' = \begin{pmatrix} 0 & 1 \\ -1 & 1 \end{pmatrix}$, we have $\begin{pmatrix} x+y & 4y \\ -y & x-y \end{pmatrix} = xI + y\alpha$ and $\begin{pmatrix} x & y \\ -y & x+y \end{pmatrix} = xI + y\alpha'$. Since α and α' satisfy respectively their characteristic equations, we have $\alpha^2 + 3I = 0$ and $\alpha'^2 - \alpha' + I = 0$. From this, it follows easily that $S \cong \mathbb{Q}(\sqrt{-3}) \cong S'$ as $\mathbb{Q}$-algebras. An explicit isomorphism from S to S' is provided by sending

$\begin{pmatrix} x+y & 4y \\ -y & x-y \end{pmatrix} \in S$ to

$$\begin{pmatrix} 1 & 0 \\ 1 & 2 \end{pmatrix}\begin{pmatrix} x+y & 4y \\ -y & x-y \end{pmatrix}\begin{pmatrix} 1 & 0 \\ 1 & 2 \end{pmatrix}^{-1} = \begin{pmatrix} x-y & 2y \\ -2y & x+y \end{pmatrix}$$
$$= \begin{pmatrix} x' & y' \\ -y' & x'+y' \end{pmatrix},$$

where $x' = x - y$, $y' = 2y$.

(g) The following subsets are subrings of $\mathbb{M}_2(\mathbb{Q})$:

$$W_1 = \left\{\begin{pmatrix} a & b \\ c & d \end{pmatrix} : a, c, d \in \mathbb{Z},\ b \in n\mathbb{Z}\right\} \quad \left(\text{abbreviated} \begin{pmatrix} \mathbb{Z} & n\mathbb{Z} \\ \mathbb{Z} & \mathbb{Z} \end{pmatrix}\right).$$

$$W_2 = \left\{\begin{pmatrix} a & b \\ c & d \end{pmatrix} : a, b, c, d \in \mathbb{Z},\ a \equiv d \ (\mathrm{mod}\ n),\ b \equiv c \ (\mathrm{mod}\ n)\right\}.$$

$$W_3 = \left\{\begin{pmatrix} a & b \\ c & d \end{pmatrix} : a, d \in \mathbb{Z},\ a \equiv d \ (\mathrm{mod}\ n),\ b, c \in n\mathbb{Z}\right\}.$$

(h) The following subsets are subrings of $\mathbb{M}_2(\mathbb{R}(x))$:

$$\begin{pmatrix} \mathbb{Z} & \mathbb{Q} \\ 0 & \mathbb{Z} \end{pmatrix}, \quad \begin{pmatrix} \mathbb{Z} & \mathbb{R} \\ 0 & \mathbb{Q} \end{pmatrix}, \quad \begin{pmatrix} \mathbb{Z} & \mathbb{R}[x] \\ 0 & \mathbb{R} \end{pmatrix}, \quad \begin{pmatrix} \mathbb{Q} & \mathbb{R}(x) \\ 0 & \mathbb{Q}[x] \end{pmatrix}.$$

(1.14) Example (Triangular Rings). The rings listed in (h) above as well as the ring $\begin{pmatrix} \mathbb{Z} & \mathbb{Z}/2\mathbb{Z} \\ 0 & \mathbb{Z} \end{pmatrix}$ considered earlier are all special cases of a more general construction. Let R, S be two rings, and let M be an (R, S)-bimodule. This means that M is a left R-module and a right S-module such that $(rm)s = r(ms)$ for all $r \in R$, $m \in M$, and $s \in S$. Given such a bimodule M, we can form

$$A = \begin{pmatrix} R & M \\ 0 & S \end{pmatrix} = \left\{\begin{pmatrix} r & m \\ 0 & s \end{pmatrix} : r \in R,\ m \in M,\ s \in S\right\},$$

and define a multiplication on A by using formal matrix multiplication:

$$\textbf{(1.15)} \qquad \begin{pmatrix} r & m \\ 0 & s \end{pmatrix}\begin{pmatrix} r' & m' \\ 0 & s' \end{pmatrix} = \begin{pmatrix} rr' & rm' + ms' \\ 0 & ss' \end{pmatrix}.$$

A routine check shows that, with this multiplication (and entry-wise addition), A becomes a ring. (The bimodule property $(rm)s = r(ms)$ on M is not needed in the above definition, but is needed in verifying the *associativity* of the multiplication in A.) This construction of the (so-called) triangular ring A clearly covers all the examples mentioned at the beginning of (1.14).

In the ring theory literature, many surprising examples and counterexamples have been produced via the triangular ring construction, by varying the choices of R, S and M. What makes this possible is the fact that the left, right and 2-sided ideal structures in A turn out to be quite tractable. In the following, we shall try to describe completely the left, right and 2-sided ideals in A.

First, it is convenient to identify R, S and M as subgroups in A (in the obvious way) and to think of A as $R \oplus M \oplus S$. In terms of this decomposition, the multiplication in A may be described by the following chart:

(1.16)

	R	M	S
R	R	M	0
M	0	0	M
S	0	0	S

From this, it is immediately clear that R is a left ideal, S is a right ideal, and M is a (square zero) ideal in A. Moreover, $R \oplus M$ and $M \oplus S$ are both ideals of A, with $A/(R \oplus M) \cong S$ and $A/(M \oplus S) \cong R$. Finally, $R \oplus S$ is a subring of A.

(1.17) Proposition.

(1) *The left ideals of A are of the form $I_1 \oplus I_2$, where I_2 is a left ideal in S, and I_1 is a left R-submodule of $R \oplus M$ containing MI_2.*

(2) *The right ideals of A are of the form $J_1 \oplus J_2$, where J_1 is a right ideal in R, and J_2 is a right S-submodule of $M \oplus S$ containing J_1M.*

(3) *The ideals of A are of the form $K_1 \oplus K_0 \oplus K_2$, where K_1 is an ideal in R, K_2 is an ideal in S, and K_0 is an (R, S)-subbimodule of M containing $K_1M + MK_2$.*

Proof. The fact that such $I_1 \oplus I_2$ is a left ideal, $J_1 \oplus J_2$ is a right ideal, and $K_1 \oplus K_0 \oplus K_2$ is an ideal is immediately clear from the multiplication table (1.16). Conversely, let I be any *left* ideal of A. If $\begin{pmatrix} r & m \\ 0 & s \end{pmatrix}$ belongs to I, then so do

$$\begin{pmatrix} 1 & 0 \\ 0 & 0 \end{pmatrix}\begin{pmatrix} r & m \\ 0 & s \end{pmatrix} = \begin{pmatrix} r & m \\ 0 & 0 \end{pmatrix}$$

and

$$\begin{pmatrix} 0 & 0 \\ 0 & 1 \end{pmatrix}\begin{pmatrix} r & m \\ 0 & s \end{pmatrix} = \begin{pmatrix} 0 & 0 \\ 0 & s \end{pmatrix}.$$

Therefore we have $I = I_1 \oplus I_2$, where $I_1 = I \cap (R \oplus M)$ and $I_2 = I \cap S$. Clearly, I_1 is a left R-submodule of $R \oplus M$, and I_2 is a left ideal of S. Lastly,

$$MI_2 = M(I \cap S) \subseteq I \cap M \subseteq I \cap (R \oplus M) = I_1.$$

This proves (1), and (2) is proved similarly. If K is an ideal of A, then, whenever $\begin{pmatrix} r & m \\ 0 & s \end{pmatrix}$ belongs to K, so do

$$\begin{pmatrix} r & m \\ 0 & s \end{pmatrix}\begin{pmatrix} 1 & 0 \\ 0 & 0 \end{pmatrix} = \begin{pmatrix} r & 0 \\ 0 & 0 \end{pmatrix},$$
$$\begin{pmatrix} 0 & 0 \\ 0 & 1 \end{pmatrix}\begin{pmatrix} r & m \\ 0 & s \end{pmatrix} = \begin{pmatrix} 0 & 0 \\ 0 & s \end{pmatrix},$$

and hence also $\begin{pmatrix} 0 & m \\ 0 & 0 \end{pmatrix}$. This shows that $K = K_1 \oplus K_0 \oplus K_2$, where

$$K_1 = K \cap R, \quad K_0 = K \cap M, \quad \text{and} \quad K_2 = K \cap S.$$

Since K and M are ideals, we must have $K_1M + MK_2 \subseteq K \cap M = K_0$, and the other required properties of K_0, K_1, K_2 are clear. QED

According to this proposition, the left and right ideal structures in A are closely tied to, respectively, the left R-module structure on M and the right S-module structure on M. Often, these two module structures on M can be arranged to be quite different. In such a situation, the ring A will exhibit drastically different behavior between its left ideals and its right ideals. To illustrate this point, we shall use the triangular formation to construct some rings below which are left noetherian (resp., artinian) but not right noetherian (resp., artinian).

First let us recall a few standard definitions. A family of subsets $\{C_i\colon i \in I\}$ in a set C is said to satisfy the *Ascending Chain Condition* (ACC) if there does not exist an infinite strictly ascending chain

$$C_{i_1} \subsetneq C_{i_2} \subsetneq \cdots$$

in the family. Two equivalent formulations of this condition are the following:

(1) For any ascending chain $C_{i_1} \subseteq C_{i_2} \subseteq \cdots$ in the family, there exists an integer n such that $C_{i_n} = C_{i_{n+1}} = C_{i_{n+2}} = \cdots$.

(2) Any nonempty subfamily of the given family has a maximal member (with respect to inclusion).

The *Descending Chain Condition* (DCC) for a family of subsets of C is defined similarly, and the obvious analogues of (1), (2) can be used as its equivalent formulations.

Let R be a ring and let M be either a left or a right R-module. We say that M is *noetherian* (resp., *artinian*) if the family of all submodules of M satisfies ACC (resp., DCC). [More briefly, we can say: M has ACC (resp., DCC) on submodules.] The following are three easy, but important, facts, which the reader should have seen from a graduate course in algebra.

(1.18) *M is noetherian iff every submodule of M is finitely generated.*

(1.19) *M is both noetherian and artinian iff M has a (finite) composition series.*

(1.20) *Let N be a submodule of M. Then M is noetherian (resp., artinian) iff N and M/N are both noetherian (resp., artinian). In particular, the direct sum of two noetherian (resp., artinian) modules is noetherian (resp., artinian).*

A ring R is said to be *left* (resp., *right*) *noetherian* if R is noetherian when viewed as a left (resp., right) R-module. If R is both left and right noetherian, we shall say that R is *noetherian*. The examples we shall present below will show that "left noetherian" and "right noetherian" are independent conditions, so a ring being noetherian is indeed a stronger condition than its being one-sided noetherian. By the preceding discussion, we see that R is *left noetherian iff every left ideal of R is finitely generated, iff any nonempty family of left ideals in R has a maximal member.*

A ring R is said to be *left* (resp., *right*) *artinian* if R is artinian when viewed as a left (resp., right) R-module. If R is both left and right artinian, we say that R is *artinian*. Again, we shall see that this is stronger than R being only one-sided artinian.

Needless to say, the nomenclature above honors, respectively, Emmy Noether and Emil Artin, who initiated the study of ascending and descending chain conditions for (one-sided) ideals and submodules. To complete our review of basic facts on chain conditions, let us also recall the following Proposition about finitely generated modules over rings satisfying chain conditions.

(1.21) Proposition. *If M is a finitely generated left module over a left noetherian (resp., artinian) ring, then M is a noetherian (resp., artinian) module.*

One of the most lovely results in ring theory is the fact that *a left (resp., right) artinian ring is always left (resp., right) noetherian.* This fact was apparently unknown to both Noether and Artin when they wrote their pioneering papers on chain conditions in the 1920's. Rather, it was proved only some years later by Levitzki and Hopkins. (We note, incidentally, that "artinian $\Rightarrow$ noetherian" works only for one-sided ideals, but not for modules!) Since this is a highly nontrivial result, we shall not assume it in the balance of this section. A full proof of the Hopkins–Levitzki Theorem will be given in §4 in conjunction with our study of the Jacobson radical of a ring.

As an application of (1.17), we shall prove the following useful result about triangular rings.

(1.22) Theorem. *Let $A = \begin{pmatrix} R & M \\ 0 & S \end{pmatrix}$ be as in (1.17). Then A is left (resp., right) noetherian iff R and S are left (resp., right) noetherian, and M as a left R-module (resp., right S-module) is noetherian. The same statement holds if we replace throughout the word "noetherian" by "artinian."*

Proof. It suffices to treat the "left noetherian" case, for the arguments in the other cases are the same. First assume A is left noetherian. Since R and S are quotient rings of A, they are also left noetherian. If $M_1 \subseteq M_2 \subseteq \cdots$ is an ascending chain of left R-submodules of M, then, passing to the $\begin{pmatrix} 0 & M_i \\ 0 & 0 \end{pmatrix}$'s, we get an ascending chain of left ideals of A. Thus $M_1 \subseteq M_2 \subseteq \cdots$ must become stationary, so M as a left R-module is noetherian. Conversely, assume that R, S are left noetherian, and that M as a left R-module is noetherian. Consider an ascending chain $I^{(1)} \subseteq I^{(2)} \subseteq \cdots$ of left ideals in A. The contraction of this chain to S must become stationary, since S is left noetherian. On the other hand, the contraction of the chain to $R \oplus M$ must also become stationary, since (by (1.20)) the left R-module $R \oplus M$ is noetherian. Recalling that

$$I^{(i)} = (I^{(i)} \cap S) \oplus (I^{(i)} \cap (R \oplus M)),$$

we see that $I^{(i)} \subseteq I^{(2)} \subseteq \cdots$ becomes stationary, so we have proved that A is left noetherian. QED

(1.23) Corollary. *Let S be a commutative noetherian domain which is not equal to its field of fractions, R. Then $A = \begin{pmatrix} R & R \\ 0 & S \end{pmatrix}$ is left noetherian and not right noetherian, and A is neither left nor right artinian.*

Proof. In view of the theorem, it suffices to show that (1) S is not artinian, and that (2) R as a (right) S-module is not noetherian. For (1), simply note that if $s \neq 0$ is a nonunit in S, then we have

$$(s) \supsetneqq (s^2) \supsetneqq (s^3) \supsetneqq \cdots.$$

For (2), assume instead that R is a noetherian S-module. Then R is, in particular, a finitely generated S-module, so there would exist a common denominator $s \in S$ for all fractions in R. But then $1/s^2 = s'/s$ for some $s' \in S$, so $s \in U(S)$, contradicting $S \neq R$. QED

The following can also be deduced immediately from (1.22).

(1.24) Corollary. *Let $S \subseteq R$ be fields such that $\dim_S R = \infty$. Then $A = \begin{pmatrix} R & R \\ 0 & S \end{pmatrix}$ is left noetherian and left artinian, but neither right noetherian nor right artinian.*

We can make two more useful remarks about the ring A in the last Corollary. First, as a left module over itself, A has a composition series of length 3, namely

$$A \supsetneqq \begin{pmatrix} 0 & R \\ 0 & S \end{pmatrix} \supsetneqq \begin{pmatrix} 0 & R \\ 0 & 0 \end{pmatrix} \supsetneqq (0).$$

The fact that this chain of left ideals cannot be further refined follows from (1.17) (1) (or from an *ad hoc* calculation). This, of course, shows directly that A is left noetherian and left artinian, in view of (1.19). Secondly, since $dim_S\, R = \infty$, we can easily construct an infinite direct sum $\bigoplus_{i=1}^{\infty} M_i$ of nonzero (right) S-subspaces in R. By passing to the $\begin{pmatrix} 0 & M_i \\ 0 & 0 \end{pmatrix}$'s, we obtain then an infinite direct sum of nonzero right ideals in A. But, of course, the fact that A is left noetherian implies that there cannot exist an infinite direct sum of nonzero left ideals in A. Using terminology to be introduced later in *Lectures*, we have in A an example of a ring which is left Goldie but not right Goldie.

Of course there are other methods for constructing rings which are noetherian on one side but not on the other. We conclude this section with two more such constructions.

(1.25) Example. *Let σ be an endomorphism of a division ring k which is not an automorphism. Then $R = k[x; \sigma]$ is left noetherian but not right noetherian.* Indeed, if I is any nonzero left ideal of R, then, choosing a monic left polynomial $f \in I$ of the least degree, the usual Euclidean algorithm argument implies that $I = R \cdot f$. Thus, every left ideal of R is principal (we say that R is a *principal left ideal domain*); in particular, R is left noetherian. On the other hand, fix an element $b \in k \backslash \sigma(k)$. We claim that $\sum_{i=0}^{\infty} x^i bxR$ is a direct sum of right ideals, which will imply that R is not right noetherian. Assume, for the moment, that there exists an equation

$$x^n bxf_n(x) + \cdots + x^{n+m} bxf_{n+m}(x) = 0,$$

where the first and the last terms are nonzero. Since R is a domain, this gives $bxf_n(x) = xg(x)$ for some $g(x) \in R$. If $f_n(x)$ has highest-degree term $c_r x^r$ $(c_r \neq 0)$ and $g(x) = \sum a_i x^i$, a comparison of the coefficients of x^{r+1} gives $b\sigma(c_r) = \sigma(a_r)$, which contradicts $b \notin \sigma(k)$. Incidentally, R is also neither left nor right artinian, since there are infinite descending chains

$$Rx \supsetneq Rx^2 \supsetneq \cdots \quad \text{and} \quad xR \supsetneq x^2R \supsetneq \cdots.$$

(1.26) Example (Dieudonné). *Let $R = \mathbb{Z}\langle x, y\rangle/(y^2, yx)$. Then R is left noetherian, but not right noetherian.* To work with R, we shall confuse x, y with their images in R. Thus, we view R as generated by x, y, with the relations $y^2 = 0$ and $yx = 0$. Then R has a direct sum decomposition $R = \mathbb{Z}[x] \oplus \mathbb{Z}[x]y$. Here $\mathbb{Z}[x]$ is a subring, and $\mathbb{Z}[x]y$ is an ideal. We shall assume the Hilbert Basis Theorem, which implies that $\mathbb{Z}[x]$ is a noetherian ring. By (1.21), $R = \mathbb{Z}[x] \oplus \mathbb{Z}[x]y$ is noetherian as a left $\mathbb{Z}[x]$-module, and hence as a left R-module. This shows that R is left noetherian. To show that R is not right noetherian, it suffices to show that $I = \mathbb{Z}[x]y$ is not finitely generated as a right R-module. Since both x and y act trivially on the right of I, if I were finitely generated as a right R-module, it would be finitely generated as an abelian group. This is clearly not the case, as

$$I = \mathbb{Z}[x]y = \bigoplus_{i=0}^{\infty} \mathbb{Z} \cdot x^i y.$$

Incidentally, the ring R in this example is neither left nor right artinian, since I is an ideal in R, and $R/I \cong \mathbb{Z}[x]$ is not an artinian ring.

Exercises for §1

Ex. 1.1. Let $(R, +, \times)$ be a system satisfying all axioms of a ring with identity, except possibly $a + b = b + a$. Show that $a + b = b + a$ for all $a, b \in R$, so R is indeed a ring.

Ex. 1.2. It was mentioned in the text that a nonzero ring R is a division ring iff every $a \in R \backslash \{0\}$ is right-invertible. Supply a proof for this statement.

Ex. 1.3. Show that the characteristic of a domain is either 0 or a prime number.

Ex. 1.4. True or False: "If ab is a unit, then a, b are units"? Show the following for any ring R:
(a) If a^n is a unit in R, then a is a unit in R.
(b) If a is left-invertible and not a right 0-divisor, then a is a unit in R.
(c) If R is a domain, then R is Dedekind-finite.

Ex. 1.4*. Let $a \in R$. (1) Show that if a has a left inverse, then a is not a left 0-divisor. (2) Show that the converse holds if $a \in aRa$.

Ex. 1.5. Give an example of an element x in a ring R such that $Rx \subsetneq xR$.

Ex. 1.6. Let a, b be elements in a ring R. If $1 - ba$ is left-invertible (resp. invertible), show that $1 - ab$ is left-invertible (resp. invertible), and construct a left inverse (resp. inverse) for it explicitly. (**Hint.** $R(1 - ab)$ contains $Rb(1 - ab) = R(1 - ba)b = Rb$, so it also contains 1. This proof lends itself easily to an explicit construction of the needed (left) inverse.)

Ex. 1.7. Let $B_1, \ldots, B_n$ be left ideals (resp. ideals) in a ring R. Show that $R = B_1 \oplus \cdots \oplus B_n$ iff there exist idempotents (resp. central idempotents) $e_1, \ldots, e_n$ with sum 1 such that $e_ie_j = 0$ whenever $i \neq j$, and $B_i = Re_i$ for all i. In the case where the B_i's are ideals, if $R = B_1 \oplus \cdots \oplus B_n$, then each B_i is a ring with identity e_i, and we have an isomorphism between R and the direct product of rings $B_1 \times \cdots \times B_n$. Show that any isomorphism of R with a finite direct product of rings arises in this way.

Ex. 1.8. Let $R = B_1 \oplus \cdots \oplus B_n$, where the B_i's are ideals of R. Show that any left ideal (resp. ideal) I of R has the form $I = I_1 \oplus \cdots \oplus I_n$ where, for each i, I_i is a left ideal (resp. ideal) of the ring B_i.

Ex. 1.9. Show that for any ring R, the center of the matrix ring $\mathbb{M}_n(R)$ consists of the diagonal matrices $r \cdot I_n$, where r belongs to the center of R.

Ex. 1.10. Let p be a fixed prime.
(a) Show that any ring (with identity) of order p^2 is commutative.
(b) Show that there exists a noncommutative ring *without identity* of order p^2.
(**Hint.** Try the multiplication $(a, b)(c, d) = ((a + b)c, (a + b)d)$ on $(\mathbb{Z}/p\mathbb{Z})^2$.)
(c) Show that there exists a noncommutative ring (with identity) of order p^3.

Ex. 1.11. Let R be a ring possibly without an identity. An element $e \in R$ is called a left (resp. right) identity for R if $ea = a$ (resp. $ae = a$) for every $a \in R$.
(a) Show that a left identity for R need not be a right identity.
(b) Show that if R has a unique left identity e, then e is also a right identity.
(**Hint.** For (b), consider $(e + ae - a)c$ for arbitrary a, $c \in R$.)

Ex. 1.12. A left R-module M is said to be *hopfian* (after the topologist H. Hopf) if any surjective R-endomorphism of M is an automorphism.
(1) Show that any noetherian module M is hopfian.
(2) Show that the left regular module ${}_RR$ is hopfian iff R is Dedekind-finite. (In particular, R being hopfian is a left-right symmetric notion.)
(3) Deduce from (1), (2) that any left noetherian ring R is Dedekind-finite.

Ex. 1.13. Let A be an algebra over a field k such that every element of A is algebraic over k. (Such A is called an *algebraic k-algebra.*)
(a) Show that A is Dedekind-finite.
(b) Show that a left 0-divisor of A is also a right 0-divisor.
(c) Show that a nonzero element of A is a unit iff it is not a 0-divisor.
(d) Let B be a subalgebra of A, and $b \in B$. Show that b is a unit in B iff it is a unit in A.

Ex. 1.14. (Kaplansky) Suppose an element a in a ring has a right inverse b but no left inverse. Show that a has infinitely many right inverses. (In particular, if a ring is finite, it must be Dedekind-finite.)

Ex. 1.15. Let $A = \mathbb{C}[x; \sigma]$, where σ denotes complex conjugation on $\mathbb{C}$.
(a) Show that $Z(A) = \mathbb{R}[x^2]$.
(b) Show that $A/A \cdot (x^2 + 1)$ is isomorphic to $\mathbb{H}$, the division ring of real quaternions.
(c) Show that $A/A \cdot (x^4 + 1)$ is isomorphic to $\mathbb{M}_2(\mathbb{C})$.
(**Hint.** For (c), define a ring homomorphism $\varphi: A \to \mathbb{M}_2(\mathbb{C})$ by taking $\varphi(x) = \begin{pmatrix} 0 & i \\ 1 & 0 \end{pmatrix}$ and $\varphi(a) = \begin{pmatrix} a & 0 \\ 0 & \sigma(a) \end{pmatrix}$ for $a \in \mathbb{C}$.)

Ex. 1.16. Let K be a division ring with center k.
(1) Show that the center of the polynomial ring $R = K[x]$ is $k[x]$.
(2) For any $a \in K \backslash k$, show that the ideal generated by $x - a$ in $K[x]$ is the unit ideal.
(3) Show that any ideal $I \subseteq R$ has the form $R \cdot h$ where $h \in k[x]$.

Ex. 1.17. Let x, y be elements in a ring R such that $Rx = Ry$. Show that there exists a right R-module isomorphism $f: xR \to yR$ such that $f(x) = y$.

Ex. 1.18. For any ring k, let

$$A = \left\{ \begin{pmatrix} a & b \\ c & d \end{pmatrix} : \ a + c = b + d \in k \right\}.$$

Show that A is a subring of $\mathbb{M}_2(k)$, and that it is isomorphic to the ring R of 2×2 lower triangular matrices over k.

Ex. 1.19. Let R be a domain. If R has a minimal left ideal, show that R is a division ring. (In particular, a left artinian domain must be a division ring.)

Ex. 1.20. Let $E = End_R(M)$ be the ring of endomorphisms of an R-module M, and let nM denote the direct sum of n copies of M. Show that $End_R(nM)$ is isomorphic to $\mathbb{M}_n(E)$ (the ring of $n \times n$ matrices over E).

Ex. 1.21. Let R be a finite ring. Show that there exists an infinite sequence $n_1 < n_2 < n_3 < \cdots$ of natural numbers such that, for any $x \in R$, we have $x^{n_1} = x^{n_2} = x^{n_3} = \cdots$.

Ex. 1.22. For any ring k, let $A = \mathbb{M}_n(k)$ and let R (resp. S) denote the ring of $n \times n$ upper (resp. lower) triangular matrices over k.
(1) Show that $R \cong S$.
(2) Suppose k has an anti-automorphism (resp. involution). Show that the same is true for A, R and S.
(3) Under the assumption of (2), show that $R, S, R^{\text{op}}, S^{\text{op}}$ are all isomorphic.

Ex. 1.23. For a fixed $n \geq 1$, let $R = \begin{pmatrix} \mathbb{Z} & n\mathbb{Z} \\ \mathbb{Z} & \mathbb{Z} \end{pmatrix}$ and $S = \begin{pmatrix} \mathbb{Z} & \mathbb{Z} \\ n\mathbb{Z} & \mathbb{Z} \end{pmatrix}$. Show that $R \cong S$, and that these are rings with involutions.

Ex. 1.24. Let R be the ring defined in Exercise 23, where $n \geq 1$ is fixed.
(1) Show that $m \in \mathbb{Z}$ is a square in R iff m is a square in $\mathbb{Z}/n\mathbb{Z}$.
(2) Let $R = \begin{pmatrix} \mathbb{Z} & 2p\mathbb{Z} \\ \mathbb{Z} & \mathbb{Z} \end{pmatrix}$ where p is an odd prime. Show that $2p \in R^2$, $p \in R^2$, but $2 \in R^2$ iff 2 is a square in $\mathbb{Z}/p\mathbb{Z}$.

Ex. 1.25. (Vaserstein) Let a, b, c be such that $ab + c = 1$ in a ring R. If there exists $x \in R$ such that $a + cx \in U(R)$, show that there exists $y \in R$ such that $b + yc \in U(R)$.
(**Hint.** Set $u = a + cx \in U(R)$, and check that the element $y = (1 - bx)u^{-1}$ works. For this choice of y, an inverse for $b + yc$ is given by $a + x(1 - ba)$. The calculations are tricky, and have to be carried out carefully.)

Ex. 1.26. For any right ideal A in a ring R, the *idealizer* of A is defined to be

$$\mathbb{I}_R(A) = \{r \in R : \ rA \subseteq A\}.$$

(1) Show that $\mathbb{I}_R(A)$ is the largest subring of R that contains A as an ideal.
(2) The ring $\mathbb{E}_R(A) := \mathbb{I}_R(A)/A$ is known as the *eigenring* of the right ideal A. Show that $\mathbb{E}_R(A) \cong End_R(R/A)$ as rings. (Note that, in a way, this "computes" the endomorphism ring of an arbitrary cyclic module over any ring.)

Ex. 1.27. Let $R = \mathbb{M}_n(k)$ where k is a ring, and let A be the right ideal of R consisting of matrices whose first r rows are zero. Compute the idealizer $\mathbb{I}_R(A)$ and the eigenring $\mathbb{E}_R(A)$.

Ex. 1.28. Do the same for the right ideal $A = xR$ in the free k-ring $R = k\langle x, y\rangle$.

§2. Semisimplicity

After studying some of the basic examples in the last section, we shall now begin more systematically the study of noncommutative rings. The main focus of the present chapter is on a very important class of rings, called *semisimple* rings. In this section, we shall first study the main properties of these semisimple rings; in the next section, we shall present the basic structure theory for these rings, due to Wedderburn and Artin. Much of this material now lies in the foundations of the theory of noncommutative rings. In fact, it is perhaps not an exaggeration to say that the Wedderburn–Artin structure theory of semisimple rings marked the beginning of the modern phase of development of noncommutative ring theory in the twentieth century.

There are several possible approaches to the notion of semisimple rings. We shall follow a somewhat more modern approach, using the convenient language of modules. We assume the reader is familiar with the rudiments of the theory of modules; in particular, we shall use freely in the text basic facts about submodules, direct sums, composition factors, homomorphisms and exact sequences, etc. If R is a ring and M is a left (resp., right) R-module, we shall often write M as ${}_RM$ (resp., M_R); this suggests that the scalars of R act on the elements of M from the left (resp., right). If R, S are rings and M is an (R, S)-bimodule (as defined in (1.14)), we shall similarly write M as ${}_RM_S$.

Fundamental to the study of the theory of modules are the following two definitions.

(2.1) Definitions. Let R be a ring, and M be a (left) R-module.

(a) M is called a *simple* (or *irreducible*) R-module if $M \neq 0$, and M has no R-submodules other than (0) and M.

(b) M is called a *semisimple* (or *completely reducible*) R-module if every R-submodule of M is an R-module direct summand of M.

Note that, according to these definitions, the zero module is semisimple, but not simple. A direct application of the definition (b) above leads to the following.

(2.2) Remark. Any submodule (resp., quotient module) of a semisimple R-module is semisimple.

Clearly, if $_RM$ is simple, then it is also semisimple. To understand more precisely the relationship between simplicity and semisimplicity, we first prove the following intermediate fact.

(2.3) Lemma. *Any nonzero semisimple left R-module M contains a simple submodule.*

Proof. Let m be a fixed nonzero element of M. In view of (2.2), it suffices to treat the case when $M = R \cdot m$. By Zorn's Lemma, there exists a submodule N of M maximal with respect to the property that $m \notin N$. Take a (necessarily nonzero) submodule N' such that $M = N \oplus N'$. We finish by showing that N' is simple. Indeed if N'' is a nonzero submodule of N', then $N \oplus N''$ must contain m (by the maximality of N), and so $N \oplus N'' = M$, which clearly implies that $N'' = N'$, as desired. QED

The Lemma above enables us to give two other characterizations of semisimple modules. Often these characterizations are used as alternative definitions for semisimplicity.

(2.4) Theorem. *For an R-module $M = {}_RM$, the following three properties are equivalent:*

(1) *M is semisimple.*

(2) *M is the direct sum of a family of simple submodules.*

(3) *M is the sum of a family of simple submodules.*

(*Convention*: The sum and direct sum of an empty family of submodules are both understood to be the zero module. This convention makes the following argument valid in all cases, including the case $M = (0)$.)

Proof of (2.4). (1) ⇒ (3). Let M_1 be the sum of all simple submodules in M, and write $M = M_1 \oplus M_2$, where M_2 is a suitable R-submodule. If $M_2 \neq (0)$, the Lemma implies that M_2 contains a simple R-submodule. But the latter must lie in M_1, a contradiction. Thus, $M_2 = (0)$; i.e., $M_1 = M$.

(3) ⇒ (1). Write $M = \sum_{i \in I} M_i$, where M_i's are simple submodules of M. Let $N \subseteq M$ be a given submodule. To show that N is a direct summand of $_RM$, consider subsets $J \subseteq I$ with the following properties:

(a) $\sum_{j \in J} M_j$ is a *direct* sum.

(b) $N \cap \sum_{j \in J} M_j = (0)$.

It is routine to check that Zorn's Lemma applies to the family of all such J's, with respect to ordinary inclusion. (This is a nonempty family as it contains the empty set.) Thus, we can pick a J to be maximal. For this J, let

$$M' := N + \sum_{j \in J} M_j = N \oplus \bigoplus_{j \in J} M_j.$$

We finish by showing that $M' = M$ (for then N is a direct summand of ${}_RM$). For this, it suffices to show that $M' \supseteq M_i$ for all $i \in I$. But if some $M_i \nsubseteq M'$, the simplicity of M_i implies that $M' \cap M_i = (0)$. From this we have

$$M' + M_i = N \oplus \left(\bigoplus_{j \in J} M_j \right) \oplus M_i,$$

in contradiction to the maximality of J.

(3) $\Rightarrow$ (2) follows from the argument above applied to $N = (0)$.

(2) $\Rightarrow$ (3) is a tautology. QED

We are now ready to introduce the notion of a (left) semisimple ring.

(2.5) Theorem and Definition. *For a ring R, the following statements are equivalent*:

(1) *All short exact sequences of left R-modules split.*

(2) *All left R-modules are semisimple.*

(3) *All finitely generated left R-modules are semisimple.*

(4) *All cyclic left R-modules are semisimple.*

(5) *The left regular R-module ${}_RR$ is semisimple.*

If any of these conditions holds, R is said to be a left semisimple ring.

Note. By using right modules instead, we can similarly define the notion of a *right semisimple ring*. We shall see later, however, that *a ring is left semisimple iff it is right semisimple.* After we prove this fact, we shall be at liberty to drop the adjectives "left" and "right" and just talk about semisimple rings.

Proof of (2.5). Note that (1), (2) are clearly equivalent, and we have a sequence of trivial implications

$$(2) \Rightarrow (3) \Rightarrow (4) \Rightarrow (5).$$

Therefore, it suffices to prove that (5) $\Rightarrow$ (2). Let M be any left R-module where R satisfies (5). In view of (2.2), (5) implies that any cyclic submodule $R \cdot m$ of M is semisimple. Since $M = \sum_{m \in M} R \cdot m$, it follows (from the characterization (2.4)(3)) that M is semisimple. QED

Let R be a left semisimple ring. Using the characterization (2.5)(5), we have a decomposition $R = \bigoplus_{i \in I} \mathfrak{A}_i$ into simple left R-modules $\mathfrak{A}_i$, which are just minimal left ideals in R. Since $1 \in R$, this direct sum is easily seen to be a *finite* direct sum. From this finite decomposition, we can write down a composition series for ${}_RR$ with composition factors $\{{}_R\mathfrak{A}_i\}$. It follows from (1.19) that ${}_RR$ satisfies both the ACC and the DCC for R-submodules.

(2.6) Corollary. *A left semisimple ring R is both left noetherian and left artinian.*

The characterization (2.5)(1) of a left semisimple ring in terms of split short exact sequences enables us to give a homological interpretation of left semisimplicity. This is done by using the important notion of a projective module. Recall that a left R-module P is *R-projective* (or projective for short) if, for any surjective R-homomorphism $f: A \to B$ between left R-modules A, B, and any R-homomorphism $g: P \to B$, there exists an R-homomorphism $h: P \to A$ such that $f \circ h = g$.

The following well-known Proposition in homological algebra offers two alternative characterizations of projective modules. The easy proof of this Proposition will be left to the reader.

(2.7) Proposition. *A (left) R-module P is projective iff P is (isomorphic to) a direct summand of a left free R-module, iff any surjective R-homomorphism from any left R-module onto P splits.*

Using this Proposition, we can now state the homological characterization of the class of (left) semisimple rings.

(2.8) Theorem. *The following conditions on a ring R are equivalent:*

(1) *R is left semisimple;*

(2) *All left R-modules are projective;*

(3) *All finitely generated left R-modules are projective;*

(4) *All cyclic left R-modules are projective.*

Proof. (1) $\Leftrightarrow$ (2) follows from (2.5) and (2.7), and (2) $\Rightarrow$ (3) $\Rightarrow$ (4) is obvious. We finish by showing that (4) $\Rightarrow$ (1). To check (1), we shall verify that the left regular module ${}_R R$ is semisimple. Consider any left ideal $\mathfrak{A} \subseteq R$. By (4) the left R-module $R/\mathfrak{A}$ is projective, so by (2.7) the short exact sequence

$$0 \to \mathfrak{A} \to R \to R/\mathfrak{A} \to 0$$

splits. This implies that $\mathfrak{A}$ is an R-module direct summand of ${}_R R$, as desired. QED

There is also the notion of an injective module which is directly dual to the notion of a projective module. We say that a left R-module I is *R-injective* (or injective for short) if, for any injective R-homomorphism $f: A \to B$ between left R-modules A, B, and any R-homomorphism $g: A \to I$, there exists an R-homomorphism $h: B \to I$ such that $h \circ f = g$. As is easily shown, the second part of (2.7) admits the following dual: *I is injective iff any injective R-homomorphism from I to any left R-module splits.* From this characterization of injective modules, we deduce the following partial analogue of (2.8).

(2.9) Theorem. *The following conditions on a ring R are equivalent:*

(1) *R is left semisimple;*

(2) *All left R-modules are injective.*

Actually, the full analogue of Theorem 2.8 does hold, so one could have added to (2.9) two more equivalent conditions:

(3) *All finitely generated left R-modules are injective.*

(4) *All cyclic left R-modules are injective.*

Of course, we have (1) $\Rightarrow$ (2) $\Rightarrow$ (3) $\Rightarrow$ (4), but the implication (4) $\Rightarrow$ (1) (due to B. Osofsky) is much harder. For a proof of this implication, we refer the reader to pp. 223–224 in *Lectures*.

There are many more characterizations of semisimple rings besides the ones mentioned here. For an exhaustive list of 23 characterizations, see p. 496 (Vol. I) of Rowen [88].

Exercises for §2

Ex. 2.1. Is any subring of a left semisimple ring left semisimple? Can any ring be embedded as a subring of a left semisimple ring?

Ex. 2.2. Let $\{F_i: i \in I\}$ be a family of fields. Show that the direct product $R = \prod_i F_i$ is a semisimple ring iff the indexing set I is finite.

Ex. 2.3. What are semisimple $\mathbb{Z}$-modules?

Ex. 2.4. Let R be the (commutative) ring of all real-valued continuous functions on $[0, 1]$. Is R a semisimple ring?

Ex. 2.5. Let R be a (left) semisimple ring. Show that, for any right ideal I and any left ideal J in R, $IJ = I \cap J$. If I, J, K are ideals in R, prove the following two distributive laws:

$$I \cap (J + K) = (I \cap J) + (I \cap K),$$
$$I + (J \cap K) = (I + J) \cap (I + K).$$

Ex. 2.6. Let R be a right semisimple ring. For $x, y \in R$, show that $Rx = Ry$ iff $x = uy$ for some unit $u \in U(R)$. (**Hint.** Assume $Rx = Ry$. By Exercise 1.17, there exists a right R-homomorphism $f: xR \rightarrow yR$ such that $f(x) = y$. Now extend f to an automorphism of R_R.)

Ex. 2.7. Show that for a semisimple module M over any ring R, the following conditions are equivalent:
(1) M is finitely generated;
(2) M is noetherian;
(3) M is artinian;
(4) M is a finite direct sum of simple modules.

Ex. 2.8. Let M be a semisimple (say, left) module over a ring. Let $\{V_i: i \in I\}$ be a complete set of nonisomorphic simple modules which occur as submodules of M. Let M_i be the sum of all submodules of M which are isomorphic to V_i. It is easy to show that $M = \bigoplus_i M_i$: the M_i's are called the *isotypic components* of M. In the following, we assume that each M_i is finitely generated. By Exercise 7, this means that each $M_i \cong m_i V_i$ for suitable integers m_i. Let N be any submodule of M. Show that $N \cong \bigoplus_i n_i V_i$ for suitable $n_i \le m_i$, and that $M/N \cong \bigoplus_i (m_i - n_i) V_i$.

§3. Structure of Semisimple Rings

In the last section, we studied various characterizing properties of left semisimple rings. In this section, we shall present the full structure theory of this important class of rings, due to J.H.M. Wedderburn (1907) and E. Artin (1927). In essence, the Wedderburn–Artin Theorem enables one to determine completely the class of (left) semisimple rings starting from the more elementary class of division rings. This theorem, regarded by many as the first major result in the abstract structure theory of rings, has remained as important today as in the earlier part of the twentieth century when it was first discovered.

As we have mentioned earlier, the definition of (left) semisimple rings we adopted in §2 is somewhat different from the one Wedderburn originally used. In developing the structure theory of finite-dimensional algebras (or "systems of hypercomplex numbers," as they were called at the beginning of the century), Wedderburn defined the *radical* of such an algebra A to be the largest nilpotent ideal in A. If this (Wedderburn) radical happens to be zero, A is called a semisimple algebra. About twenty years later, Artin extended Wedderburn's methods to the class of rings satisfying the DCC on left ideals.[2] For these rings A (now called left artinian rings), Artin showed that there is also a largest nilpotent ideal, so the Wedderburn radical of A is still defined. If this radical is zero, the left artinian ring A is said to be semisimple. In his 1927 paper, Artin obtained the structure theory of these semisimple (left artinian) rings, in full generalization of Wedderburn's earlier structure theory of semisimple finite-dimensional algebras.

In the next section, we shall show that the Wedderburn–Artin definition of semisimple rings agrees with the definition we gave in §2 (cf. (4.14)). In our exposition, we do not emphasize the Wedderburn radical since it is defined only for certain classes of rings, instead of for all rings. (In the next chapter, we shall study more generally the Jacobson radical, which is defined for all rings, and which agrees with the Wedderburn radical for left artinian rings.) The definition of (left) semisimple rings we gave in (2.5) has the advantage that it is independent of the notion of the radical; for this reason, it is more convenient for our exposition. Actually, this definition of semisimplicity is also quite natural, and very much in keeping with the spirit of the work of E. Noether and H. Weyl in representation theory. We shall continue to use this definition throughout this section.

---

[2] Actually, Artin worked with rings satisfying both DCC and ACC on left ideals, without realizing that the former implies the latter, which was later proved by Levitzki and Hopkins.

Before we proceed to the formulation of the Wedderburn–Artin Theorem, it will be useful to first construct some examples of left semisimple rings. The only obvious example so far is that of a division ring. If D is a division ring, then its left modules are the left vector spaces over D. It is well-known that any short exact sequence of vector spaces splits, so D is left (and right) semisimple. To produce more examples, we shall use the construction of (finite) matrix rings. First let us prove the following elementary result on the classification of ideals in a full matrix ring.

(3.1) Theorem. *Let R be a ring and $\mathbb{M}_n(R)$ be the ring of $n \times n$ matrices over R. Then any ideal I of $\mathbb{M}_n(R)$ has the form $\mathbb{M}_n(\mathfrak{A})$ for a uniquely determined ideal $\mathfrak{A}$ of R. In particular, if R is a simple ring, so is $\mathbb{M}_n(R)$.*

Proof. If $\mathfrak{A}$ is an ideal in R, clearly $\mathbb{M}_n(\mathfrak{A})$ is an ideal in $\mathbb{M}_n(R)$. If $\mathfrak{A}$, $\mathfrak{B}$ are both ideals in R, it is also clear that $\mathfrak{A} = \mathfrak{B}$ iff $\mathbb{M}_n(\mathfrak{A}) = \mathbb{M}_n(\mathfrak{B})$. Now let I be any ideal in $\mathbb{M}_n(R)$, and let $\mathfrak{A}$ be the set of all the $(1, 1)$-entries of matrices in I. This $\mathfrak{A}$ is easily seen to be an ideal in R, and we are done if we can show that $I = \mathbb{M}_n(\mathfrak{A})$. For any matrix $M = (m_{ij})$, we have an identity

$$E_{ij} M E_{k\ell} = m_{jk} E_{i\ell}, \tag{3.2}$$

where $\{E_{ij}\}$ denote the matrix units. Assume $M \in I$. Letting $i = \ell = 1$, the equation above shows that $m_{jk} E_{11} \in I$, and so $m_{jk} \in \mathfrak{A}$ for all j, k. This shows that $I \subseteq \mathbb{M}_n(\mathfrak{A})$. Conversely, take any $(a_{ij}) \in \mathbb{M}_n(\mathfrak{A})$. To show that $(a_{ij}) \in I$, it is enough to show that $a_{i\ell} E_{i\ell} \in I$ for all i, ℓ. Find a matrix $M = (m_{ij}) \in I$ such that $a_{i\ell} = m_{11}$. Then, for $j = k = 1$, (3.2) gives

$$a_{i\ell} E_{i\ell} = m_{11} E_{i\ell} = E_{i1} M E_{1\ell} \in I.$$

The last conclusion of the Theorem is now clear. QED

In the next theorem, we study in detail the properties of a matrix ring over a division ring.

(3.3) Theorem. *Let D be a division ring, and let $R = \mathbb{M}_n(D)$. Then*

(1) *R is simple, left semisimple, left artinian and left noetherian.*

(2) *R has (up to isomorphism) a unique left simple module V. R acts faithfully on V, and ${}_R R \cong n \cdot V$ as R-modules.*[3]

(3) *The endomorphism ring $\mathrm{End}({}_R V)$, viewed as a ring of right operators on V, is isomorphic to D.*

Before we proceed to the proof, a few words on notation are in order here, concerning (3). As a rule, when we consider modules over a ring, endomorphisms will be written *opposite* the scalars. Thus, for a *left* module, we'll write endomorphisms on the *right* of the arguments, and consequently, we'll

[3] $n \cdot V$ (or sometimes nV) denotes the direct sum of n copies of the R-module V.

use the right-hand rule for mapping composition. This convention means, in essence, that we'll be using sometimes the left-hand rule and sometimes the right-hand rule. Whichever rule is being used, however, should always be clear from the context, as it is dictated by the side on which the mappings are written. The adoption of this convention enables us to completely get rid of the formation of opposite rings in the formulation of the Wedderburn–Artin theory. Another considerable notational advantage is the following: If R is a ring and V is a left R-module, with $E = End(_RV)$ viewed as a ring of *right* operators on V, then $V = {_RV_E}$ becomes as (R, E)-bimodule.

We now return to give the

Proof of (3.3). Since D is a simple ring, the simplicity of R follows from (3.1). We may view $R = \mathbb{M}_n(D)$ as a left D-vector space, and, as such, R has finite D-dimension n^2. Since the left ideals of R are D-subspaces of R, it is clear that they must satisfy the *DCC* as well as the *ACC*.

Let V be the n-tuple column space D^n, viewed as a right D-vector space. The ring $R = \mathbb{M}_n(D)$ acts on the left of V by matrix multiplication, so we can view V as a left R-module. In fact R may be identified with $End(V_D)$ by using the usual matrix representation of linear transformations. This shows that $_RV$ is a faithful R-module, and facts in linear algebra (over a division ring) imply that it is a simple R-module. (Alternatively, one can check by a direct matrix calculation that, for any $v \neq 0$ in V, $R \cdot v = V$. This clearly implies the simplicity of $_RV$.)

Now consider the direct sum decomposition

$$R = \mathfrak{A}_1 \oplus \cdots \oplus \mathfrak{A}_n,$$

where $\mathfrak{A}_i$ $(1 \leq i \leq n)$ is the left ideal of R consisting of matrices all of whose columns other than the ith are equal to zero. As a left R-module, $\mathfrak{A}_i$ is clearly isomorphic to $_RV$, so $_RR \cong n \cdot V$ is semisimple. This shows that the ring R is left semisimple. To show the uniqueness of V, let V' be another simple left R-module. Since $V' \cong R/\mathfrak{m}$ for some maximal left ideal $\mathfrak{m} \subset R$, V' is a composition factor of $_RR$. By the Jordan–Hölder Theorem, it follows that $V' \cong V$.

Finally, let us compute $E := End(_RV)$. We have a natural ring homomorphism $\Delta: D \rightarrow E$ defined by

$$v \cdot \Delta(d) = v \cdot d \quad (v \in V, d \in D).$$

The proof will be complete if we can show that Δ is an isomorphism. The injectivity of Δ is clear since D acts faithfully on V_D. To prove the surjectivity of Δ, consider $f \in E$. Writing

$$\begin{pmatrix} 1 \\ 0 \\ \vdots \\ 0 \end{pmatrix} f = \begin{pmatrix} d \\ * \\ \vdots \\ * \end{pmatrix} \quad (d \in D),$$

we have

$$\begin{pmatrix} a_1 \\ \vdots \\ a_n \end{pmatrix} f = \left(\begin{pmatrix} a_1 & 0 & \cdot & 0 \\ \vdots & \vdots & \vdots & \vdots \\ a_n & 0 & \cdot & 0 \end{pmatrix} \begin{pmatrix} 1 \\ 0 \\ \vdots \\ 0 \end{pmatrix} \right) f = \begin{pmatrix} a_1 & 0 & \cdot & 0 \\ \vdots & \vdots & \vdots & \vdots \\ a_n & 0 & \cdot & 0 \end{pmatrix} \begin{pmatrix} d \\ * \\ \vdots \\ * \end{pmatrix}$$

$$= \begin{pmatrix} a_1 d \\ \vdots \\ a_n d \end{pmatrix} = \begin{pmatrix} a_1 \\ \vdots \\ a_n \end{pmatrix} \Delta(d).$$

Hence $f = \Delta(d)$. QED

In order to produce more examples of semisimple rings, we make the following observation on finite direct products of rings.

(3.4) Proposition. *Let $R_1, \ldots, R_r$ be left semisimple rings. Then their direct product $R = R_1 \times \cdots \times R_r$ is also a left semisimple ring.*

Proof. Let $R_i = \mathfrak{A}_{i1} \oplus \cdots \oplus \mathfrak{A}_{im_i}$, where each $\mathfrak{A}_{ij}$ is a minimal left ideal of R_i. Viewing R_i as an ideal in R, $\mathfrak{A}_{ij}$ is also a minimal left ideal of R. From

$${}_RR = R_1 \oplus \cdots \oplus R_r = \bigoplus_{i,j} \mathfrak{A}_{ij},$$

we conclude that R is left semisimple. QED

From (3.3) and (3.4), it follows that, if $D_1, \ldots, D_r$ are division rings, then for arbitrary natural numbers $n_1, \ldots, n_r$,

$$\mathbb{M}_{n_1}(D_1) \times \cdots \times \mathbb{M}_{n_r}(D_r)$$

is a left semisimple ring. This gives a good stock of examples of left semisimple rings. Remarkably, it turns out that these are all the examples! This is the content of the following celebrated result.

(3.5) Wedderburn–Artin Theorem. *Let R be any left semisimple ring. Then $R \cong \mathbb{M}_{n_1}(D_1) \times \cdots \times \mathbb{M}_{n_r}(D_r)$ for suitable division rings $D_1, \ldots, D_r$ and positive integers $n_1, \ldots, n_r$. The number r is uniquely determined, as are the pairs $(n_1, D_1), \ldots, (n_r, D_r)$ (up to a permutation). There are exactly r mutually nonisomorphic left simple modules over R.*

Before we give the proof of this theorem, let us first prove another classical result, due to Issai Schur.

(3.6) Schur's Lemma. *Let R be any ring, and ${}_RV$ be a simple left R-module. Then $End({}_RV)$ is a division ring.*

Proof. Let $0 \neq f \in End(_RV)$. Then $im(f) \neq 0$ and $ker(f) \neq V$. Since $im(f)$ and $ker(f)$ are both submodules of V, it follows that $im(f) = V$ and $ker(f) = 0$; i.e., f is invertible in $End(_RV)$. QED

To prove the Wedderburn–Artin Theorem, let R be a left semisimple ring. Decompose $_RR$ into a finite direct sum of minimal left ideals. Grouping these according to their isomorphism types as left R-modules, we can write

$$\text{(A)} \qquad _RR \cong n_1 V_1 \oplus \cdots \oplus n_r V_r,$$

where $V_1, \ldots, V_r$ are mutually nonisomorphic simple left R-modules. If V is any simple left R-module, V is isomorphic to a quotient of $_RR$ and hence (by the Jordan–Hölder Theorem) isomorphic to some V_i. Therefore $\{V_1, \ldots, V_r\}$ is a full set of mutually nonisomorphic left simple R-modules.

Let us now compute the R-endomorphism rings of the two modules in (A), using the convention that endomorphisms of left modules are written on the right. For $_RR$, the R-endomorphisms are given by right multiplications by elements of R, so $End(_RR) \cong R$. (This is the analogue of (1.12) for the left regular module.) To compute $End(n_1 V_1 \oplus \cdots \oplus n_r V_r)$, let $D_i = End(V_i)$. By Schur's Lemma, each D_i is a division ring, and by Exercise 1.20, $End(n_i V_i) \cong \mathbb{M}_{n_i}(D_i)$. Since there is no nonzero homomorphism from V_i to V_j for $i \neq j$, we have

$$\begin{aligned} End(n_1 V_1 \oplus \cdots \oplus n_r V_r) &\cong End(n_1 V_1) \times \cdots \times End(n_r V_r) \\ &\cong \mathbb{M}_{n_1}(D_1) \times \cdots \times \mathbb{M}_{n_r}(D_r). \end{aligned}$$

Thus, we get a ring isomorphism $R \cong \mathbb{M}_{n_1}(D_1) \times \cdots \times \mathbb{M}_{n_r}(D_r)$.

To prove the uniqueness of this decomposition, suppose we have another isomorphism

$$R \cong \mathbb{M}_{n_1'}(D_1') \times \cdots \times \mathbb{M}_{n_s'}(D_s'),$$

where $D_1', \ldots, D_s'$ are division rings. Let V_i' be the unique simple left module over $\mathbb{M}_{n_i'}(D_i')$. We can also view V_i' as a simple left module over R; clearly $V_i' \ncong V_j'$ as R-modules if $i \neq j$. By (3.3) and (3.4) we have

$$\text{(A}'\text{)} \qquad _RR \cong n_1' V_1' \oplus \cdots \oplus n_s' V_s'.$$

By the Jordan–Hölder Theorem, we see from (A) and (A′) that $r = s$ and that (upon reindexing) $n_i = n_i'$, and $V_i \cong V_i'$ for all i. Writing $R_i = \mathbb{M}_{n_i}(D_i')$, we have by (3.3)(3):

$$D_i' \cong End_{R_i}(V_i') \cong End_R(V_i') \cong End_R(V_i) = D_i$$

for all i. QED

Since $\mathbb{M}_{n_1}(D_1) \times \cdots \times \mathbb{M}_{n_r}(D_r)$ is right semisimple as well as left semisimple, we have the following interesting consequence of (3.5).

(3.7) Corollary. *A left semisimple ring R is always right semisimple (and conversely).*

From now on, we may, therefore, speak of "semisimple rings" without the adjectives "left" or "right."

Note that the existence of an isomorphism

$$R \cong \mathbb{M}_{n_1}(D_1) \times \cdots \times \mathbb{M}_{n_r}(D_r)$$

in (3.5) amounts to the fact that there are ideals $B_1, \ldots, B_r$ in R such that $R = B_1 \oplus \cdots \oplus B_r$, and that, as a ring, each B_i is isomorphic to the simple ring $\mathbb{M}_{n_i}(D_i)$. The following general lemma on direct decompositions of a ring implies that the B_i's are not only determined up to isomorphism as rings, but they are, in fact, uniquely determined as ideals in R. In the sequel, we shall call them the *simple components* (or *Wedderburn components*) of the (semisimple) ring R.

(3.8) Lemma. *Let R be a ring with nonzero ideals $B_1, \ldots, B_r$ and $C_1, \ldots, C_s$ such that*

$$R = B_1 \oplus \cdots \oplus B_r = C_1 \oplus \cdots \oplus C_s$$

and such that each B_i as well as each C_j is indecomposable as an ideal (i.e., not a direct sum of two nonzero subideals). Then $r = s$, and after a permutation of indices, $B_i = C_i$ for $1 \leq i \leq r$.

Proof. Viewing the B_i's as rings, $R \cong B_1 \times \cdots \times B_r$. Under this isomorphism, the ideal C_1 in R corresponds to an ideal $I_1 \times \cdots \times I_r$ where each I_i is an ideal in B_i (see Exercise 1.8). Since C_1 is indecomposable as an ideal, all but one I_i must be zero. After a permutation of indices, we may assume $I_2 = \cdots = I_r = 0$, and so $C_1 \subseteq B_1$. Similarly, we have $B_1 \subseteq C_i$ for some i. But then $C_1 \subseteq C_i$ implies that $i = 1$; hence $C_1 = B_1$. Repeating the same argument for the other C_j's, we obtain the desired conclusion. QED

Let R be a semisimple ring. By (3.5) and (3.8), we know that R has a finite number of uniquely determined simple components: they are minimal (two-sided) ideals in R, and R is their direct sum. One may ask the following natural question: Is it possible to give a *more intrinsic* construction of these simple components? We shall show that this is indeed possible, even independently of the proof we have given for (3.5). In particular, the construction below provides a second route to the existence of the "simple decomposition" of R.

For any minimal left ideal $\mathfrak{A}$ in a ring R, let $B_{\mathfrak{A}}$ be the sum of all the minimal left ideals of R which are isomorphic to $\mathfrak{A}$ as left R-modules. The following general properties of $B_{\mathfrak{A}}$ are valid without any assumptions on the ring R.

(3.9) Lemma. (1) *$B_{\mathfrak{A}}$ is an ideal of R.* (2) *If $\mathfrak{A}$, $\mathfrak{A}'$ are minimal left ideals which are not isomorphic as left R-modules, then $B_{\mathfrak{A}} \cdot B_{\mathfrak{A}'} = 0$.*

Proof. For (1), it is enough to show that, if $\mathfrak{B}$ is a minimal left ideal with $\mathfrak{B} \cong \mathfrak{A}$ (as left R-modules), then $\mathfrak{B} \cdot r \subseteq B_{\mathfrak{A}}$ for any $r \in R$. But $\mathfrak{B} \cdot r$ (as a left

R-module) is a homomorphic image of $\mathfrak{B}$, so we can only have $\mathfrak{B} \cdot r = 0$ or $\mathfrak{B} \cdot r \cong \mathfrak{B} \cong \mathfrak{A}$. In either case, $\mathfrak{B} \cdot r \subseteq B_{\mathfrak{A}}$. For (2), it is enough to show that $\mathfrak{A} \cdot \mathfrak{A}' = 0$. Assume, on the contrary, that $\mathfrak{A} \cdot a' \neq 0$ for some $a' \in \mathfrak{A}'$. Since $\mathfrak{A}'$ is a minimal left ideal, we must have $\mathfrak{A} \cdot a' = \mathfrak{A}'$. But then $\mathfrak{A} \cong \mathfrak{A} \cdot a' = \mathfrak{A}'$ (as left R-modules), a contradiction. QED

Now assume R is a semisimple ring. We can decompose ${}_RR$ as

$$\mathfrak{A}_1 \oplus \cdots \oplus \mathfrak{A}_r \oplus \cdots \oplus \mathfrak{A}_n$$

where each $\mathfrak{A}_i$ is a minimal left ideal. We may assume the indexing is arranged in such a way that $\mathfrak{A}_1, \ldots, \mathfrak{A}_r$ are pairwise nonisomorphic (as left R-modules), and that each $\mathfrak{A}_i$ is isomorphic to (exactly) one of $\mathfrak{A}_1, \ldots, \mathfrak{A}_r$. Let $B_i = B_{\mathfrak{A}_i}$ $(1 \leq i \leq r)$. By (3.9)(1), these are ideals in R; their sum includes all $\mathfrak{A}_i$ $(1 \leq i \leq n)$, and so must be equal to R. By (3.9)(2), we also have $B_i \cdot B_j = 0$ for $i \neq j$; from this, we see easily that

$$(*) \qquad R = B_1 \oplus \cdots \oplus B_r.$$

Note that for any minimal left ideal $\mathfrak{A}$ of R, we have (by the Jordan–Hölder Theorem) $\mathfrak{A} \cong \mathfrak{A}_i$ for some $i \leq r$, and so $B_{\mathfrak{A}} = B_i$ for the same i. (The B_i's are just the isotypic components of ${}_RR$ in the sense of Exercise 2.8. Thus, the direct sum decomposition in $(*)$ would also have followed from that exercise.) Finally, we claim that, *for each i, B_i is a simple, left artinian ring.* The fact that B_i is left artinian is clear, since R itself is left artinian (by (2.6)) and B_i is a homomorphic image of R. To see that B_i is simple, let $I \neq 0$ be an ideal of B_i. Then I is also an ideal in R, and so I contains a minimal left ideal $\mathfrak{A}$ of R. By what we said earlier, $B_{\mathfrak{A}}$ is one of $B_1, \ldots, B_r$; since $B_{\mathfrak{A}}$ contains $\mathfrak{A} \subseteq B_i$, we must have $B_{\mathfrak{A}} = B_i$. We finish by showing that any minimal left ideal $\mathfrak{B} \cong \mathfrak{A}$ is contained in I, for then we must have $I \supseteq B_{\mathfrak{A}} = B_i$. Fix an R-module isomorphism $\varphi \colon \mathfrak{A} \to \mathfrak{B}$. Since $\mathfrak{A}$ is a direct summand of ${}_RR$, we have $\mathfrak{A} = R \cdot e$ for some idempotent $e \in \mathfrak{A}$ (see Exercise 1.7). Then $\mathfrak{A} \cdot e = (R \cdot e)e = \mathfrak{A}$, and so

$$\mathfrak{B} = \varphi(\mathfrak{A}) = \varphi(\mathfrak{A} \cdot e) = \mathfrak{A} \cdot \varphi(e) \subseteq I,$$

as desired.

Note that, by the above method, we have decomposed the semisimple ring R into a finite number of simple components, independently of the proof of (3.5). As a by-product of this construction, we have the following variation of the Wedderburn–Artin Theorem, characterizing the class of simple left artinian rings.

(3.10) Theorem. *Let R be a simple ring. The following statements are equivalent:*

(1) *R is left artinian.*

(2) *R is (left) semisimple.*

(3) *R has a minimal left ideal.*

(4) *$R \cong \mathbb{M}_n(D)$ for some natural number n and some division ring D.*

Proof. Since R is simple, (2) $\Leftrightarrow$ (4) follows from (3.5). Thus we need only show the equivalence of (1), (2), and (3). Now (1) $\Rightarrow$ (3) is obvious, and (2) $\Rightarrow$ (1) is done in (2.6), so the only implication left is (3) $\Rightarrow$ (2). Let $\mathfrak{A}$ be a minimal left ideal of R and consider the ideal $B_{\mathfrak{A}}$. Since R is simple, $B_{\mathfrak{A}} = R$. But then ${}_RR$ is a semisimple R-module, so (2) follows. QED

From the equivalences in (3.10), it follows that if a simple ring R is left artinian, then it is also right artinian (and conversely, by symmetry). For brevity, we shall henceforth refer to such a ring R as a *simple artinian ring*, without the adjectives "left" or "right." Note that, by (2.6), such a ring is also left and right noetherian.

The classification of simple artinian rings as matrix rings over division rings is usually regarded as a part of the Wedderburn–Artin Theorem. There are several possible ways to arrive at this classification. We have described one approach in the foregoing. Because of the great importance of the Wedderburn–Artin Theorem, it will be worthwhile to describe another approach, due to M. Rieffel. This approach is no more complicated than the one we have already described, and yet it gives a more general result which is meaningful for any simple ring (not just the ones satisfying the descending chain condition). The result is the following.

(3.11) Theorem. *Let R be a simple ring, and $\mathfrak{A}$ be a nonzero left ideal. Let $D = End({}_R\mathfrak{A})$ (viewed as a ring of right operators on $\mathfrak{A}$). Then the natural map $f\colon R \to End(\mathfrak{A}_D)$ is a ring isomorphism.*

(In the literature, this property of ${}_R\mathfrak{A}$ is sometimes referred to as the "Double Centralizer Property.")

Proof (M. Rieffel). By definition, the natural map f takes $r \in R$ to the left multiplication by r on $\mathfrak{A}$. This f is a ring homomorphism into $End(\mathfrak{A}_D)$ (the latter being viewed as a ring of *left* operators on $\mathfrak{A}$). Since R is simple, f is injective; we finish by showing that f is surjective. First let us show that, for $r \in \mathfrak{A}$ and $h \in E := End(\mathfrak{A}_D)$, we have

$$\textbf{(3.12)} \qquad h \cdot f(r) = f(h(r)) \in E.$$

In fact, for any $a \in \mathfrak{A}$, *right* multiplication by a on $\mathfrak{A}$ gives an element in D, so $h(ra) = h(r)a$. From this, we have

$$(h \cdot f(r))(a) = h(ra) = h(r)a = f(h(r))(a),$$

hence (3.12). From (3.12), it follows that $E \cdot f(\mathfrak{A}) \subseteq f(\mathfrak{A})$. Since R is simple and $\mathfrak{A} \neq 0$, we have $\mathfrak{A} \cdot R = R$ and so $f(R) = f(\mathfrak{A})f(R)$. But then

$$E \cdot f(R) = E \cdot f(\mathfrak{A})f(R) \subseteq f(\mathfrak{A})f(R) = f(R),$$

so $f(R)$ is a left ideal in E. Since $f(R)$ clearly contains the identity of E, this implies that $f(R) = E$. QED

Using the theorem above, it is easy to give another proof of the structure theorem for simple left artinian rings (or simple rings with minimal left ideals).

(3.13) Corollary. *Let R be a simple ring with a minimal left ideal. Then $R \cong \mathbb{M}_n(D)$ for some n and some division ring D, both of which are uniquely determined.*

Proof. The uniqueness of n and D follows from (3.3). To prove their existence, let $\mathfrak{A}$ be any minimal left ideal of R. By Schur's Lemma, $D := End({}_R\mathfrak{A})$ is a division ring, so $\mathfrak{A}$ is a right vector space over D. By (3.11), we have $R \cong E := End(\mathfrak{A}_D)$. Thus, E is simple. If $dim_D\mathfrak{A}$ is infinite, the set of endomorphisms of finite rank in E would give an ideal different from (0) and E. Thus, we must have $n := dim_D\mathfrak{A} < \infty$, and hence $R \cong E \cong \mathbb{M}_n(D)$, as desired. QED

Now let R be any semisimple ring. Using the "$B_{\mathfrak{A}}$" construction, we have a unique decomposition of R into its simple components, say

$$R = B_1 \times \cdots \times B_r.$$

Each B_i is a simple left artinian ring, so by (3.13), $B_i \cong \mathbb{M}_{n_i}(D_i)$ for some integer n_i and some division ring D_i. This gives a second proof of the Wedderburn–Artin Theorem.

In later chapters (and in *Lectures*), when we go more deeply into the study of the structure of rings, we will find several results which are of the same genre as the Wedderburn–Artin Theorem. For instance, Jacobson's structure theorem on left primitive rings and Morita's theorem on the equivalence of module categories are both more powerful results than the Wedderburn–Artin Theorem, and may be regarded as generalizations of it. To prove these more powerful results, we shall need to use certain new ideas and methods. In the special case of simple (left) artinian rings, these higher methods will yield two more proofs of the structure theorem (3.13) for such rings. However, in order to preserve the historical perspective of this structure theorem, we have refrained in this section from using the more powerful methods of primitive rings or the tools of category equivalences. In this way, we have tried to make the exposition of the Wedderburn–Artin Theorem as elementary as possible. Later, when we study the more general theorems of Jacobson and Morita, we will be able to see how the main idea of the Wedderburn–Artin Theorem evolved into its various higher forms.

To conclude our discussion of the Wedderburn–Artin Theorem, let us remark on two of its special cases. Let k be a field and let R be a finite-dimensional k-algebra. Since any left (or right) ideal of R is a k-subspace of R, any chain of such ideals has bounded length. In particular, R is left (and right) noetherian and artinian. If R is a simple algebra, we have by (3.13) $R \cong \mathbb{M}_n(D)$ for some n and some division ring D. Since D is characterized as the R-endomorphism ring of the unique left simple R-module V, D has also the structure of a (finite-dimensional) k-algebra. Similarly, if R is a finite-dimensional semisimple k-algebra, then

$$R \cong \mathbb{M}_{n_1}(D_1) \times \cdots \times \mathbb{M}_{n_r}(D_r),$$

where each D_i is a finite-dimensional k-division algebra. We shall refer to this fact as the Wedderburn Theorem, since this was the case originally treated by Wedderburn in his 1907 paper. If the ground field k is algebraically closed, each D_i must be k itself and we have

$$R \cong \mathbb{M}_{n_1}(k) \times \cdots \times \mathbb{M}_{n_r}(k).$$

In the special case when $k = \mathbb{C}$, this fact goes back even earlier to the Estonian mathematician T. Molien.

If we work in the category of commutative rings, the Wedderburn–Artin Theorem is basically an easy result. Since simple commutative rings are just fields, the decomposition argument following (3.9) shows that *any semisimple commutative ring is a finite direct product of fields* (*and conversely*). Rings of this type occur, for instance, in linear algebra. Recall that a linear operator T on a finite-dimensional vector space V over a field k is said to be semisimple iff any T-invariant subspace of V has a T-invariant complement. Assuming some future results (cf. (11.1) and (11.7) (1)), it can be seen that *T is semisimple iff the subalgebra $k[T]$ of $\mathrm{End}_k(V)$ is semisimple.* Now if $m(x)$ is the minimal polynomial of T, and

$$m(x) = m_1(x)^{e_1} \cdots m_r(x)^{e_r} \in k[x],$$

where the $m_i(x)$'s are distinct irreducible factors, then

(3.14) $$k[T] \cong \frac{k[x]}{(m(x))} \cong \prod_{i=1}^{r} \frac{k[x]}{(m_i(x)^{e_i})}.$$

This gives a decomposition of $k[T]$ into a finite direct sum of indecomposable ideals. From (3.8), it follows that $k[T]$ is semisimple iff each

$$k[x]/(m_i(x)^{e_i})$$

is a field; i.e., iff each $e_i = 1$. This is a well-known characterization of semisimple operators in linear algebra. Note that, in the special case when k is algebraically closed, this characterization boils down to: T is semisimple iff it is diagonalizable.

Since we have completely determined the structure of artinian simple rings, one may wonder how much of this structure theory can be extended to nonartinian simple rings. For left noetherian simple rings, there does exist a fairly substantial structure theory (see, for instance, Cozzens–Faith [75]); however, the structure of general simple rings remains difficult. We shall not go into this subject in this book. To close this section, we shall, instead, give some examples of nonartinian simple rings. This will serve to beef up our stock of nontrivial examples of rings started in §1.

The first example to be constructed is based on the following observation. Suppose we have a chain of rings

$$R_0 \subseteq R_1 \subseteq R_2 \subseteq \cdots$$

which share the same identity; if each R_i is a simple ring, then so is the union

$$R = \bigcup_{i \geq 0} R_i.$$

In fact, let I be a nonzero ideal in R. Then $I \cap R_i \neq 0$ for some i, and $I \cap R_i$ is an ideal in R_i. Since R_i is simple, $I \cap R_i$ contains 1, and so $I = R$. To construct an explicit example, let D be a fixed division ring and consider $R_i = \mathbb{M}_{2^i}(D)$ $(i \geq 0)$. We shall regard R_i as a subring of R_{i+1} by identifying a $2^i \times 2^i$ matrix M with the $2^{i+1} \times 2^{i+1}$ matrix $\begin{pmatrix} M & 0 \\ 0 & M \end{pmatrix}$. In this way, we have a chain of simple rings

$$R_0 \subseteq R_1 \subseteq R_2 \subseteq \cdots,$$

where $R_0 = D$. Let R be their union. This is a simple ring; however, we shall show below that it is not (left) artinian. For $i \geq 0$, let e_i be the matrix unit in R_i with 1 in the $(1, 1)$-position, and zeros elsewhere. Each e_i is to be viewed as an element of R by using the embeddings

$$R_i \subseteq R_{i+1} \subseteq \cdots \subseteq R.$$

It is easy to see that, for every i, $e_{i+1} = e_{i+1}e_i \in R_{i+1}$. Hence we have a descending chain

$$Re_0 \supseteq Re_1 \supseteq Re_2 \supseteq \cdots$$

of left ideals in R. We shall show that this is a *strictly* descending chain by showing that, for each i, $e_i \notin Re_{i+1}$. This will complete the proof that R is a nonartinian simple ring. Assume, for the moment, that $e_i \in Re_{i+1}$. Then $e_i \in R_je_{i+1}$ for some $j > i$, so $e_i = Me_{i+1} \in R_j$ for some $2^j \times 2^j$ matrix M. However, the $(2^i + 1, 2^i + 1)$ entry of Me_{i+1} is 0, and the $(2^i + 1, 2^i + 1)$ entry of e_i (viewed as a matrix in R_j) is 1. This gives the desired contradiction. We leave it to the reader to show that the ring R here is also not left (or right) noetherian.

To get a second example of a nonartinian simple ring, let D be a division ring as before, and let

$$V = \bigoplus_{i \geq 1} e_iD$$

be a right D-vector space of countably infinite dimension. Let $E = End(V_D)$ and let I be the ideal of E consisting of endomorphisms of finite rank. *We claim that the quotient $R := E/I$ is a simple ring.* To see this, let $\mathfrak{A}$ be any ideal of R properly containing I; we fix an endomorphism $g \in \mathfrak{A} \backslash I$. It suffices to show that, for suitable endomorphisms $f, h \in E$, we have $fgh = 1$, for then $\mathfrak{A}$ must be the unit ideal. Write $V = \ker g \oplus U$ and let $\{u_1, u_2, \ldots\}$ be a basis for U. Then $\{g(u_1), g(u_2), \ldots\}$ are linearly independent, so there exists $f \in E$ which sends each $g(u_i)$ to e_i. Finally, let $h \in E$ be the endomorphism which sends each e_i to u_i. Then $fgh(e_i) = f(g(u_i)) = e_i$ for any i, so we have

$fgh = 1$, as desired. Next, we claim that *the simple ring R is not left noetherian*. To see this, we choose another basis of V consisting of the vectors

$$\{w_{ij}: \ 1 \leq i < \infty, \ 1 \leq j < \infty\},$$

and, for $n \geq 1$, let

$$\mathfrak{A}_n = \{f \in E: \ f(w_{ij}) = 0 \quad \text{for } i \geq n \text{ and } j \geq 1\}.$$

Then, $0 = \mathfrak{A}_1 \subseteq \mathfrak{A}_2 \subseteq \cdots$ is a chain of left ideals in E. We finish by showing that

$$\mathfrak{A}_1 + I \subseteq \mathfrak{A}_2 + I \subseteq \mathfrak{A}_3 + I \subseteq \cdots$$

is a *strictly* increasing chain. To show that

$$\mathfrak{A}_n + I \subsetneq \mathfrak{A}_{n+1} + I,$$

consider an endomorphism $f \in \mathfrak{A}_{n+1}$ with the property that $f(w_{nj}) = w_{nj}$ for all j. If $f = h + k$ where $h \in \mathfrak{A}_n$ and $k \in I$, then

$$w_{nj} = f(w_{nj}) = h(w_{nj}) + k(w_{nj}) = k(w_{nj})$$

for all j. But then $k(V) \supseteq \sum_{j=1}^{\infty} w_{nj}D$, contradicting the fact that k is an endomorphism of finite rank. This shows that R is not left noetherian, and hence also not left artinian, by (3.3) and (3.10). More explicitly, a strictly decreasing chain of left ideals containing I can be constructed by using the same idea: we simply take

$$\mathfrak{B}_n = \{f \in E: \ f(w_{ij}) = 0 \quad \text{for } i \leq n \text{ and } j \geq 1\}$$

and show, as before, that

$$\mathfrak{B}_1 + I \supsetneq \mathfrak{B}_2 + I \supsetneq \cdots.$$

For further properties of the ring E, see Exercises 14 and 15 below.

To conclude this chapter, we shall present some more constructions of nonartinian simple rings by using skew polynomials and skew Laurent polynomials. We first consider the case of the differential polynomial ring $k[x;\delta]$, where δ is a derivation on the ring k. Let us call an ideal $\mathfrak{A}$ of k a *δ-ideal* if $\delta(\mathfrak{A}) \subseteq \mathfrak{A}$. We shall say k is *δ-simple* if $k \neq (0)$ and the only δ-ideals in k are (0) and k. Using these notions, we have the following characterization for $k[x;\delta]$ to be a simple ring, in the case when k is a $\mathbb{Q}$-algebra.

(3.15) Theorem. *For any $\mathbb{Q}$-algebra k with a derivation δ, $R = k[x;\delta]$ is a simple ring iff k is δ-simple and δ is not an inner derivation on k.*

Proof. First assume δ is inner, so that, for some $c \in k$, $\delta(b) = cb - bc$ for all $b \in k$. Then, as we have observed in (1.9), $R = k[t]$ for $t = x - c$, so R is clearly not simple. Next assume k has a δ-ideal $\mathfrak{A} \neq (0)$, k. Then

$$I := \left\{\sum a_i x^i \in R: \ \text{all } a_i \in \mathfrak{A}\right\} = \mathfrak{A} \cdot R$$

is an ideal of R since

$$x\sum a_i x^i = \sum a_i x^{i+1} + \sum \delta(a_i)x^i \in I$$

if all $a_i \in \mathfrak{A}$. Since $I \neq (0), R$, again R is not simple. This proves the "only if" part of the theorem. To prove the converse, assume k is δ-simple, but R is not simple, with, say, an ideal $I \neq (0), R$. We finish by showing that δ must be inner. To see this, let n be the minimum degree for the nonzero (left) polynomials in I, and let $\mathfrak{A}$ be the set of leading coefficients a of polynomials $f \in I$ of degree n, together with 0. Observing that

$$xf - fx = \delta(a)x^n + \cdots,$$

we see easily that $\mathfrak{A}$ is a (nonzero) δ-ideal of k, so $1 \in \mathfrak{A}$; i.e., there exists a polynomial $g = x^n + dx^{n-1} + \cdots$ in I. Since $I \neq R, n > 0$. For any $b \in k$, we see easily by induction that $x^n b = bx^n + n\delta(b)x^{n-1} + \cdots$, so

$$bg - gb = (bd - db - n\delta(b))x^{n-1} + (\text{lower-degree terms}).$$

Since $bg - gb \in I$ (and $\mathbb{Q} \subseteq k$), we see that

$$\delta(b) = b\left(\frac{d}{n}\right) - \left(\frac{d}{n}\right)b$$

for every $b \in k$, so δ is an inner derivation. QED

(3.16) Corollary (Amitsur). *Let k be any simple ring of characteristic* 0. *Then for any non-inner derivation δ on k, $R = k[x;\delta]$ is a nonartinian simple ring.*

Proof. The center C of k is a field (by Exercise 3.4 below), so char $k = 0$ implies that $\mathbb{Q} \subseteq C$; i.e., k is a $\mathbb{Q}$-algebra. Obviously k is δ-simple, so the theorem applies. The descending chain of left ideals

$$Rx \supsetneq Rx^2 \supsetneq \cdots$$

shows that R is not (left) artinian. QED

(3.17) Corollary. *Let k_0 be any simple ring (resp., domain) of characteristic* 0. *Then the Weyl algebras $A_n(k_0)$ $(n \geq 1)$ are all nonartinian simple rings (resp., domains).*

Proof. Since $A_n(k_0) = A_1(A_{n-1}(k_0))$, it suffices to prove the Corollary for $n = 1$. The fact that $A_1(k_0)$ is nonartinian follows as in the last corollary. To show that $A_1(k_0)$ is simple, we use the identification $A_1(k_0) = k_0[y][x;\delta]$, where $\delta = d/dy$ on $k = k_0[y]$. Since y is in the center of $k_0[y]$ but $\delta(y) = 1$, we see that δ is not an inner derivation on k. We finish by showing that, *though k is not simple, it is δ-simple.* Let $\mathfrak{A}$ be any nonzero δ-ideal of k. If $f = ay^n + \cdots$ has minimal degree n among the nonzero (left) polynomials in

$\mathfrak{A}$, then

$$\delta(f) = \frac{df}{dy} = nay^{n-1} + \cdots \in \mathfrak{A}$$

implies that $na = 0$. Since $a \neq 0$ and $\mathbb{Q} \subset k_0$, we must have $n = 0$. Thus $f = a \in \mathfrak{A} \cap k_0$, and the fact that k_0 is a simple ring implies that $1 \in \mathfrak{A} \cap k_0$, so $\mathfrak{A} = k_0[y]$. QED

Though the $A_n(k_0)$'s are not artinian, it can be shown that they are noetherian if the ground ring k_0 is noetherian. Thus, (3.17) provides an interesting class of noetherian simple rings (resp., domains). We note, however, that the assumption on the characteristic of k_0 is essential in (3.17). In fact, if k_0 has characteristic $p > 0$, then in

$$A_1(k_0) = k_0\langle x, y\rangle/(xy - yx - 1),$$

we have (by induction on m) $x^m y - yx^m = mx^{m-1}$, and hence x^p commutes with y (as well as with x). Therefore $A_1(k_0)x^p$ is a nonzero ideal $\subsetneq A_1(k_0)$, so $A_1(k_0)$ is not simple. Similarly, the higher Weyl algebras are also not simple.

Next we shall present the analogue of (3.15) for skew Laurent polynomials. Here, we start with a base ring k equipped with an automorphism σ, and consider the ring $R = k[x, x^{-1}; \sigma]$ of skew Laurent polynomials defined in (1.8). The elements of R have the form $\sum_{i=r}^{s} a_i x^i$ where $a_i \in k$, $r \leq s$ in $\mathbb{Z}$, and multiplication for elements of R is induced by the rule $x^i a = \sigma^i(a)x^i$. The *inner order* of σ is the smallest natural number n such that σ^n is an inner automorphism of k. If no such natural number n exists, we say that σ has infinite inner order. In analogy to the case of derivations considered before, we call an ideal $\mathfrak{A} \subseteq k$ a *σ-ideal* if $\sigma(\mathfrak{A}) = \mathfrak{A}$, and we say that k is *σ-simple* if $k \neq (0)$, and the only σ-ideals of k are (0) and k. Using this terminology, we can give an explicit criterion for $R = k[x, x^{-1}; \sigma]$ to be simple, in analogy to (3.15). Our presentation here follows a paper of D.A. Jordan [84].

(3.18) Theorem. *For k and $R = k[x, x^{-1}; \sigma]$ as above, the following statements are equivalent*:

(1) *R is a simple ring.*

(2) *k is σ-simple and σ has infinite inner order.*

(3) *k is σ-simple, and there is no natural number m for which σ^m is an inner automorphism of k induced by a unit of k fixed by σ.*

Proof. First we prove (2) $\Leftrightarrow$ (3), which can be done without reference to the ring R. Of course, we need only prove (3) $\Rightarrow$ (2). Suppose, for some $n \geq 1, \sigma^n$ is an inner automorphism, say $\sigma^n(r) = brb^{-1}$ for all r, where $b \in U(k)$. Let

$a = b \cdot \sigma(b) \cdot \sigma^2(b) \cdots \sigma^{n-1}(b) \in U(k)$. Since $\sigma^n(b) = b$, we have

$$\begin{aligned} \sigma(a) &= \sigma(b)\sigma^2(b) \cdots \sigma^n(b) \\ &= \sigma(\sigma^n b)\sigma^2(\sigma^n b) \cdots \sigma^n(\sigma^n b) \\ &= \sigma^n(\sigma(b)\sigma^2(b) \cdots \sigma^n(b)) \\ &= b(\sigma(b)\sigma^2(b) \cdots \sigma^n(b))b^{-1} \\ &= a\sigma^n(b)b^{-1} = a. \end{aligned}$$

Now consider the inner automorphism of k induced by the unit $\sigma^i(b)$ $(i \geq 0)$. For $c \in k$, we have

$$\begin{aligned} \sigma^i(b)c\sigma^i(b)^{-1} &= \sigma^i(b\sigma^{-i}(c)b^{-1}) \\ &= \sigma^i(\sigma^n(\sigma^{-i}(c))) \\ &= \sigma^n(c). \end{aligned}$$

Thus, the inner automorphism induced by $\sigma^i(b)$ (for any $i \geq 0$) is σ^n. It follows that the inner automorphism induced by $a = b \cdot \sigma(b) \cdots \sigma^{n-1}(b)$ is σ^{n^2}. Since $\sigma(a) = a$, this gives what we want.

Next we shall prove (1) $\Rightarrow$ (3). Assuming (1), let $\mathfrak{A} \neq 0$ be a σ-ideal in k. Then $\mathfrak{A}[x, x^{-1}; \sigma]$ is easily seen to be a (nonzero) ideal in R. Therefore we must have $\mathfrak{A}[x, x^{-1}; \sigma] = R$, and so $\mathfrak{A} = k$. Next, assume that, for some $n \geq 1$, there is a unit $a \in U(k)$ fixed by σ such that $\sigma^n(c) = aca^{-1}$ for all $c \in k$. We claim that $a^{-1}x^n$ is in the center of R. To see this, it suffices to show that $a^{-1}x^n$ commutes with x and with any $c \in k$. This is seen as follows:

$$x \cdot a^{-1}x^n = \sigma(a^{-1})x^{n+1} = a^{-1}x^{n+1} = a^{-1}x^n \cdot x,$$

$$a^{-1}x^n \cdot c = a^{-1}\sigma^n(c)x^n = a^{-1} \cdot aca^{-1}x^n = c \cdot a^{-1}x^n.$$

Therefore, $1 + a^{-1}x^n$ is central in R, and, being a nonunit, it generates an ideal $\neq (0), R$ in R, contradicting (1).

We finish by proving now (3) $\Rightarrow$ (1). Assuming (3), let $I \neq 0$ be an ideal in R. Clearly, $I \cap k[x; \sigma] \neq 0$. The leading coefficients of the nonzero polynomials in $I \cap k[x; \sigma]$ of minimal degree (say n), together with 0, form a nonzero ideal $\mathfrak{A}$ in k. If

$$b_n x^n + \cdots + b_0 \in I \cap k[x; \sigma],$$

we also have

$$x(b_n x^n + \cdots + b_0)x^{-1} \in I \cap k[x; \sigma];$$

hence $\sigma(b_n) \in \mathfrak{A}$, and similarly $\sigma^{-1}(b_n) \in \mathfrak{A}$. This shows that $\sigma(\mathfrak{A}) = \mathfrak{A}$ and so, by (3), $\mathfrak{A} = k$. Using this, we can find a *monic* polynomial

$$f(x) = x^n + a_{n-1}x^{n-1} + \cdots + a_0 \in I \cap k[x; \sigma].$$

The polynomials $f(x) - xf(x)x^{-1}$ and $cf(x) - f(x)\sigma^{-n}(c)$ (for $c \in k$) both belong to $I \cap k[x; \sigma]$ and have degree $< n$. Hence they must be both zero, which shows that $a_i = \sigma(a_i)$ and $ca_i = a_i\sigma^{i-n}(c)$ for all $c \in k$ and $i \leq n-1$. In particular, $a_i k = ka_i$, and this is a σ-ideal in k. If some $a_i \neq 0$, then, by (3), we must have $a_i k = ka_i = k$, so $a_i \in U(k)$. But then $c = a_i\sigma^{i-n}(c)a_i^{-1}$, and hence $\sigma^{n-i}(c) = a_i c a_i^{-1}$, for all $c \in k$, in contradiction to (3). Thus, all a_i must be zero, and so $f(x) = x^n \in I$. Since $x^n \in U(R)$, we have now $I = R$, as desired. QED

The following easy consequence of (3.18) yields another large family of nonartinian simple rings.

(3.19) Corollary. *Let k be any field with an automorphism σ of infinite order. Then $R = k[x, x^{-1}; \sigma]$ is a nonartinian simple domain.*

Proof. The simplicity of the domain R follows from (3.18). The descending chain of left ideals

$$R \cdot (x+1) \supsetneq R \cdot (x+1)^2 \supsetneq \cdots$$

shows that R is not (left) artinian. (In general, a domain R can never be left or right artinian, unless it is a division ring: see Exercise 1.19.) QED

Exercises for §3

Ex. 3.1. Show that if R is semisimple, so is $\mathbb{M}_n(R)$.

Ex. 3.2. Let R be a domain. Show that if $\mathbb{M}_n(R)$ is semisimple, then R is a division ring.

Ex. 3.3. Let R be a semisimple ring.
(a) Show that any ideal of R is a sum of simple components of R.
(b) Using (a), show that any quotient ring of R is semisimple.
(c) Show that a simple artinian ring S is isomorphic to a simple component of R iff there is a surjective ring homomorphism from R onto S.

Ex. 3.4. Show that the center of a simple ring is a field, and the center of a semisimple ring is a finite direct product of fields.

Ex. 3.5. Let M be a finitely generated left R-module and $E = End({}_RM)$. Show that if R is semisimple (resp. simple artinian), then so is E.

Ex. 3.6A. Let M be a left R-module and $E = End({}_RM)$. If ${}_RM$ is a semisimple R-module, show that M_E is a semisimple E-module.

Ex. 3.6B. In the above Exercise, if M_E is a semisimple E-module, is ${}_RM$ necessarily a semisimple R-module?

Ex. 3.7. Let R be a simple ring which is finite-dimensional over its center k. (k is a field by Exercise 4 above.) Let M be a finitely generated left R-module and let $E = End(_R M)$. Show that

$$(dim_k\ M)^2 = (dim_k\ R)(dim_k\ E).$$

Ex. 3.8. For R as in Exercise 7, show that R is isomorphic to a matrix algebra over its center k iff R has a nonzero left ideal $\mathfrak{A}$ with $(dim_k\ \mathfrak{A})^2 \le dim_k\ R$.

Ex. 3.9. (a) Let R, S be rings such that $\mathbb{M}_m(R) \cong \mathbb{M}_n(S)$. Does this imply that $m = n$ and $R \cong S$?
(b) Let us call a ring A a matrix ring if $A \cong \mathbb{M}_m(R)$ for some integer $m \ge 2$ and some ring R. True or False: "A homomorphic image of a matrix ring is also a matrix ring"?

Ex. 3.10. Let R be any semisimple ring.
(1) Show that R is Dedekind-finite, i.e. $ab = 1$ implies $ba = 1$ in R.
(2) If $a \in R$ is such that $I = aR$ is an ideal in R, then $I = Ra$.
(3) Every element $a \in R$ can be written as a unit times an idempotent.
Remark. (3) expresses the fact that semisimple rings are "unit-regular": for a more general view of this, see Exercises 4.14B and 4.14C.

Ex. 3.11. Let R be an n^2-dimensional algebra over a field k. Show that $R \cong \mathbb{M}_n(k)$ (as k-algebras) iff R is simple and has an element whose minimal polynomial over k has the form $(x - a_1) \cdots (x - a_n)$ where $a_1, \ldots, a_n \in k$. (**Hint.** For the "if" part, produce a chain of left ideals of length n in R, and apply the Wedderburn-Artin Theorem.)

Ex. 3.12. For a subset S in a ring R, let $ann_\ell(S) = \{a \in R : aS = 0\}$ and $ann_r(S) = \{a \in R : Sa = 0\}$. Let R be a semisimple ring, I be a left ideal and J be a right ideal in R. Show that $ann_\ell(ann_r(I)) = I$ and $ann_r(ann_\ell(J)) = J$.

Ex. 3.13. Let R be a simple, infinite-dimensional algebra over a field k. Show that any nonzero left R-module V is also infinite-dimensional over k.

Ex. 3.14. (Over certain rings, the "rank" of a free module may not be defined.) Let D be a division ring, $V = \bigoplus_{i=1}^{\infty} e_i D$, and $E = End(V_D)$. Define $f_1, f_2 \in E$ by $f_1(e_n) = e_{2n}$, $f_2(e_n) = e_{2n-1}$ for $n \ge 1$. Show that $\{f_1, f_2\}$ form a free E-basis for E_E. Therefore, as right E-modules, $E \cong E^2$; using this, show that $E^m \cong E^n$ for any finite $m, n > 0$!

Ex. 3.15. Show that the ring E above has exactly three ideals: 0, E, and the ideal consisting of endomorphisms of finite rank.

Ex. 3.16. Generalize the exercise above to the case of $E = End(V_D)$, where $\dim_D V = \alpha$ is an arbitrary infinite cardinal. (**Hint.** For any infinite cardinal $\beta \le \alpha$, let E_β be the ideal of E consisting of all endomorphisms of rank $< \beta$. Show that the ideals of E are: 0, E, and all the E_β's. It follows, in particular, that E/E_α is a simple ring.)

Ex. 3.17. (K.A. Hirsch) Let k be a field of characteristic zero, and (a_{ij}) be an $m \times m$ skew symmetric matrix over k. Let R be the k-algebra generated by

$x_1, \ldots, x_m$ with the relations $x_i x_j - x_j x_i = a_{ij}$ for all i, j. Show that R is a simple ring iff $\det(a_{ij}) \neq 0$. In particular, R is always nonsimple if m is odd. (**Hint.** After a congruence transformation, we may assume that (a_{ij}) consists of a number of diagonal blocks $\begin{pmatrix} 0 & 1 \\ -1 & 0 \end{pmatrix}$ together with a zero block of size $r \geq 0$. If $r > 0$, x_m generates a proper ideal in R. If $r = 0$, then $m = 2n$ for some n, and R is the nth Weyl algebra $A_n(k)$ over k.)

Ex. 3.18. (Quebbemann) Let k be a field of characteristic zero, and let R be the Weyl algebra $A_1(k)$ with generators x, y and relation $xy - yx = 1$. Let $p(y) \in k[y]$ be a fixed polynomial.
(a) Show that $R \cdot (x - p(y))$ is a maximal left ideal in R, and that the simple R-module $V = R/R \cdot (x - p(y))$ has R-endomorphism ring equal to k.
(b) Show that $R \to End(V_k)$ is injective but not an isomorphism.
(**Hint.** Identify V with $k[y]$ and show that $x - p(y)$ acts as differentiation on $k[y]$. If $v(y) \in V \backslash \{0\}$ has degree m, $R \cdot v(y)$ contains $(x - p(y))^m \cdot v$, which is a nonzero constant. Finally, let $f \in End({}_R V)$ and let $f(1) = g(y)$. Then $f(x \cdot 1) = xf(1)$ shows that $dg/dy = 0$. Therefore $g \in k$ and $f(v(y)) = f(v(y) \cdot 1) = v(y)g$.)

Ex. 3.19. True or False: "If I is a minimal left ideal in a ring R, then $\mathbb{M}_n(I)$ is a minimal left ideal in $\mathbb{M}_n(R)$"?

Ex. 3.20. Let $\mathfrak{A}_i$ $(1 \leq i \leq n)$ be ideals in a ring R, and let $\mathfrak{A} = \bigcap_i \mathfrak{A}_i$. If each $R/\mathfrak{A}_i$ is semisimple, show that $R/\mathfrak{A}$ is semisimple.

Ex. 3.21. For any finitely generated left module M over a ring R, let $\mu(M)$ denote the smallest number of elements that can be used to generate M. If R is an artinian simple ring, find a formula for $\mu(M)$ in terms of $\ell(M)$, the composition length of M.

Ex. 3.22. (1) Generalize the computation of $\mu(M)$ in the above exercise to the case of a finitely generated left module M over a semisimple ring R.
(2) Show that μ is subadditive, in the sense that $\mu(M \oplus N) \leq \mu(M) + \mu(N)$ for finitely generated R-modules M, N.
(3) Show that $N \subseteq M \Rightarrow \mu(N) \leq \mu(M)$.

Ex. 3.23. Show that a nonzero ring R is simple iff each simple left R-module is faithful.

Ex. 3.24. (Jacobson) A subset S in a ring R is said to be *nil* (resp. *nilpotent*) if every $s \in S$ is nilpotent (resp. if $S^m = 0$ for some m, where S^m denotes the set of all products $s_1 \cdots s_m$ with $s_i \in S$).
(1) Let $R = \mathbb{M}_n(D)$ where D is a division ring. Let $S \subseteq R$ be a nonempty nil set which is closed under multiplication. Show that $S^n = 0$.
(2) Let R be any semisimple ring. Show that any nonempty nil set $S \subseteq R$ closed under multiplication is nilpotent.

CHAPTER 2
Jacobson Radical Theory

Historically, the notion of the radical was a direct outgrowth of the notion of semisimplicity. It may be somewhat surprising, however, to remark that the radical was studied first in the context of nonassociative rings (namely, finite-dimensional Lie algebras) rather than associative rings. In the work of E. Cartan, the radical of a finite-dimensional Lie algebra A (say over $\mathbb{C}$) is defined to be the maximal solvable ideal of A: it is obtained as the sum of all the solvable ideals in A. The Lie algebra A is semisimple iff its radical is zero, i.e., iff it has no nonzero solvable ideals. Cartan characterized the semisimplicity of a Lie algebra in terms of the nondegeneracy of its Killing form, and showed that any semisimple Lie algebra is a finite direct sum of simple Lie algebras. Moreover, he classified the finite-dimensional simple Lie algebras (over $\mathbb{C}$). Therefore, the structure theory of finite-dimensional semisimple Lie algebras is completely determined.

The theory of semisimple rings we developed in the last chapter may be viewed as the analogue of Cartan's theory in the context of associative rings. It was developed by Molien and Wedderburn for finite-dimensional (associative) algebras, and generalized later by Artin to rings satisfying the descending chain condition. In the last chapter, we based the development of this theory on the use of semisimple (or completely reducible) modules; this treatment is somewhat different from the original treatment of Wedderburn. In developing the theory of finite-dimensional algebras over a field, Wedderburn defined for every such algebra A an ideal, *rad* A, which is the largest nilpotent ideal of A, i.e., the sum of all the nilpotent ideals of A. In parallel with Cartan's theory, the (finite-dimensional) algebra A is semisimple iff its radical is zero. Such an algebra A is (uniquely) the direct product of a finite number of finite-dimensional simple algebras A_i, and each A_i is (uniquely) a matrix algebra over a finite-dimensional division algebra. This beautiful theory of Wedderburn laid the modern foundation for the study of the structure of finite-dimensional algebras. Artin extended Wedderburn's theory to rings with the minimum condition (appropriately called Artinian rings). For such rings R, the sum of all nilpotent ideals in R is nilpotent, so R has a largest nilpotent ideal *rad* R, called the *Wedderburn radical* of R. As we saw in the last Chapter, Wedderburn's theory of simple and semisimple algebras can be extended successfully to rings satisfying the descending chain condition on one-sided ideals.

What about rings which do not satisfy Artin's descending chain condition? For these rings R, the sum of all nilpotent ideals need no longer be nilpotent; thus, R may not possess a largest nilpotent ideal, and so we no longer have the notion of a Wedderburn radical (see Ex. 4.25). The problem of finding the appropriate generalization of Wedderburn's radical for arbitrary rings remained untackled for almost forty years. Finally, in a fundamental paper in 1945, N. Jacobson initiated the general notion of the radical of an arbitrary ring R: by definition, the (Jacobson) radical, *rad* R, of R is the intersection of the maximal left (or maximal right) ideals of R. For rings satisfying a one-sided minimum condition, the Jacobson radical agrees with the classical Wedderburn radical, so, in general, the former provides a good substitute for the latter. Ever since its inception, Jacobson's general theory of the radical has proved to be fundamental for the study of the structure of rings. In this chapter, we shall present the basic definition and properties of the Jacobson radical, and study the behavior of the radical under certain changes of rings. In the next chapter, we shall apply this material to the representation theory of algebras and groups, and explain the basic connections between ring theory and group representation theory, with some applications to group theory itself.

Needless to say, this chapter is a beginning, not an end. Having defined the Jacobson radical for arbitrary rings, we are led to a more general notion of semisimplicity: a ring R is called *Jacobson* (or *J*-) *semisimple* if *rad* $R = 0$. These J-semisimple rings generalize the semisimple rings in Chapter 1, and therefore should play an important role in the study of rings possibly not satisfying the descending chain condition. We shall try to develop this theme in more detail in Chapter 4. Also, there are several other radicals which can be defined for arbitrary rings, and which provide alternative generalizations of the Wedderburn radical. These other radicals may not be as fundamental as the Jacobson radical, but in one way or another, they reflect more accurately the structure of the nil (and nilpotent) ideals of the ring, so one might say that these other radicals resemble the Wedderburn radical more than does the Jacobson radical. However, we can do only one thing at a time. So, in this chapter, we focus our attention on the Jacobson radical; other kinds of radicals (upper and lower nilradicals and the Levitzki radical) will be taken up in a future chapter.

§4. The Jacobson Radical

As we mentioned in the Introduction, the *Jacobson radical* of a ring R, denoted by $rad\ R$, is defined to be the intersection of all the maximal left ideals of R. Note that if $R \neq 0$, maximal left ideals always exist by Zorn's Lemma, and so $rad\ R \neq R$. If $R = 0$, then there are no maximal left ideals; in this case, of course, we define the Jacobson radical to be zero.

In the definition of $rad\ R$ above, we used the maximal left ideals of R, so $rad\ R$ should be called the left radical of R, and we can similarly define the right radical of R (by intersecting the maximal right ideals). It turns out, fortuitously, that the left and right radicals coincide, so the distinction is, after all, unnecessary. We shall now try to prove this result: this is done by obtaining a left–right symmetric characterization of the (left) radical $rad\ R$. First we prove a lemma characterizing the elements of $rad\ R$ in terms of the left-invertible elements of R, and in terms of the simple left R-modules.

(4.1) Lemma. *For $y \in R$, the following statements are equivalent:*

(1) $y \in rad\ R$;

(2) $1 - xy$ *is left-invertible for any* $x \in R$;

(3) $yM = 0$ *for any simple left R-module M.*

Proof. (1) $\Rightarrow$ (2) Assume $y \in rad\ R$. If, for some x, $1 - xy$ is not left-invertible, then $R \cdot (1 - xy) \subsetneq R$ is contained in a maximal left ideal $\mathfrak{m}$ of R. But $1 - xy \in \mathfrak{m}$ and $y \in \mathfrak{m}$ imply that $1 \in \mathfrak{m}$, a contradiction.

(2) $\Rightarrow$ (3) Assume $ym \neq 0$ for some $m \in M$. Then we must have $R \cdot ym = M$. In particular, $m = x \cdot ym$ for some $x \in R$, so $(1 - xy)m = 0$. Using (2), we get $m = 0$, a contradiction.

(3) $\Rightarrow$ (1) For any maximal left ideal $\mathfrak{m}$, $R/\mathfrak{m}$ is a simple left R-module, so by (3), $y \cdot R/\mathfrak{m} = 0$ which implies that $y \in \mathfrak{m}$. By definition, we have $y \in rad\ R$. QED

For any left R-module M, the annihilator of M is defined to be

$$ann\ M := \{r \in R:\ rM = 0\}.$$

This is easily seen to be an ideal of R. Consider the special case of a *cyclic* module M: we can take M to be $R/\mathfrak{A}$, where $\mathfrak{A}$ is a left ideal in R. In this case

$$ann\ M = \{r \in R:\ r \cdot R/\mathfrak{A} = 0\} = \{r \in R:\ rR \subseteq \mathfrak{A}\}.$$

This is easily seen to be the largest ideal of R contained in $\mathfrak{A}$. It is sometimes called the *core* of the left ideal $\mathfrak{A}$. (If R happens to be commutative then, of course, $ann(R/\mathfrak{A}) = \mathfrak{A}$.) The Lemma (4.1) above has the following immediate consequence.

(4.2) Corollary. *rad* $R = \bigcap$ *ann* M, *where* M *ranges over all the simple left* R*-modules. In particular, rad* R *is an ideal of* R.

The next result is a refinement of (4.1). It adds a fourth condition to the list in (4.1) which is a strengthening of the condition (2) there. We could have proved all four equivalences in one stroke, but the proof below will show that it is more convenient to prove (4.1) (and (4.2)) first before adding the fourth equivalent condition.

(4.3) Lemma. *For* $y \in R$, *the following statements are equivalent*:

(1) $y \in rad\ R$;

(2)′ $1 - xyz \in U(R)$ (*the group of units of* R) *for any* $x, z \in R$.

Proof. Since (2)′ ⇒ (2) in (4.1) (by letting $z = 1$), it suffices to prove (1) ⇒ (2)′. Let $y \in rad\ R$, $x, z \in R$. By (4.2), $yz \in rad\ R$, so by (4.1), there exists $u \in R$ such that $u(1 - xyz) = 1$. Again by (4.2), $xyz \in rad\ R$, so another application of (4.1) shows that $u = 1 + u(xyz)$ is left-invertible. Since u is also right-invertible, we have $u \in U(R)$ and hence $1 - xyz \in U(R)$. QED

(4.4) Remark. Since (2) and (2)′ involve only the notion of invertible and left-invertible elements, it is perhaps not unreasonable to ask for a direct proof for (2) ⇒ (2)′, not using the notion of the radical. Such a proof can indeed be given, using an exercise from Chapter 1. Suppose $y \in R$ satisfies (2) in (4.1), and let $x, z \in R$. Then there exists $v \in R$ such that $v(1 - zxy) = 1$. Now $v = 1 + (vzx)y$ is left-invertible as well as right-invertible, so $v \in U(R)$ and therefore $1 - z(xy) \in U(R)$. By Exercise 1.6, it follows that $1 - (xy)z \in U(R)$.

Let us now record some consequences of the results above.

(4.5) Corollary. (A) *rad* R *is the largest left ideal* (*and hence the largest ideal*) $\mathfrak{A} \subseteq R$ *such that* $1 + \mathfrak{A} \subseteq U(R)$. (B) *The left radical of* R *agrees with its right radical.*

Proof. (A) follows from (4.1), (4.2) and (4.3). Since (A) gives a left–right symmetric characterization of *rad* R, the conclusion (B) follows. (Of course, (2)′ in (4.3) is another left–right symmetric characterization of *rad* R.) QED

For later reference, we record here one more property of the Jacobson radical. The proof of this is immediate, so we suppress it.

(4.6) Proposition. *Let* $\mathfrak{A}$ *be any ideal of* R *lying in rad* R. *Then* $rad(R/\mathfrak{A}) = (rad\ R)/\mathfrak{A}$.

The notion of the Jacobson radical of a ring leads to a new notion of semisimplicity, which we now introduce.

(4.7) Definition. A ring R is called *Jacobson semisimple* (or *J-semisimple* for short) if $rad\ R = 0$.

Jacobson semisimple rings are also called *semiprimitive rings* in the literature; we shall henceforth use these two terms interchangeably. Of course, the latter term may seem a little mysterious at this point, since we have not yet introduced the notion of primitive rings. After we have introduced primitive rings in a later chapter, the reader will be able to put the term "semiprimitive rings" in a better perspective. At this point, the reader should be warned that, in many books and papers, "J-semisimplicity" is taken as the definition of "semisimplicity." We will *not* adopt this convention as it will confuse the J-semisimple rings in the sense of (4.7) with the semisimple rings we studied in Chapter 1. (The precise relationship between these two notions will be clarified a little later.)

In a manner of speaking, Jacobson semisimple rings are ubiquitous: for any ring R, $R/rad\ R$ is a J-semisimple ring associated with R (see (4.6)). One might hope to study the structure of a ring R by first studying the structure of $R/rad\ R$. The two rings R and $R/rad\ R$ share certain common properties, as we shall show in the following.

(4.8) Proposition. *R and $R/rad\ R$ have the same simple left modules. An element $x \in R$ is left-invertible (resp., invertible) in R iff $\bar{x} \in \bar{R}$ is left-invertible (resp., invertible) in $\bar{R} := R/rad\ R$.*

Proof. The first statement follows easily from (4.2). For the second statement, it is enough to treat the case of left-invertibility. The "only if" part is clear. For the "if" part, take $y \in R$ such that $\bar{y}\bar{x} = 1 \in \bar{R}$. Then $1 - yx \in rad\ R$, so

$$yx \in 1 + rad\ R \subseteq U(R).$$

Clearly this implies that x has a left inverse in R. QED

Next we shall study the relationship between $rad\ R$ and the nil (resp., nilpotent) ideals of R. Let us first recall the appropriate definitions.

(4.9) Definition. A one-sided (or two-sided) ideal $\mathfrak{A} \subseteq R$ is said to be *nil* if $\mathfrak{A}$ consists of nilpotent elements; $\mathfrak{A}$ is said to be *nilpotent* if $\mathfrak{A}^n = 0$ for some natural number n.

Note that $\mathfrak{A}^n = 0$ means that $a_1 \cdots a_n = 0$ for any set of elements $a_1, \ldots, a_n \in \mathfrak{A}$. This condition is much stronger than $\mathfrak{A}$ being nil. For instance, in the (commutative) ring

$$R = \mathbb{Z}[x_1, x_2, x_3, \ldots]/(x_1^2, x_2^3, x_3^4, \ldots),$$

the ideal $\mathfrak{A}$ generated by $\bar{x}_1, \bar{x}_2, \bar{x}_3, \ldots$ is nil, but easily shown to be not nilpotent. One advantage of "nilpotent" over "nil" is seen from the following easy result.

(4.10) Lemma. *Let $\mathfrak{A}_i$ $(1 \leq i \leq m)$ be a finite set of left ideals in R. If each $\mathfrak{A}_i$ is nilpotent, then $\mathfrak{A}_1 + \cdots + \mathfrak{A}_m$ is also nilpotent.*

Proof. By induction, it is enough to handle the case $m = 2$. Changing notations, let $\mathfrak{A}$, $\mathfrak{B}$ be nilpotent left ideals, say $\mathfrak{A}^n = 0 = \mathfrak{B}^n$. For $\mathfrak{C} = \mathfrak{A} + \mathfrak{B}$, we claim that $\mathfrak{C}^{2n} = 0$. To see this, consider a product

$$(a_1 + b_1) \cdots (a_{2n} + b_{2n})$$

of $2n$ elements in $\mathfrak{C}$ $(a_i \in \mathfrak{A}, b_i \in \mathfrak{B})$. When this product is expanded, each term in it is a product of $2n$ elements, some from $\mathfrak{A}$ and some from $\mathfrak{B}$. In each of these terms, there will be at least n factors from $\mathfrak{A}$, or else at least n factors from $\mathfrak{B}$. Since $\mathfrak{A}$, $\mathfrak{B}$ are left ideals, it follows from $\mathfrak{A}^n = \mathfrak{B}^n = 0$ that such a product is zero, and so $\mathfrak{C}^{2n} = 0$ as claimed. QED

(4.11) Lemma. *If a left (resp., right) ideal $\mathfrak{A} \subseteq R$ is nil, then $\mathfrak{A} \subseteq rad\ R$.*

Proof. Let $y \in \mathfrak{A}$. Then for any $x \in R$, $xy \in \mathfrak{A}$ is nilpotent. It follows that $1 - xy$ has an inverse (given by $\sum_{i=0}^{\infty}(xy)^i$). Therefore, by (4.1), we have $y \in rad\ R$. QED

We are now ready to show that the Jacobson radical provides a good generalization of the Wedderburn radical in that, in the case of left artinian rings, the two radicals indeed coincide.

(4.12) Theorem. *Let R be a left artinian ring. Then $rad\ R$ is the largest nilpotent left ideal, and it is also the largest nilpotent right ideal.*

Proof. In view of the above lemma, we are done if we can show that $J := rad\ R$ is nilpotent. Applying the left DCC to

$$J \supseteq J^2 \supseteq J^3 \supseteq \cdots,$$

there exists an integer k such that

$$J^k = J^{k+1} = \cdots = I \quad \text{(say)}.$$

We claim that $I = 0$. Indeed, if $I \neq 0$, then, among all left ideals $\mathfrak{A}$ such that $I \cdot \mathfrak{A} \neq 0$, we can choose a minimal one, say $\mathfrak{A}_0$ (by the left DCC). Fix an element $a \in \mathfrak{A}_0$ such that $I \cdot a \neq 0$. Then

$$I \cdot (Ia) = I^2 a = Ia \neq 0,$$

so by the minimality of $\mathfrak{A}_0$, we have $I \cdot a = \mathfrak{A}_0$. Thus, $a = ya$ for some $y \in I \subseteq rad\ R$. But then $(1-y)a = 0$ implies that $a = 0$ since $1 - y \in U(R)$. This is a contradiction, so we must have $I = J^k = 0$. QED

The theorem we just proved and the lemma preceding it have the following pleasant consequence:

(4.13) Corollary. *In a left artinian ring, any nil* 1*-sided ideal is nilpotent.*

In ring theory, there are many similar results of the "nil implies nilpotent" variety. The one above is the first one we encounter in our exposition. There will be a few other such results that will be developed in future chapters.

If R is a commutative ring, then any nilpotent element of R is contained in *rad* R (since all the nilpotent elements form a nil ideal). If R is not commutative, however, this may no longer be the case. For instance, let D be any division ring, and $R = \mathbb{M}_n(D)$ $(n \geq 2)$. Using the known structure of left ideals in R (as developed in Chapter 1), it is easy to see that *rad* $R = 0$. Thus, R has no nonzero nil left ideals, but nevertheless, nilpotent elements abound.

The next theorem gives the basic connection between the semisimple rings as we have defined them in Chapter 1, and the J-semisimple rings defined in (4.7).

(4.14) Theorem. *For any ring R, the following three statements are equivalent:*

(1) *R is semisimple.*

(2) *R is J-semisimple and left artinian.*

(3) *R is J-semisimple, and satisfies DCC on principal left ideals.*

Proof. (1) $\Rightarrow$ (2). Assume R is semisimple, and let $\mathfrak{A} = rad\ R$. There exists a left ideal $\mathfrak{B}$ such that $R = \mathfrak{A} \oplus \mathfrak{B}$. If $\mathfrak{A} \neq 0$, then $\mathfrak{B}$ is contained in a maximal left ideal $\mathfrak{m}$. But then $\mathfrak{m} \not\supseteq \mathfrak{A}$, a contradiction.

(2) $\Rightarrow$ (3) is trivial, so it only remains to show that (3) $\Rightarrow$ (1). Assume R satisfies (3). We can derive the following two properties of R:

(a) *Every left ideal $\mathfrak{A} \neq 0$ contains a minimal left ideal I.* (Indeed, choose I to be a minimal member of the family of nonzero principal left ideals $\subseteq \mathfrak{A}$; then I is clearly minimal as a left ideal.)

(b) *Every minimal left ideal $\mathfrak{B}$ is a direct summand of ${}_RR$.* (In fact, since $\mathfrak{B} \neq 0 = rad\ R$, there exists a maximal left ideal $\mathfrak{m}$ not containing $\mathfrak{B}$. Then $\mathfrak{B} \cap \mathfrak{m} = 0$ and so ${}_RR = \mathfrak{B} \oplus \mathfrak{m}$.)

Now assume R is not semisimple. Take a minimal left ideal $\mathfrak{B}_1$, and write ${}_RR = \mathfrak{B}_1 \oplus \mathfrak{A}_1$. Then $\mathfrak{A}_1 \neq 0$, and so (by (a)) there exists a minimal left ideal $\mathfrak{B}_2 \subseteq \mathfrak{A}_1$. By (b), $\mathfrak{B}_2$ is a direct summand in ${}_RR$ and hence also in $\mathfrak{A}_1$, so we can write $\mathfrak{A}_1 = \mathfrak{B}_2 \oplus \mathfrak{A}_2$. Continuing in this fashion, we get a descending chain of left ideals

$$\mathfrak{A}_1 \supsetneqq \mathfrak{A}_2 \supsetneqq \mathfrak{A}_3 \supsetneqq \cdots.$$

These are direct summands of $_RR$, so they are principal left ideals of R. This contradicts (3), so R must be semisimple. QED

In the theorem above, of course, we can also show (2) $\Rightarrow$ (1) without routing through the condition (3). Assuming R is left artinian, we can write $rad\ R = \bigcap_{i=1}^{n} \mathfrak{m}_i$ for a finite number of maximal left ideals $\mathfrak{m}_1, \ldots, \mathfrak{m}_n$. If $rad\ R = 0$, then R embeds into $\bigoplus_{i=1}^{n} R/\mathfrak{m}_i$, so $_RR$ is semisimple. However, it is nice to have the extra equivalent condition (3). More importantly, the class of rings satisfying the *DCC* on principal left ideals turn out to be of independent interest: these are the *right* (not left!) perfect rings which we shall study in a later chapter. Using this terminology, (1) $\Leftrightarrow$ (3) in the theorem says that *R is semisimple iff R is J-semisimple and right perfect.* This result will be rather "clear" from the more general perspective of perfect rings from §§23–24. Also worth noting is the fact that the same result is true with "*J*-semisimple" above replaced by "semiprime": see (10.24).

When Artin proved his analogues of Wedderburn's structure theorems for left *DCC* rings in 1927, he did not seem to realize that left *DCC* in fact implies left *ACC*. Throughout his work, he assumed, in fact, that the rings in consideration satisfy *both* chain conditions. The result that left *DCC* implies left *ACC* was obtained only some years later, independently by C. Hopkins and J. Levitzki. Using the notion of the Jacobson radical, we shall now give a proof of this very important result.

(4.15) Hopkins–Levitzki Theorem (1939). *Let R be a ring for which rad R is nilpotent, and $\bar{R} = R/rad\ R$ is semisimple.* (*Such a ring R is called semiprimary.*) *Then for any R-module $_RM$, the following statements are equivalent*:

(1) *M is noetherian.*

(2) *M is artinian.*

(3) *M has a composition series.*

In particular, (A) *a ring is left artinian iff it is left noetherian and semiprimary*; (B) *any finitely generated left module over a left artinian ring has a composition series.*

Proof. By (4.12) and (4.14), a left artinian ring is semiprimary. Thus, (A) follows from the equivalence of (1) and (2), applied to the left regular module $_RR$. (B) follows from the equivalence of (2) and (3) since a finitely generated left module over a left artinian ring is also artinian.

We have observed before (cf. (1.19)) that, for any $_RM$, (3) implies (1) and (2). To complete the proof, it is therefore enough to show that (1) $\Rightarrow$ (3) and (2) $\Rightarrow$ (3) for semiprimary rings.

Assume M is either noetherian or artinian. For $J = rad\ R$, fix an integer n such that $J^n = 0$ and let $\bar{R} = R/J$. Consider the filtration

$$M \supseteq JM \supseteq J^2M \supseteq \cdots \supseteq J^nM = 0.$$

It is enough to show that each filtration factor $J^iM/J^{i+1}M$ has a composition series. But $J^iM/J^{i+1}M$ is either noetherian or artinian, as a module over $\bar{R}$. Since $\bar{R}$ is semisimple, $J^iM/J^{i+1}M$ is a direct sum of simple $\bar{R}$-modules. The chain condition on $J^iM/J^{i+1}M$ implies that this direct sum must be finite, so $J^iM/J^{i+1}M$ does have a composition series as an $\bar{R}$-module.

QED

For an example of a semiprimary ring which is neither left nor right artinian (resp., noetherian), see Exercise 26.

We now give some examples illustrating the notion of the Jacobson radical.

(1) The ring $\mathbb{Z}$, or more generally, any full ring R of algebraic integers in a number field K ($[K : \mathbb{Q}] < \infty$) is J-semisimple. In fact, if $0 \neq a \in R$, then only a finite number of prime ideals contain it. On the other hand, there are infinitely many nonzero prime ($=$ maximal) ideals in R, so a cannot be in the Jacobson radical.

(2) Let R be a commutative affine algebra over a field k. By this we mean that R is finitely generated as a k-algebra, say $R \cong k[x_1, \ldots, x_n]/\mathfrak{A}$, where $\mathfrak{A}$ is an ideal in the polynomial ring $k[x_1, \ldots, x_n]$. By Hilbert's Nullstellensatz, the radical of $\mathfrak{A}$, defined by

$$\sqrt{\mathfrak{A}} := \{f \in k[x_1, \ldots, x_n] : \ f^r \in \mathfrak{A} \text{ for some } r\},$$

coincides with the intersection of all the maximal ideals of $k[x_1, \ldots, x_n]$ containing $\mathfrak{A}$. (A full proof of this will be given in the next section in the more general context of Hilbert rings: see (5.4).) From this, it follows that $rad\ R$ is exactly the nil radical

$$Nil\ R = \{r \in R : \ r \text{ is nilpotent}\}.$$

In particular, it follows that R is J-semisimple iff R is reduced (i.e., R has no nonzero nilpotent elements).

(3) For commutative rings R in general, $rad\ R$ may not be equal to the nil radical $Nil\ R$. For instance, if R is a commutative local domain, then $Nil\ R = 0$, but $rad\ R$ is the unique maximal ideal of R, which is nonzero if R is not a field.

(4) Let V be an n-dimensional vector space over a field k, and let

$$R = \bigwedge(V) = k \oplus \bigwedge^1(V) \oplus \cdots \oplus \bigwedge^n(V)$$

be its exterior algebra. Let $\mathfrak{m}$ be the ideal $\bigwedge^1(V) \oplus \cdots \oplus \bigwedge^n(V)$. Since $v \wedge v = 0$ for any vector $v \in V$, we see easily that $\mathfrak{m}^{n+1} = 0$. Thus $\mathfrak{m} \subseteq rad\ R$. From $R/\mathfrak{m} \cong k$, it follows that $\mathfrak{m} = rad\ R$, and that $\mathfrak{m}$ is the unique maximal left (resp., right) ideal of R. (In the terminology of §19, R is a (noncommutative) local ring.)

(5) Any simple ring R is J-semisimple, since $rad\ R$, being an ideal $\subsetneq R$, must be zero.

(6) Let k be a division ring, and R be the ring of upper triangular $n \times n$ matrices with entries in k. Let J be the subset of R consisting of matrices with zeros on the main diagonal. *We claim that* $J = rad\ R$. First, it is easy to see that J is an ideal of R, and that $J^n = 0$. Thus, $J \subseteq rad\ R$. By (4.6), we have $(rad\ R)/J = rad(R/J)$. But R/J is isomorphic to the ring of $n \times n$ diagonal matrices, so

$$R/J \cong k \times \cdots \times k$$

is semisimple. This gives $rad(R/J) = 0$, and so $rad\ R = J$. From the decomposition of R/J, we see that, up to isomorphism, there are exactly n simple left R-modules, each being 1-dimensional over k. The ith one, M_i, is given by ${}_Rk$, with the action

$$\begin{pmatrix} a_{11} & & * \\ & \ddots & \\ 0 & & a_{nn} \end{pmatrix} \cdot b = a_{ii}b \quad (\text{for } b \in k).$$

There is also a natural left R-module V, given by the space of column n-tuples with the usual matrix action of R on the left. It is of interest to compute the composition factors of ${}_RV$. To do this, let V_i $(0 \le i \le n)$ be the set of column n-tuples with the last $n - i$ entries $= 0$. These are easily seen to be R-submodules of V, giving a filtration

$$0 = V_0 \subsetneq V_1 \subsetneq \cdots \subsetneq V_n = V.$$

The filtration factor V_i/V_{i-1} is 1-dimensional over k, with k-basis given by the ith unit column vector e_i. The R-action on V_i/V_{i-1} is given by

$$(a_{ij}) \cdot (e_i + V_{i-1}) = a_{ii}e_i + V_{i-1}.$$

Thus, $V_i/V_{i-1} \cong M_i$, and the composition factors of ${}_RV$ are precisely $\{M_1, \ldots, M_n\}$, each occurring with multiplicity 1. As an easy exercise, the reader can verify that $J^iV = V_{n-i}$, and that ${}_RV$ is "uniserial," that is,

$$0 = V_0 \subsetneq V_1 \subsetneq \cdots \subsetneq V_n = V$$

is the *only* composition series for ${}_RV$.

(7) *What is the Jacobson radical of a full matrix ring* $\mathbb{M}_n(R)$ *over a given ring* R? The hardly surprising answer is that

$$rad\ \mathbb{M}_n(R) = \mathbb{M}_n(rad\ R).$$

For the inclusion "$\supseteq$," it suffices to show that $a \in rad\ R$ implies that $aE_{ij} \in rad\ \mathbb{M}_n(R)$, i.e., that $N = I - M \cdot aE_{ij}$ is invertible for every

$M \in \mathbb{M}_n(R)$. (As usual, the E_{ij}'s denote the matrix units.) Write $M = \sum m_{k\ell} E_{k\ell}$. Then

$$N = I - M \cdot aE_{ij} = I - \sum_k m_{ki} a E_{kj} = I - m_{ji} a E_{jj} - \sum_{k \neq j} m_{ki} a E_{kj}.$$

Write $(1 - m_{ji}a)^{-1}$ as $1 - b$, where $b \in R$. Then $I - bE_{jj}$ is the inverse of $I - m_{ji} a E_{jj}$, so

$$(I - bE_{jj})N = I - (I - bE_{jj}) \sum_{k \neq j} m_{ki} a E_{kj} = I - \sum_{k \neq j} m_{ki} a E_{kj}.$$

Since the matrix on the *RHS* is invertible (with inverse $I + \sum_{k \neq j} m_{ki} a E_{kj}$), it follows that N is invertible. This shows that $rad(\mathbb{M}_n(R)) \supseteq \mathbb{M}_n(rad\ R)$. For the reverse inclusion, write $rad\ \mathbb{M}_n(R) = \mathbb{M}_n(\mathfrak{A})$, where $\mathfrak{A}$ is a suitable ideal in R (see (3.1)). For $a \in \mathfrak{A}$, we have then $a \cdot I \in rad\ \mathbb{M}_n(R)$ so $I - b \cdot aI = (1 - ba)I$ is invertible for any $b \in R$. This clearly implies that $1 - ba$ is invertible in R and so, by (4.1), $a \in rad\ R$. Thus, $\mathfrak{A} \subseteq rad\ R$, from which we have $rad\ \mathbb{M}_n(R) \subseteq \mathbb{M}_n(rad\ R)$. (A more general method for deriving the equation $rad\ \mathbb{M}_n(R) = \mathbb{M}_n(rad\ R)$ will be given later by using the theory of idempotents; see (21.14).)

(8) *Let R be a ring such that $S := U(R) \cup \{0\}$ is a division ring. Then R is J-semisimple.* To see this, note that $S \cap rad\ R$ is an ideal in S, so it is zero. Now let $y \in rad\ R$. Then $1 + y \in U(R) \subseteq S$. Subtracting 1, we see that $y \in S \cap rad\ R = 0$.

(4.16) Corollary. *Any ring R freely generated by a set of indeterminates $\{x_i\}$ over a division ring k is J-semisimple.*

Proof. By an easy degree argument, a polynomial in the indeterminates $\{x_i\}$ is invertible iff it is a nonzero constant in k. Thus $U(R) \cup \{0\} = k$, so the Corollary follows from the preceding observation (8). QED

By exactly the same degree argument, we can also deduce from (8):

(4.17) Corollary. *Let k be any division ring. Then any polynomial ring $k[\{x_i\}]$ in the commuting variables $\{x_i\}$ is J-semisimple. The skew polynomial rings $k[x; \sigma]$ (σ any endomorphism of k) and $k[x; \delta]$ (δ any derivation of k) are also J-semisimple.*

(9) Let k be a field, and R be a k-algebra. An element $x \in R$ is said to be *algebraic over k* if it satisfies a nontrivial polynomial equation with coefficients in k. We have the following interesting description of the algebraic elements in the Jacobson radical of R.

(4.18) Proposition. *Let $x \in rad\ R$, where R is a k-algebra. Then x is algebraic over k iff x is nilpotent.*

Proof. The "if" part is obvious. For the converse, let $x \in rad\ R$ be algebraic over k. Write down a polynomial equation for x in ascending degrees, say

$$x^r + a_1 x^{r+1} + \cdots + a_n x^{r+n} = 0$$

where $a_i \in k$. Since

$$1 + a_1 x + \cdots + a_n x^n \in 1 + rad\ R \subseteq U(R),$$

it follows that $x^r = 0$ so we must have $r \geq 1$ and x is nilpotent. QED

A k-algebra R is said to be an *algebraic algebra* if every element $x \in R$ is algebraic over k. The Proposition we just proved, together with (4.11), imply the following:

(4.19) Corollary. *Let R be an algebraic algebra over k. Then rad R is the largest nil ideal of R.*

At this point, we ought to mention some examples of algebraic algebras. First, any finite-dimensional algebra over a field k is clearly an algebraic k-algebra. In general, then, an algebraic algebra is just a k-algebra which is a union of its finite-dimensional k-subalgebras. For some infinite-dimensional examples, we can take algebraic field extensions K/k with $[K : k] = \infty$. Further examples are given by group algebras kG over groups G which are locally finite (any finitely generated subgroup of G is finite): here, any element $\alpha = a_1 g_1 + \cdots + a_n g_n$ of kG belongs to the finite-dimensional k-subalgebra kH where H is the (finite) group generated by $g_1, \ldots, g_n$. In particular, if G is any abelian torsion group, then kG is an algebraic k-algebra.

There are some further consequences of (4.18) which involve the consideration of the cardinal numbers, $|k|$ and $dim_k\ R$. We shall give some such consequences in the following

(4.20) Theorem (Amitsur). *Suppose $dim_k\ R < |k|$ (as cardinal numbers), where R is a k-algebra. Then rad R is the largest nil ideal of R.*

Proof. It suffices to show that *rad* R is nil. First suppose k is a finite field. The hypothesis implies that R is a finite ring. In particular R is left artinian, so by (4.12), *rad* R is, in fact, nilpotent. In the following, we may therefore assume that k is infinite. To show that *rad* R is nil, it suffices (by (4.18)) to show that every $r \in rad\ R$ is algebraic over k. For any $a \in k^* = k \backslash \{0\}$, $a - r = a(1 - a^{-1} r) \in U(R)$. Since $dim_k\ R < |k| = |k^*|$, the elements $\{(a - r)^{-1} : a \in k^*\}$ cannot be k-linearly independent. Therefore, there exist distinct elements $a_1, \ldots, a_n \in k^*$ such that there is a dependence relation

$$\sum_{i=1}^{n} b_i (a_i - r)^{-1} = 0,$$

where $b_i \in k$ are not all zero. Clearing denominators, we have

$$\sum_{i=1}^{n} b_i(a_1 - r) \cdots \widehat{(a_i - r)} \cdots (a_n - r) = 0,$$

where, as usual, *wedge* means omission of a factor. Therefore, r is a root of the k-polynomial

$$f(x) = \sum b_i(a_1 - x) \cdots \widehat{(a_i - x)} \cdots (a_n - x).$$

Since $f(a_i) = b_i \prod_{j \neq i}(a_j - a_i)$ is nonzero at least for some i, f is not the zero polynomial. Therefore, r is algebraic over k, as claimed. QED

(4.21) Corollary. *Let R be a countably generated algebra over an uncountable field k. Then rad R is the largest nil ideal of R.*

Proof. As a k-vector space, R has a countable basis, so the hypothesis $dim_k\ R < |k|$ in the Theorem is fulfilled. QED

To conclude this section, we shall prove the following result which is of fundamental importance in the theory of rings and modules.

(4.22) Nakayama's Lemma. *For any left ideal $J \subseteq R$, the following statements are equivalent:*

(1) *$J \subseteq rad\ R$.*

(2) *For any finitely generated left R-module M, $J \cdot M = M$ implies that $M = 0$.*

(3) *For any left R-modules $N \subseteq M$ such that M/N is finitely generated, $N + J \cdot M = M$ implies that $N = M$.*

Proof. (1) $\Rightarrow$ (2). Assume $M \neq 0$. Then, among all submodules $\subsetneq M$, there is a maximal one, say M'. (This M' exists by Zorn's Lemma, in view of the finite generation of M.) Then M/M' is simple, and so $J \cdot (M/M') = 0$; i.e., $J \cdot M \subseteq M'$. In particular, $J \cdot M \neq M$.

(2) $\Rightarrow$ (3) follows by applying (2) to the quotient module M/N.

(3) $\Rightarrow$ (1). Suppose some element $y \in J$ is not in *rad* R. Then $y \notin \mathfrak{m}$ for some maximal left ideal $\mathfrak{m}$ of R. We have $\mathfrak{m} + J = R$ so, a fortiori, $\mathfrak{m} + J \cdot R = R$. From (3) it follows that $\mathfrak{m} = R$, a contradiction. QED

Remark. (1) $\Rightarrow$ (2) can also be proved *without* Zorn's Lemma, as follows. If $J \cdot M = M$ and $M \neq 0$, let $m_1, \ldots, m_k$ be a minimal set of generators ($k \geq 1$). We can then write $m_1 = r_1 m_1 + \cdots + r_k m_k$ for suitable $r_i \in J$. Since $1 - r_1$ is a unit, this implies that $m_1 \in Rm_2 + \cdots + Rm_k$, a contradiction.

While we have called (4.22) Nakayama's Lemma, the idea behind this lemma originated from the work of more than one mathematician. In the commutative case and when M itself is an ideal of R, (1) $\Rightarrow$ (2) was discovered and used effectively by W. Krull. The module-theoretic formulation of (2), (3) above is due to G. Azumaya and T. Nakayama. When Nakayama himself was asked what would be the correct attribution of (4.22) (see Nagata [62], p. 213), he suggested modestly that it should be Krull–Azumaya in the commutative case, and Jacobson–Azumaya in the noncommutative case. Since this is obviously too complicated, we shall follow the majority of mathematicians and call (4.22) Nakayama's Lemma. Most often, this Lemma is used in the form (3) for $J = rad\ R$.

We close this section by discussing a very important class of rings which is "between" semisimple rings and J-semisimple rings. These are the von Neumann regular rings, discovered (around 1935) by John von Neumann in connection with his work on continuous geometry and operator algebras. The following memorable result is "Part II, Chapter 2, Theorem 2.2" in von Neumann's book "Continuous Geometry."

(4.23) Theorem. *For any ring R, the following are equivalent:*

(1) *For any $a \in R$, there exists $x \in R$ such that $a = axa$.*

(2) *Every principal left ideal is generated by an idempotent.*

(2)′ *Every principal left ideal is a direct summand of ${}_RR$.*

(3) *Every finitely generated left ideal is generated by an idempotent.*

(3)′ *Every finitely generated left ideal is a direct summand of ${}_RR$.*

Since the condition (1) is left–right symmetric, we see that the same theorem also holds with the word "left" replaced by "right" in the last four conditions. In general, an element $a \in R$ is said to be *von Neumann regular* if $a \in aRa$. If every $a \in R$ is von Neumann regular, we say that R is a *von Neumann regular ring*. The conditions (2), (2)′, (3), (3)′ above (and their right analogues) are therefore characterizations of such rings.

Proof of (4.23). (2) $\Leftrightarrow$ (2)′ and (3) $\Leftrightarrow$ (3)′ are easy (see Exercise 1.7). Let us prove (1) $\Leftrightarrow$ (2). Assume (1), and consider a principal left ideal $R \cdot a$. Choose $x \in R$ such that $axa = a$. Then

$$e := xa = xaxa = e^2,$$

and $e \in R \cdot a$ while $a = axa = ae \in R \cdot e$, so $R \cdot a = R \cdot e$. Conversely, assume (2) and let $a \in R$. Writing $R \cdot a = R \cdot e$ where $e = e^2$, we have $e = xa$ and $a = ye$ for some $x, y \in R$. Then

$$axa = ye \cdot e = ye = a.$$

Since (3) obviously implies (2), it only remains to show that (2) $\Rightarrow$ (3). By induction, it suffices to show that, for any two idempotents $e, f, I = Re + Rf$

is generated by an idempotent. Now $I = Re + Rf(1-e)$ and $Rf(1-e) = Re'$ for some idempotent e', for which $e'e \in Rf(1-e)e = 0$. Thus, $e'(e'+e) = e'$, which leads easily to

$$I = Re + Re' = R(e' + e).$$

Therefore, $I = Re''$ for some idempotent e''. (An explicit choice for e'' is $e + e' - ee'$, as the reader may check.) QED

(4.24) Corollary. *Semisimple $\Rightarrow$ von Neumann regular $\Rightarrow$ J-semisimple.*

Proof. The first implication follows from the characterization $(2)'$ of von Neumann regular rings. (In view of this characterization, we see, in fact, that von Neumann regularity is a very natural weakening of semisimplicity.) The second implication follows from the general observation that, if $a \in aRa$ in any ring R, then $a \in rad\ R \Rightarrow a = 0$. Indeed, if $a = axa$ where $x \in R$, then $a(1 - xa) = 0$ implies $a = 0$ since $1 - xa \in U(R)$.

The earlier result (4.14) that semisimple rings are exactly the left (resp., right) artinian J-semisimple rings has the following good analogue.

(4.25) Theorem. *Semisimple rings are exactly the left (resp., right) noetherian von Neumann regular rings.*

Proof. We have already seen that semisimple rings are left noetherian and von Neumann regular ((2.6) and (4.24)). Conversely, if a ring R is left noetherian and von Neumann regular, then every left ideal of R is finitely generated and hence a direct summand of $_RR$, by using the characterization $(3)'$ of (4.23). Therefore, R is (left) semisimple. QED

(4.26) Corollary. *If a von Neumann regular ring is left noetherian, then it is noetherian and artinian.*

Note that direct products and quotient rings of von Neumann regular rings are all von Neumann regular. Any Boolean ring (a ring in which every element is idempotent) is von Neumann regular. More generally, any ring in which every element a satisfies $a^{n(a)} = a$ for some $n(a) \geq 2$ is von Neumann regular.

For an element a in a von Neumann regular ring R, there exist usually more than one $x \in R$ such that $a = axa$. Any such x may be thought of as a kind of pseudo-inverse of a. (If $a \in U(R)$, then of course x is unique and $x = a^{-1}$.) The idea of such pseudo-inverses is best illustrated by the proof of the following Proposition, which provides a big class of examples of von Neumann regular rings.

(4.27) Proposition. *Let M be any semisimple (right) module over a ring k. Then $R = End(M_k)$ is von Neumann regular.*

Proof. Consider any $f \in R$, with $K := ker\, f$. Fix a k-submodule $N \subseteq M$ such that $M = K \oplus N$. Then f maps N isomorphically onto $N' := f(N)$, and we can find another k-submodule $K' \subseteq M$ such that $M = K' \oplus N'$. Now define $g \in R$ to be such that $g(K') = 0$ and $g|N'$ is the inverse of $f|N$. Then clearly $fgf = f$! QED

In the above, if k is a division ring, then M_k is always semisimple. In the special case when M has finite dimension n over k, the Proposition shows that $\mathbb{M}_n(k)$ is von Neumann regular. By taking finite direct products of such matrix rings, we have a round-about way of seeing that semisimple rings are von Neumann regular. But we could have taken infinite direct products to get nonsemisimple examples. And of course, if $dim\, M_k$ is infinite, we get nonsemisimple, non-Dedekind-finite examples as well.

Finally, we note that the two rings R constructed on p. 40 are both *simple non-noetherian* von Neumann regular rings.

Exercises for §4

In this book we deal only with rings with an identity element. In particular, the theory of the Jacobson radical was developed in the text for rings with an identity. However, by doing things a little more carefully, the whole theory can be carried over to rings possibly without an identity. In Exercises 1–7 below, we sketch the steps necessary in developing this more general theory; in these exercises, R denotes a ring possibly without 1.

Ex. 4.1. In R, define $a \circ b = a + b - ab$. Show that this binary operation is associative, and that $(R, \circ)$ is a monoid with zero as the identity element.

Ex. 4.2. An element $a \in R$ is called *left* (resp. *right*) *quasi-regular* if a has a left (resp. right) inverse in the monoid $(R, \circ)$ with identity. If a is both left and right quasi-regular, we say that a is *quasi-regular*.
(1) Show that if ab is left quasi-regular, then so is ba.
(2) Show that any nilpotent element is quasi-regular.
(3) Show that, if R has an identity 1, the map φ: $(R, \circ) \to (R, \times)$ sending a to $1 - a$ is a monoid isomorphism. In this case, an element a is left (right) quasi-regular iff $1 - a$ has a left (resp. right) inverse with respect to ring multiplication.

Ex. 4.3. A set $I \subseteq R$ is called quasi-regular (resp. left or right quasi-regular) if every element of I is quasi-regular (resp. left or right quasi-regular). Show that if a left ideal $I \subseteq R$ is left quasi-regular, then it is quasi-regular.

Ex. 4.4. Define the Jacobson radical of R by

$$rad\ R = \{a \in R:\ Ra \text{ is left quasi-regular}\}.$$

Show that rad R is a quasi-regular ideal which contains every quasi-regular left (resp. right) ideal of R. (In particular, rad R contains every nil left or right ideal of R.) Show that, if R has an identity, the definition of *rad* R here agrees with the one given in the introduction to this section.

Ex. 4.5. A left ideal $I \subseteq R$ is said to be *modular* (or *regular*[1]) if there exists $e \in R$ which serves as a "*right* identity mod I"; i.e. $re \equiv r \pmod{I}$ for every $r \in R$.
(a) Show that if $I \subsetneq R$ is a modular left ideal, then I can be embedded in a modular maximal left ideal of R.
(b) Show that rad R is the intersection of all modular maximal left (resp. right) ideals of R. (**Hint.** If $e \in R$ is *not* left quasi-regular, then $I = \{r - re\colon\ r \in R\}$ is a modular left ideal not containing e.)

Ex. 4.6. A left R-module M is said to be *simple* (or *irreducible*) if $R \cdot M \neq 0$ and M has no R-submodules other than (0) and M. Show that $_RM$ is simple iff $M \cong R/\mathfrak{m}$ (as left R-modules) for a suitable modular maximal left ideal $\mathfrak{m} \subset R$. Show that *rad* R is the intersection of the annihilators of all simple left R-modules.

Ex. 4.7. Show that $rad(R/rad\ R) = 0$, and that, if I is an ideal in R, then, viewing I as a ring, $rad\ I = I \cap rad\ R$. This shows, in particular, that a ring R may be equal to its Jacobson radical: if this is the case, R is said to be a *radical ring*. Show that R is a radical ring iff it has no simple left (resp. right) modules.

In the following problems, we return to our standing assumption that all rings to be considered have an identity element.

Ex. 4.8. An ideal $I \subsetneq R$ is called a *maximal ideal* of R if there is no ideal of R strictly between I and R. Show that any maximal ideal I of R is the annihilator of some simple left R-module, but not conversely. Defining $rad'R$ to be the intersection of all maximal ideals of R, show that $rad\ R \subseteq rad'R$, and give an example to show that this may be a strict inclusion. ($rad'R$ is called the *Brown-McCoy radical* of R.)

Ex. 4.9. Let R be a J-semisimple domain and a be a nonzero central element of R. Show that the intersection of all maximal left ideals not containing a is zero.

Ex. 4.10. Show that if $f\colon R \to S$ is a surjective ring homomorphism, then $f(rad\ R) \subseteq rad\ S$. Give an example to show that $f(rad\ R)$ may be smaller than $rad\ S$.

Ex. 4.11. If an ideal $I \subseteq R$ is such that R/I is J-semisimple, show that $I \supseteq rad\ R$. (Therefore, $rad\ R$ is the smallest ideal $I \subseteq R$ such that R/I is J-semisimple.)

---

[1] We mention this alternate term only because it is sometimes used in the literature. Since "regular" has too many meanings, we shall avoid using it altogether.

Ex. 4.12A. Let $\mathfrak{A}_i$ $(i \in I)$ be ideals in a ring R, and let $\mathfrak{A} = \bigcap_i \mathfrak{A}_i$. True or False: "If each $R/\mathfrak{A}_i$ is J-semisimple, then so is $R/\mathfrak{A}$"?

Ex. 4.12B. Show that, for any direct product of rings $\prod R_i$, $rad(\prod R_i) = \prod rad\ R_i$. (Therefore, any direct product of J-semisimple rings is J-semisimple.)

Ex. 4.13. Let R be the ring of all continuous real-valued functions on a topological space. Show that R is J-semisimple, but "in most cases" not von Neumann regular.

Ex. 4.14. (Generalization of (4.27).) Show that a ring R is von Neumann regular iff $IJ = I \cap J$ for every right ideal I and every left ideal J in R.

Ex. 4.14A. Let $R = End_k(M)$ where M is a right module over a ring k. Show that an element $f \in R$ is von Neumann regular iff $ker(f)$ and $im(f)$ are both direct summands of M_k.

Ex. 4.14B. For any ring R, show that the following are equivalent:
(1) For any $a \in R$, there exists a unit $u \in U(R)$ such that $a = aua$.
(2) Every $a \in R$ can be written as a unit times an idempotent.
(2)′ Every $a \in R$ can be written as an idempotent times a unit.
If R satisfies (1), it is said to be *unit-regular*.
(3) Show that any unit-regular ring R is Dedekind-finite.

Ex. 4.14C. (Ehrlich, Handelman) Let M be a right module over a ring k such that $R = End_k(M)$ is von Neumann regular. Show that R is unit-regular iff, whenever $M = K \oplus N = K' \oplus N'$ (in the category of k-modules), $N \cong N'$ implies $K \cong K'$.

Ex. 4.14D. Let M be a semisimple right k-module. Show that $R = End_k(M)$ in unit-regular iff the isotypic components M_i of M (as defined in Exercise 2.8) are all finitely generated.

Ex. 4.15. For a commutative ring R, show that the following are equivalent:
(1) R has Krull dimension 0.[2]
(2) $rad\ R$ is nil and $R/rad\ R$ is von Neumann regular.
(3) For any $a \in R$, the descending chain $Ra \supseteq Ra^2 \supseteq \cdots$ stabilizes.
(4) For any $a \in R$, there exists $n \geq 1$ such that a^n is regular (i.e. such that $a^n \in a^n R a^n$).

Specializing the above result, show that the following are also equivalent:
(A) R is reduced (no nonzero nilpotents), and $K\text{-}dim\ R = 0$.
(B) R is von Neumann regular.
(C) The localizations of R at its maximal ideals are all fields.

Ex. 4.16. (Cf. Exercise 1.12) A left R-module M is said to be *cohopfian* if any injective R-endomorphism of M is an automorphism.

---

[2] Recall that the *Krull dimension* of a commutative ring R is the supremum of the lengths of chains of prime ideals in R. In particular, $K\text{-}dim\ R = 0$ means that all prime ideals in R are maximal ideals.

(1) Show that any artinian module M is cohopfian.
(2) Show that the left regular module ${}_RR$ is cohopfian iff every non right-0-divisor in R is a unit. In this case, show that ${}_RR$ is also hopfian.
Remark. The fact that ${}_RR$ is cohopfian $\Longrightarrow {}_RR$ is hopfian may be viewed as an analogue of the fact that ${}_RR$ is artinian $\Longrightarrow {}_RR$ is noetherian. It is, however, easy to see that ${}_RR$ is hopfian need not imply that ${}_RR$ is cohopfian (e.g. take $R = \mathbb{Z}$).

Ex. 4.17. Let R be a ring in which all descending chains

$$Ra \supseteq Ra^2 \supseteq Ra^3 \supseteq \cdots \quad (\text{for } a \in R)$$

stabilize. Show that R is Dedekind-finite, and every non right-0-divisor in R is a unit. (**Comment.** Rings satisfying the descending chain condition above are known as *strong π-regular* rings. For more details, see Exercise 23.5.)

Ex. 4.18. The *socle* $soc(M)$ of a left module M over a ring R is defined to be the sum of all simple submodules of M. Show that

$$soc(M) \subseteq \{m \in M : \ (rad\ R) \cdot m = 0\},$$

with equality if $R/rad\ R$ is an artinian ring.

Ex. 4.19. Show that for any ring R, $soc({}_RR)$ (= sum of all minimal left ideals of R) is an ideal of R. Using this, give a new proof for the fact that if R is a simple ring which has a minimal left ideal, then R is a semisimple ring.

Ex. 4.20. For any left artinian ring R with Jacobson radical J, show that

$$soc({}_RR) = \{r \in R : \ Jr = 0\} \quad \text{and} \quad soc(R_R) = \{r \in R : \ rJ = 0\}.$$

Using this, construct an artinian ring R in which $soc({}_RR) \neq soc(R_R)$.

Ex. 4.21. For any ring R, let $GL_n(R)$ denote the group of units of $\mathbb{M}_n(R)$. Show that for any ideal $I \subseteq rad\ R$, the natural map $GL_n(R) \to GL_n(R/I)$ is surjective. (**Hint.** First prove this for $n = 1$.)

Ex. 4.22. Using the definition of $rad\ R$ as the intersection of the maximal left ideals, show directly that $rad\ R$ is an ideal. (**Hint.** For $y \in rad\ R$, $r \in R$, and $\mathfrak{m}$ any maximal left ideal, we must have $yr \in \mathfrak{m}$. We may assume that $r \notin \mathfrak{m}$, so $\mathfrak{m} + Rr = R$. Look at $R \to R/\mathfrak{m}$ given by right multiplication by r, and show that the kernel is a maximal left ideal.)

Ex. 4.23. (Herstein) In commutative algebra, it is well known (as a consequence of Krull's Intersection Theorem) that, for any commutative noetherian R, $\bigcap_{n \geq 1} (rad\ R)^n = 0$. Show that this need not be true for non-commutative right noetherian rings.

Ex. 4.24. For any ring R, we know that

$$rad(R) \subseteq \{r \in R : \ r + U(R) \subseteq U(R)\}.$$

Give an example to show that this need not be an equality. (**Hint.** Let $I(R)$ be the set on the RHS, and consider the case $R = A[t]$ where A is a commutative domain. Here, $rad(R) = 0$, but $U(R) = U(A)$ shows that $I(R)$

contains $rad(A)$, which may not be zero. Of course, equality does hold sometimes; see, for instance, Exercise 20.10B.)

Ex. 4.25. Let R be the commutative $\mathbb{Q}$-algebra generated by $x_1, x_2, \ldots$ with the relations $x_n^n = 0$ for all n. Show that R does not have a largest nilpotent ideal (so there is no "Wedderburn radical" for R).

Ex. 4.26. Let R be the commutative $\mathbb{Q}$-algebra generated by $x_1, x_2, \ldots$ with the relations $x_i x_j = 0$ for all i, j. Show that R is semiprimary, but neither artinian nor noetherian. (For a noncommutative example, see Exercise 20.5.)

§5. Jacobson Radical Under Change of Rings

The main problem we shall consider in this section is the following: suppose S is a ring, and R is a subring of S, what kind of relations hold between the Jacobson radical of R and the Jacobson radical of S? In general, we certainly cannot expect that one radical would "determine" the other, but if we are given more specific information on the pair of rings R and S, it is reasonable to expect that certain inclusion relations hold between $rad\ R$ and $R \cap rad\ S$, or between $rad\ S$ and $S \cdot (R \cap rad\ S)$. This section is devoted to results of this general sort. In particular, we shall consider the behavior of the Jacobson radical under polynomial extensions of rings, and under scalar extensions of algebras over fields.

To begin this section, we shall first work with *commutative* rings. To determine the behavior of the Jacobson radical under a polynomial extension turns out to be fairly straightforward in the commutative case. In the following, let $T = \{t_i \colon i \in I\}$ be a (nonempty) set of commuting independent variables over a commutative ring R. Recall that $Nil\ R$ denotes the ideal of nilpotent elements of R. The following theorem gives the complete determination of the Jacobson radical of $R[T] = R[t_i \colon i \in I]$.

(5.1) Theorem (E. Snapper). *Let R be a commutative ring and let $R[T]$ be a polynomial ring over R. Then $rad\ R[T] = Nil(R[T]) = (Nil\ R)[T]$.*

Proof. Recall that a ring is called *reduced* if it has no nonzero nilpotent elements. Since $R/Nil\ R$ is reduced, it is easy to see that $(R/Nil\ R)[T]$ is reduced. But

$$(R/Nil\ R)[T] \cong R[T]/(Nil\ R)[T],$$

so it follows that $(Nil\ R)[T] = Nil(R[T])$. Also, $Nil(R[T]) \subseteq rad(R[T])$, so it only remains to show the reverse inclusion. For this, we may assume that T is a singleton, say t. Let

$$f(t) = r_0 + \cdots + r_n t^n \in rad(R[t]).$$

Then

$$1 + tf(t) = 1 + r_0 t + \cdots + r_n t^{n+1} \in U(R[t]).$$

Let $\mathfrak{p}$ be any prime ideal in R. Then the invertibility of the polynomial above in $(R/\mathfrak{p})[t]$ implies that each $r_i \in \mathfrak{p}$. Since this holds for all prime ideals

$\mathfrak{p} \subset R$, we have $r_i \in Nil\ R$ by a standard theorem in commutative algebra. Thus, $f(t) \in (Nil\ R)[t]$. QED

(5.2) Corollary. *Let R and T be as above. Then $R[T]$ is Jacobson semisimple iff R is reduced.*

Next we shall obtain some general results about a class of commutative rings called *Hilbert rings.* These results will be strong enough for us to deduce the classical Hilbert Nullstellensatz which we alluded to in Example (2) of §4. First we prove the following result on finitely generated commutative ring extensions.

(5.3) Theorem. *Let $R \subseteq A$ be commutative domains such that A is finitely generated as an R-algebra, and R is J-semisimple. Then A is also J-semisimple.*

Proof. It suffices to treat the case where $A = R[a]$. We may assume that a is algebraic over the quotient field K of R, for otherwise we are done by Snapper's Theorem above. Assume that there exists a nonzero element $b \in rad\ A$. Then a and b are both algebraic over K. Let

$$\sum_{i=0}^{n} r_i t^i, \quad \sum_{i=0}^{m} s_i t^i \in R[t]$$

be polynomials of the smallest possible degrees $n, m \geq 1$, satisfied, respectively, by a and b. Since A is a domain,

$$s_0 = -\sum_{i=1}^{m} s_i b^i \in rad\ A$$

is not zero, and so $r_n s_0 \neq 0$. From $rad\ R = 0$, we can find a maximal ideal $\mathfrak{m}$ of R such that $r_n s_0 \notin \mathfrak{m}$. Upon localizing at $S = R \backslash \mathfrak{m}$, r_n becomes a unit, so a satisfies a monic equation over $S^{-1}R$; in particular, $S^{-1}A = (S^{-1}R)[a]$ is finitely generated as a *module* over $S^{-1}R$. By Nakayama's Lemma (4.22),

$$(rad\ S^{-1}R) \cdot S^{-1}A \subsetneq S^{-1}A.$$

In particular, $\mathfrak{m} \cdot A \subsetneq A$. Let $\mathfrak{m}'$ be a maximal ideal of A containing $\mathfrak{m} \cdot A$. Then clearly $\mathfrak{m}' \cap R = \mathfrak{m}$, and so $s_0 \notin \mathfrak{m}$ implies that $s_0 \notin \mathfrak{m}'$, contradicting the fact that $s_0 \in rad\ A$. QED

It is now convenient to define Hilbert rings. A commutative ring (resp., domain) R is called a *Hilbert ring* (resp., *Hilbert domain*) if every prime ideal in R is an intersection of maximal ideals. If R is Hilbert, clearly so is every homomorphic image $R/\mathfrak{A}$; moreover, $rad\ R$ is the intersection of all prime ideals of R, so $rad\ R = Nil\ R$. In particular, if R is a Hilbert domain, then R is J-semisimple.

(5.4) Corollary. *Let $R \subseteq A$ be commutative rings such that A is finitely generated as an R-algebra and R is a Hilbert ring. Then A is also a Hilbert ring. (In particular, $rad\ A = Nil\ A$.)*

Proof. Let $\mathfrak{p} \subset A$ be a prime ideal. Then $A/\mathfrak{p}$ is a finitely generated domain over $R/\mathfrak{p} \cap R$. The latter is a Hilbert domain so $rad(R/\mathfrak{p} \cap R) = 0$. By the Theorem, $rad(A/\mathfrak{p}) = 0$, which means precisely that $\mathfrak{p}$ is the intersection of all the maximal ideals containing $\mathfrak{p}$. Therefore, A is Hilbert. QED

Taking R above to be a field k, it follows from (5.4) that any commutative k-affine algebra A is Hilbert. This proves the assertions made about A in Example (2) of §4. To complete our discussion of Hilbert's Nullstellensatz, we give one more result on commutative J-semisimple domains.

(5.5) Theorem. *Let $R \subseteq A$ be commutative domains such that $rad\ R = 0$ and A is finitely generated as an R-algebra. If A is a field, then so is R, and A/R is a finite (algebraic) field extension.*

Proof. First let us treat the "monogenic" case: $A = R[a]$. Clearly a must be algebraic over the quotient field of R. Let

$$\sum_{i=0}^{n} r_i t^i \in R[t]$$

be a polynomial (with $r_n \neq 0$) satisfied by a, and let $\mathfrak{m}$ be a maximal ideal in R with $r_n \notin \mathfrak{m}$. (Such a maximal ideal exists because $rad\ R = 0$.) As we saw in the proof of (5.3), $\mathfrak{m} \cdot A \subsetneq A$. Since A is a field, the ideal $\mathfrak{m} \cdot A$ must be zero. Therefore, $\mathfrak{m}$ itself is zero, which implies that R is a field. To treat the general case, let $A = R[a_1, \ldots, a_m]$ and write $R' = R[a_1]$. By (5.3), $rad\ R' = 0$. Invoking an inductive hypothesis (on m), we see that R' is a field and each a_i $(2 \leq i \leq m)$ is algebraic over R'. By the monogenic case, we conclude that R is a field and a_1 is algebraic over R. It follows that each a_i is algebraic over R, and that A/R is a finite (algebraic) field extension.

QED

In the special case when R is already *assumed* to be a field, Theorem (5.5) is known as Zariski's Lemma in commutative algebra. Stated in another form, this says that, *for any field R, and any maximal ideal $\mathfrak{m}$ of the polynomial ring $R[x_1, \ldots, x_m]$, the quotient ring $R[x_1, \ldots, x_m]/\mathfrak{m}$ is a finite field extension of R*. In case R is algebraically closed, this implies, in particular, that $\mathfrak{m}$ has the form $(x_1 - b_1, \ldots, x_m - b_m)$ for suitable $b_1, \ldots, b_m \in R$. It follows that *any proper ideal $\mathfrak{A} \subsetneq R[x_1, \ldots, x_m]$ has a zero in R^m*: this is the so-called *Weak Nullstellensatz*. The *Strong Nullstellensatz* is essentially the geometrical translation of the fact that $rad\ A = Nil\ A$ for any (commutative) affine algebra A over a field. In (5.3), (5.4), (5.5), we have succeeded in extending these basic results in commutative algebra to a somewhat more general setting by using the notion of the Jacobson radical. Our treatment here follows the ideas of Eagon [67].

We shall now leave commutative rings and consider again general rings. Our next goal will be to extend the result (5.1) to the noncommutative case. The main trouble in this case is that the nilpotent elements of R need no longer form an ideal (or even an additive group), so we first need to find a substitute for the expression $(Nil\ R)[T]$ in Snapper's Theorem. Also, since the proof of this theorem depends heavily on the commutativity of R, some new ideas are needed to extend the proof to the case of noncommutative rings.

Before we proceed to the consideration of $R[T]$, it will be convenient to collect a few general facts about the behavior of the radical under a change of rings. In the first result below, we consider a pair of rings $R \subseteq S$ and study sufficient conditions on R and S from which we can draw the conclusion $R \cap rad\ S \subseteq rad\ R$.

(5.6) Proposition. *Let $R \subseteq S$ be two rings. Assume either*

(1) *as a left R-module, ${}_R R$ is a direct summand of ${}_R S$, or*

(2) *there is a group G of automorphisms of the ring S such that R is the subring of fixed points $S^G := \{s \in S :\ g(s) = s\ \forall g \in G\}$.*

Then $R \cap rad\ S \subseteq rad\ R$.

Proof. First assume (1). Write $S = R \oplus T$ where T is a suitable left R-submodule of ${}_R S$. We are done if we can show that $r_0 \in R \cap rad\ S \Rightarrow 1 - r_0$ is right-invertible in R. Let

$$1 = (1 - r_0)(r + t) = (1 - r_0)r + (1 - r_0)t,$$

where $r \in R$ and $t \in T$. Since $S = R \oplus T$ and $1 \in R$, this implies that $(1 - r_0)r = 1$, as desired.

Next assume that $R = S^G$, where G is as in (2). We proceed as in the proof of (1). For $r_0 \in R \cap rad\ S$, let $(1 - r_0)s = 1$ where $s \in S$. Clearly s is fixed under the action of G. Hence $s \in S^G = R$ so $1 - r_0$ is right-invertible in R.

QED

Next, let i: $R \to S$ be a ring homomorphism. We shall investigate sufficient conditions under which we can conclude that $i(rad\ R) \subseteq rad\ S$. One such sufficient condition which readily comes to mind is that i: $R \to S$ be surjective. Under this assumption, an easy application of (4.1) shows that $i(rad\ R) \subseteq rad\ S$. (This was Exercise 4.10.) In the following, we shall try to develop a more general sufficient condition. Note that, via the homomorphism i: $R \to S$, we can view S as a bimodule ${}_R S_R$. For ease of notation, we shall denote the left and right R-actions on S by multiplication: for instance, if $r \in R$ and $s \in S$, $r \cdot s$ shall mean $i(r)s$.

(5.7) Proposition. *Let i: $R \to S$ be as above. Assume that*

$$S = R \cdot x_1 + \cdots + R \cdot x_n$$

where each x_j commutes (elementwise) with $i(R)$. Then $i(rad\ R) \subseteq rad\ S$.

Proof. Note that if M is any left S-module, we can view M also as a left R-module via i. Let $J := rad\ R$. To prove that $i(J) \subseteq rad\ S$, it suffices to show that J annihilates every simple left S-module M (see (4.1)). Write $M = S \cdot a$ for some $a \in M$. Then

$$M = (R \cdot x_1 + \cdots + R \cdot x_n)a = R \cdot x_1 a + \cdots + R \cdot x_n a,$$

so ${}_R M$ is finitely generated. Consider $J \cdot M$. This is an S-submodule of M since

$$x_j(J \cdot M) = x_j J \cdot M = J \cdot x_j M \subseteq J \cdot M.$$

Since $M \neq 0$, Nakayama's Lemma (4.22) implies that $J \cdot M \subsetneq M$. Recalling that ${}_S M$ is a simple module, we have $J \cdot M = 0$, as desired. QED

(5.8) Remark. The proof above shows that the conclusion $i(rad\ R) \subseteq rad\ S$ is already valid under the following considerably weaker hypothesis on x_j $(1 \leq j \leq n)$: For each x_j, there exists a ring automorphism σ_j of R such that $x_j \cdot r = \sigma_j(r) \cdot x_j$ (for every $r \in R$). For, under this hypothesis, we have already $x_j \cdot J \subseteq \sigma_j(J) \cdot x_j = J \cdot x_j$ ($J = rad\ R$ being clearly invariant under all automorphisms of R), and the rest of the proof goes over without any change.

The Proposition above is most often used in the following somewhat simpler form:

(5.9) Corollary. *Let R be a commutative ring and S be an R-algebra such that S is finitely generated as an R-module. Then $(rad\ R) \cdot S \subseteq rad\ S$.*

After the above preliminaries, we shall now resume the consideration of a polynomial extension, $R \subset S := R[T]$, where $T = \{t_i: i \in I\}$ is a nonempty set of independent (commuting) variables. (Of course, we shall not assume that R is commutative.) The following nice result of Amitsur [56] describes the structure of the radical of $S = R[T]$, thus providing a noncommutative analogue of Snapper's Theorem (5.1).

(5.10) Amitsur's Theorem. *Let R be any ring, and $S = R[T]$. Let $J = rad\ S$ and $N = R \cap J$. Then N is a nil ideal in R, and $J = N[T]$. In particular, if R has no nonzero nil ideal, then S is Jacobson semisimple.*

The proof of this important theorem will be presented in several steps. In the following, the notations introduced in (5.10) will be fixed. Let us first prove the easier part.

(5.10A) Proposition. *N is a nil ideal in R.*

Proof. Let $a \in N$ and $t = t_{i_0}$ be one of the variables. Then $1 - at$ is invertible in $R[T]$, say $(1 - at)g(T) = 1$. Setting all variables t_i $(i \neq i_0)$ equal to zero, we have

$$(1 - at)(a_0 + a_1 t + \cdots + a_n t^n) = 1$$

for some $a_j \in R$. Comparing coefficients, we have

$$a_0 = 1, \quad a_1 = aa_0 = a, \quad \ldots, \quad a_n = aa_{n-1} = a^n,$$

and $0 = aa_n = a^{n+1}$, as desired. QED

The next two results establish the truth of (5.10) in the case of one variable, $T = \{t\}$. Our treatment here is modeled upon that of Passman [91] (p. 192), following an earlier idea of G. Bergman (cf. Exercise 8 below).

(5.10B) Proposition. *Let $S = R[t]$, $J = rad\ S$, and $a_0, \ldots, a_n \in R$. If $f(t) = a_0 + a_1 t + \cdots + a_n t^n \in J$, then $a_i t^i \in J$ for all i.*

Proof. The conclusion is clearly true for $n = 0$. By induction, we may assume the truth of the conclusion (for *all* rings R) for smaller n. Let p be any prime number $> n$, and let R_1 be the ring

$$R[\zeta]/(1 + \zeta + \cdots + \zeta^{p-1}).$$

To simplify notations, we shall write ζ for the image of ζ in R_1; then $\zeta^p = 1$ in R_1. Note that, for any positive integer $j < p$, we have

$$p \in (\zeta^j - 1)R_1. \tag{5.11}$$

In fact, in the quotient ring $R_1/(\zeta^j - 1)R_1$ we have $\bar{\zeta}^j = 1$ and hence $\bar{\zeta} = 1$. Therefore, $1 + \bar{\zeta} + \cdots + \bar{\zeta}^{p-1} = 0$ implies that $\bar{p} = 0$. Now let $S_1 = R_1[t]$ and $J_1 = rad\ S_1$. Since

$$S_1 = S \oplus \zeta S \oplus \cdots \oplus \zeta^{p-2} S$$

and ζ is central in S_1, we have $J_1 \cap S = J$ by (5.6) and (5.7). Applying the automorphism $t \mapsto \zeta t$ on S_1, $f(t) \in J \subseteq J_1$ leads to $f(\zeta t) \in J_1$ and hence

$$\zeta^n f(t) - f(\zeta t) = a_0(\zeta^n - 1) + a_1(\zeta^n - \zeta)t + \cdots + a_{n-1}(\zeta^n - \zeta^{n-1})t^{n-1} \in J_1.$$

Invoking the inductive hypothesis (over R_1), we have $a_i(\zeta^n - \zeta^i)t^i \in J_1$ and hence $a_i(\zeta^{n-i} - 1)t^i \in J_1$ for any $i \leq n - 1$. Using (5.11), we see that $pa_i t^i \in J_1 \cap S = J$. Applying this argument to another prime $q > n$, we have also $qa_i t^i \in J$; therefore, $a_i t^i \in J$ for all $i \leq n - 1$. Since $f(t) \in J$, it follows that $a_n t^n \in J$ as well. QED

(5.10C) Proposition. *In the notation of* (5.10B), *if $f(t) \in J$, then $a_i \in J$ for all i.*

Proof. Applying the automorphism $t \mapsto t + 1$ on $R[t]$, the earlier conclusion $a_i t^i \in J$ leads to

$$a_i(1 + t)^i = a_i + ia_i t + \cdots + a_i t^i \in J.$$

Applying (5.10B) again, we see that $a_i \in J$. QED

For another proof of (5.10C) using somewhat different ideas, in the case when R is an algebra over a field, see Exercises (3) and (4) below.

Coming back now to Amitsur's Theorem (5.10), our remaining task is to deduce the general case of many variables from the case of one variable settled in (5.10C).

Proof of (5.10). The desired conclusion $J = N[T]$ $(N = J \cap R)$ means that, if a polynomial $f(T) \in J$, then all of its coefficients must belong to J. To see this, we induct on the number m of variables appearing in f. If $m = 0$, this is clear. If $m > 0$, fix a variable t appearing in f. Write $T = T_0 \dot{\cup} \{t\}$ (disjoint union) and $f(T) = \sum_i a_i(T_0)t^i$. Applying (5.10C) to $R[T] = R[T_0][t]$, we see that $a_i(T_0) \in J$ for all i. Since the number of variables actually appearing in each $a_i(T_0)$ is $\leq m - 1$, the induction proceeds. QED

At this point, let us make some comments about Amitsur's Theorem (5.10). One drawback of this theorem is that it does not "determine" what $N = R \cap \text{rad}\, R[T]$ really is, other than that it is a nil ideal. Now in any ring R, there is always a largest nil ideal, since the sum of all nil ideals is nil. (The easy proof of this can be found in (10.25).) Let us denote this largest nil ideal by $\text{Nil}^*\, R$; it is called the *upper nilradical* of R. An interesting problem is to determine whether the N in Amitsur's Theorem is indeed equal to $\text{Nil}^*\, R$, so that $\text{rad}\, R[T] = (\text{Nil}^*\, R)[T]$. In other words:

(5.12) Problem. *If I is a nil ideal in R, is $I[T] \subseteq \text{rad}\, R[T]$?*

If I is in fact nilpotent, say $I^n = 0$, then clearly $I[T]^n = 0$ and we can conclude from (4.11) that $I[T] \subseteq \text{rad}\, R[T]$. But if I is only nil, and R is noncommutative, (5.12) has remained a difficult unsolved problem in ring theory. In fact, (5.12) turns out to be equivalent to another famous problem in ring theory called Köthe's Conjecture. For more information on this, see §10 and Exercise 10.25.

In the case when R is an algebra over a field k, it is possible to obtain an analogue of (5.10) for the algebra $R(T)$. Here, $R(T)$ is defined to be the scalar extension of the k-algebra R when we enlarge the scalars from k to $k(T)$, the rational function field over k in the set of variables T. Since $k(T)$ is the quotient field of $k[T]$ and $R \otimes_k k[T] = R[T]$, $R(T)$ is the localization of $R[T]$ at the central multiplicative set $k[T]\backslash\{0\}$. For the structure of the radical of $R(T)$, we have the following analogue of (5.10), also due to Amitsur.

(5.13) Theorem. *Let $J' = \text{rad}\, R(T)$ and $N' = R \cap J'$. Then N' is a nil ideal in R, and $J' = N'(T) := N' \otimes_k k(T)$. In particular, if R has no nonzero nil ideals, then $R(T)$ is Jacobson semisimple.*

Proof. The proof here runs along the same line as in the case of a polynomial extension. Repeating the earlier arguments, we can show that, in the case of one variable, whenever $a_i \in R$:

$$(\dagger) \qquad \sum_i a_i t^i \in rad\ R(t) \Longrightarrow \text{all } a_i \in rad\ R(t).$$

In the case of many variables, consider $f(T)/g(T) \in J'$, where $f(T) \in R[T]$ and $0 \neq g(T) \in k[T]$. We have $f(T) \in g(T)J' \subseteq J'$, so it suffices to show that $f(T) \in J'$ implies that all coefficients of f are in J'. Inducting on the number of variables appearing in f, write $f(T) = \sum a_i(T_0)t^i$ as in the proof of (5.10), where $T = T_0 \dot\cup \{t\}$. Let $R^* = R(T_0)$, viewed as an algebra over $k^* = k(T_0)$. We can make the following identifications:

$$R(T) = R \otimes_k k(T_0) \otimes_{k(T_0)} k(T_0, t) = R^* \otimes_{k^*} k^*(t) = R^*(t).$$

From $\sum a_i(T_0)t^i \in J'$, we conclude from (†) (applied to R^*) that $a_i(T_0) \in J'$ for all i, so we are done by induction as before.

To complete the proof of (5.13), we still have to show that N' is a nil ideal in R. In fact a somewhat stronger statement is true for scalar extensions of algebras. We shall now investigate, in more detail, the behavior of the radical of algebras under scalar extensions. Then result (5.15) we prove below will, in particular, imply that the ideal N' in (5.13) is nil.

Let R be a k-algebra where k is a field, and let $K \supseteq k$ be a field extension. We can form the algebra $R^K := R \otimes_k K$, in which multiplication is defined by

$$(r \otimes a)(r' \otimes a') = rr' \otimes aa' \quad \text{for} \quad r, r' \in R \quad \text{and} \quad a, a' \in K.$$

The K-algebra R^K is called the *scalar extension* of R to the new scalar field K. The subring $R \otimes k = R \otimes 1$ of R^K is isomorphic to R, so we shall identify it with R. Note that we do not need to impose any condition on the extension $K \supseteq k$; thus, this may be a transcendental extension as well as an algebraic extension. The following sequence of results describes the relationship between the two radicals $rad\ R$ and $rad\ R^K$.

(5.14) Theorem. *For any k-algebra R and any field extension K/k, we have $R \cap rad\ R^K \subseteq rad\ R$. If K/k is an algebraic extension, or if $dim_k\ R < \infty$, then $R \cap rad\ R^K = rad\ R$. If $[K : k] = n < \infty$, then*

$$(rad\ R^K)^n \subseteq (rad\ R)^K (= (rad\ R) \otimes_k K).$$

Proof. Let $\{e_i\}$ be a basis of K as a k-vector space, with, say $e_{i_0} = 1$. Then

$$R^K = R \oplus \bigoplus_{i \neq i_0} R \cdot e_i$$

is a direct sum decomposition of R^K as a left R-module. Therefore, $R \cap rad\ R^K \subseteq rad\ R$ by (5.6). If $dim_k\ R < \infty$, then by (4.12) $rad\ R$ is nilpotent, so $(rad\ R)^K \subseteq R^K$ is also nilpotent. Therefore $(rad\ R)^K \subseteq rad\ R^K$ and hence $rad\ R \subseteq R \cap rad\ R^K$. To prove the same inclusion for K/k algebraic (and for arbitrary R), an easy direct limit argument reduces the consideration to the case $[K : k] = n < \infty$. In this case, the direct sum decomposition for R^K above is finite, and each e_i centralizes R. By (5.7), we have $rad\ R \subseteq R \cap rad\ R^K$, as desired. Finally, to see that $(rad\ R^K)^n \subseteq (rad\ R) \otimes K$, let V be any simple right R-module. Then $V^K = V \otimes_k K$ is a right R^K-module. Viewed as an R-module,

$$V^K \cong \bigoplus_{i=1}^{n} (V \otimes e_i)$$

has composition length n. Therefore, viewed as an R^K-module, V^K has composition length $\leq n$. Thus, for any $z \in (rad\ R^K)^n$, $V^K \cdot z = 0$. Write $z = \sum r_i \otimes e_i$ where $r_i \in R$. For any $v \in V$, we have

$$0 = (v \otimes 1)\left(\sum r_i \otimes e_i\right) = \sum vr_i \otimes e_i \Longrightarrow vr_i = 0 \quad (1 \leq i \leq n).$$

Therefore $V \cdot r_i = 0$ and so $r_i \in rad\ R$ for all i, from which we have $z = \sum r_i \otimes e_i \in (rad\ R) \otimes K$. QED

If K/k is not an algebraic extension, the inclusion $R \cap rad\ R^K \subseteq rad\ R$ may no longer be an equality. In fact, we shall show below that in this case, $R \cap rad\ R^K$ is always a nil ideal. So, if $rad\ R$ itself is not nil, we have a strict inclusion $R \cap rad\ R^K \subsetneq rad\ R$.

(5.15) Proposition. *Let K/k be a field extension which is not algebraic. Then for any k-algebra R, $R \cap rad\ R^K$ is a nil ideal in R. (This shows, in particular, that the ideal N' in (5.13) is nil.)*

Proof. Let $a \in R \cap rad\ R^K$ and let $t \in K$ be transcendental over k. Applying the first part of (5.14) to the extension $k(t) \subseteq K$, we see that $a \in R \cap rad\ R^{k(t)}$. To see that a is nilpotent, it suffices therefore to assume that $K = k(t)$. Since $1 - at \in U(R^K)$, there exists an equation

$$(1 - at) \cdot f(t)/g(t) = 1,$$

where

$$f(t) = b_0 + b_1 t + \cdots + b_m t^m \in R[t] \quad (b_m \neq 0),$$

$$g(t) = c_0 + c_1 t + \cdots + c_{m+1} t^{m+1} \in k[t].$$

Comparing coefficients, we get $c_0 = b_0$, $c_i = b_i - ab_{i-1}$ $(1 \leq i \leq m+1)$, with the convention that $b_{m+1} = 0$. Solving the b_i's in terms of the c_j's, we have, inductively,

$$b_i = a^i c_0 + a^{i-1} c_1 + \cdots + c_i.$$

For $i = m$, the fact that $b_m \neq 0$ implies that $c_0, \ldots, c_m$ are not all zero. For $i = m+1$, therefore, the equation

$$0 = b_{m+1} = a^{m+1} c_0 + \cdots + a c_m + c_{m+1}$$

shows that a is algebraic over k. Since $a \in rad\ R^{k(t)}$, it follows from (4.18) that a is nilpotent. QED

From here on we shall focus our attention on scalar extensions of algebras when K/k is a separable algebraic extension. In this case, one can obtain very specific information: see (5.17) below.

(5.16) Lemma. *Let R be a k-algebra, and K/k be a separable algebraic extension. If R is J-semisimple, then so is R^K.*

Proof. If $z \in rad\ R^K$, then, for some $L \subseteq K$ of finite degree over k, we have $z \in R^L \cap rad\ R^K \subseteq rad\ R^L$, by (5.14). Therefore we may assume that K/k is finite. Let E be the normal hull of K over k; then E/k is finite Galois. By (5.14) (applied to E/K),

$$rad\ R^K \subseteq rad(R^K)^E = rad\ R^E.$$

Therefore it suffices to show that $rad\ R^E = 0$. Let $e_1, \ldots, e_n$ be a k-basis for E, and let G be the Galois group of E/k. We can extend the G-action to $R^E = R \otimes_k E$ by identifying $\sigma \in G$ with $1 \otimes \sigma$. Given any element

$$z = \sum_i r_i \otimes e_i \in rad\ R^E,$$

we have, for any $\sigma \in G$ and any index j:

$$\sigma(ze_j) = \sigma\left(\sum_i r_i \otimes e_i e_j\right) = \sum_i r_i \otimes \sigma(e_i e_j).$$

These elements belong to $rad\ R^E$ since $rad\ R^E$ is an ideal of R^E which is invariant under all automorphisms. Summing these elements over $\sigma \in G$ and writing "tr" for the field trace of E/k, we get

$$\begin{aligned}\sum_i r_i \otimes \sum_{\sigma \in G} \sigma(e_i e_j) &= \sum_i r_i \otimes tr(e_i e_j)\\ &= \sum_i r_i\, tr(e_i e_j) \otimes 1.\end{aligned}$$

This element belongs to $R \cap rad\ R^E \subseteq rad\ R = 0$ (see (5.14)); hence $\sum_i r_i\, tr(e_i e_j) = 0$ for all j. Since E/k is separable, the trace form

$$(x, y) \mapsto tr(xy)$$

is nondegenerate; equivalently, the k-matrix $(tr(e_i e_j))$ is invertible. From the linear equations $\sum_i r_i\, tr(e_i e_j) = 0$, we conclude, therefore, that $r_i = 0$ for all i, and so $z = 0$. QED

(5.17) Theorem. *Let R be a k-algebra, and K/k be a separable algebraic extension. Then $rad(R^K) = (rad\ R)^K$.*

Proof. By (5.14), we have $(rad\ R)^K \subseteq rad(R^K)$. Moreover,

$$R^K/(rad\ R)^K \cong (R/rad\ R)^K.$$

Since $R/rad\ R$ is J-semisimple, the Lemma implies that $(R/rad\ R)^K$ is J-semisimple. By (4.6), it follows that $rad(R^K) = (rad\ R)^K$. QED

Note that the hypothesis that K/k be separable is essential in (5.16) and (5.17). If K contains inseparable elements over k, counterexamples are easy to find. The following is perhaps everyone's favorite counterexample. Let

k be a field of characteristic $p > 0$ and suppose $a \in k \backslash k^p$. Let $K = k(\alpha)$ where $\alpha^p = a$, so K is a purely inseparable extension of k. As a k-algebra, $R := K$ is of course semisimple (it is a field!), but the scalar extension $R^K = R \otimes_k K = K \otimes_k K$ is isomorphic to

$$K \otimes_k \frac{k[t]}{(t^p - a)} \cong \frac{K[t]}{(t^p - a)} = \frac{K[t]}{(t^p - \alpha^p)} = \frac{K[t]}{(t - \alpha)^p}.$$

This is not semisimple; in fact its radical is clearly the nilpotent ideal $(t-\alpha)/(t-\alpha)^p$. Tracing back through the isomorphism, we have that $rad(K \otimes_k K)$ is generated by the nilpotent element $1 \otimes \alpha - \alpha \otimes 1$. In fact, with the notation above, A^K is never J-semisimple for *any* k-algebra A containing K in its center, since $A \otimes_k K$ has a central nilpotent element $1 \otimes \alpha - \alpha \otimes 1$.

Exercises for §5

Ex. 5.0. (This exercise refines some of the ideas used in the proof of (5.6).) For any subring $R \subseteq S$, consider the following conditions:
(1) ${}_R R$ is a direct summand of ${}_R S$ and R_R is a direct summand of S_R.
(2) R is a *full subring* of S in the sense that $R \cap U(S) \subseteq U(R)$.
(3) $R \cap rad\ S \subseteq rad\ R$.
(For examples of full subrings, see, for instance, Exercises 1.13 and 6.3.)
(A) Show that (1) $\Rightarrow$ (2) $\Rightarrow$ (3).
(B) Deduce from the above that, if $C = Z(S)$ (the center of S), then $C \cap rad\ S \subseteq rad\ C$.
(C) Does equality hold in general in (B)?

Ex. 5.1. Let R be a commutative domain and $S^{-1}R$ be the localization of R at a multiplicative set S. Determine if any of the following inclusion relations holds:
(a) $rad\ R \subseteq R \cap rad\ S^{-1}R$,
(b) $R \cap rad\ S^{-1}R \subseteq rad\ R$,
(c) $rad\ S^{-1}R \subseteq S^{-1}(rad\ R)$.

Ex. 5.2. Give an example of a ring R with $rad\ R \neq 0$ but $rad\ R[t] = 0$.

In the following two exercises, we sketch another proof of (5.10C) in the case when R is an algebra over a field k. As in (5.10C), we let $J = rad\ R[t]$ and $N = R \cap J$.

Ex. 5.3. Assume k is an infinite field. Show that $J = N[t]$. (**Hint.** First show that $J \neq 0 \Longrightarrow N \neq 0$. Let $f(t) = a_0 + a_1 t + \cdots + a_n t^n \in J \backslash \{0\}$ with n chosen minimal. For any $\alpha \in k$, $f(t + \alpha) - f(t) \in J$ implies that $f(t+\alpha) = f(t)$. Setting $t = 0$ gives $a_n \alpha^n + \cdots + a_1 \alpha = 0$ for any $\alpha \in k$. Since k is infinite, this implies that $a_n = \cdots = a_1 = 0$.)

Ex. 5.4. Assume k is a finite field. Show that $J = N[t]$. (**Hint.** Let $\tilde{k}$ be the algebraic closure of k, and $\tilde{R} = R \otimes_k \tilde{k}$, $\tilde{J} = rad(\tilde{R}[t])$. By the first part of

(5.14), $J = R[t] \cap \tilde{J}$. Let $f(t) = a_0 + a_1 t + \cdots + a_n t^n \in J$. Then $f(t) \in \tilde{J}$, and, since $\tilde{k}$ is infinite, Exercise 3 yields $a_i \in \tilde{J}$ for all i. But then $a_i \in R \cap \tilde{J} = R \cap (R[t] \cap \tilde{J}) = R \cap J = N$.)

Ex. 5.5. Let R be any ring whose additive group is torsion-free. Show (without using Amitsur's Theorem) that $J = rad\ R[t] \neq 0$ implies that $R \cap J \neq 0$. (**Hint.** Let $f(t) \in J \backslash \{0\}$ be of minimal degree and consider $f(t+1) - f(t) \in J$.)

Ex. 5.6. For any ring R, show that the Jacobson radical of the power series ring $A = R[[t]]$ is given by

$$P := \{a + tf(t) : \ a \in rad\ R,\ f(t) \in A\}.$$

Ex. 5.7. For any k-algebra R and any finite field extension K/k, show that *rad* R is nilpotent iff *rad* R^K is nilpotent.

Ex. 5.8. (This problem, due to G. Bergman, is the origin of the proof of (5.10B).) Let R be a graded ring, i.e.

$$R = R_0 \oplus R_1 \oplus \cdots$$

where the R_i's are additive subgroups of R such that $R_i R_j \subseteq R_{i+j}$ (for all i, j) and $1 \in R_0$. Show that $J = rad\ R$ is a *graded ideal* of R, in the sense that J has a decomposition

$$J = J_0 \oplus J_1 \oplus \cdots,$$

where $J_i = J \cap R_i$. (**Hint.** Use the ideas in the proof of (5.10B).)

Ex. 5.9. Let $A = R[T]$, where T is an infinite set of commuting indeterminates. Show that *rad* A is a nil ideal.

Ex. 5.10. Let us call a *commutative* ring R "rad-nil" if its Jacobson radical is nil, that is, if $rad(R) = Nil(R)$. [Examples include: (commutative) J-semisimple rings, algebraic algebras and affine algebras over fields, etc.] Show that:

(1) a commutative ring is Hilbert iff all of its quotients are rad-nil;
(2) any commutative artinian ring is Hilbert;
(3) any commutative ring is a quotient of a rad-nil ring.
(4) Construct a commutative noetherian rad-nil ring that is not Hilbert.

§6. Group Rings and the J-Semisimplicity Problem

In §1 we have explained the formation of a group ring. This not only gives a nice source of examples of noncommutative as well as commutative rings, but also provides the basic connection between ring theory and the theory of group representations. Classically, group rings of finite groups over the complex numbers provided some of the earliest nontrivial examples of semisimple rings. For infinite groups, the associated group rings are no longer semisimple; the study of this class of rings can be used, therefore, as a guide to the general study of nonsemisimple rings. On the other hand, ideas and results in ring theory have had very important impact on the development of the theory of representations of groups. In the 1920's, Emmy Noether initiated the viewpoint that representations of groups amount to modules over the associated group rings. From this viewpoint, the Wedderburn structure theory of finite-dimensional algebras has natural interpretations in the framework of finite group representations. In particular, this enabled Noether to re-prove effectively many of the classical results of Frobenius and Schur in representation theory from a ring-theoretic perspective. This section will be devoted to the elementary study of group rings, with special emphasis on the question of semisimplicity (and J-semisimplicity). The major application of this material to representation theory will be given in §8 of the next chapter.

We begin by recalling the basic relationship between group representations and modules over group rings. To simplify matters, the base coefficients are assumed to form a field k. For any (multiplicative) group G, let $R = kG$ be the group ring (or group algebra) of G over k. An n-dimensional representation of G over k is defined, classically, to be a homomorphism D from G to $GL_n(k)$, the general linear group of invertible $n \times n$ matrices over k. (D stands for Darstellung, the German word for representation.) Given such a homomorphism, G then acts as a group of linear transformations on k^n; we denote this action by $(\sigma, v) \mapsto \sigma \cdot v$, for $\sigma \in G$ and $v \in k^n$. If we extend this action to the group ring $R = kG$ by taking

$$\left(\sum_{\sigma \in G} a_\sigma \sigma\right) \cdot v = \sum_{\sigma \in G} a_\sigma(\sigma \cdot v),$$

the vector space k^n becomes a (left) R-module. Conversely, if we are given a left R-module V such that $dim_k\, V = n$, then G acts as a group of linear transformations on V, and, by fixing a k-basis on V, we obtain a representation D of G by $n \times n$ invertible matrices. If we use a different k-basis for V, a routine computation shows that the resulting representation D' of G differs from D by an inner automorphism of $GL_n(k)$. Two representations D, D' which differ by an inner automorphism of $GL_n(k)$ are said to be *equivalent*. Conversely, it is easy to show that two equivalent representations give rise to a pair of isomorphic (left) kG-modules. Thus, equivalence classes of finite-dimensional representations of G may be identified with isomorphism classes of left kG-modules finite-dimensional over k. In this sense, the study of representations of G and the study of (left) kG-modules are essentially equivalent. The k-representations of G afforded by irreducible kG-modules are called the *irreducible representations* of G over k.

Classically, the representation theory of finite groups over fields of characteristic zero is of special importance. The most basic ring-theoretic result in this setting is that the associated group ring is a semisimple ring (in the sense of §2). This famous result is due to H. Maschke in 1898. Maschke's Theorem is also valid for fields k whose characteristic does not divide the order of G (as pointed out by Dickson), but not valid otherwise. We state below this theorem of Maschke, with its modern embellishments.

(6.1) Theorem. *Let k be any ring and G be a finite group. Then $R = kG$ is semisimple iff k is semisimple and $|G| \cdot 1$ is a unit in k.*

Proof. For the "if" part, let W be an R-submodule of a left R-module V. We want to show that W is an R-module direct summand of V. Fix a k-homomorphism $f\colon V \to W$ such that $f|W$ is the identity. (Such a map exists since W is a k-module direct summand of V.) We shall modify f into a map g with the same properties as f, but such that g is a homomorphism of R-modules. If such a g can be found, then $V = W \oplus ker(g)$ gives what we want. We define $g\colon V \to V$ by the following "averaging" device:

$$g(v) := |G|^{-1} \sum_{\sigma \in G} \sigma^{-1} f(\sigma v), \quad v \in V.$$

Since $g(v) \in |G|^{-1} \sum_{\sigma \in G} \sigma^{-1} \cdot W \subseteq W$, we may view the k-homomorphism g as from V to W. If $v \in W$, then

$$g(v) = |G|^{-1} \sum_{\sigma \in G} \sigma^{-1}(\sigma v) = v,$$

so g is the identity on W. Finally, the following computation shows that g is an R-homomorphism: for any $\tau \in G$,

$$\begin{aligned} g(\tau v) &= |G|^{-1} \sum_{\sigma \in G} \sigma^{-1}(f(\sigma\tau \cdot v)) \\ &= |G|^{-1} \sum_{\sigma' \in G} \tau\sigma'^{-1} f(\sigma' v) \\ &= \tau g(v). \end{aligned}$$

For the "only if" part of the theorem, assume now that $R = kG$ is semisimple. We have a ring homomorphism (the augmentation map)

$$\varepsilon\colon\ kG \longrightarrow k$$

defined by taking $\varepsilon|k = Id_k$ and $\varepsilon(G) = 1$. Therefore, as a homomorphic image of kG, k is semisimple (cf. Exercise 3.3). We finish by showing that *any prime p dividing $|G|$ is a unit in k.* By Cauchy's Theorem in group theory, there exists an element $\sigma \in G$ of order p. Since the semisimple ring R is von Neumann regular (see (4.24)), there exists an element $\alpha \in R$ such that $(1 - \sigma)\alpha(1 - \sigma) = 1 - \sigma$, from which

$$[1 - (1 - \sigma)\alpha] \cdot (1 - \sigma) = 0.$$

By the Lemma below, we can write

$$1 - (1 - \sigma)\alpha = \beta \cdot (1 + \sigma + \cdots + \sigma^{p-1})$$

for some $\beta \in R$. Applying the augmentation map ε, we have $1 = \varepsilon(\beta) \cdot p$, so $p = p \cdot 1$ is invertible in k, as desired. QED

We now supply the missing link in the argument above.

(6.2) Lemma. *For $r \in R = kG$, and $\sigma \in G$ of order p, $r \cdot (1 - \sigma) = 0$ iff $r \in R \cdot (1 + \sigma + \cdots + \sigma^{p-1})$.*

Proof. We need only prove the "only if" part. Let $r = \sum_{\tau \in G} r_\tau \tau$. We shall induct on the number n of τ's occurring in r with nonzero coefficients. If $n = 0$, then $r = 0$ and we are done. Otherwise, look at some τ with $r_\tau \neq 0$. Since

$$r = r \cdot \sigma = r \cdot \sigma^2 = \cdots,$$

the group elements $\tau, \tau\sigma, \ldots, \tau\sigma^{p-1}$ all appear in r with the same coefficient r_τ. Therefore

$$r = r_\tau(\tau + \tau\sigma + \cdots + \tau\sigma^{p-1}) + \begin{pmatrix} k\text{-combination of} \\ \text{other group elements} \end{pmatrix}$$

$$= r_\tau \tau(1 + \sigma + \cdots + \sigma^{p-1}) + r' \quad \text{(say)}.$$

Since $r \cdot (1 - \sigma) = 0$ implies that $r' \cdot (1 - \sigma) = 0$, the proof proceeds by induction. QED

In Maschke's Theorem, we have considered only finite groups. The following supplement to the theorem explains why.

(6.3) Proposition. *Let $k \neq 0$ be any ring, and G be an infinite group. Then the group ring $R = kG$ is never semisimple.*

Proof. For the augmentation map $\varepsilon\colon kG \to k$ defined above, let $\mathfrak{A} := ker(\varepsilon)$ be the "augmentation ideal." Assuming $R = kG$ is semisimple, we have $R = \mathfrak{A} \oplus \mathfrak{B}$ where $\mathfrak{B} \subset R$ is a suitable left ideal. Write

$$\mathfrak{A} = R \cdot e \quad \text{and} \quad \mathfrak{B} = R \cdot f,$$

where e, f are idempotents such that $e + f = 1$ (see Exercise 1.7). Clearly, e, f are not zero. We have $\mathfrak{A} \cdot f = Re \cdot f = 0$, so $(\sigma - 1)f = 0$, i.e., $f = \sigma f$, for any $\sigma \in G$. Let $\tau \in G$ be a group element which appears in f with a nonzero coefficient. Then $\sigma\tau$ appears in f with the same coefficient, for any $\sigma \in G$. This means that f involves *all* group elements of G; since G is infinite, this contradicts the definition of a group ring. QED

After Jacobson introduced the Jacobson radical in 1945, one obtains a new notion of semisimplicity for rings possibly not satisfying any chain conditions: a ring R is called *Jacobson semisimple* if *rad* $R = 0$. In view of Maschke's classical result that the group ring of a finite group over a field of characteristic zero is semisimple, a natural question would be to ask whether the group ring of an arbitrary group over a field of characteristic zero is always J-semisimple. The earliest result in this direction is due to C. Rickart [50] who used Banach algebra methods to show that, for any group G, the complex and real group algebras $\mathbb{C}G$ and $\mathbb{R}G$ are, indeed, J-semisimple. Later, Amitsur and Herstein showed independently that kG is J-semisimple for k any uncountable field of characteristic zero. Improving this result further, Amitsur [59] showed that, for any field k of characteristic zero, kG is J-semisimple except perhaps when k is a field of algebraic numbers (i.e., when k is an algebraic extension of $\mathbb{Q}$). In this case, a number of difficulties remain, and the problem seems to be still unsolved as of this date. Considerable work has also been devoted to special classes of groups, e.g., ordered groups, abelian groups, solvable groups, and certain linear groups. For all the special classes considered so far, the known answers to the J-semisimplicity problem have been uniformly affirmative. Analogous results have also been obtained for fields of characteristic p.

In the following, we shall give an exposition of the results on the J-semisimplicity problem described above. We shall follow the chronological order of these results in spite of the (obvious) fact that the later results are stronger than the earlier ones. The motivation for this approach is, in part, that we want to preserve the historical perspective of the problem: it was Rickart's pioneering result on the J-semisimplicity of $\mathbb{C}G$ and $\mathbb{R}G$ that gave the main impetus to the effort for solving the same problem over arbitrary fields of characteristic zero. A second reason for including Rickart's result is that Rickart's proof used certain interesting ideas from the theory of topological algebras. It is hoped that a study of this proof will provide a glimpse of the interaction between the purely ring-theoretic methods and the methods of functional analysis and complex analysis.

(6.4) Rickart's Theorem. *For any group G, the complex group algebra $\mathbb{C}G$ is J-semisimple.* (*In view of* (5.7), *this implies that the real group algebra $\mathbb{R}G$ is also J-semisimple.*)

For the sake of a self-contained exposition, we shall present Rickart's proof in a somewhat disguised form, following Passman [77], pp. 269–271. In this version of the proof, one does not need to introduce the general terminology of Banach algebras used in Rickart's paper. Nevertheless, this proof will be sufficient to convey the general flavor of Rickart's analytic methods.

Let us first give an overview of the proof. For any element $\alpha = \sum \alpha_g g$ in the complex group algebra $\mathbb{C}G$, we define the trace of α to be $tr(\alpha) = \alpha_1 \in \mathbb{C}$. The proof of (6.4) consists of two main parts.

(6.5) Lemma (The Algebraic Part). *If $rad(\mathbb{C}G) \neq 0$, there exists an element $\alpha \in rad(\mathbb{C}G)$ such that, for any $m \geq 1$,*

$$tr(\alpha^{2^m}) \in \mathbb{R} \quad and \quad tr(\alpha^{2^m}) \geq 1.$$

(6.6) Lemma (The Analytic Part). *For any $\alpha \in rad(\mathbb{C}G)$,*

$$\lim_{n \to \infty} tr(\alpha^n) = 0 \quad in\ \mathbb{C}.$$

Obviously, these two lemmas give the desired conclusion $rad(\mathbb{C}G) = 0$. Now we must prove these lemmas! Since the algebraic part is the easier one, we shall do it first.

We begin by recalling the idea of an involution. By an *involution* on a ring k, we mean an additive homomorphism $*: k \to k$ such that $a^{**} = a$ and

$(ab)^* = b^*a^*$ for all $a, b \in k$. (The latter implies that $1^* = 1$.) For instance, if k is a commutative ring, the identity map $a \mapsto a^* = a$ gives an involution on k, and the transpose map on matrices gives an involution on $\mathbb{M}_n(k)$. The usual complex conjugation map gives an involution on $\mathbb{C}$.

If k is a ring with an involution $^-$, we can define on any group ring kG:

$$\left(\sum \alpha_g g\right)^* = \sum \bar{\alpha}_g g^{-1}.$$

It is easy to check that this gives an involution on kG extending the given involution on k. In the following, we shall take k to be $\mathbb{C}$, with "bar" given by complex conjugation. The group algebra $R = \mathbb{C}G$ then has an involution *, as defined above.

Proof of (6.5). For any $\alpha = \sum \alpha_g g \in R$, we have

$$tr(\alpha^*\alpha) = \sum_{g \in G} \bar{\alpha}_g \alpha_g = \sum_{g \in G} |\alpha_g|^2 \geqq |\alpha_1|^2 = |tr\ \alpha|^2,$$

where $|z|$ denotes the modulus of a complex number z. In particular, if α is a *-symmetric element (i.e., $\alpha^* = \alpha$), then by induction on $m \geq 1$ we have $tr(\alpha^{2^m}) \geq |tr\ \alpha|^{2^m}$ in $\mathbb{R}$. If $rad(R)$ contains a nonzero element $\beta = \sum \beta_g g$, then $tr(\beta^*\beta) = \sum |\beta_g|^2 \neq 0$ and we have

$$\alpha := \beta^*\beta / tr(\beta^*\beta) \in rad\ R$$

since $rad\ R$ is an ideal. Clearly, $\alpha^* = \alpha$, $tr(\alpha) = 1$, so by what we said above, $tr(\alpha^{2^n})$ is a real number ≥ 1 for all n. QED

Next, we shall try to do the "Analytic Part" (Lemma 6.6). To begin the proof, let us first set up some notations. For $\alpha = \sum \alpha_g g \in R$, define $|\alpha| = \sum |\alpha_g| \in \mathbb{R}$. One checks easily the following properties:

(1) $|tr\ \alpha| \leq |\alpha|$,

(2) $|\alpha + \beta| \leq |\alpha| + |\beta|$,

(3) $|\alpha\beta| \leq |\alpha| \cdot |\beta|$.

In view of (2), if we define a distance function on R by

$$dist(\alpha, \beta) := |\alpha - \beta|,$$

R becomes a metric space. Thus, we can talk about continuity of functions to or from R. For instance, by property (1) above, $tr: R \to \mathbb{C}$ is a continuous ($\mathbb{C}$-linear) function.

In the following, let α be a *fixed* element of $rad\ R$. Then for any $z \in \mathbb{C}$, $1 - z\alpha \in U(R)$, so we can define a function $\varphi: \mathbb{C} \to R$ by

$$\varphi(z) = (1 - z\alpha)^{-1} \in R.$$

The proof of (6.6) depends on the following three crucial properties of φ.

(A) *φ is continuous at every $z \in \mathbb{C}$.*

(B) *φ is differentiable at every $z \in \mathbb{C}$.*

(C) *If $|z|$ is sufficiently small, then $\varphi(z) = \sum_{n=0}^{\infty} \alpha^n z^n \in R$.*

Here, (A) is a consequence of the fact that the inverse map on R is continuous. To give a more detailed proof, consider two points $y, z \in \mathbb{C}$. Since $\varphi(y)$ and $\varphi(z)$ commute, we have

$$\textbf{(6.7)} \qquad \begin{aligned} \varphi(y) - \varphi(z) &= [(1 - z\alpha) - (1 - y\alpha)](1 - y\alpha)^{-1}(1 - z\alpha)^{-1} \\ &= (y - z)\, \alpha\, \varphi(y)\varphi(z). \end{aligned}$$

This implies that, for a given z, $|\varphi(y)|$ is bounded in a suitable neighborhood of z. In fact, from (6.7), we have

$$|\varphi(y)| \leq |\varphi(z)| + |y - z| \cdot |\varphi(y)| \cdot |\alpha\, \varphi(z)|,$$

so

$$|\varphi(y)| \cdot (1 - |y - z| \cdot |\alpha\, \varphi(z)|) \leq |\varphi(z)|.$$

If y is sufficiently close to z, we can make the expression in parentheses to be $\geq 1/2$, so $|\varphi(y)| \leq 2|\varphi(z)|$, and (6.7) gives

$$|\varphi(y) - \varphi(z)| \leq 2|\alpha| \cdot |y - z| \cdot |\varphi(z)|^2.$$

This implies (A), and also shows that

$$\textbf{(6.8)} \qquad \varphi'(z) := \lim_{y \to z} \frac{\varphi(y) - \varphi(z)}{y - z} = \lim_{y \to z} \alpha\, \varphi(y)\varphi(z) = \alpha\, \varphi(z)^2,$$

for every $z \in \mathbb{C}$. Now let z be such that $|z| \cdot |\alpha| < 1$. Then $|z\alpha| = |z| \cdot |\alpha| < 1$, and the usual geometric series argument shows that $\varphi(z) = \sum_{n=0}^{\infty} z^n \alpha^n$. In fact, for any integer N:

$$\varphi(z) - \sum_{n=0}^{N} z^n \alpha^n = \varphi(z) \left\{ 1 - (1 - z\alpha) \sum_{n=0}^{N} z^n \alpha^n \right\} = \varphi(z)(z\alpha)^{N+1},$$

so

$$\left| \varphi(z) - \sum_{n=0}^{N} z^n \alpha^n \right| \leq |\varphi(z)| \cdot |z\alpha|^{N+1}.$$

From this estimate, we see that

$$\left| \varphi(z) - \sum_{n=0}^{N} z^n \alpha^n \right| \to 0 \quad \text{as } N \to \infty,$$

proving (C).

Proof of (6.6). For any group element $g \in G$, we can define $tr_g \colon R \to \mathbb{C}$ by sending any $\alpha = \sum_{h \in G} \alpha_h h \in R$ to $\alpha_g \in \mathbb{C}$. This map is clearly continuous and

$\mathbb{C}$-linear, and our earlier trace map "tr" is just tr_1. For any fixed $\alpha \in rad\ R$ and $g \in G$, *we shall show more generally that*

$$\lim_{n \to \infty} tr_g(\alpha^n) = 0.$$

Let

$$f = tr_g \circ \varphi : \mathbb{C} \longrightarrow \mathbb{C}; \quad \text{i.e., } f(z) = tr_g(1 - z\alpha)^{-1}.$$

By (B), f is an entire function (with $f'(z) = tr_g(\alpha\, \varphi(z)^2)$), and by (C):

(6.9)
$$f(z) = tr_g\left(\sum_{n=0}^{\infty} \alpha^n z^n\right) = \sum_{n=0}^{\infty} tr_g(\alpha^n) z^n$$

for sufficiently small $|z|$. In particular, (6.9) gives the Taylor expansion of f at the origin. Since f is entire, a well-known theorem in complex analysis guarantees that this Taylor series has *infinite* radius of convergence, and converges to f everywhere in $\mathbb{C}$. At the point $z = 1$, this yields

$(\ast)$
$$f(1) = \sum_{n=0}^{\infty} tr_g(\alpha^n).$$

But then the nth term of this series must converge to 0, as desired. QED

(6.10) Remark. By definition, $f(1)$ is $tr_g(1 - \alpha)^{-1}$. If we knew that $|\alpha| < 1$, then $(1 - \alpha)^{-1} = \sum_{n=0}^{\infty} \alpha^n$ would hold in R, and we would have the equation $(\ast)$ by taking trace. But of course, we did not know that $|\alpha| < 1$, nor that $\sum_{n=0}^{\infty} \alpha^n$ would converge in R. Therefore, we have to invoke a basic theorem on entire functions to justify the key equation $(\ast)$.

While the proof of (6.6) was designed to use a minimum amount of analysis, a few remarks on its hidden connections to Banach algebras are in order. In Rickart's original proof, G is thought of as a locally compact group with the discrete topology, and one considers the $\mathbb{C}$-algebra of L^1-integrable functions from G to $\mathbb{C}$ with respect to the Haar measure, the multiplication of functions being given by "convolution." This algebra, traditionally denoted by $L^1(G)$, may be called the "analyst's version" of the group algebra: it is a Banach algebra over $\mathbb{C}$. The discrete group algebra $\mathbb{C}G$ can be embedded as a dense subring of $B := L^1(G)$ by identifying $\sum \alpha_g g$ with the function $g \mapsto \alpha_g$. For any element α in the Banach algebra B, the *resolvent set* of α is

$$\{z \in \mathbb{C} : \ z - \alpha \in U(B)\},$$

and on this resolvent set, the "resolvent function"

$$\psi(z) = (z - \alpha)^{-1}$$

is analytic, with values in B. The function φ used in our proof of (6.6) is just a slight variant of ψ, namely $\varphi(z) = \psi(z^{-1})/z$. The argument in (6.6), applied in the general setting of analytic functions with values in a Banach space, shows that the radical of a complex Banach algebra B is "topologically nil," i.e., any element α in it has $\alpha^n \to 0$. It follows easily from this that *rad B is the largest topologically nil* (*1-sided or 2-sided*) *ideal in B.* This is, therefore, the general fact which underlies the analytic lemma (6.6). On the other hand, the algebraic lemma (6.5) made crucial use of the *-involution on $\mathbb{C}G$; this amounts to exploiting the standard C^*-algebra structure in $L^1(G)$. The argument on the 2^n-powers of *-symmetric elements essentially shows that, in a C^*-algebra, the only topologically nil 1-sided ideal is the zero ideal. In particular, this argument suffices to show that *any C^*-algebra over $\mathbb{C}$ is Jacobson semisimple.*

Having thus explained the proof of Rickart's Theorem, we shall now return to more conventional ring theory. Our next goal is to discuss Amitsur's results on the J-semisimplicity of group rings in characteristic zero and Passman's analogous results in characteristic p. The methods we use to establish these results will be purely algebraic. We begin by proving a result on nil one-sided ideals in group rings of characteristic zero. The idea used in the proof of the Proposition below is very similar to that used for Lemma 6.5.

(6.11) Proposition. *Let k be a ring with an involution * such that*

$$\sum \alpha_i^* \alpha_i = 0 \Longrightarrow \text{all } \alpha_i = 0 \quad \text{in } k.$$

Then, for any group G, the group ring $R = kG$ has no nonzero nil left ideals. In particular, this conclusion holds in the following two cases:

(a) *k is a commutative ring which is formally real, in the sense that $\sum \alpha_i^2 = 0 \Longrightarrow$ all $\alpha_i = 0$.*

(b) *k is an algebraically closed field of characteristic zero.*[3]

Proof. For $\alpha = \sum \alpha_g g \in R$, we define $tr(\alpha) = \alpha_1$ as before, and we extend * to an involution on R by defining α^* to be $\sum \alpha_g^* g^{-1}$. Since

$$tr(\alpha^*\alpha) = \sum \alpha_g^* \alpha_g,$$

the hypothesis on $(k, {}^*)$ amounts to:

$$tr(\alpha^*\alpha) = 0 \Longrightarrow \alpha = 0 \quad \text{in } R.$$

Assume that R has a nonzero nil left ideal $\mathfrak{B}$, say with $0 \neq \beta \in \mathfrak{B}$. Then $0 \neq \gamma := \beta^*\beta \in \mathfrak{B}$, and $\gamma^* = \gamma$. Choose $n \geq 1$ so that $\gamma^n \neq 0 = \gamma^{n+1}$. For $\alpha := \gamma^n$, we have $\alpha^*\alpha = \alpha^2 = \gamma^{2n} = 0$, but by what we said above, $\alpha = 0$, a contradiction. If k is a formally real commutative ring, we can take the involution * to be the identity on k, and the argument above applies. Now let

[3] The conclusion that kG has no nonzero nil left ideals actually holds for all fields k of characteristic zero, without the assumption that k be algebraically closed. However, we shall not prove this more general result here.

k be an algebraically closed field of characteristic zero. By a basic theorem in field theory, we know that $k = k_0[i]$ where k_0 is a real-closed field $\subset k$, and $i = \sqrt{-1}$. (A real-closed field is a formally real field with no formally real proper algebraic extensions.) Defining $(a+bi)^* = a - bi$ on k $(a, b \in k_0)$, the fact that k_0 is formally real gives:

$$0 = \sum (a_j + b_j i)^* (a_j + b_j i) = \sum a_j^2 + \sum b_j^2 \Longrightarrow a_j = b_j = 0 \quad \forall j.$$

Therefore, the first part of the Proposition applies to kG. QED

Note that the last part of the Proposition can be used to give another proof of the characteristic zero case of Maschke's Theorem. Indeed, let k be a field of characteristic 0, and let G be a finite group. Since kG is artinian, $(rad\ kG)^n = 0$ for some n, and hence $((rad\ kG) \cdot \bar{k})^n = 0$ for $\bar{k}$ the algebraic closure of k. By the Proposition, we have $(rad\ kG) \cdot \bar{k} = 0$ and hence $rad\ kG = 0$, showing that kG is semisimple. (In the case of characteristic p, a similar remark can be made after the proof of (6.13).) However, the argument in (6.11) works for *any* group G, and our present goal is to get theorems on the J-semisimplicity of kG for possibly infinite groups. In fact, we are now ready to prove the following result of Amitsur [59], which extends Rickart's Theorem from the complex base field to "almost" any field of characteristic zero.

(6.12) Amitsur's Theorem. *Let K be a nonalgebraic field extension of $\mathbb{Q}$. Then for any group G, the group ring KG is J-semisimple.*

Proof. Let $F = \mathbb{Q}(\{x_i\})$, where $\{x_i\}$ is a (nonempty) transcendence basis for $K/\mathbb{Q}$. Note that the scalar extension $\mathbb{Q}G \otimes_{\mathbb{Q}} F$ is just FG. Let $J = rad\ FG$. By (5.13), $N = J \cap \mathbb{Q}G$ is a nil ideal of $\mathbb{Q}G$ and $J = N \otimes_{\mathbb{Q}} F = N \cdot F$. However, by Proposition 6.11(a), $\mathbb{Q}G$ has no nonzero nil left ideals; hence $N = 0$ and so $J = 0$. This shows that FG is J-semisimple. Since we are in characteristic zero, K/F is a *separable* algebraic extension. Therefore, by (5.16), the scalar extension $FG \otimes_F K = KG$ is also J-semisimple. QED

Next we shall try to obtain the characteristic p analogue of the above result. We first need a characteristic p analogue of the nonexistence of nonzero one-sided nil ideals. To do this, we introduce the following group-theoretic notion: for a prime p, we say that *a group G is a p'-group if G has no element of order p.* Note that, by Cauchy's Theorem, if G is a finite group, this condition simply means that p does not divide the order of G.

The characteristic p analogue of (6.11), due independently to D.S. Passman and I.G. Connell, is as follows.

(6.13) Proposition. *Let k be a commutative reduced ring of prime characteristic $p > 0$. Let G be any p'-group. Then $R = kG$ has no nonzero nil left ideals.*

Proof. Assume R has a nonzero nil left ideal $\mathfrak{B}$, say

$$0 \neq \beta = \sum \beta_g g \in \mathfrak{B}.$$

After left multiplying β with a suitable group element, we may assume that $tr(\beta) = \beta_1 \neq 0$. We claim that $tr(\beta^p) = (tr(\beta))^p$. If so, then by iteration,

$$tr(\beta^{p^n}) = (tr(\beta))^{p^n} \neq 0$$

for every n, and we get the desired contradiction. To show our claim, note that

$$tr(\beta^p) = tr\Big(\Big(\sum \beta_g g\Big)^p\Big) = \sum \beta_{g_1}\beta_{g_2}\cdots\beta_{g_p},$$

where the sum is over the set S of (ordered) p-tuples $(g_1, \ldots, g_p)$ of group elements such that $g_1 \cdots g_p = 1$. The cyclic group $H = \langle\sigma\rangle$ of order p acts on S by

$$\sigma * (g_1, \ldots, g_p) = (g_2, \ldots, g_p, g_1).$$

The H-orbits on S have cardinality either 1 or p. For an orbit of cardinality p, since all the p-tuples in the orbit make the same contribution to $tr(\beta^p)$, the total contribution is a multiple of p, and therefore is zero. Now look at a singleton orbit $H * (g_1, \ldots, g_p)$. We must have $g_1 = g_2 = \cdots = g_p$ and hence $g_1^p = 1$. Since G is a p'-group, we have $g_1 = \cdots = g_p = 1$. Therefore, there is a unique singleton orbit in S, and its contribution to $tr(\beta^p)$ is β_1^p, as claimed. QED

In order to prove the characteristic p analogue of (6.12), we need the following intermediate result.

(6.14) Proposition. *Let K/F be an algebraic extension of fields of characteristic p, and let G be a p'-group. If FG is J-semisimple, so is KG.*

Proof. First let us assume that $[K : F] = n < \infty$. By (5.14),

$$(rad(KG))^n \subseteq (rad\ FG) \cdot K,$$

and, by hypothesis, $rad\ FG = 0$. Thus, $rad\ KG$ is a nilpotent ideal. By the above Proposition, therefore, $rad\ KG = 0$. Now drop the assumption that $[K : F] < \infty$. Given any element $\alpha \in rad\ KG$, we can find a field $K_0 \subseteq K$ of finite degree over F such that $\alpha \in K_0G$. According to (5.14), we have

$$\alpha \in K_0G \cap rad\ KG \subseteq rad\ K_0G.$$

But by the case we have already treated, $rad\ K_0G = 0$ and so $\alpha = 0$.

QED

The following analogue of Amitsur's Theorem (6.12) was proved by Passman in 1962.

(6.15) Passman's Theorem. *Let K be a nonalgebraic field extension of $\mathbb{F}_p$ (the field of p elements). Then for any p'-group G, the group ring KG is J-semisimple.*

Proof. As before, let $\{x_i\}$ be a (nonempty) transcendence basis for $K/\mathbb{F}_p$, and let $F = \mathbb{F}_p(\{x_i\})$. By (6.13), $\mathbb{F}_pG$ has no nonzero nil left ideals. Arguing as in the proof of (6.12), we see that FG is J-semisimple. Applying (6.14) to the algebraic extension K/F, it follows that KG is also J-semisimple. QED

It is perhaps not unreasonable to conjecture that (6.12) and (6.15) both remain true in the case when K is algebraic over its prime field. To prove this conjecture, it would suffice to show that, for any group G, $\mathbb{Q}G$ is J-semisimple, and that, in case G is a p'-group, $\mathbb{F}_pG$ is J-semisimple. Once these cases are known, the general case can be deduced affirmatively from (5.16) (since a prime field is perfect so all of its algebraic extensions are separable). In spite of considerable effort, however, this problem has remained unsolved. Surprising as it might seem, it is the case when K is a prime field that presents the so far insurmountable difficulties.

In the case of characteristic p, one may also ask for *necessary* conditions for a group ring KG to be J-semisimple. One rather obvious necessary condition is that any *finite* normal subgroup $H \subseteq G$ must be a p'-group. For, if there exists such an H with $|H|$ divisible by p, then the element $\alpha = \sum_{h\in H} h$ is in the center of KG with $\alpha^2 = |H|\alpha = 0$. But then $KG \cdot \alpha$ is a nonzero ideal with square zero and so KG is not J-semisimple. However, for KG to be J-semisimple, it is not necessary that G itself be a p'-group. For instance, if G is the infinite dihedral group, then G has elements of order 2, but Wallace has shown that KG *is* J-semisimple for all fields K of characteristic 2 (cf. Exercise 14). If G is the group consisting of permutations of an infinite set S moving only finitely many elements of S, then G has elements of order p for all primes p, but Formanek has shown that KG is J-semisimple for *all* fields, independently of their characteristics. Many other similar examples are now known, leaving not much of a clue as to precisely when group algebras are J-semisimple over fields of characteristic p.

We now finish this section by studying some other problems concerning group rings which are related to the J-semisimplicity problem. One of these is the unit problem, and the other is the zero-divisor problem. These problems are of interest mainly for the class of torsion-free groups, i.e., groups without nonidentity elements of finite order.

First let us define the notion of "nontrivial units." In any group ring kG over a ring k, we always have the units $a \cdot g$, where a is a unit in k and $g \in G$. These are called the *trivial units* of kG; other units of kG are called *nontrivial units*. As an example, for any group G with at most four elements, the integral group ring $\mathbb{Z}G$ has only trivial units. To see this, first assume $|G| = 2$. Then $\mathbb{Z}G \cong \mathbb{Z}[t]/(t^2-1)$, which is isomorphic to

$$\{(a,b) \in \mathbb{Z} \times \mathbb{Z}:\ a \equiv b \pmod 2\}.$$

Thus, $\mathbb{Z}G$ has exactly four units, which are necessarily trivial. The computations for $|G| = 3, 4$ are similar and are left to the reader. On the other hand, if $|G| = 5$, we have seen in (1.4) that $\mathbb{Z}G$ has nontrivial units.

For infinite groups, one simple example to keep in mind is that of the infinite cyclic group $G = \langle x \rangle$. For any domain k, kG is the ring of Laurent polynomials $k[x, x^{-1}]$, and an easy degree argument shows that kG has only trivial units, and is a domain. Repeating this argument, one sees that the same holds for any finitely generated free abelian group, and hence also for any *torsion-free abelian* group.

In general, for any torsion-free group G and any domain k, the following are important problems for the study of group rings:

(6.16) Problem U. *Are units of kG all trivial?*

(6.17) Problem R. *Is kG always reduced?*

(6.18) Problem D. *Is kG always a domain?*

(6.19) Problem J. *If $G \neq \{1\}$, is kG J-semisimple?*

Of course, Problem J is related to the earlier material on J-semisimplicity in this section, except that we now allow k to be a domain instead of a field. Note that since G is assumed to be torsion-free, we need not impose the p'-group assumption on G, even if k has characteristic $p > 0$. Note also that, for Problem D to have an affirmative answer, the torsion-freeness of G is a necessary condition, for if G has an element x of finite order $n > 1$, then

$$(x-1)(x^{n-1} + \cdots + x + 1) = 0 \quad \text{in } kG$$

shows that kG has zero-divisors. Indeed, the idea of Problem D is that, if no such element of finite order is available, then perhaps kG will not have zero-divisors.

The four problems raised in (6.16)–(6.19) are somewhat interconnected. The known relationship between the four can be summarized as follows:

(6.20)
$$\begin{array}{ccc} & & R \Longleftrightarrow D \\ & \nearrow & \\ U & & \\ & \searrow & \\ & & J \end{array}$$

Here, $U \Rightarrow R$ means that if the answer to Problem U is "yes," then so is the answer to Problem R, etc. Since any domain is reduced, the implication $D \Rightarrow R$ is trivial. The reverse implication $R \Rightarrow D$ is rather deep and requires substantial work for its proof. We shall postpone this implication, and first prove the following Proposition which gives the two other implications $U \Rightarrow R$, $U \Rightarrow J$, under much weaker assumptions on k and G.

(6.21) Proposition. *Let $k \neq 0$ be a ring and $G \neq \{1\}$ be a group such that $A = kG$ has only trivial units.*

(1) *If k is reduced and G has no element of order 2, then A is reduced.*

(2) *A is J-semisimple except when $|k| = |G| = 2$.*

Proof. (1) It suffices to show that, for $\alpha \in A$, $\alpha^2 = 0 \Rightarrow \alpha = 0$. From $\alpha^2 = 0$, we have

$$(1-\alpha)(1+\alpha) = 1 - \alpha^2 = 1,$$

so $1-\alpha$ is a unit, and we have $1 - \alpha = ag$ for some $a \in U(k)$ and $g \in G$. If $g \neq 1$, the equation

$$0 = \alpha^2 = (1-ag)^2 = 1 - 2ag + a^2g^2$$

gives a contradiction since $g \neq 1 \neq g^2$ and hence there is no term to "cancel out" 1. Thus we must have $g = 1$, whence $\alpha = 1 - a \in k$. But then $\alpha^2 = 0 \Rightarrow \alpha = 0$ since k is reduced. For (2), note that if $|k| = |G| = 2$, say $G = \langle g \rangle$, then $U(A) = G$, but $rad\ A = \{0, g-1\} \neq 0$, so this case is an exception. Now assume we are not in this case and let $\alpha \in rad\ A$. Then the unit $1 - \alpha$ has the form ag, where $a \in U(k)$ and $g \in G$. *We claim again that* $g = 1$. Assume, instead, that $g \neq 1$. If $|k| \geq 3$, there exists $b \in k \backslash \{0, 1\}$ and

$$1 - \alpha b = 1 - b + abg$$

is a nontrivial unit. If $|G| \geq 3$, there exits $h \in G \backslash \{1, g^{-1}\}$ and now

$$1 - \alpha h = 1 - h + agh$$

is a nontrivial unit. Therefore, we must have $g = 1$, and so $\alpha = 1 - a \in k$. Now for any $h \neq 1$ in G, the unit $1 + \alpha h$ must be trivial, hence $\alpha = 0$.

QED

Next we shall try to give an account for the implication "$R \Rightarrow D$." The usefulness of this lies in the fact that, combined with (6.21)(2), it gives the implication "$U \Rightarrow D$" which is not so easy to obtain otherwise. For the proof of "$R \Rightarrow D$," we need several group-theoretic lemmata.

(6.22) Lemma. *Let G be a group such that the center $H = Z(G)$ has finite index n in G. Then G is n-abelian; i.e., $(ab)^n = a^n b^n$ for all $a, b \in G$.*

Proof (Sketch). The easiest proof of this makes use of the transfer homomorphism φ: $G \to H/H'$, where H' denotes the commutator subgroup of H. Since H is abelian, $H' = \{1\}$ here, and using the definition of φ, it can be seen that $\varphi(a) = a^n \in H$. Since φ is a homomorphism, we have

$$(ab)^n = \varphi(ab) = \varphi(a)\varphi(b) = a^n b^n \in H. \qquad \text{QED}$$

For any group G, we define $\Delta(G)$ to be the subgroup of G consisting of elements $g \in G$ with only finitely many distinct conjugates. $\Delta(G)$ is a subgroup since, for $a, b \in \Delta(G)$,

$$x^{-1}(ab)x = (x^{-1}ax)(x^{-1}bx) \quad (x \in G)$$

can take only finitely many values in G. Note that

(6.23) $$\Delta(G) = \{g \in G:\ [G : C_G(g)] < \infty\},$$

and that $\Delta(G)$ is a characteristic subgroup of G, with $\Delta(\Delta(G)) = \Delta(G)$. For instance, if G is the infinite dihedral group

$$\langle a, b:\ b^2 = 1,\ bab^{-1} = a^{-1}\rangle,$$

then $\Delta(G)$ is the characteristic subgroup $\langle a\rangle$.

A group H is called an *f.c. group* (finite conjugate group) if every element of H has only finitely many conjugates in H. Clearly, any subgroup and any quotient group of an f.c. group are also f.c. From what we said above, it follows that, for any group G, $\Delta(G)$ is a normal f.c. subgroup of G. Let us record the following important consequence of (6.22) for f.c. groups.

(6.24) Corollary. *Any torsion-free f.c. group G is abelian.*

Proof. For $x, y \in G$, we want to show that $xy = yx$. Since any subgroup of G is f.c., we may assume that G is generated by x and y. Then $Z(G) = C_G(x) \cap C_G(y)$ has finite index, say n, in G. By (6.22),

$$(x^{-1}y^{-1}xy)^n = (x^{-1})^n(y^{-1})^n x^n y^n = (x^n)^{-1}(y^n)^{-1}x^n y^n = 1,$$

since $x^n \in Z(G)$. But G is torsion-free, so $x^{-1}y^{-1}xy = 1$. QED

A second group-theoretic lemma needed is the following.

(6.25) Neumann's Lemma. *Let $H_1, \ldots, H_m$ be subgroups in a group G. Suppose there are finitely many elements a_{ij} $(1 \le i \le m, 1 \le j \le n)$ in G such that*

$$G = \bigcup_{i=1}^{m} \bigcup_{j=1}^{n} H_i a_{ij}.$$

Then some H_i is of finite index in G.

Proof. We proceed by induction on m, the case $m = 1$ being clear. For $m \ge 2$, we may assume that $[G : H_1]$ is infinite, so there is a right coset $H_1 b$ disjoint from $\bigcup_{j=1}^{n} H_1 a_{1j}$. Then

$$H_1 b \subseteq \bigcup_{i=2}^{m} \bigcup_{j=1}^{n} H_i a_{ij},$$

and right multiplication by $b^{-1}a_{1k}$ gives

$$H_1 a_{1k} \subseteq \bigcup_{i=2}^{m} \bigcup_{j=1}^{n} H_i a_{ij} b^{-1} a_{1k}.$$

This shows that G is covered by a finite number of right cosets of $H_2, \ldots, H_m$, so we are done by induction. QED

Let k be any ring and Δ be a subgroup of a group G. We can define a projection map $\pi = \pi_\Delta: kG \to k\Delta$ by:

(6.26) $$\pi\left(\sum_{g \in G} a_g g\right) = \sum_{g \in \Delta} a_g g.$$

(6.27) Proposition. (a) *π is a homomorphism of $(k\Delta, k\Delta)$-bimodules, i.e.*

$$\pi\left(\alpha \cdot \sum_{g \in G} a_g g \cdot \beta\right) = \alpha\, \pi\left(\sum_{g \in G} a_g g\right)\beta$$

for all $\alpha, \beta \in k\Delta$.

(b) *Let $\mathfrak{A}$ be a right ideal of kG. Then $\mathfrak{A} \subseteq \pi(\mathfrak{A}) \cdot kG$. In particular, $\mathfrak{A} \neq 0 \Rightarrow \pi(\mathfrak{A}) \neq 0$.*

Proof. (a) follows directly from the definition (6.26). For (b), pick right coset representatives g_i $(i \in I)$ such that $G = \bigcup_{i \in I} \Delta g_i$. Then

$$kG = \bigoplus_{i \in I} k\Delta \cdot g_i.$$

For $\gamma \in \mathfrak{A}$, write $\gamma = \sum_{i \in I} \alpha_i g_i$, where $\alpha_i \in k\Delta$. Multiplying this from the right by g_j^{-1}, we see that $\alpha_j = \pi(\gamma g_j^{-1}) \in \pi(\mathfrak{A})$ (since $\mathfrak{A}$ is a right ideal), and this shows that $\gamma \in \pi(\mathfrak{A}) \cdot kG$. QED

The crucial point in the proof of "$R \Rightarrow D$" is the following ingenious result of Passman.

(6.28) Proposition. *Let k be a ring and G be any group. Let $\Delta = \Delta(G)$ and $\pi = \pi_\Delta$. Then for any two elements $\gamma, \gamma' \in kG$, we have*

$$\gamma\, kG\, \gamma' = 0 \Longrightarrow \pi(\gamma)\pi(\gamma') = 0 \in k\Delta.$$

Before we prove this Proposition, let us first show how it can be used to ascertain the implication "$R \Rightarrow D$" in (6.20).

Proof of "$R \Rightarrow D$". In this proof, k denotes a domain, and G denotes a torsion-free group. We assume that kG is reduced, and proceed to prove that kG is a domain. Assume, instead, that there exist nonzero $\gamma, \gamma' \in kG$ such that $\gamma'\gamma = 0$. For any $\tau \in kG$,

$$(\gamma\tau\gamma')^2 = \gamma\tau\gamma'\gamma\tau\gamma' = 0,$$

so $\gamma\tau\gamma' = 0$, since kG is reduced. This shows that $\gamma\, kG\, \gamma' = 0$. Let $\pi = \pi_{\Delta(G)}$. By (6.27)(b), $\pi(\gamma\, kG) \neq 0$, so $\pi(\gamma\tau) \neq 0$ for some $\tau \in kG$. Similarly, $\pi(\tau'\gamma') \neq 0$ for some $\tau' \in kG$. On the other hand, $\gamma\tau\, kG\, \tau'\gamma' \subseteq \gamma\, kG\, \gamma' = 0$, so (6.28) implies that $\pi(\gamma\tau)\pi(\tau'\gamma') = 0$ in $k\Delta$. But $\Delta = \Delta(G)$ is f.c. and torsion-free, so by (6.24) it is abelian. By the observation made before (6.16) (or by

(6.29) below), we know that $k\Delta$ is a domain, in contradiction to the equation $\pi(\gamma\tau)\pi(\tau'\gamma') = 0$. QED

To conclude our arguments, we now present

Proof of (6.28). Here, k and G are arbitrary, and $\Delta = \Delta(G)$. Given $\gamma\, kG\, \gamma' = 0$, it suffices to show that $\pi(\gamma)\gamma' = 0$, for (6.27)(a) will then give $0 = \pi(\pi(\gamma)\gamma') = \pi(\gamma)\pi(\gamma')$. Write $\gamma = \gamma_0 + \gamma_1$, where

$$\gamma_0 = a_1u_1 + \cdots + a_ru_r \quad (u_\ell \in \Delta),$$
$$\gamma_1 = b_1v_1 + \cdots + b_mv_m \quad (v_i \notin \Delta), \quad \text{and}$$
$$\gamma' = c_1w_1 + \cdots + c_nw_n \quad (w_j \in G).$$

Then the subgroup $C := \bigcap C_G(u_\ell)$ has finite index in G since each $C_G(u_\ell)$ does. Assume that $0 \neq \pi(\gamma)\gamma' = \gamma_0\gamma'$ and *fix* an element $g \in G$ which appears with a nonzero coefficient in $\gamma_0\gamma'$. If v_i happens to be conjugate to gw_j^{-1} in G, we also fix an element $g_{ij} \in G$ so that $g_{ij}^{-1}v_ig_{ij} = gw_j^{-1}$. If v_i is not conjugate to gw_j^{-1}, we just take $g_{ij} = 1$. The hypothesis $\gamma\, kG\, \gamma' = 0$ implies, in particular, that $x^{-1}\gamma x\gamma' = 0$ for every $x \in C$. Since x commutes with each u_ℓ, this gives

$$\begin{aligned}\gamma_0\gamma' &= (a_1u_1 + \cdots + a_ru_r)\gamma' \\ &= -x^{-1}(b_1v_1 + \cdots + b_mv_m)\, x\, (c_1w_1 + \cdots + c_nw_n).\end{aligned}$$

Since g appears on the *LHS*, we must have $g = x^{-1}v_ixw_j$ for some i,j. Thus, v_i is conjugate to gw_j^{-1}, so we have

$$x^{-1}v_ix = gw_j^{-1} = g_{ij}^{-1}v_ig_{ij};$$

that is, $x \in C_G(v_i)g_{ij}$. Here, $x \in C$ is arbitrary, so

$$C \subseteq \bigcup_{i,j} C_G(v_i)g_{ij}.$$

Since $[G : C] < \infty$, it follows that G itself is covered by a finite number of right cosets of $C_G(v_1), \ldots, C_G(v_m)$. This contradicts Neumann's Lemma, since $v_i \notin \Delta$ implies that each $C_G(v_i)$ has *infinite* index in G. QED

Now that we have established the interconnections (6.20) between the Problems U, R, D, and J, we see that the strongest possible "theorem" would be an affirmative answer to Problem U, for this would imply affirmative answers to all the other problems (for k a domain and G a torsion-free group). We shall now try to find some classes of groups for which Problem U has indeed an affirmative answer.

The best known class of such groups is the class of ordered groups. We say that a multiplicative group G is *ordered* if there is given a total ordering "$<$" on the elements of G such that, for any $x, y, z \in G$, we have

$$x < y \Longrightarrow xz < yz \quad \text{and} \quad zx < zy.$$

A prototype for ordered groups is the multiplicative group $\dot{\mathbb{R}}^+$ of positive real numbers, with its usual ordering. The additive groups $\mathbb{Z}$, $\mathbb{Q}$, and $\mathbb{R}$ are also ordered groups with respect to their usual orderings; each of these can be order-embedded into $\dot{\mathbb{R}}^+$ by the exponential map $a \mapsto e^a$ where $e > 1$. In fact, the exponential map gives an order-isomorphism from $\mathbb{R}$ onto $\dot{\mathbb{R}}^+$.

Given an ordering "$<$" on a multiplicative group G, its *positive cone* P is defined to be the set of elements $x \in G$ with $x > 1$. This cone has the following properties:

(1) $P \cdot P \subseteq P$;

(2) $G\backslash\{1\}$ is the disjoint union of P and $P^{-1} = \{x^{-1} : \ x \in P\}$;

(3) $zPz^{-1} \subseteq P$ for any $z \in G$.

Conversely, if we are given a set $P \subseteq G$ satisfying these three properties, we can define "$<$" by:

$$x < y \Longleftrightarrow x^{-1}y \in P \Longleftrightarrow yx^{-1} \in P.$$

It is straightforward to check that "$<$" makes G into an ordered group, with positive cone P. Because of this, it is often more convenient to define orderings on groups by specifying their positive cones.

Note that any ordered group $(G, <)$ is always torsion-free. For, if $g > 1$, then

$$1 < g < g^2 < \cdots,$$

and if $g < 1$, then

$$1 > g > g^2 > \cdots,$$

so g^n is never equal to 1. However, there exist torsion-free groups which *cannot* be ordered. For instance, the group G generated by two elements x, y with the relation $yxy^{-1} = x^{-1}$ is an extension of $\mathbb{Z}(\cong \langle x \rangle)$ by $\mathbb{Z}(\cong \langle y \rangle)$ so it is torsion-free. But G cannot be ordered since the positive cone of any ordering on G has to contain one, and hence both, of x, x^{-1}, which is impossible. For abelian groups, the situation is much better: we shall show a little later that *an abelian group G can be ordered iff G is torsion-free.*

We show next that all the problems raised in (6.16)–(6.19) have affirmative answers for the class of ordered groups.

(6.29) Theorem. *Let k be a domain and $(G, <)$ be an ordered group. Then $A = kG$ has only trivial units and is a domain. If $G \neq \{1\}$, A is J-semisimple.*

Proof. Consider a product $\alpha\beta$ where

$$\alpha = a_1 g_1 + \cdots + a_m g_m, \quad g_1 < \cdots < g_m, \quad a_i \neq 0 \quad (1 \leq i \leq m),$$

$$\beta = b_1 h_1 + \cdots + b_n h_n, \quad h_1 < \cdots < h_n, \quad b_j \neq 0 \quad (1 \leq j \leq n).$$

We have $g_1h_1 \leq g_ih_j$, with equality iff $i = j = 1$. Thus, the "smallest" group element appearing in $\alpha\beta$ is g_1h_1 (with nonzero coefficient a_1b_1), and similarly the "largest" one is g_mh_n (with nonzero coefficient a_mb_n). In particular, $\alpha\beta \neq 0$, and if $\alpha\beta = \beta\alpha = 1$, we must have $m = n = 1$, so $\alpha = a_1g_1$, $\beta = b_1h_1$, with $a_1b_1 = b_1a_1 = 1$ in k and $g_1h_1 = 1$ in G. This proves that A is a domain, and that A has only trivial units. The last statement of the theorem now follows from (6.21)(2). (Note that for this proof, we do not need to use the implication "$U \Rightarrow D$" in (6.20).) QED

As an application of (6.29), we shall give the following complete solution to the J-semisimplicity Problem for abelian group algebras over fields.

(6.30) Theorem. *Let k be a field and G be an abelian group.*

(1) *If char $k = 0$, then $A = kG$ is J-semisimple.*

(2) *If char $k = p$, then $A = kG$ is J-semisimple iff G is a p'-group.*

Proof. First assume that *char* $k = p$ and A is J-semisimple. Then G must be a p'-group, for if $x \in G$ has order p, then $((x-1)A)^p = 0$ and *rad* $A \supseteq (x-1)A \neq 0$. Now assume that *char* $k = 0$, or that *char* $k = p$ and G is a p'-group. To show that kG is J-semisimple, we may assume that G is finitely generated. For, if $\alpha \in$ *rad* kG, there exists a finitely generated subgroup $G_0 \subseteq G$ such that $\alpha \in kG_0$. Then $\alpha \in kG_0 \cap$ *rad* $kG \subseteq$ *rad* kG_0 (see Exercise 3), so it suffices to show that *rad* $kG_0 = 0$. If G is finitely generated, we can write $G = G_t \times H$ where G_t is the torsion subgroup of G and H is a free abelian group of finite rank. As is easily verified, $A = kG$ is isomorphic to the group ring RH where $R = kG_t$. Since G_t is finite and *char* k does not divide its order, Maschke's Theorem (6.1) implies that R is semisimple, and so $R \cong k_1 \times \cdots \times k_m$ where the k_i's are suitable fields. We have an isomorphism

$$A \cong RH \cong (k_1 \times \cdots \times k_m)H \cong k_1H \times \cdots \times k_mH,$$

so it suffices to show that each k_iH is J-semisimple. But H is an ordered group since $H \cong \mathbb{Z} \times \cdots \times \mathbb{Z}$ can be given the lexicographic ordering. Therefore, the J-semisimplicity of k_iH follows from (6.29). QED

In view of (6.29), it is of interest to know more examples of ordered groups. In the proof above, we have seen that a free abelian group of finite rank can be ordered (say lexicographically). In what follows, we shall prove two standard results on the orderability of groups:

(1) An abelian group can be ordered iff it is torsion-free (Theorem of Levi);

(2) Any free group can be ordered (Theorem of Birkhoff, Iwasawa and Neumann).

These results lead to more examples of group rings which are J-semisimple domains. The fact that free groups can be ordered will also prove to be useful later when we study the problem of embedding free algebras into division rings; see (14.25).

(6.31) Theorem. *Let G be either a torsion-free abelian group or a free group. Then G can be ordered. In particular, for any domain k, kG has only trivial units, and is a J-semisimple domain if $G \neq \{1\}$.*

Proof. First assume G is torsion-free and abelian. Then G embeds into $G_1 = \mathbb{Q} \otimes_{\mathbb{Z}} G$ which is a $\mathbb{Q}$-vector space. Clearly, it suffices to construct a positive cone P for G_1. To do this, we use again the idea of the lexicographic ordering. Choose a $\mathbb{Q}$-basis $\{g_i \colon i \in I\}$ for G_1, and fix a total ordering "$<$" on the indexing set I. Using the additive notation for G_1, we can define P to be the set of elements

$$a_1 g_{i_1} + \cdots + a_n g_{i_n},$$

where $i_1 < \cdots < i_n$ in I and $a_1 > 0$ in $\mathbb{Q}$. It is easy to check that P satisfies the axioms of a positive cone. Therefore P induces an ordered group structure on G_1 (and hence on G).

Next assume G is a free group. We shall construct a positive cone on G; however, this construction will not be entirely self-contained. We have to invoke the following known properties of the free group G:

(6.32) Magnus–Witt Theorem. *Define the lower central series*

$$G \supseteq G^{(1)} \supseteq G^{(2)} \supseteq \cdots$$

of a free group G by $G^{(1)} = [G, G]$ (the commutator group) and $G^{(n+1)} = [G, G^{(n)}]$ $(n \geq 1)$. Then $\bigcap_{n \geq 1} G^{(n)} = \{1\}$, and each $G^{(n)}/G^{(n+1)}$ is free abelian.

For a proof of this, we refer the reader to the book "Combinatorial Group Theory" (Sec. 5.7) of Magnus, Karrass, and Solitar (Dover, 1976), or the book "Finite Groups II" (pp. 380–383) by Huppert and Blackburn (Springer, 1982). Granting this result, we can construct a positive cone P on the free group G as follows. Since $G^{(n)}/G^{(n+1)}$ is free abelian, there exists a positive cone P_n in $G^{(n)}/G^{(n+1)}$ defining on it the structure of an ordered abelian group. Now let P be the subset of G consisting of all elements $g \neq 1$ with the property that, if n is the (unique) integer such that $g \in G^{(n)} \backslash G^{(n+1)}$ then the coset $gG^{(n+1)}$ belongs to P_n. Clearly, G is the disjoint union of $\{1\}$, $\dot{P}$ and $\dot{P}^{-1}$. It is also easy to check that, for any $z \in G$, $z^{-1}Pz \subseteq P$. For, if $g \in G$ is such that $g \in G^{(n)} \backslash G^{(n+1)}$ and $gG^{(n+1)} \in P_n$, then $z^{-1}gz \in G^{(n)} \backslash G^{(n+1)}$ and

$$z^{-1}gz = g \cdot g^{-1}z^{-1}gz \in g \cdot [G, G^{(n)}] = gG^{(n+1)},$$

so $z^{-1}gzG^{(n+1)} \in P_n$. To complete the proof that P is a positive cone on G, it only remains to show that $P \cdot P \subseteq P$. Let g, h be elements in P, with

$$g \in G^{(n)} \backslash G^{(n+1)}, \quad h \in G^{(m)} \backslash G^{(m+1)}, \quad \text{and}$$
$$gG^{(n+1)} \in P_n, \quad hG^{(m+1)} \in P_m.$$

To show that $gh \in P$, we may assume that $m \geq n$. If $m > n$, then $h \in G^{(m)} \subseteq G^{(n+1)}$; in this case $gh \in G^{(n)} \backslash G^{(n+1)}$ and

$$ghG^{(n+1)} = gG^{(n+1)} \in P_n,$$

so by definition $gh \in P$. If $m = n$, then $gG^{(n+1)}$, $hG^{(n+1)} \in P_n$ imply that $ghG^{(n+1)} \in P_n$; in particular, $gh \in G^{(n)} \backslash G^{(n+1)}$, so again $gh \in P$. QED

In the literature on group rings, there are many other classes of groups G for which one or more of the Problems (6.16)–(6.19) have been shown to have affirmative answers, especially in the case when k is a field. For an excellent treatise of these and other related problems over various classes of groups, we refer the reader to Passman [77].

Exercises for §6

In the following exercises, k denotes a field and G denotes a group, unless otherwise specified.

Ex. 6.1. Let V be a kG-module and H be a subgroup in G of finite index n not divisible by *char* k. Modify the proof of Maschke's Theorem to show the following: If V is semisimple as a kH-module, then V is semisimple as a kG-module.

Ex. 6.2. Let A be a normal elementary p-subgroup of a finite group G such that the index of the centralizer $C_G(A)$ is prime to p. Show that for any normal subgroup B of G lying in A, there exists another normal subgroup C of G lying in A such that $A = B \times C$. (**Hint.** Consider the conjugation action of G on A and apply Maschke's Theorem with $k = \mathbb{F}_p$.)

Ex. 6.3. Let G be a finite group whose order is a unit in a ring k, and let $W \subseteq V$ be left kG-modules.
(1) If W is a direct summand of V as k-modules, show that W is a direct summand of V as kG-modules.
(2) If V is projective as a k-module, show that V is projective as a kG-module.

Ex. 6.4. (This exercise is valid for any ring k.) For any subgroup H of a group G, show that

$$(*) \qquad kH \cap U(kG) \subseteq U(kH) \quad \text{and} \quad kH \cap \mathit{rad}\, kG \subseteq \mathit{rad}\, kH.$$

Deduce that, if kH is J-semisimple for any finitely generated subgroup H of G, then kG itself is J-semisimple.

Ex. 6.5. (Amitsur, Herstein) If k is an uncountable field, show that, for any group G, *rad* kG is a nil ideal. (**Hint.** Use Exercise 4 above and (4.21).)

Ex. 6.6. Let H be a normal subgroup of G. Show that $I = kG \cdot rad\ kH$ is an ideal of kG. If $rad\ kH$ is nilpotent, show that I is also nilpotent. (In particular, if H is finite, I is always nilpotent.)

Ex. 6.7. (For this Exercise, we assume Wedderburn's Theorem that finite division rings are commutative. A proof of this theorem will be given in (13.1).) Show that if k_0 is any finite field and G is any finite group, then $(k_0G/rad\ k_0G) \otimes_{k_0} K$ is semisimple for any field extension $K \supseteq k_0$.

Ex. 6.8. Let $k \subseteq K$ be two fields and G be a finite group. Show that

$$rad(KG) = (rad\ kG) \otimes_k K.$$

(**Hint.** It is enough to treat the case where k has characteristic $p > 0$. From Exercise 7, deduce that, in this case, $rad(kG) = (rad\ \mathbb{F}_pG) \otimes_{\mathbb{F}_p} k$, and $rad(KG) = (rad\ \mathbb{F}_pG) \otimes_{\mathbb{F}_p} K$.)

Ex. 6.9. Let $k \subseteq K$ and G be as above. Show that a kG-module M is semisimple iff the KG-module $M^K = M \otimes_k K$ is semisimple.

Ex. 6.10. Let k be a commutative ring and G be any group. If kG is left noetherian (resp. left artinian), show that kG is right noetherian (resp. right artinian).

Ex. 6.11. (Hölder's Theorem) An ordered group $(G, <)$ is said to be *archimedean* if, for any $a, b > 1$ in G, we have $a < b^n$ for some integer $n \geq 1$. Show that if $(G, <)$ is archimedean, then G is commutative and $(G, <)$ is order-isomorphic to an additive subgroup of $\mathbb{R}$ with the usual ordering.

Ex. 6.12. Assume $char(k) = 3$, and let $G = S_3$ (symmetric group on three letters).
(1) Compute the Jacobson radical $J = rad(kG)$, and the factor ring kG/J.
(2) Determine the index of nilpotency for J, and find a k-basis for J^i for each i.
(The case where $char(k) = 2$ will be covered by the first part of the next exercise.)

Ex. 6.13. (Passman) Assume $char\ k = 2$. Let A be an abelian $2'$-group and let G be the semidirect product of A and a cyclic group $\langle x \rangle$ of order 2, where x acts on A by $a \mapsto a^{-1}$.
(1) If $|A| < \infty$, show that $rad\ kG = k \cdot \sum_{g \in G} g$, and $(rad\ kG)^2 = 0$.
(2) If A is infinite, show that kG has no nonzero nil ideals.
(**Hint.** Any element in kG can be expressed in the form $\alpha + \beta x$, $\alpha, \beta \in kA$. For $\alpha = \sum_{a \in A} \alpha_a a \in kA$, let $\alpha^* = \sum \alpha_a a^{-1}$; then $x\alpha = \alpha^* x$. Let I be any nil ideal in kG. If $\alpha + \beta x \in I$, show that $(\alpha + \beta x)(\alpha^* + x\beta^*) = \alpha\alpha^* + \beta\beta^*$, and conclude that $\alpha\alpha^* = \beta\beta^*$. Then show that $(\alpha + \beta x)^m = (\alpha + \alpha^*)^{m-1}(\alpha + \beta x)$ and deduce $\alpha = \alpha^*$.)

Ex. 6.14. (Wallace) Assume $char\ k = 2$, and let $G = A \cdot \langle x \rangle$ as in Exercise 13, where A is the infinite cyclic group $\langle y \rangle$. (G is the infinite dihedral group.) Show that $R = kG$ is J-semisimple (even though G has an element of order

2). (**Hint.** Let $H_i = \langle y^{3^i} \rangle$ $(1 \le i < \infty)$ and note that $G_i = G/H_i$ is a dihedral group of order $2 \cdot 3^i$. Show that $\varphi : kG \to \prod_{i=1}^{\infty} kG_i$ is injective. By part (1) of Exercise 13, $(rad\ kG_i)^2 = 0$. Deduce that $(rad\ kG)^2 = 0$ and conclude from part (2) of Exercise 13 that $rad\ kG = 0$.)

Ex. 6.15. (Dietzmann's Lemma) Let G be a group generated by $x_1, \ldots, x_n$ where each x_i has finite order and has only finitely many conjugates in G. Show that G is a finite group.

Ex. 6.16. (1) Let G be a group such that $[G : Z(G)] < \infty$. Show that the commutator subgroup $[G, G]$ is finite.
(2) Let G be an f. c. group, i.e. each $g \in G$ has only finitely many conjugates in G. Show that $[G, G]$ is torsion. If, moreover, G is finitely generated, show that $[G, G]$ is finite.

Ex. 6.17. For any group G, let

$$\Delta(G) = \{g \in G : \ [G : C_G(g)] < \infty\}, \quad \text{and}$$
$$\Delta^+(G) = \{g \in \Delta(G) : \ g \text{ has finite order}\}.$$

(1) Show that $\Delta^+(G)$ is a characteristic subgroup of G and that $\Delta^+(G)$ is the union of all finite normal subgroups of G. (**Hint.** For $a, b \in \Delta^+(G)$, Exercise 15 shows that the conjugates of a and b generate a finite normal subgroup of G.)
(2) (B.H. Neumann) Show that $\Delta(G)/\Delta^+(G)$ is torsion-free abelian. (**Hint.** $\Delta(G)/\Delta^+(G)$ is torsion-free and f.c. Apply (6.24).)

Ex. 6.18. A total ordering "$<$" of the elements of a group G is said to be a *right ordering* of G if $x < y \Rightarrow xz < yz$ for any $x, y, z \in G$. Show that Theorem (6.29) remains valid as long as the group G can be right-ordered.

Ex. 6.19. For any von Neumann regular ring k, show that any finitely generated submodule M of a projective k-module P is a direct summand of P (and hence also a projective k-module). (**Hint.** Reduce to the case where P is a free module $e_1 k \oplus \cdots \oplus e_n k$. Map M to $e_n k$ by coordinate projection, and induct on n.)

Ex. 6.20. Show that the conclusion of the last exercise is equivalent to the fact that, if k is a von Neumann regular ring, then so is $\mathbb{M}_n(k)$ for any $n \ge 1$.

Ex. 6.21. (Auslander, McLaughlin, Connell) For any nonzero ring k and any group G, show that the group ring kG is von Neumann regular iff k is von Neumann regular, G is locally finite (that is, any finite subset in G generates a finite subgroup), and the order of any finite subgroup of G is a unit in k. (**Hint.** For the "only if" part, use the fact that, if $h_1, \ldots, h_n$ generate a subgroup $H \subseteq G$, then

$$\sum_{h \in H} kG \cdot (h - 1) = \sum_{i=1}^{n} kG \cdot (h_i - 1).$$

The "if" part can be deduced from Exercise 19 and (1) of Exercise 3.)

CHAPTER 3

Introduction to Representation Theory

After studying the J-semisimplicity problem for the group ring in the last chapter, a natural topic to discuss next will be the representation theory of groups. We have already explained, in the introduction to §6, how ring theory may be brought to bear on group representation theory by viewing representations as modules over group rings. From this viewpoint, many facts in the representation theory and character theory of finite groups can be deduced from facts concerning modules over finite-dimensional algebras. This ring-theoretic approach to group representation theory was first effectively used by Emmy Noether, and subsequently greatly popularized by her disciples and followers.

In this chapter, we shall give a short introduction to representation theory, from the ring-theoretic (and module-theoretic) perspective. Our goal will be to illustrate the role played by the methods of ring theory in the development of the representation theory of finite groups. In the beginning section, we first study more generally modules over finite-dimensional algebras, and establish the basic facts about irreducible modules, scalar extensions and splitting fields. In the second section, we specialize these facts to group algebras and develop the rudiments of group representation theory and character theory. The results obtained from representation theory are, in turn, applied to obtain ring-theoretic information about group algebras, e.g., the structure of their unit groups and idempotents. In the final section of the chapter, we make a short excursion into the theory of linear groups over a field k. Here, the groups G studied may no longer be finite, but the role of the group algebra kG may often be replaced by that of the finite-dimensional k-algebra spanned by G in an appropriate matrix algebra. Thus, the tools of ring theory still prove to be effective in analyzing the structure of G. We shall study in this section some classical results of Burnside, Schur, Lie and

Kolchin, using a combination of techniques from group theory, ring theory and linear algebra. Throughout this exposition, the notion of the Jacobson radical of a ring plays a fundamental role.

§7. Modules Over Finite-Dimensional Algebras

The module theory we shall develop in this section is essentially a refinement or elaboration of the general structure theory of semisimple rings given in Chapter 1. In the category of finite-dimensional algebras, certain aspects of this structure theory can be made somewhat more explicit. Therefore, we can often expect to get sharper results for finite-dimensional algebras than for artinian rings. It is true that many such results can eventually be generalized, in one form or another, to artinian rings. However, at this early stage, it seems best not to worry about these generalizations so that we can focus our attention on finite-dimensional algebras (and their modules).

Throughout this section, k denotes a field whose characteristic is arbitrary, unless stated otherwise. We shall consider *finite-dimensional* k-algebras, usually denoted by R, with Jacobson radical $rad\ R$. Note that subalgebras and quotient algebras of R are all finite-dimensional; hence they are left and right artinian rings. In particular, by (4.14), $\bar{R} := R/rad\ R$ is semisimple, so the structure theory developed in §3 applies to $\bar{R}$.

(7.1) Notation. Let $\bar{R} = B_1 \times \cdots \times B_r$ be the decomposition of $\bar{R}$ into simple components, and let M_i $(1 \leq i \leq r)$ be the unique simple left module over B_i. Then $M_1, \ldots, M_r$ form a complete set of (isomorphism classes of) simple left $\bar{R}$-modules; by (4.8), they form a complete set of simple left R-modules. Let

$$D_i = End({}_{B_i}M_i) = End({}_RM_i),$$

and $n_i = dim_{D_i}\ M_i$. By the Wedderburn theory, $B_i \cong End(M_i)_{D_i} \cong \mathbb{M}_{n_i}(D_i)$. (Recall that elements of D_i are composed as right operators on M_i so that we can avoid forming opposite rings in applying the Wedderburn theory.) We have then

$$\bar{R} \cong \mathbb{M}_{n_1}(D_1) \times \cdots \times \mathbb{M}_{n_r}(D_r), \quad \text{and}$$

$${}_{\bar{R}}\bar{R} \cong n_1 M_1 \oplus \cdots \oplus n_r M_r.$$

The important thing to note here is that each of the objects above has a natural structure as a k-vector space, and as such, it is finite-dimensional. In particular, B_i and D_i are finite-dimensional k-algebras. For later reference, we record the following useful facts.

(7.2) Proposition. *In the notation of* (7.1), *we have*

(1) $dim_k\ M_i = n_i\ dim_k\ D_i$.

(2) $dim_k\ R = dim_k\ rad\ R + \sum_{i=1}^{r} n_i^2\ dim_k\ D_i$.

(3) *The natural map* $R \to End(M_i)_{D_i}$ *is onto.*

In the special case when $D_i = k$, the last part of the Proposition amounts to the following classical fact.

(7.3) Burnside's Lemma. *Let M be a finite-dimensional right k-vector space and A be a k-subalgebra of $End(M_k)$ such that M is simple as a left A-module. If $End(_AM) = k$, then $A = End(M_k)$.*

In general, the D_i's are (finite-dimensional) k-division algebras, by Schur's Lemma. The extent to which we can determine the structure of D_i would largely depend on the nature of k. *The easiest case is when k is an algebraically closed field.* In this case D_i must be k itself (for, if $d \in D_i$, then $k[d]$ is a finite field extension of k, and hence d must be in k). Consequently, we have

$$\bar{R} \cong \mathbb{M}_{n_1}(k) \times \cdots \times \mathbb{M}_{n_r}(k),$$

and (7.2) simplifies to $dim_k\ M_i = n_i$ and

$$dim_k\ R = dim_k\ rad\ R + \sum_{i=1}^{r} n_i^2.$$

Also, in (7.3), the condition $End(_AM) = k$ is automatic and therefore can be removed: this gives in fact Burnside's Theorem in its original form.

Without any condition on k and M_i, of course D_i may *not* be equal to k. To give an example, let $k = \mathbb{Q}$ and consider $R = \mathbb{Q}G$ where $G = \langle g \rangle$ is the cyclic group of order 3. Let $M = \mathbb{Q}(\omega)$ where ω is a primitive cubic root of unity. Via the surjection $\mathbb{Q}G \to \mathbb{Q}(\omega)$ (sending g to ω), M becomes a (left) simple R-module. Clearly $D = End(_RM)$ is given by right multiplications of $\mathbb{Q}(\omega)$, so $D \cong \mathbb{Q}(\omega) \supsetneq \mathbb{Q}$.

We might also consider the converse question: Let M be a left R-module; *if $End(_RM) = k$, is $_RM$ necessarily simple*? In general, the answer to this question is "no." For instance, let R be the ring of upper triangular matrices $\left\{\begin{pmatrix} a & b \\ 0 & c \end{pmatrix}\right\}$ over k, and let $M = k^2$, viewed as a left R-module by matrix multiplication. If a k-endomorphism on k^2 given by $\begin{pmatrix} x & y \\ z & w \end{pmatrix}$ commutes with all $\begin{pmatrix} a & b \\ 0 & c \end{pmatrix}$, an easy computation shows that $y = z = 0$ and $x = w$. Therefore $End(_RM) = k$, but $_RM$ is *not* simple, as $\left\{\begin{pmatrix} d \\ 0 \end{pmatrix} : d \in k\right\}$ is a proper R-submodule. The question raised above has an affirmative answer only if we are willing to add more conditions. For instance, if $_RM$ is semisimple, the answer becomes "yes." In fact, if $M = M' \oplus M''$ with $M' \neq 0 \neq M''$, then the endomorphism of M obtained by projecting M to M' is clearly not a scalar multiplication.

The simple R-modules M with the property that $End({}_RM) = k$ are of special significance. Our next goal will be to characterize these modules. We first need some notations. Let $K \supseteq k$ be an extension of the ground field k. Then we have the K-algebra $R^K := R \otimes_k K$ obtained by "scalar extension." For any (left) R-module M, we also have a scalar extension $M^K := M \otimes_k K$, which is a (left) R^K-module via the action

$$(r \otimes a)(m \otimes b) = rm \otimes ab.$$

The following fact about homomorphisms between extended modules is of fundamental importance.

(7.4) Lemma. *Let R be a k-algebra (not necessarily of finite dimension over k) and let M, N be left R-modules, with $dim_k\ M < \infty$. Then the natural map*

$$\theta\colon\ (Hom_R(M,N))^K \longrightarrow Hom_{R^K}(M^K, N^K)$$

is an isomorphism of K-vector spaces.

Proof. Let $\{a_i\colon i \in I\}$ be a k-basis of K. Then M^K has a decomposition $\bigoplus_i (M \otimes a_i)$, and similarly for N^K. Consider an R^K-homomorphism $f\colon M^K \to N^K$. Then

$$f(m \otimes 1) = \sum_i g_i(m) \otimes a_i$$

where $g_i\colon M \to N$ are uniquely determined k-linear maps. We claim that each g_i is an R-homomorphism. To see this, let $r \in R$ and $m \in M$. On the one hand, we have

$$f(rm \otimes 1) = \sum g_i(rm) \otimes a_i;$$

on the other hand,

$$\begin{aligned} f(rm \otimes 1) &= f((r \otimes 1)(m \otimes 1)) \\ &= (r \otimes 1)\Big(\sum g_i(m) \otimes a_i\Big) \\ &= \sum r g_i(m) \otimes a_i. \end{aligned}$$

This implies that $g_i(rm) = rg_i(m)$, so $g_i \in Hom_R(M,N)$ for all i. Also, since $dim_k\ M < \infty$, it is easy to see that only a finite number of the g_i's can be nonzero. Thus, $g := \sum g_i \otimes a_i$ makes sense, and g is an element of $(Hom_R(M,N))^K$ which maps to f under the natural map θ. This shows the surjectivity of θ. For the injectivity of θ, note that any element in $(Hom_R(M,N))^K$ can be written in the form $\sum_i f_i \otimes a_i$ where $f_i \in Hom_R(M,N)$. If this maps to zero under θ, then

$$0 = \sum_i \theta(f_i \otimes a_i)(m \otimes 1) = \sum_i f_i(m) \otimes a_i$$

implies that $f_i(m) = 0$ for all $m \in M$, and therefore $\sum_i f_i \otimes a_i = 0$. (This argument does not depend on the finite dimensionality of M.) QED

We are now ready to characterize the simple R-modules M with $End({}_RM) = k$.

(7.5) Theorem. *Let R be a k-algebra (not necessarily finite-dimensional), and let M be a simple left R-module with $\dim_k M < \infty$. The following statements are equivalent:*

(1) $End({}_RM) = k$.

(2) *The map $R \to End(M_k)$ expressing the R-action on M is surjective.*

(3) *For any field extension $K \supseteq k$, M^K is a simple R^K-module.*

(4) *There exists an algebraically closed field $E \supseteq k$ such that M^E is a simple R^E-module.*

If one (and hence all) of these conditions holds, we say that M is an absolutely simple (or absolutely irreducible) R-module.

Proof. Clearly (3) $\Rightarrow$ (4), so it suffices to show

$$(4) \Rightarrow (1) \Rightarrow (2) \Rightarrow (3).$$

Assume (4) holds. Since E is algebraically closed, this implies that $Hom_{R^E}(M^E, M^E) \cong E$ and therefore, by (7.4), $Hom_R(M, M) \cong k$. (Actually this uses only the trivial injectivity part of (7.4). The nontrivial surjectivity part will be used a bit later.) Next, (1) $\Rightarrow$ (2) follows from Burnside's Lemma (7.3). Now assume (2). To prove (3), we may replace R by $End(M_k)$. If $M_k = k^n$, we can then identify R with the full matrix algebra $\mathbb{M}_n(k)$. With these identifications, we have $M^K = K^n$ and $R^K = \mathbb{M}_n(K)$, so M^K is a (left) simple R^K-module, as desired. QED

Next we introduce the important notion of a splitting field. *The algebras to be considered in the rest of this section will be assumed to be finite-dimensional.*

(7.6) Definition. Let R be a (finite-dimensional) k-algebra. We say that a field $K \supseteq k$ is a *splitting field* for R (or that R *splits over* K) if every left irreducible R^K-module is absolutely irreducible.

Strictly speaking, the K above should be called a *left* splitting field, since we use left R-modules in its definition. The following characterization, however, shows that this notion is left-right symmetric, so the distinction will not be necessary.

(7.7) Theorem. *In the notation of* (7.6), *K is a (left) splitting field for R iff $R^K/rad(R^K)$ is a finite direct product of matrix algebras over K.*

Proof. To simplify the notation, we may as well take $K = k$. If each irreducible R-module M_i is absolutely irreducible, then by (7.5), each $D_i = End_R(M_i)$ is k so $R/rad\ R$ is a finite direct product of matrix algebras over k. The converse is proved similarly. QED

In a similar vein, we also have the following characterization whose easy proof is omitted.

(7.8) Corollary. *A k-algebra R splits over k iff*

$$dim_k\ R = dim_k(rad\ R) + \sum (dim_k\ M_i)^2,$$

where $\{M_i\}$ is a full set of simple left R-modules.

Note that any k-algebra R has a splitting field: in fact, R always splits over E, the algebraic closure of k. The following proposition reduces the consideration of splitting fields to the case of semisimple algebras.

(7.9) Proposition. *An extension $K \supseteq k$ is a splitting field for R iff it is a splitting field for $\bar{R} := R/rad\ R$.*

Proof. Since we assume $dim_k\ R < \infty$, (5.14) implies that

$$(rad\ R)^K \subseteq rad(R^K).$$

Therefore, the radical of

$$\bar{R}^K \cong R^K/(rad\ R)^K$$

is given by $rad(R^K)/(rad\ R)^K$, and so the simple left modules over $\bar{R}^K$ are the same as those over $R^K/rad(R^K)$, or those of R^K. The Proposition now follows immediately from the definition of a splitting field. QED

In applications, it is often convenient to work with splitting fields which are finite-dimensional over k. The following Proposition guarantees that such splitting fields do exist if the ground field k is a perfect field.

(7.10) Proposition. *Any algebra R over a perfect field k splits over some finite extension K of k.*

Proof. By (7.9), we may assume that R is semisimple. Since k is perfect, the algebraic closure E of k is separable over k and therefore, by (5.16), R^E is semisimple. By Wedderburn's Theorem, R^E is a finite direct product of matrix algebras over E. The matrix units defining such a decomposition, being finite in number, lie in R^K for some $K \subseteq E$ of finite degree over k. Therefore, R^K is a finite direct product of matrix algebras over K, and so K is a splitting field for R. QED

(7.11) Remark. In the proof above, the hypothesis that k be perfect was only used to see that R^E is semisimple. Therefore, the Proposition is valid for any semisimple k-algebra R for which R^E remains semisimple, where E is the algebraic closure of k. Such algebras are called *separable* k-algebras; we shall not digress to discuss their further properties here.

In elementary field theory we have the notion of a splitting field for a polynomial. Our definition (7.6) for the splitting field of an algebra is, in fact, a generalization of this. To see the connection, let $f \in k[t]$ be a polynomial over k, and let R be the quotient ring $k[t]/(f)$, viewed, as usual, as a finite-dimensional k-algebra.

(7.12) Proposition. *An extension $K \supseteq k$ is a splitting field for the algebra $R = k[t]/(f)$ iff K is a splitting field for f in the sense of field theory.*

Proof. Let $f = f_1^{n_1} \cdots f_r^{n_r}$ be a factorization of f into irreducible factors in $K[t]$. Then

$$R \otimes_k K \cong K[t]/(f_1^{n_1} \cdots f_r^{n_r}) \cong \prod_{i=1}^{r} K[t]/(f_i^{n_i})$$

and so

$$R^K/rad(R^K) \cong \prod_{i=1}^{r} K[t]/(f_i).$$

Here $K_i := K[t]/(f_i)$ are field extensions of K with $[K_i : K] = deg\, f_i$. By (7.7), K is a splitting field for R iff $K_i = K$ for all i. This is the case iff all f_i's have degree 1, i.e., iff f splits completely in K. QED

Next we prove a result relating the simple modules of an algebra R to those of the extended algebra R^K.

(7.13) Proposition. *Let R be a k-algebra and $K \supseteq k$ be a field extension. Then:*

(1) *any simple left R^K-module V is a composition factor of M^K for some simple left R-module M; and*

(2) *if M_1, M_2 are non-isomorphic simple left R-modules, then M_1^K and M_2^K cannot have a common composition factor.*

Proof. (2) Let $R/rad\, R = B_1 \times \cdots \times B_r$ be the decomposition of $R/rad\, R$ into its simple components. We may assume that M_i is the unique simple module of B_i $(i = 1, 2)$. Let $a_1 \in R$ be chosen such that $\bar{a}_1$ in $R/rad\ R$ is the unit element of B_1. Then $a_1 \otimes 1$ acts as the identity on any composition factor V_1 of M_1^K, and acts as zero on any composition factor V_2 of M_2^K. Clearly, this implies that $V_1 \not\cong V_2$ as R^K-modules. To prove (1), let

$$0 = I_0 \subsetneq \cdots \subsetneq I_r = R$$

be a composition series of the left regular module ${}_RR$. Then

$$0 = I_0^K \subsetneq \cdots \subsetneq I_r^K = R^K$$

is a filtration of the left regular module R^K. Any simple left R^K-module V is a composition factor of the latter, so V is a composition factor of some $I_{i+1}^K / I_i^K \cong (I_{i+1}/I_i)^K$. QED

The following Proposition gives some of the nice properties of a splitting field.

(7.14) Proposition. *Let $K \supseteq k$ be a splitting field for the k-algebra R. Let $\{V_1, \ldots, V_m\}$ be a full set of left simple R^K-modules. Then for any field $L \supseteq K$, $\{V_1^L, \ldots, V_m^L\}$ is a full set of left simple R^L-modules. In particular, L is also a splitting field for R.*

Proof. By the definition of a splitting field, $V_1^L, \ldots, V_m^L$ are simple modules over R^L, and by (7.13)(2), these are mutually nonisomorphic. (Another way to see the latter is to use (7.4).) By (7.13)(1), each simple R^L-module is isomorphic to one of $V_1^L, \ldots, V_m^L$. This proves the first statement in the Proposition. Since each V_i^L is clearly absolutely irreducible, it follows that L is a splitting field for R. QED

Next we would like to derive some general results on the number of simple modules over an algebra R. These results will have nice applications in the next section when we try to determine the number of irreducible representations of a finite group over a splitting field. We shall, however, formulate our results with sufficient generality so that they are meaningful over arbitrary fields (see (7.17)).

First we give a general definition. For any ring R, and elements $a, b \in R$, let $[a, b]$ denote the element $ab - ba$: this is called the *additive commutator* (or sometimes *Lie product*) of the elements a, b. The additive subgroup of R generated by all $[a, b]$ is denoted by $[R, R]$. If R is a k-algebra where k is a commutative ring, then $[R, R]$ is a k-submodule of R. In general, $[R, R]$ is neither a left nor a right ideal of R. Some properties of $[R, R]$ are as follows.

(7.15) Lemma. *Let R be a ring of characteristic p, where p is a prime. Let $S = [R, R]$. Then*

(1) *For any* $a_1, \ldots, a_n \in R$, $(a_1 + \cdots + a_n)^{p^r} \equiv a_1^{p^r} + \cdots + a_n^{p^r} \pmod{S}$ *for all* $r \geq 0$.

(2) $s \in S \Longrightarrow s^{p^r} \in S$ *for all* $r \geq 0$.

Proof. Clearly it suffices to prove (1), (2) for $r = 1$. If S were an ideal, (1) would follow easily by working in the quotient ring R/S of characteristic p. Since S may not be an ideal, we have to argue more carefully. Regarding $a_1, \ldots, a_n$ as (noncommuting) symbols, the expansion of $(a_1 + \cdots + a_n)^p$ consists of n^p distinct words of length p in $a_1, \ldots, a_n$. The cyclic group G of order p acts on these words by cyclic permutations. Using the definition of S, we see easily that two words in the same orbit are always congruent modulo S. Now there are exactly n singleton orbits, consisting, respectively, of $a_1^p, \ldots, a_n^p$. The other orbits are of cardinality p.[1] Summing over each orbit in R, we get modulo S an element of $p \cdot R = 0$. Thus,

$$(a_1 + \cdots + a_n)^p \equiv a_1^p + \cdots a_n^p \pmod{S}.$$

Finally, let $s = \sum (a_i b_i - b_i a_i) \in S$. Modulo S, we have

$$\begin{aligned} s^p &\equiv \sum (a_i b_i - b_i a_i)^p \\ &\equiv \sum ((a_i b_i)^p - (b_i a_i)^p) \\ &\equiv \sum (a_i \cdot (b_i a_i)^{p-1} b_i - (b_i a_i)^{p-1} b_i \cdot a_i). \end{aligned}$$

Therefore, $s^p \in S$. QED

(7.16) Lemma. *Let* $R = \mathbb{M}_n(k)$ *where* k *is a commutative ring. Then*

$$[R, R] = \{M \in \mathbb{M}_n(k) : \ tr(M) = 0\}.$$

Proof. The inclusion "$\subseteq$" follows from the observation that $tr(MN) = tr(NM)$. To prove the reverse inclusion, let $S = [R, R]$, and let $\{E_{ij}\}$ be the matrix units. If $i \neq j$, we have

$$E_{ij} = E_{ii} E_{ij} - E_{ij} E_{ii} \in S, \quad \text{and}$$

$$E_{ii} - E_{jj} = E_{ij} E_{ji} - E_{ji} E_{ij} \in S.$$

Noting that S is a k-module, we have, for any $M = (a_{ij})$, the following congruence:

$$M = \sum_{i,j} a_{ij} E_{ij} \equiv \sum_i a_{ii} E_{ii} \equiv \sum_i a_{ii} E_{11} \pmod{S}.$$

In particular, if $tr(M) = 0$, this implies $M \in S$. QED

---

[1] The number of nonsingleton orbits is $(n^p - n)/p$. This, incidentally, gives another proof of Fermat's Little Theorem: $n^p \equiv n \pmod{p}$.

Using (7.16) (but not (7.15)), we can derive the following result on the number of simple modules over a k-algebra R, in the case when k is a splitting field for R.

(7.17) Theorem. *For a finite-dimensional k-algebra R, let*

$$T(R) = rad\ R + [R, R].$$

If R splits over k, then the number of left simple R-modules (up to isomorphism) is $dim_k\ R/T(R)$; moreover, $T(R)$ contains all nilpotent elements of R.

Proof. Let $\bar{R} = R/rad\ R$. Since $[R, R]$ maps onto $[\bar{R}, \bar{R}]$ under the projection map, we have

$$dim_k\ R/T(R) = dim_k\ \bar{R}/[\bar{R}, \bar{R}].$$

By (7.7),

$$\bar{R} \cong A_1 \times \cdots \times A_r$$

where each A_i is a matrix algebra over k, and r is the number of simple left R-modules. Clearly $[\bar{R}, \bar{R}] \cong \prod_{i=1}^{r} [A_i, A_i]$, so

$$(\dagger) \qquad \bar{R}/[\bar{R}, \bar{R}] \cong \prod_{i=1}^{r} A_i/[A_i, A_i].$$

Each factor on the *RHS* has k-dimension 1 by (7.16). Therefore, taking k-dimensions in ($\dagger$), we have $r = dim_k\ \bar{R}/[\bar{R}, \bar{R}]$. The last conclusion follows from (7.16) since any nilpotent matrix over k has trace zero. QED

From (7.17), we can deduce the following result which is valid over any field k.

(7.18) Corollary. *Let R be a k-algebra and $K \supseteq k$ be a splitting field for R. Let r (resp., r') be the number of simple left R-modules (resp., R^K-modules). Then $r \leq r' \leq dim_k\ R/T(R)$.*

Proof. The first inequality follows from (7.13)(2). For the second inequality, note that, by the theorem above, $r' = dim_K\ R^K/T(R^K)$. Since

$$[R, R]^K \subseteq [R^K, R^K] \quad \text{and} \quad (rad\ R)^K \subseteq rad(R^K),$$

we have $T(R)^K \subseteq T(R^K)$ and so

$$r' = dim_K\ R^K/T(R^K) \leq dim_K\ R^K/T(R)^K = dim_k\ R/T(R). \qquad \text{QED}$$

To conclude this section, we shall discuss briefly the notion of characters. For any (finite-dimensional) k-algebra R and any (left) R-module M of finite dimension over k, we can associate a function $\chi_M\colon R \to k$ defined by $\chi_M(a) = tr(a)$, where, for any $a \in R$, $tr(a)$ denotes the trace of the linear transformation on M given by left multiplication of a. Clearly, χ_M is a k-linear functional on R: it is called the *character* associated with the R-module M. Note that if we have an exact sequence of R-modules

$$0 \to M' \to M \to M'' \to 0$$

each of which is finite-dimensional over k, then, computing the traces using suitable bases on the three modules, we get $\chi_M = \chi_{M'} + \chi_{M''}$. In particular, the composition factors of a module M, counted with multiplicities, completely determine the character χ_M. The converse of this is also true, in the characteristic zero case: this is the next result.

(7.19) Theorem. *Let R be a k-algebra where char $k = 0$, and let M be a left R-module with $\dim_k M < \infty$. Then χ_M completely determines the composition factors of M, counted with multiplicities. More precisely, if M, M' are R-modules (of finite k-dimensions) with $\chi_M = \chi_{M'}$, then M and M' have the same composition factors, counted with multiplicities; if M and M' are both semisimple, then in fact $M \cong M'$.*

Proof. Using the notation of (7.1), suppose M_i $(1 \le i \le r)$ occurs as a composition factor for M with multiplicity $m_i \ge 0$. Then

$$\chi_M = \sum_{i=1}^{r} m_i \chi_{M_i},$$

and our job is to show that χ_M determines the integers $\{m_1, \ldots, m_r\}$. Let $a_i \in R$ $(1 \le i \le r)$ be chosen such that $\bar{a}_i$ in $R/rad\ R$ is the identity of the ith simple component of $R/rad\ R$. Then a_i acts as zero on M_j for $j \neq i$, and acts as the identity on M_i. Computing the character χ_M on a_i, we have $\chi_M(a_i) = m_i \dim_k M_i$. Since *char* $k = 0$, this gives

$$m_i = \chi_M(a_i)/\dim_k M_i$$

in k, and therefore the same equation holds in $\mathbb{Q}$, for $1 \le i \le r$. The rest is clear. QED

Without the assumption that M be semisimple, χ_M will not determine the isomorphism class of M, even in characteristic 0. For instance, over the two-dimensional k-algebra $R = k[t]/(t^2)$, the module $M = ke_1 \oplus ke_2$ with the t-action $t(e_1) = e_2$, $t(e_2) = 0$ has character χ_M given by $\chi_M(a + bt) = 2a$ for $a, b \in k$. The module $M' = k^2$ with the trivial t-action has the same character, but clearly $M \not\cong M'$. If the characteristic of k is $p > 0$, then χ_M need not determine the composition factors of M; in fact, it may not even determine $\dim_k M$. For instance, let M_1, M_2 be nonisomorphic simple R-modules. Then

$$M = M_1 \oplus \cdots \oplus M_1,$$

$$M' = M_2 \oplus \cdots \oplus M_2$$

(p copies each) both have zero characters, but they may have different dimensions, and they do not even have a composition factor in common.

We do have the following result which holds in any characteristic.

(7.20) Theorem. *Let M, M' be modules over the k-algebra R, with M absolutely irreducible over R. Assume either* (1) $\dim_k M = \dim_k M'$ *or* (2) *M' is irreducible. Then $M \cong M'$ iff $\chi_M = \chi_{M'}$.*

Proof ("If" part). Using again the notation of (7.1), let us assume $M = M_1$ and that, for any i, M_i occurs as a composition factor of multiplicity m_i in M'. Since M is absolutely irreducible, the map

$$R \to B_1 = End(M_1)_k$$

is a surjection. Pick $a \in R$ such that a projects to a k-endomorphism of trace 1 in B_1 and projects to zero in $B_i = End(M_i)_{D_i}$ for $i \geq 2$. Then

$$1 = \chi_M(a) = \chi_{M'}(a) = \sum_{i=1}^{r} m_i \chi_{M_i}(a) = m_1 \chi_M(a) = m_1 \cdot 1$$

in k. In particular, as an integer, $m_1 \geq 1$. If we assume either (1) or (2), this clearly forces $m_2 = \cdots = m_r = 0$ and $m_1 = 1$. Therefore $M' \cong M_1 = M$.

QED

(7.21) Corollary. *Let R be a k-algebra which splits over k. Then two simple R-modules are isomorphic iff they have the same character.*

Throughout this section, we have focused our attention essentially on irreducible (and absolutely irreducible) modules over algebras. If the algebra in question happens to be semisimple, this would give a fairly complete picture of the module theory over the algebra. But if the algebra is not semisimple, what we did in this section will certainly *not* be enough to reveal the general behavior of the modules over the algebra. In the nonsemisimple case, the role of the irreducible modules is to be replaced by that of the indecomposable modules. (A nonzero module is said to be *indecomposable* if it is not a direct sum of two nonzero submodules.) The classification of indecomposable modules over an algebra is an extremely complicated task which is beyond the scope of this book. We shall, however, study some of the basic facts on indecomposable modules in a later section, in the more general setting of artinian rings. To conclude this section, let us give some examples of indecomposable modules over finite-dimensional algebras. In particular, we shall see that such algebras may have infinitely many mutually nonisomorphic indecomposable modules.

First we consider 2-dimensional algebras (over a field k). Such an algebra R has the form $k[x]/(q(x))$ where $q(x)$ is a monic quadratic polynomial. If $q(x)$ is irreducible in $k[x]$, R is a quadratic field extension of k, so the only indecomposable R-module is ${}_RR$. If $q(x) = (x-a)(x-b)$ where $a \neq b$ in k, then $R \cong k \times k$ is semisimple, and the indecomposable R-modules are just the simple R-modules, of which there are two, up to isomorphism. Finally, if $q(x) = (x-a)^2$ where $a \in k$, then

$$R = k[x]/(x-a)^2 \cong k[x]/(x^2).$$

We shall see below that R also has precisely two indecomposable modules.

Consider more generally $R = k[x]/(x^r)$. A finitely generated R-module is just a finite-dimensional k-vector space equipped with an endomorphism λ with $\lambda^r = 0$. By the Jordan Canonical Form Theorem, any such R-module is isomorphic to $n_1M_1 \oplus \cdots \oplus n_rM_r$, where

$$M_i = R/(\bar{x}^i) = k[x]/(x^i) \quad (1 \leq i \leq r),$$

and the nonnegative integers $n_1, \ldots, n_r$ are uniquely determined. In particular, $M_1, \ldots, M_r$ are the only indecomposable R-modules.

While a 2-dimensional algebra can have at most two indecomposable modules, algebras of dimension ≥ 3 may have infinitely many indecomposable modules. In the following, we shall work with the algebra R with two commuting generators α, β, subject to the relations

$$\alpha^2 = \beta^2 = \alpha\beta = 0.$$

We have $dim_k\, R = 3$, as $\{1, \alpha, \beta\}$ is a k-basis of R. Define an R-module $M = M_{2n+1}$ of k-dimension $2n+1$ by $M = U \oplus V$, where U has basis $u_0, \ldots, u_n$, V has basis $v_1, \ldots, v_n$, with

$$\alpha(U) = 0, \quad \beta(U) = 0,$$

$$\alpha(v_i) = u_i \quad \text{and} \quad \beta(v_i) = u_{i-1} \quad \text{for } i \geq 1.$$

Similarly, define an R-module $\overline{M} = M_{2n}$ by $\overline{M} = W \oplus V$, where W has basis $w_1, \ldots, w_n$, V has basis $v_1, \ldots, v_n$, with

$$\alpha(W) = 0, \quad \beta(W) = 0, \quad \alpha(v_i) = w_i,$$

$$\beta(v_i) = w_{i-1} \quad \text{for } i \geq 2, \quad \text{and} \quad \beta(v_1) = 0.$$

Here we used the notation $\overline{M}$ for M_{2n}, since $\overline{M}$ is easily seen to be isomorphic to M/ku_0.

(7.22) Proposition. *$M = M_{2n+1}$ and $\overline{M} = M_{2n}$ are both indecomposable modules over the 3-dimensional commutative algebra*

$$R = k\langle \alpha, \beta : \alpha^2 = \beta^2 = \alpha\beta = \beta\alpha = 0 \rangle.$$

In particular, R has indecomposable modules in every positive dimension.

Proof. To prove the indecomposability of M, we first make the following observation:

(7.23) *If S is any nonzero k-subspace of V, then $\beta S \not\subseteq \alpha S$.*

In fact, let m be the largest integer such that $S \subseteq kv_m + \cdots + kv_n$. Then there exists a vector $s = a_m v_m + \cdots + a_n v_n$ with $a_m \neq 0$. But then

$$\beta s = a_m u_{m-1} + \cdots + a_n u_{n-1} \notin \alpha S,$$

as $\alpha S \subseteq ku_m + \cdots + ku_n$. Next, for any nonzero R-module $N \subseteq M$, we claim that

$$\dim N \geq 1 + 2 \dim \pi(N),$$

where π is the projection of M onto V with respect to the decomposition $M = U \oplus V$. To prove this, we may assume that $S := \pi(N) \neq 0$. Since α, β act as zero on U, we have $\alpha S = \alpha N \subseteq N$ and $\beta S = \beta N \subseteq N$. On the other hand, $\alpha S + \beta S \subseteq U$, so π induces a surjection $N/(\alpha S + \beta S) \to S$. Therefore, by (7.23):

$$\begin{aligned} \dim N &\geq \dim(\alpha S + \beta S) + \dim S \\ &\geq 1 + \dim \alpha S + \dim S \\ &= 1 + 2 \dim S, \end{aligned}$$

as claimed. If $M = N \oplus N'$ where N, N' are nonzero R-submodules, then $S + S' = V$ for $S' := \pi(N')$, and we have

$$\dim N + \dim N' \geq 2 + 2(\dim S + \dim S') \geq 2 + 2n,$$

a contradiction. Next we try to prove the indecomposability of $\overline{M} = M_{2n} = W \oplus V$. Here we let $\bar{\pi}$ be the projection of $\overline{M}$ onto V with respect to this k-decomposition. Suppose we have a decomposition $\overline{M} = N \oplus N'$, where N, N' are nonzero R-modules. Let $S = \bar{\pi}(N)$ and $S' = \bar{\pi}(N')$. If either $\beta S \not\subseteq \alpha S$, or $\beta S' \not\subseteq \alpha S'$, or $\dim S + \dim S' > n$, we will get a contradiction as before. Therefore, we may assume that $\beta S \subseteq \alpha S$, $\beta S' \subseteq \alpha S'$, and that $V = S \oplus S'$. We claim that S, S' are both nonzero. Indeed, if say, $S' = 0$ and $S = V$, then

$$N' \subseteq W = \alpha V = \alpha S = \alpha N \subseteq N,$$

a contradiction. Therefore, $W = \alpha V = \alpha S \oplus \alpha S'$ is a decomposition of W into *nonzero* k-subspaces. Let λ be the k-endomorphism of W defined by $\lambda(w_i) = w_{i-1}$ for $i \geq 2$, and $\lambda(w_1) = 0$. Then $\lambda(\alpha S) = \beta S \subseteq \alpha S$ and $\lambda(\alpha S') = \beta S' \subseteq \alpha S'$. Therefore $W = \alpha S \oplus \alpha S'$ is a decomposition of W into a direct sum of two nonzero $k[\lambda]$-modules. This is impossible since $k[\lambda] \cong k[x]/(x^n)$, and, as $k[\lambda]$-module, $W \cong k[\lambda]$ is indecomposable.

QED

Next we shall construct a family of 2-dimensional indecomposable modules over R. For any element $a \in k$, define the R-module $M(a)$ as follows:

$M(a) = kw \oplus kv$, with action

$$\alpha w = 0, \quad \alpha v = w, \quad \text{and} \quad \beta w = 0, \quad \beta v = aw.$$

In matrix notation, we have $\alpha \mapsto \begin{pmatrix} 0 & 1 \\ 0 & 0 \end{pmatrix}$ and $\beta \mapsto \begin{pmatrix} 0 & a \\ 0 & 0 \end{pmatrix}$ with respect to the ordered basis $\{w, v\}$. Note that the module M_2 considered in (7.22) is just $M(0)$.

(7.24) Proposition. *For any $a \in k$, $M(a)$ is an indecomposable R-module. If $a \neq a'$ in k, then $M(a) \not\cong M(a')$ as R-modules.*

Proof. The first assertion is clear, since α and β must act trivially on an R-module of k-dimension 1. Now assume h: $M(a) \to M(a')$ is an R-module isomorphism. We represent $M(a')$ as $kw' \oplus kv'$ with

$$\alpha w' = 0, \quad \alpha v' = w', \quad \beta w' = 0, \quad \text{and} \quad \beta v' = a'w'.$$

Say

$$h(w) = bw' + cv' \quad \text{and} \quad h(v) = dw' + ev'.$$

Then $h(\alpha w) = \alpha h(w)$ gives $0 = \alpha(bw' + cv') = cw'$, so $c = 0$. With this, $h(\alpha v) = \alpha h(v)$ gives $bw' = \alpha(dw' + ev') = ew'$, so $b = e$. Finally, $h(\beta v) = \beta h(v)$ gives

$$abw' = \beta(dw' + ev') = ea'w'.$$

Since $\begin{pmatrix} b & d \\ c & e \end{pmatrix} = \begin{pmatrix} b & d \\ 0 & e \end{pmatrix}$ is invertible, we have $b = e \neq 0$, and hence $a = a'$. QED

The Proposition above shows that the cardinality of isomorphism types of indecomposable 2-dimensional R-modules is at least as large as the cardinality of k. In particular, if k is an infinite field, R has infinitely many mutually nonisomorphic 2-dimensional indecomposable modules. Note that M_{2n+1}, M_{2n} and the $M(a)$'s may also be viewed as modules over the finite-dimensional commutative algebras $k[x, y]/(x^r, y^s)$ for $r, s \geq 2$. Now consider the elementary p-group G with two generators f, g, and let k be a field of characteristic p. Then

$$kG \cong k[X, Y]/(X^p - 1, Y^p - 1) \cong k[X, Y]/((X-1)^p, (Y-1)^p).$$

Since (by a "change of variables") this is isomorphic to $k[x, y]/(x^p, y^p)$, it follows that kG has indecomposable modules in every positive dimension, and that it has mutually nonisomorphic indecomposable 2-dimensional modules $M(a)$ given by the matrix representations:

$$f \mapsto \begin{pmatrix} 1 & 1 \\ 0 & 1 \end{pmatrix}, \quad \text{and} \quad g \mapsto \begin{pmatrix} 1 & a \\ 0 & 1 \end{pmatrix},$$

where $a \in k$.

The above examples serve to show that there is a lot to be said about the determination and classification of indecomposable modules over finite-dimensional algebras. In the representation theory of algebras, an algebra R is said to be of *finite representation type* if R has only finitely many finite-dimensional indecomposable modules (up to isomorphism). Otherwise, R is said to be of *infinite representation type*. For instance, $k[x]/(x^r)$ has finite representation type, but $k[x, y]/(x^r, y^s)$ for $r, s \geq 2$ has infinite representation type. The *Brauer–Thrall Conjecture* in representation theory stated that *if an algebra R has infinite representation type, then it has indecomposable modules of arbitrarily large dimension*. This conjecture was proved by Roiter in 1968, and subsequently extended by M. Auslander to artinian rings. A stronger version of the Brauer–Thrall Conjecture stated that *if an algebra R has infinite representation type, then there is an infinite sequence $d_1 < d_2 < \cdots$ such that R has infinitely many indecomposable modules at each dimension d_i*. So far this stronger conjecture has been proved by Nazarova and Roiter over algebraically closed fields, and subsequently by Ringel over perfect fields. The modern representation theory of algebras is a very rich subject on which there has been a lot of exciting research. However, due to space limitation, we will not be able to go into the details of this theory. For a good introduction to this subject, see Pierce [82].

Exercises for §7

In the following exercises, k denotes a field.

Ex. 7.1. Let M, N be finite-dimensional modules over a finite-dimensional k-algebra R. For any field $K \supseteq k$, show that M^K and N^K have a common composition factor as R^K-modules iff M and N have a common composition factor as R-modules. (**Hint.** Use (7.13)(2).)

Ex. 7.2. Let R be a finite-dimensional k-algebra which splits over k. Show that, for any field $K \supseteq k$, $rad(R^K) = (rad\ R)^K$.

Ex. 7.3. Let R be a finite-dimensional k-algebra, M be an R-module and $E = End_R\ M$. Show that if $f \in E$ is such that $f(M) \subseteq (rad\ R)M$, then $f \in rad\ E$. (**Hint.** Show that the set of such f's form a nilpotent ideal in E.)

Ex. 7.4. Let R be a left artinian ring and C be a subring in the center $Z(R)$ of R. Show that $Nil\ C = C \cap rad\ R$. If R is a finite-dimensional algebra over a subfield $k \subseteq C$, show that $rad\ C = C \cap rad\ R$.

Ex. 7.5. Let R be a finite-dimensional k-algebra which splits over k. Show that any k-subalgebra $C \subseteq Z(R)$ also splits over k.

Ex. 7.6. For a finite-dimensional k-algebra R, let $T(R) = rad\ R + [R, R]$, where $[R, R]$ denotes the subgroup of R generated by $ab - ba$ for all $a, b \in R$.

Assume that k has characteristic $p > 0$. Show that

$$T(R) \subseteq \{a \in R\colon\ a^{p^m} \in [R, R] \quad \text{for some } m \geq 1\},$$

with equality if k is a splitting field for R.

Ex. 7.7. Using (4.1) and (7.13), give another proof for the fact (already established in (5.14)) that for any finite-dimensional k-algebra R and any field extension $K \supseteq k$, $(rad\ R)^K \subseteq rad(R^K)$.

Ex. 7.8. Let R be a finite-dimensional k-algebra and let $L \supseteq K \supseteq k$ be fields. Assume that L is a splitting field for R. Show that K is a splitting field for R iff, for every simple left R^L-module M, there exists a (simple) left R^K-module U such that $U^L \cong M$.

Ex. 7.9. If $K \supseteq k$ is a splitting field for a finite-dimensional k-algebra R, does it follow that K is also a splitting field for any quotient algebra $\bar{R}$ of R?

Ex. 7.10. (Suggested by I. Emmanouil) Give a basis-free proof for (7.4) in the case when M is a finitely presented left module over R. (**Hint.** Note that (7.4) is true when $M = R^n$. Now apply the left exact functor $Hom_R(-, N)$ to a finite presentation $R^m \to R^n \to M \to 0$.)

§8. Representations of Groups

At the beginning of §6, we have commented on the fundamental connections between ring theory and the representation theory of groups. For a field k and a group G, if we view the k-representations of G as afforded by modules over the group algebra kG, the study of the representation theory of G then becomes a special case of the study of modules over algebras. This ring-theoretic perspective of representation theory stems from a classical paper of E. Noether [29], who based her ideas in part on the earlier work of T. Molien. Since we have already developed in the last section the basic facts on simple modules and splitting fields of algebras, it is now easy to specialize this information to group algebras and deduce the standard classical results in group representations. This is done in the first half of this section. The second half goes on to study the theory of group characters, focusing on the semisimple case. We shall relate the arithmetic properties of the characters of the irreducible representations of a group G to the ring structure of the group algebra kG. This enables us to give a couple of nice ring-theoretic applications of the theory of characters.

As we have explained before, it will be convenient to view representations of G over k as given by kG-modules. Unless specified otherwise, G will be assumed to be a finite group, and all representations considered will be finite-dimensional over k. Thus the kG-modules affording the representations will be finite-dimensional k-vector spaces. Unless stated otherwise, the characteristic of k will be arbitrary. If *char* k does not divide $|G|$ (including the case when *char* $k = 0$), kG is semisimple by Maschke's Theorem (6.1) and so all representation modules are semisimple. On the other hand, if *char* k divides $|G|$ then kG is no longer semisimple; the representations in this case are classically known as *modular representations* (after L.E. Dickson).

The fundamental objects used in studying the k-representations of G can be laid down as in (7.1). Recapitulating, let $M_1, \ldots, M_r$ be a complete set of simple left kG-modules; let $D_i = End({}_{kG}M_i)$ and $n_i = dim_{D_i} M_i$. Then we have

(8.1) Theorem.

(1) $kG/rad\, kG \cong \mathbb{M}_{n_1}(D_1) \times \cdots \times \mathbb{M}_{n_r}(D_r)$.

(2) *As a left kG-module,* $kG/rad\, kG \cong n_1 M_1 \oplus \cdots \oplus n_r M_r$.

(3) $dim_k\, M_i = n_i\, dim_k\, D_i$.

(4) $|G| = dim_k(rad\, kG) + \sum_{i=1}^{r} n_i^2\, dim_k\, D_i$.

In the special case when G is abelian, this result simplifies further, for then all the n_i's are 1 and all the D_i's are commutative (so they are finite field extensions of k). In this case

$$kG/rad\, kG \cong D_1 \times \cdots \times D_r,$$

and the D_i's, viewed as ideals in $kG/rad\, kG$, afford the r different irreducible representations of G over k.

Based on what we did in the last section, there is a natural notion of a splitting field for a group G.

(8.2) Definition. We say that a field k is a *splitting field for* G if the group algebra kG splits over k (in the sense of (7.6)).

Using our earlier result (7.10), we shall establish the following basic fact in representation theory.

(8.3) Theorem. *Let k be any field, and G be any group. Then some finite extension $K \supseteq k$ is a splitting field for G.*

Proof. Let k_0 be the prime field of k, and $\bar{k}$ be the algebraic closure of k. Since k_0 is perfect, (7.10) implies that a finite extension $k_1 \supseteq k_0$ in $\bar{k}$ is a splitting field for G. Let $K = k \cdot k_1$, the field compositum being formed in $\bar{k}$. Clearly K is a finite extension of k, and by (7.14), k_1 being a splitting field for G implies the same for K. QED

In the proof above, we only knew that the splitting field k_1 exists, but we were not able to describe more concretely the construction of k_1 or to control its size. At this point, it is appropriate to mention without proof a famous result of Brauer. Let m be the exponent of G, i.e., the smallest integer such that $g^m = 1$ for every $g \in G$. According to Brauer's Theorem, in case $k_0 = \mathbb{Q}$, the splitting field k_1 may be taken to be $\mathbb{Q}(\zeta_m)$ where ζ_m is a primitive mth root of unity. And, in case $k_0 = \mathbb{F}_p$, k_1 may be taken to be $\mathbb{Z}[\zeta_m]/\mathfrak{p}$ where $\mathfrak{p}$ is any prime ideal of $\mathbb{Z}[\zeta_m]$ containing p. For proofs of these results, we refer the reader to Curtis and Reiner [62] (p. 292 and p. 592).

Over a splitting field, of course, the representations of a group are much nicer to deal with, and they also have more definitive meanings. Take, again, the case of an abelian group G. Over a splitting field k, we have all $D_i = k$ as well as all $n_i = 1$, so all $dim_k\ M_i = 1$. Thus, the irreducible k-representations are 1-dimensional; they correspond to the different homomorphisms of G into the multiplicative group k^* of the field k. (For a converse of this in the semisimple case, see Exercise 2 in this section.)

We shall now give some more examples, first in the characteristic 0 case, then in the modular case.

Let $G = \langle g \rangle$ be a cyclic group of order n, and let $R = kG$. We shall determine the simple R-modules (i.e., irreducible representations of G over k) for $k = \mathbb{Q}$, $\mathbb{R}$, and $\mathbb{C}$ respectively.

First, for $k = \mathbb{C}$ or $\mathbb{Q}(\zeta)$, where ζ is a primitive nth root of unity, we have the decomposition

$$R \cong \frac{k[x]}{(x^n - 1)} \cong \prod_{i=1}^{n} \frac{k[x]}{(x - \zeta^i)} \cong \prod_{i=1}^{n} k.$$

Thus, k is a splitting field for G and there are n simple R-modules $M_i = k$ $(1 \le i \le n)$, with g acting via multiplication by ζ^i on M_i.

Secondly, for $k = \mathbb{Q}$, we have the following decomposition of R into its simple components:

$$R \cong \frac{\mathbb{Q}[x]}{(x^n - 1)} = \frac{\mathbb{Q}[x]}{(\prod_{d|n} \Phi_d(x))} \cong \prod_{d|n} \frac{\mathbb{Q}[x]}{(\Phi_d(x))} \cong \prod_{d|n} \mathbb{Q}(\zeta_d).$$

Here ζ_d is a primitive dth root of unity, and $\Phi_d(x)$ is the dth cyclotomic polynomial, i.e., the minimal polynomial of ζ_d over $\mathbb{Q}$, where d denotes a positive divisor of n. A typical simple R-module is $N_d = \mathbb{Q}(\zeta_d)$, on which g acts via multiplication by ζ_d. When N_d is tensored up to $\mathbb{C}$, it splits into $\varphi(d)$ 1-dimensional representations of G which map g into the different primitive dth roots of unity in $\mathbb{C}^*$. Here, φ denotes the Euler function, and (8.1)(4) recovers the familiar formula $n = \sum_{d|n} \varphi(d)$.

Finally, let $k = \mathbb{R}$. Consider first the case $n = 2m + 1$. Then

$$R \cong \frac{\mathbb{R}[x]}{((x-1)\prod_{j=1}^{m}(x - \zeta^j)(x - \bar{\zeta}^j))} \cong \mathbb{R} \times \prod_{j=1}^{m} \frac{\mathbb{R}[x]}{((x - \zeta^j)(x - \bar{\zeta}^j))}.$$

Therefore, there is a unique 1-dimensional R-module (with trivial G-action), and there are m two-dimensional simple R-modules

$$V_j = \mathbb{C} = \mathbb{R}^2 \quad (1 \le j \le m)$$

with g acting via multiplication by ζ^j, i.e., as a rotation by the angle $2\pi j/n$ (clockwise or counterclockwise). Upon scalar extension, $V_j \otimes_{\mathbb{R}} \mathbb{C} \cong M_j \oplus M_{n-j}$. The case $n = 2m$ is analyzed similarly, with

$$R \cong \mathbb{R} \times \mathbb{R} \times \prod_{j=1}^{m-1} \frac{\mathbb{R}[x]}{((x - \zeta^j)(x - \bar{\zeta}^j))}.$$

Here there are *two* 1-dimensional R-modules supported by $\mathbb{R}$, with g acting as I and $-I$ respectively, and there are $m - 1$ two-dimensional simple R-modules

$$V_j = \mathbb{C} = \mathbb{R}^2 \quad (1 \le j \le m - 1),$$

with g acting via multiplication by ζ^j. Again,

$$V_j \otimes_{\mathbb{R}} \mathbb{C} \cong M_j \oplus M_{n-j}.$$

This example gives a good illustration of (7.13) relating the simple modules of $\mathbb{Q}G$, $\mathbb{R}G$ and $\mathbb{C}G$.

To give a nonabelian example, let G be the quaternion group of order 8, generated by two elements a, b with relations

$$a^4 = 1, \quad b^2 = a^2 \quad \text{and} \quad b^{-1}ab = a^{-1}.$$

We first try to construct the irreducible representations of G over $\mathbb{Q}$. Here, the commutator subgroup G' is $\langle a^2 \rangle$, and G/G' is the Klein 4-group. The latter has four homomorphisms into $\{\pm 1\}$ which lead to four different $\mathbb{Q}G$-modules M_1, M_2, M_3, M_4, each 1-dimensional over $\mathbb{Q}$. The corresponding simple components are $\cong \mathbb{Q}$, by (8.1)(3). To construct one more $\mathbb{Q}G$-representation, consider D, the division ring of rational quaternions. As is well-known, we can realize G as a subgroup of D^* by identifying a with i and b with j, so that

$$G = \{\pm 1, \pm i, \pm j, \pm k\}.$$

Thus, D may be viewed as a (left) $\mathbb{Q}G$-module M_5. Since G spans D as a $\mathbb{Q}$-vector space, this $\mathbb{Q}G$-module is irreducible. (A $\mathbb{Q}G$-submodule would amount to a left ideal of D, but D is a division ring.) The corresponding simple component is just D. In the formula

$$8 = |G| = \sum_{i=1}^{r} n_i \, dim \, M_i,$$

we already have $dim\, M_1 = \cdots = dim\, M_4 = 1$ and $dim\, M_5 = 4$. Hence all $n_i = 1$, $r = 5$, and the simple decomposition of $\mathbb{Q}G$ is

$$\mathbb{Q} \times \mathbb{Q} \times \mathbb{Q} \times \mathbb{Q} \times D,$$

so $\mathbb{Q}$ is not a splitting field for G. If we replace the ground field $\mathbb{Q}$ by $\mathbb{R}$, the same analysis remains valid, with D replaced by the real quaternions. (Incidentally, the computations above hint at interesting connections between representation theory of groups and the study of finite multiplicative subgroups of division rings.)

What happens in the example above if we choose the ground field to be $K = \mathbb{Q}(\sqrt{-1})$? The modules $M_1, \ldots, M_4$, being one-dimensional, clearly remain irreducible when we extend the scalars from $\mathbb{Q}$ to K; however, M_5 does not. The simple component D of $\mathbb{Q}G$, upon tensoring up to K, becomes $\mathbb{M}_2(K)$ as is easily verified. The unique left $\mathbb{M}_2(K)$-module M_5' occurs twice in the decomposition of $\mathbb{M}_2(K)$ into minimal left ideals. Therefore the left regular module KG decomposes as

$$M_1' \oplus M_2' \oplus M_3' \oplus M_4' \oplus 2M_5',$$

where $M_i' = M_i \otimes_{\mathbb{Q}} K$ for $i \leq 4$, and we have $n_1' = n_2' = n_3' = n_4' = 1$, $n_5' = 2$. By working explicitly with matrices, the irreducible 2-dimensional KG-representation afforded by M_5' is seen to be:

$$a \mapsto \begin{pmatrix} 0 & 1 \\ -1 & 0 \end{pmatrix}, \quad b \mapsto \begin{pmatrix} 0 & \sqrt{-1} \\ \sqrt{-1} & 0 \end{pmatrix} \quad \text{(up to equivalence)}.$$

Clearly, the quadratic extension K of $\mathbb{Q}$ is a splitting field for G.

Let us now give an example to illustrate the relations between representations in characteristic zero and those in characteristic p. The group under study will be the symmetric group $G = S_3$, and we shall consider the three fields $\mathbb{Q}$, $\mathbb{F}_2$, and $\mathbb{F}_3$. Over $k = \mathbb{Q}$, we have two 1-dimensional (irreducible) representations: the trivial representation M_1, and the sign representation M_1'. There is also a natural 2-dimensional representation M_2, given by

$$ke_1 \oplus ke_2 \oplus ke_3 / k \cdot (e_1 + e_2 + e_3),$$

where G acts by permuting the unit vectors e_1, e_2, e_3. This is easily seen to be irreducible, and since $6 = |G| = 1^2 + 1^2 + 2^2$, $\{M_1, M_1', M_2\}$ is a full set of simple (left) $\mathbb{Q}G$-modules and $\mathbb{Q}$ is a splitting field for G.

Next, let $k = \mathbb{F}_2$. Here we have the trivial representation $M_1 = M_1'$, and we have the 2-dimensional representation M_2 which is easily checked to be irreducible. By explicit computation, we see that $D_2 = End({}_{kG}M_2) = k$, so $n_2 = 2$ and M_2 is in fact absolutely irreducible. From the equation

$$\begin{aligned} 6 = |G| &= dim_k(rad\ kG) + \sum_{i=1}^{r} n_i^2 \ dim_k\ D_i \\ &= dim_k(rad\ kG) + 1 + 4 + \cdots, \end{aligned}$$

we see that $r = 2$, $dim_k(rad\ kG) = 1$, and $k = \mathbb{F}_2$ is a splitting field for G. Incidentally, this enables us to determine $rad\ kG$ explicitly. In fact, let $\sigma = \sum_{g \in G} g \in kG$. Clearly $k \cdot \sigma$ is an ideal of square zero. Therefore, the computation above implies that *rad* $kG = k \cdot \sigma$. The composition factors of ${}_{kG}kG$ are $\{M_1, M_1, M_2, M_2\}$.

Finally, consider $k = \mathbb{F}_3$. Here, we have the two *distinct* 1-dimensional representations M_1, M_1', but the 2-dimensional representation M_2 is no longer irreducible. (Check that $e_1 - e_2$ generates a proper kG-submodule!) In the following, we shall prove Theorem (8.4) which implies that the normal 3-Sylow group $\langle(123)\rangle$ of G must act trivially on any irreducible kG-representation. Thus, the irreducible k-representations of G are the same as those of $G/\langle(123)\rangle$, so M_1, M_1' are the only possibilities. The equation (8.1)(4) implies that $\dim_k$ *rad* $kG = 4$, and that $k = \mathbb{F}_3$ is a splitting field for G. We leave it as an exercise for the reader to show that the composition factors of kG are $\{M_1, M_1, M_1, M_1', M_1', M_1'\}$.

We shall now give the theorem which was used in the argument above.

(8.4) Theorem. *Let k be a field of characteristic $p > 0$, and G be a finite group. Then any normal p-subgroup $H \subseteq G$ acts trivially on any simple left kG-module. Thus, simple left kG-modules are the same as simple left $k[G/H]$-modules. (In particular, if G is a p-group, then the only simple left kG-module is k, with trivial G-action.)*

One of the main tools used for the proof of the theorem is the following.

(8.5) Clifford's Theorem. *Let k be any field, and H be a normal subgroup of a (possibly infinite) group G. If V is a simple left kG-module, then ${}_{kH}V$ is a semisimple kH-module.*

Proof. Let M be a simple kH-submodule of V. For any $g \in G$, $g \cdot M$ is also a kH-submodule of V since

$$h(gM) = g(h^g M) = gM,$$

where $h^g = g^{-1}hg \in H$. Moreover, gM is a *simple* kH-module because if M' is a kH-submodule of it, $g^{-1}M'$ would be a kH-submodule of M. Now consider $V' := \sum_{g \in G} gM$. This is a semisimple kH-module, and since it is also a kG-submodule of V, we have $V' = V$. QED

Proof of (8.4). In view of Clifford's Theorem, we are reduced to proving that H acts trivially on any simple kH-module M. We do this by induction on $|H|$, the case $|H| = 1$ being trivial. If $|H| > 1$, let $h \neq 1$ be an element in the center of H, say of order p^n. Since

$$(h-1)^{p^n} = h^{p^n} - 1 = 0 \in kH,$$

$h - 1$ acts as a nilpotent transformation on M, so its kernel is nonzero. Let $M_0 = \{m \in M : hm = m\} \neq 0$. This is a kH-submodule of M and so $M_0 = M$. Therefore, M may be viewed as a (simple) $k[H/\langle h\rangle]$-module, and we are done by induction. QED

In any finite group G, let $O_p(G)$ denote the intersection of all its p-Sylow groups. Clearly $O_p(G)$ is the largest normal p-subgroup of G. We have the following nice characterization of $O_p(G)$ in terms of the modular representations of G.

(8.6) Corollary. *Let k be a field of characteristic $p > 0$, and G be a finite group. For $h \in G$, the following are equivalent:*

(1) $h \in O_p(G)$.

(2) *h acts trivially on all simple left kG-modules.*

(3) $h - 1 \in rad(kG)$.

(Note that, although (2) and (3) ostensibly depend on the ground field k, the condition (1) does not.)

Proof. (2) $\Leftrightarrow$ (3) is trivial, and (1) $\Rightarrow$ (2) follows from (8.4) since $O_p(G)$ is normal. Now assume (2). By considering a composition series of the left regular module kG, we see that $h - 1$ acts as a nilpotent transformation on kG. Thus, for a sufficiently large integer n,

$$(h-1)^{p^n} = h^{p^n} - 1 = 0 \in kG;$$

i.e., the order of h is a power of p. How let H be the set of all elements h of G satisfying (2). This H is easily seen to be a normal subgroup of G, and what we did above shows that H is a p-group. Thus $H \subseteq O_p(G)$ and we are done. QED

(8.7) Corollary (Wallace). *Let k be a field of characteristic p, and G be a finite group with a normal p-Sylow group H. Then*

$$rad\ kG = \sum_{h \in H} kG \cdot (h-1)$$

with $dim_k\ rad\ kG = [G : H]\ (|H| - 1)$.

Proof. Since $(h-1)g = g(h^g - 1)$ for any $g \in G$, the left ideal

$$\mathfrak{A} := \sum_{h \in H} kG \cdot (h-1)$$

is in fact an ideal, and this lies in $rad\ kG$ by (8.4). The quotient $kG/\mathfrak{A}$ is easily seen to be isomorphic to $k[G/H]$. (We have a natural map from $kG/\mathfrak{A} \to k[G/H]$, and an inverse can be easily constructed.) Since $p = char\ k$ is not a divisor of $|G/H|$, $k[G/H]$ is semisimple by Maschke's Theorem. Therefore (from (4.6)) we have $rad\ kG = \mathfrak{A}$ and

$$dim_k\ rad\ kG = |G| - |G/H| = [G : H]\ (|H| - 1). \quad \text{QED}$$

(8.8) Corollary. *Let k be a field of characteristic p, and G be a finite p-group. Then $J = rad\ kG$ equals the augmentation ideal of kG, and we have $J^{|G|} = 0$. If G is generated as a group by $\{g_1, \ldots, g_n\}$, then J is generated as a left ideal by $\{g_1 - 1, \ldots, g_n - 1\}$.*

Proof. The first conclusion follows from (8.7) since the augmentation ideal is generated as a k-space by $\{g - 1 : g \in G\}$. For the second conclusion, note that ${}_{kG}kG$ has $|G|$ composition factors. Since J acts trivially on each, we have $J^{|G|} = 0$. The third conclusion is left as an exercise. QED

For a prime p, an element g in a finite group G is said to be *p-regular* if p does not divide the order of g. A conjugacy class $\mathscr{C}$ of G is said to be *p-regular* if one, and hence all, elements of $\mathscr{C}$ are p-regular. Actually, in these definitions, it is convenient to allow for the possibility that p be zero: by definition, any element and any conjugacy class are always 0-regular. In the results (8.4), (8.6) above, we saw that certain group elements whose orders are powers of p essentially play no role in determining the structure of the irreducible representations at characteristic p. Thus the group elements which govern the behavior of these representations ought to be the p-regular elements. In 1935, Brauer gave more substance to this viewpoint by proving the following beautiful result.

(8.9) Brauer's Theorem. *Let G be a finite group, and k be a splitting field of G of characteristic $p \geq 0$. Then the number of irreducible kG-representations is equal to the number of p-regular conjugacy classes of G.*

In (7.17), the number of irreducible kG-representations was determined to be $dim_k\ R/T(R)$ where $R = kG$ and $T(R) = rad\ R + [R, R]$. The idea of the proof of (8.9) is that, for $R = kG$, $dim_k\ R/T(R)$ can further be computed in purely group-theoretic terms. To begin this computation, we first give another characterization of $[R, R]$. The following characterization is valid for any group G and for any commutative ring k.

(8.10) Lemma. *An element $\alpha \in R = kG$ belongs to $[R, R]$ iff the sum of its coefficients over each conjugacy class of G is zero.*

Proof. For the "only if" part, we may assume that $\alpha = ab - ba$ where $a, b \in kG$. Writing $a = \sum a_g g$ and $b = \sum b_h h$, we have

$$\alpha = \sum_{g,h} a_g b_h (gh - hg),$$

so it suffices to deal with $\alpha_0 = gh - hg$. Since gh, hg are conjugate in G, the desired conclusion in this case is clear. For the converse, note that if $g_1, g_2 \in G$ are conjugate, then we can write $g_1 = gh$, $g_2 = hg$ for suitable $g, h \in G$, so $g_1 \equiv g_2 \pmod{[R, R]}$. Consider a conjugacy class $\mathscr{C} = \{g_1, \ldots, g_n\}$ and an element

$$\alpha = \varepsilon_1 g_1 + \cdots + \varepsilon_n g_n \in kG.$$

By the foregoing,

$$\alpha \equiv (\varepsilon_1 + \cdots + \varepsilon_n) g_1 \pmod{[R, R]}.$$

Thus, if $\varepsilon_1 + \cdots + \varepsilon_n = 0$, we have $\alpha \in [R, R]$. QED

In the following, let $\{a_i : i \in I\}$ be a complete set of representatives of the conjugacy classes of a group G. The lemma, together with its proof, clearly imply the following.

(8.11) Corollary. *Let $R = kG$, where G is any group and k is any commutative ring. Then $R/[R, R]$ is a free k-module with basis*

$$\{a_i + [R, R] : \ i \in I\}.$$

Now let J be the subset of I such that $\{a_j : j \in J\}$ is a complete set of representatives of the p-regular conjugacy classes of G. (From here on, we assume again G is finite and k is a field.) We now complete the proof of Brauer's Theorem by proving the following analogue of (8.11).

(8.12) Lemma. *Let k be a splitting field for G of characteristic $p \geq 0$, and let $R = kG$. Then $R/T(R)$ is a k-vector space with basis*

$$\{a_j + T(R) : \ j \in J\}.$$

Proof. If $p = 0$, we have $rad\ R = 0$ so $T(R) = [R, R]$. In this case $I = J$ and we are done by (8.11). In the following, we may, therefore, assume $p > 0$. Let $g \in G$. By standard group theory, there is a factorization $g = ab = ba$ where a is p-regular and $b^{p^n} = 1$ for some n. Since a and b commute, we have

$$(ab - a)^{p^n} = a^{p^n} b^{p^n} - a^{p^n} = 0 \in R,$$

so $ab - a$ is nilpotent. By the second conclusion of (7.17), we get $g = ab \equiv a \pmod{T(R)}$. We have also observed before that, if g, g' are conjugate, then $g \equiv g' \pmod{[R, R]}$. Therefore, the a_j's $(j \in J)$ span $R/T(R)$ as a k-space. To finish, we need to show that

$$\sum_{j \in J} \varepsilon_j a_j \in T(R) \Longrightarrow \text{all } \varepsilon_j = 0 \quad \text{in } k.$$

Write $\sum_{j \in J} \varepsilon_j a_j = c + d$ where $c \in rad\ R$, $d \in [R, R]$. Let m be the LCM of the orders of the a_j's. Then p is a unit mod m, so for some N, $p^N \equiv 1 \pmod{m}$. Choose N large enough so that $c^{p^N} = 0$; this is possible since $c \in rad\ R$ is nilpotent. Let $q = p^N$. Then $a_j^q = a_j$ for each $j \in J$, and, by

(7.15)(2), $d^q \in [R, R]$. Hence, by (7.15)(1),

$$0 = c^q = \left(\sum \varepsilon_j a_j - d\right)^q \equiv \sum \varepsilon_j^q a_j^q - d^q \equiv \sum \varepsilon_j^q a_j \pmod{[R, R]}.$$

Now by (8.11), all $\varepsilon_j^q = 0$ and hence all $\varepsilon_j = 0$. QED

(8.13) Corollary. *For any field k of characteristic $p \geq 0$, the number of irreducible kG-representations is bounded by the number of p-regular conjugacy classes in G.*

Proof. This follows from Brauer's Theorem and (7.18). QED

It is worthwhile to point out that, in the *semisimple* case, there is another, more or less "dual," method by which one can deduce the two results (8.9) and (8.13). In fact, assume $p = char\, k$ does not divide $|G|$. Let $\mathscr{C}_1, \ldots, \mathscr{C}_s$ be the (necessarily p-regular) conjugacy classes of G, and let

$$C_i := \sum_{g \in \mathscr{C}_i} g \in kG.$$

As an analogue of (8.11), one can show easily that $C_1, \ldots, C_s$ form a k-basis of $Z(kG)$, the center of kG.[2] Let

$$kG \cong \mathbb{M}_{n_1}(D_1) \times \cdots \times \mathbb{M}_{n_r}(D_r),$$

where the D_i's are k-division algebras, and r is the number of irreducible kG-representations. Taking centers, we have

$$\textbf{(8.14)} \qquad \begin{aligned} Z(kG) &\cong Z(\mathbb{M}_{n_1}(D_1)) \times \cdots \times Z(\mathbb{M}_{n_r}(D_r)) \\ &\cong Z(D_1) \times \cdots \times Z(D_r), \end{aligned}$$

and so $s = \sum_{i=1}^r dim_k\, Z(D_i) \geq r$. If k happens to be a splitting field, then $D_i = k$ for all i, and the inequality becomes an equality.

Our next goal is to discuss characters of group representations. Recall that, for any module M over $R = kG$ of finite dimension over k, there is an associated k-linear character χ_M: $R \to k$. Since the elements of G form a k-basis of R, we can think of χ_M as a function from G to k, extended by linearity to R. Note that, for any $g, h \in G$,

$$\chi_M(ghg^{-1}) = \chi_M(h);$$

i.e., χ_M is constant on each conjugacy class of G. In the following, we shall show that we can get a lot more specific information on characters for representations of groups than for representations of general algebras. Our intention, however, is not to give a full treatment of the character theory

[2] This statement is, in fact, valid for any commutative ring k.

of groups. Since our primary interest is in ring theory, we shall focus our attention only on those aspects of character theory which have repercussions on the structure of group rings. To illustrate the interplay between character theory and ring theory, we shall derive a few ring-theoretic applications of group characters at the end of our discussion. The applications of group characters to the structure of groups is also an important topic; this, however, lies beyond the scope of this book, and will not be explored in detail here.

To simplify the exposition, *we shall assume in the following that char k does not divide* $|G|$, *and that k is a splitting field for G*. The notations used in (7.1) for the simple modules over $R = kG$ will be fixed. Since k is a splitting field, all D_i's are equal to k, and $n_i = dim_k\ M_i$. We write $\chi_i = \chi_{M_i}$ and let $\{e_1, \ldots, e_r\}$ be the central idempotents in R giving the Wedderburn decomposition of R into its simple components. For any $g \in G$, let $C_g \in R$ denote the sum of the group elements in the conjugacy class of g. Now consider $Z(kG)$. This has two different k-bases:

$$\{e_i :\ 1 \le i \le r\}, \quad \text{and} \quad \{C_g :\ g \in G\}.$$

(The latter set, of course, has cardinality r, not $|G|$.) Our first result gives explicit formulas expressing one basis in terms of the other.

(8.15) Proposition. (1) $e_i = |G|^{-1} n_i \sum_{g \in G} \chi_i(g^{-1}) g$. *(Since conjugate elements have the same coefficients in the summation, the* RHS *is a k-linear combination of the* C_g*'s.) In particular, char k does not divide* n_i *for any i.*

(2) $C_g = m_g \sum_i \chi_i(g) e_i / n_i$, *where* m_g *is the cardinality of the conjugacy class of* g.

Proof. Let χ_{reg} denote the character of the left regular representation, afforded by ${}_R R$. Clearly, $\chi_{reg}(1) = |G|$, and $\chi_{reg}(g) = 0$ if $g \ne 1$. To prove (1), write $e_i = \sum_h a_{i,h} h$ where $a_{i,h} \in k$. Then, for any $g \in G$,

$$\chi_{reg}(e_i g^{-1}) = \sum_h a_{ih} \chi_{reg}(h g^{-1}) = a_{ig} |G|.$$

On the other hand, $\chi_{reg} = \sum_j n_j \chi_j$, so

$$a_{ig} = \frac{1}{|G|} \chi_{reg}(e_i g^{-1}) = \frac{1}{|G|} \sum_j n_j \chi_j(e_i g^{-1}).$$

Since e_i acts as δ_{ij} (Kronecker deltas) on M_j, $\chi_j(e_i g^{-1}) = \delta_{ij} \chi_j(g^{-1})$. Therefore,

$$a_{ig} = \frac{n_i}{|G|} \chi_i(g^{-1}),$$

as claimed in (1). Since $e_i \ne 0$, the formula in (1) implies that $n_i \ne 0 \in k$; i.e., *char k* does not divide n_i.

To prove (2), write $C_g = \sum_i b_{g,i} e_i$. Applying χ_j leads to

$$m_g \, \chi_j(g) = \sum_i b_{g,i} \, \chi_j(e_i) = b_{g,j} \, n_j,$$

and so $b_{g,j} = m_g \chi_j(g)/n_j$, as claimed in (2). QED

We can now derive the following two famous character formulas due to G. Frobenius.

(8.16) Theorem (First and Second Orthogonality Relations).

(A) $\sum_g \chi_i(g^{-1})\chi_j(g) = \delta_{ij}|G|$.

(B) $\sum_i \chi_i(g)\chi_i(h^{-1}) = \delta |C_G(g)|$ *where* $\delta = 1$ *if* g, h *are conjugate, and* $\delta = 0$ *if* g, h *are not conjugate, and* $C_G(g)$ *denotes the centralizer of* g *in* G.

Proof. (A) follows by applying χ_j to (8.15)(1) and canceling n_i. For (B), we plug (8.15)(1) into (8.15)(2) to get

$$\begin{aligned} C_g &= m_g \sum_i \frac{\chi_i(g)}{n_i} \cdot \frac{n_i}{|G|} \sum_h \chi_i(h^{-1})h \\ &= \frac{m_g}{|G|} \sum_h \left(\sum_i \chi_i(g)\chi_i(h^{-1}) \right) h. \end{aligned}$$

Noting that $m_g = [G : C_G(g)]$, the formula (B) follows by comparing coefficients of h on the two sides of the equation. QED

We should perhaps explain the term "orthogonality relations" used for the two formulas in (8.16). A function μ: $G \to k$ is called a *class function* if μ is constant on each conjugacy class of G. The set $F_k(G)$ of all class functions on G forms a k-vector space of dimension r (the number of conjugacy classes of G). On $F_k(G)$, we can introduce an inner product, defined by

$$[\mu, \nu] = \frac{1}{|G|} \sum_g \mu(g^{-1})\nu(g).$$

Similarly, in the r-dimensional k-space $Z(kG)$, we can introduce an inner product, defined by

$$[\alpha, \beta] = \frac{1}{|G|} \sum_i \chi_i(\alpha)\chi_i(\bar{\beta}),$$

where, for $\beta = \sum \beta_g g$, $\bar{\beta}$ means $\sum \beta_g g^{-1}$. With respect to these inner products, (8.16) amounts to (1) and (2) of the Corollary below.

(8.17) Corollary. *Under the hypotheses made in the paragraph preceding* (8.15):

(1) *The r irreducible characters* $\{\chi_i\}$ *form an orthonormal basis of* $F_k(G)$.

(2) *The r class sums* $\{C_g:\ g \in G\}$ *form an orthogonal basis of* $Z(kG)$, *with* $[C_g, C_g] = m_g$.

(3) $F_k(G)$ *and* $Z(kG)$ *are dual spaces of each other, under the pairing* $(\mu, \alpha) \mapsto \mu(\alpha)$. *Moreover,* $\{n_i^{-1}\chi_i\}$ *and* $\{e_j\}$ *are dual bases of one another.*

In view of the last conclusion above, it is of interest to record the values of $n_i^{-1}\chi_i$ on the other basis $\{C_g\}$ of $Z(kG)$, which are

$$\frac{1}{n_i}\chi_i(C_g) = \frac{m_g}{n_i}\chi_i(g).$$

By (8.15)(2), these are precisely the "coordinates" of C_g in terms of the basis $\{e_i\}$ of $Z(kG)$, as we could have predicted from (8.17)(3). In characteristic zero, we have the following important arithmetic information.

(8.18) Theorem. *Assume that* char $k = 0$ *and let A be the ring of algebraic integers in k (i.e., elements of k which satisfy a monic polynomial equation over* $\mathbb{Z}$*). Then*

(1) *For any* $g \in G$, $C_g \in \sum A \cdot e_i$. *(In other words,* $\chi_i(C_g)/n_i = \chi_i(g)m_g/n_i \in A$ *for all i.)*

(2) *For any i,* $|G|e_i/n_i \in \sum A \cdot C_g$.

Proof. Look at $Z(\mathbb{Z}G) \subseteq Z(kG)$. Since the commutative ring $Z(\mathbb{Z}G) = \sum \mathbb{Z} \cdot C_g$ is finitely generated as an abelian group, each $C_g \in Z(\mathbb{Z}G)$ is integral over $\mathbb{Z}$. Therefore, the coordinates of C_g with respect to the decomposition $\prod_{i=1}^r k \cdot e_i$ are also integral over $\mathbb{Z}$. This proves (1). For (2), note that for any $g \in G$ and any (finite-dimensional) kG-module M, $\chi_M(g) \in A$. In fact, let T be the matrix of the action of g with respect to a k-basis of M. Then $T^n = I$ for $n =$ order of g. Therefore, all eigenvalues of T (in the algebraic closure of k) are nth roots of unity. Since $\chi_M(g)$ is the sum of these eigenvalues, we have $\chi_M(g) \in A$. Thus, (2) follows from (8.15)(1). QED

From (1) and (2) above, we have

$$\frac{|G|}{n_i}e_i \in \sum_g A \cdot \left(\sum_j A \cdot e_j\right) \subseteq \sum_j A \cdot e_j.$$

Thus, $|G|/n_i \in A \cap \mathbb{Q} = \mathbb{Z}$. This shows that, in characteristic zero, *the degrees* $\{n_i\}$ *of the irreducible kG-representations are divisors of* $|G|$. By arguing a little more carefully, we can further refine this result, as follows.

(8.19) Theorem. *Let k and A be as in* (8.18) *and let* $H = Z(G)$ *(the center of* G*). Then*

(1) (Schur) *Each* n_i *divides* $[G : H]$.

(2) *Each* $n_i \leq \sqrt{[G : H]}$.

(3) *If* G *is a p-group, then each* n_i^2 *divides* $[G : H]$.

Proof. To simplify the notation, let $\chi = \chi_i$, $n = n_i$ and $M = M_i$. We may assume that G acts faithfully on M. (If K is the kernel of the G-action, we may replace G by $\bar{G} = G/K$. Since there is a surjection $G/Z(G) \to \bar{G}/Z(\bar{G})$, $[\bar{G} : Z(\bar{G})]$ is a divisor of $[G : Z(G)]$.)

Since $End({}_{kG}M) = k$, the center $H = Z(G)$ acts on M by scalar multiplications; i.e., $h \cdot m = \mu(h)m$ where $h \in H$, $m \in M$, and μ is a homomorphism from H to $k \backslash \{0\}$. Since the H-action on M is also faithful, μ is in fact a monomorphism. Note that, for $g \in G$ and $h \in H$, we have $\chi(gh) = \chi(hg) = \mu(h)\chi(g)$.

Now define an equivalence relation "$\sim$" on G by declaring

$$g \sim a \quad \text{iff} \quad a = g_1 h_1 \quad \text{where } g_1 \text{ is conjugate to } g, \text{ and } h_1 \in H.$$

Then $\chi(g^{-1})\chi(g)$ is constant on each $\sim$-equivalence class. For, if $a = g_1 h_1$ is as above, then

$$\begin{aligned} \chi(a^{-1})\chi(a) &= \chi(h_1^{-1} g_1^{-1})\chi(g_1 h_1) \\ &= \mu(h_1^{-1})\mu(h_1)\chi(g_1^{-1})\chi(g_1) \\ &= \chi(g^{-1})\chi(g). \end{aligned}$$

For $g \in G$, let $C\ell(g)$ denote the $\sim$-equivalence class of g. We claim that, if $\chi(g) \neq 0$, then $|C\ell(g)| = m_g|H|$ where m_g is the number of elements conjugate to g. To see this, it suffices to show that, for any $a \sim g$, the factorization $a = g_1 h_1$ in the foregoing notation is unique. In fact, if $a = g_2 h_2$ is another such factorization, then $\chi(g_1)\mu(h_1) = \chi(g_2)\mu(h_2)$. Since $\chi(g_1) = \chi(g_2) = \chi(g) \neq 0$, we have $\mu(h_1) = \mu(h_2)$ and so $h_1 = h_2$, $g_1 = g_2$.

Now, changing notations, let $\{g_j\}$ be a complete set of representatives for the $\sim$-equivalence classes $C\ell(g)$ with $\chi(g) \neq 0$. Then

$$\begin{aligned} |G| &= \sum_g \chi(g^{-1})\chi(g) \\ &= \sum_j |C\ell(g_j)| \cdot \chi(g_j^{-1})\chi(g_j) \\ &= |H| \cdot \sum_j m_{g_j}\chi(g_j^{-1})\chi(g_j). \end{aligned}$$

And so, using (8.18)(1),

$$\frac{[G : H]}{n} = \sum_j \chi(g_j^{-1}) \cdot m_{g_j}\chi(g_j)/n \in A \cap \mathbb{Q} = \mathbb{Z}.$$

This proves (1). For (2), let $D: kG \to End(M_k)$ be the representation associated with M. By (7.3), D is surjective; i.e., $\{D(g): g \in G\}$ spans $End(M_k)$. But for $h \in H$, $D(gh) = \mu(h)D(g)$. Thus, $End(M_k)$ is already spanned by $\{D(t_j)\}$ where $\{t_j\}$ is a complete set of coset representatives modulo H. Comparing dimensions, we get $n^2 \leq [G:H]$, as asserted in (2), If G is a p-group, then n_i^2 and $[G:H]$ are both powers of p, and (2) amounts to $n_i^2 \mid [G:H]$. QED

Remark. Itô has further improved Schur's result (8.19)(1) by showing that n above actually divides $[G:H]$ for any abelian normal subgroup H of G.

If we fix a set of representatives $\{a_j\colon 1 \leq j \leq r\}$ for the conjugacy classes of a group G, we can form an $r \times r$ matrix whose (i,j)-entry is $\chi_i(a_j)$. This matrix is called the *character table* of G (with respect to the splitting field k). With the quantitative results on characters obtained so far, it is an easy and rather pleasant task to compute the character tables of groups of small order. The irreducible representations of S_3 and the quaternion group of order 8 have been worked out before, so it is trivial to write down their character tables. To avoid repetitions, we deal here with S_4, A_4 and A_5. For convenience, we will take $k = \mathbb{C}$, although, as we shall see, a much smaller field will suffice.

For $G = S_4$, the five conjugacy classes are represented by

$$(1),\ (12),\ (123),\ (1234) \quad \text{and} \quad (12)(34),$$

and their cardinalities are 1, 6, 8, 6, and 3. We have exactly two linear characters (characters associated with 1-dimensional representations): χ_1, the trivial character, and χ_2, the sign character. We have also a standard 3-dimensional representation, afforded by the module

$$M_4 = ke_1 \oplus ke_2 \oplus ke_3 \oplus ke_4/k \cdot (e_1 + e_2 + e_3 + e_4),$$

on which G acts by permuting the e_i's. It is easy to see that M_4 is irreducible, by verifying that its character χ_4 satisfies

$$\sum \chi_4(g^{-1})\chi_4(g) = |G|$$

(see Exercise 10). We get another 3-dimensional representation $M_5 = M_2 \otimes_k M_4$, with G acting diagonally. Its character χ_5 is obtained by "twisting" χ_4 with the sign character χ_2. Since $1^2 + 1^2 + 3^2 + 3^2 = 20$, the only "missing" irreducible module is a certain M_3 of dimension 2. This can be constructed by using the well-known fact that

$$H = \{1, (12)(34), (13)(24), (14)(23)\}$$

is a normal subgroup in S_4 with $S_4/H \cong S_3$. Thus M_3 may be obtained by taking the irreducible 2-dimensional representation of S_3, and lifting it to S_4.

The character table for S_4 is therefore as follows (the first column being the degrees of the various irreducible representations):

$G = S_4$

	(1)	(12)	(123)	(1234)	(12)(34)
χ_1	1	1	1	1	1
χ_2	1	-1	1	-1	1
χ_3	2	0	-1	0	2
χ_4	3	1	0	-1	-1
χ_5	3	-1	0	1	-1

For the tetrahedral group $G = A_4$ of order 12, we have the normal subgroup $H \lhd A_4$ (constructed above), with A_4/H cyclic of order 3. Thus we obtain three linear characters χ_1, χ_2, χ_3 by lifting those of A_4/H. The kS_4-module M_4, viewed as a kA_4-module, affords a character χ_4, and can be shown to be irreducible (over kA_4) by using the same idea as in the kS_4 case. Since $1^2 + 1^2 + 1^2 + 3^2 = 12 = |A_4|$, the determination of the irreducible characters is complete. Taking 1, (12)(34), (123), and (132) as the representatives for the conjugacy classes (with cardinalities 1, 3, 4, and 4), we arrive at the following character table:

$G = A_4$

	(1)	(12)(34)	(123)	(132)
χ_1	1	1	1	1
χ_2	1	1	ω	ω^2
χ_3	1	1	ω^2	ω
χ_4	3	-1	0	0

Here, ω denotes a primitive cubic root of unity. One can verify that the irreducible 3-dimensional representation used above is equivalent to the representation of A_4 as the rotation group of the regular tetrahedron.

For the alternating group $G = A_5$, the computations become even more interesting. This is the smallest noncyclic simple group; in particular, $[G, G] = G$ and G has only one linear character, χ_1. Next, we have a 4-dimensional standard representation

$$M_4 = \bigoplus_{i=1}^{5} k \cdot e_i / k \cdot (e_1 + \cdots + e_5),$$

which can be shown to be irreducible over A_5 by a character computation (cf. Exercise 10). For the conjugacy classes, we can take as representatives

$$\{(1),\ (12)(34),\ (123),\ (12345),\ (13524)\};$$

the class cardinalities are 1, 15, 20, 12, and 12 respectively. We have

$$1^2 + n_2^2 + n_3^2 + 4^2 + n_5^2 = 60,$$

so $n_2^2 + n_3^2 + n_5^2 = 43$. The only possibility is (up to a permutation) $n_2 = n_3 = 3$, $n_5 = 5$. Here, M_2 and M_3 can be obtained by the two different ways of representing A_5 as the rotation group of the icosahedron. Their characters χ_2, χ_3 are computed by using the fact that a matrix representing a rotation of degree θ in 3-space has trace $1 + 2\cos\theta$. For the two icosahedral representations at hand, θ can only be 0°, 180°, ±120°, or multiples of 72°. This enables us to determine χ_2 and χ_3 as in the following table. (We leave the details to the reader.) Note that χ_2 and χ_3 are conjugate under the nontrivial automorphism of $\mathbb{Q}(\sqrt{5})$. Alternatively, we can also obtain M_3 from M_2 by "twisting" M_2 with the automorphism $g \mapsto (12)g(12)^{-1}$ of A_5.

$\boxed{G = A_5}$

	(1)	(12)(34)	(123)	(12345)	(13524)
χ_1	1	1	1	1	1
χ_2	3	−1	0	$(1+\sqrt{5})/2$	$(1-\sqrt{5})/2$
χ_3	3	−1	0	$(1-\sqrt{5})/2$	$(1+\sqrt{5})/2$
χ_4	4	0	1	−1	−1
χ_5	5	1	−1	0	0

There are various ways to construct explicitly the representation module M_5. By identifying A_5 with

$$PSL_2(\mathbb{F}_5) = SL_2(\mathbb{F}_5)/(\pm I),$$

we get a permutation representation of A_5 on the six points $\{e_1, e_2, \ldots, e_6\}$ of the projective line over $\mathbb{F}_5$. Then M_5 can be obtained as the module

$$\bigoplus_{i=1}^{6} k \cdot e_i / k(e_1 + \cdots + e_6),$$

where A_5 acts on $\{e_1, \ldots, e_6\}$ by the permutation representation just described. Another way to get this permutation representation is to let A_5 act by conjugation on the six 5-Sylow-groups of S_5. If we assume the knowledge of M_5 and its character χ_5, then the two characters χ_2, χ_3 associated with the icosahedral representations can in turn be determined by the orthogonality relations.

Of course, our brief treatment of group representations so far has only barely scratched the surface of a vast and very beautiful subject. For more comprehensive treatments of the subject, we refer the reader to the books of Feit, Curtis and Reiner, and Isaacs. Note that our treatment has particularly stressed the ring-theoretic perspective, so that we saw how the notions of modules, idempotents, and Wedderburn decomposition played a role in the development of representation theory. To close this section, we shall give two ring-theoretic applications of the methods developed so far. Note that, while we have assumed throughout that k is a splitting field for G, this does not necessarily limit the applications to the splitting field case since we can always "go up" to a splitting field k, apply results over k, and try to pull back the information. In fact, both of the applications presented below concern group rings over arbitrary rings of algebraic integers. We shall obtain interesting information on the units and idempotents of such integral group rings by using character methods. The following elementary fact about complex numbers will be needed; its easy proof is left to the reader.

(8.20) Lemma. *Let $\omega_1, \ldots, \omega_n \in \mathbb{C}$ be such that $|\omega_1| = \cdots = |\omega_n| = 1$. Then $|\omega_1 + \cdots + \omega_n| \leq n$, with equality iff $\omega_1 = \cdots = \omega_n$.*

The following result determines the central units of finite (multiplicative) order in certain integral group rings.

(8.21) Theorem. *Let k be an arbitrary field of algebraic numbers, and A be its ring of algebraic integers. Let α be a central unit of finite order in AG, where G is a finite group. Then $\alpha = \omega \cdot g$ for some $g \in Z(G)$ (the center of G), and some root of unity $\omega \in A$.*

In the case when G is abelian, this result was first proved by G. Higman. Here, we generalize Higman's method of proof by using general character theory so that we get the version (8.21) above which is meaningful for any finite group G.

We should note, however, that there are more general results in the literature dealing with (not necessarily central) units of finite order in AG where G is any group and A is any integral domain of characteristic zero. Theorem 8.21 above is only a special case of these more general results.

Proof of (8.21). We may clearly assume that $[k : \mathbb{Q}] < \infty$. To prove the theorem, we are also free to replace k by any algebraic extension $k' \supseteq k$. In fact, let A' be the ring of algebraic integers in k'. Suppose we know the result for $A'G$. Then, since α remains central in $A'G$, we can write $\alpha = \omega' \cdot g$ for some $g \in Z(G)$ and some root of unity $\omega' \in A'$. But then $\omega' = \alpha g^{-1} \in A' \cap AG = A$.

After suitably enlarging k, we may therefore assume that k is Galois over $\mathbb{Q}$, and is a splitting field for G. We shall now use the general notations in (8.15) through (8.19). Since $\alpha \in Z(kG) \cap AG$, we can write

$$\alpha = \sum b_i e_i = \sum a_g g,$$

where $b_i \in k$ and $a_g \in A$. Fix an integer m such that $\alpha^m = 1$. Then, computing in $Z(kG) = \prod_{i=1}^{r} k \cdot e_i$, we have $b_i^m = 1$; in particular, $|b_i| = 1$ (for all i). (We think of k as a subfield of $\mathbb{C}$; the absolute value refers to the modulus of a complex number.) Expressing the e_i's in terms of the g's by (8.15)(1), we get

$$a_g = \frac{1}{|G|} \sum_i n_i b_i \chi_i(g^{-1}).$$

Now fix a $g \in G$ such that $a_g \neq 0$, and write

$$\chi_i(g^{-1}) = \omega_{i1} + \cdots + \omega_{in_i},$$

where $\{\omega_{ij}\}$ are the characteristic roots of a matrix giving the g^{-1}-action on M_i. As we have observed before, the ω_{ij}'s are roots of unity. Therefore, by (8.20),

(8.22) $$|a_g| = \frac{1}{|G|} \cdot \left| \sum_{i,j} n_i b_i \omega_{ij} \right| \leq \frac{1}{|G|} \sum_i n_i^2 = 1.$$

For any $\sigma \in Gal(k/\mathbb{Q})$, a similar argument yields $|a_g^\sigma| \leq 1$ and so

$$|N_{k/\mathbb{Q}}(a_g)| = \prod_\sigma |a_g^\sigma| \leq 1.$$

But since $a_g \in A$, we have $N_{k/\mathbb{Q}}(a_g) \in \mathbb{Z}$ and so $|N_{k/\mathbb{Q}}(a_g)| = 1$. In particular, the inequality in (8.22) must be an equality. By (8.20), all the $b_i\omega_{ij}$'s must be equal, say

$$b_i \omega_{ij} = \omega \quad (1 \leq i \leq r,\ 1 \leq j \leq n_i).$$

Then $\chi_i(g^{-1}) = n_i\omega_{ij} = n_i\omega/b_i$, and

$$a_g = \frac{1}{|G|} \sum n_i^2 \omega = \omega.$$

We also need an expression for $\chi_i(g)$. Since the g-action on M_i has characteristic roots $\omega_{i1}^{-1}, \ldots, \omega_{in_i}^{-1}$, we have

$$\chi_i(g) = \omega_{i1}^{-1} + \cdots + \omega_{in_i}^{-1} = n_i b_i / \omega,$$

so $n_i b_i = \omega \chi_i(g)$. Now consider any $h \in G$ which is not conjugate to g. Then

$$a_h = \frac{1}{|G|} \sum_i n_i b_i \chi_i(h^{-1}) = \frac{\omega}{|G|} \sum_i \chi_i(g) \chi_i(h^{-1}) = 0$$

by the Second Orthogonality Relation (8.16)(B). Therefore, we have $\alpha = \omega C_g$. On the other hand, we have

$$|C_G(g)| = \sum_i \chi_i(g)\chi_i(g^{-1}) = \sum n_i^2 = |G|.$$

This implies that $g \in Z(G)$ and so $\alpha = \omega g$, as desired. QED

(8.23) Remark. The proof above actually showed more than was asserted in (8.21). Instead of requiring that $\alpha \in AG$, all we need is that $\alpha = \sum a_g g$ has at least one nonzero coefficient $a_g \in A$. If, however, all $a_g \notin A$, then $\alpha^m = 1$ may not imply that α is a k-multiple of a group element (see Exercise 27).

Let us now record a few consequences of (8.21).

(8.24) Corollary. *If G is a finite group with a trivial center, then any central unit of finite order in AG is a root of unity in A. If G is a finite abelian group and A is the ring of algebraic integers in a number field k which is not totally imaginary (i.e., k has at least one real embedding), then any unit of finite order in AG has the form $\pm g$ where $g \in G$.*

(8.25) Corollary (G. Higman, S.D. Berman). *Let A be as in* (8.21), *and G, H be finite groups. If AG and AH are isomorphic as rings, then $Z(G)$ and $Z(H)$ are isomorphic as groups.*

Proof. If $AG \cong AH$ as rings, then their groups of central units of finite order are also isomorphic. By (8.21), we get $U \times Z(G) \cong U \times Z(H)$, where U is the group of roots of unity in A. By the uniqueness part of the Fundamental Theorem on Finite Abelian Groups, it follows easily that $Z(G) \cong Z(H)$.

QED

We conclude with the following result on idempotents in integral group rings which can be proved by applying the same kind of arithmetic methods used above.

(8.26) Theorem. *Let A and k be as in* (8.21), *and G be any finite group. Then AG has no idempotents except* 0 *and* 1.

Proof. Again we use the general notations set up earlier in this section. Let $e = \sum a_g g \in AG$ be an idempotent, and let

$$e' = 1 - e = \sum a'_g g \in AG$$

be its complementary idempotent. Let $\theta_{i1}, \ldots, \theta_{in_i}$ be the characteristic values of the action of e on M_i. We see easily that each θ_{ij} is an idempotent. Therefore $\theta_{ij} \in \{0, 1\}$ and

$$\chi_i(e) = \theta_{i1} + \cdots + \theta_{in_i} \in \{0, 1, \ldots, n_i\}.$$

Applying the character $\chi_{reg} = \sum n_i \chi_i$, we get

$$a_1 = \frac{1}{|G|}\chi_{reg}\left(\sum a_g g\right) = \frac{1}{|G|}\sum_{i=1}^{r} n_i \chi_i(e) \in \mathbb{Q}, \tag{8.27}$$

and so

$$0 \leq a_1 \leq \frac{1}{|G|} \sum n_i^2 = 1.$$

Since $a_1 \in A \cap \mathbb{Q} = \mathbb{Z}$, this implies that $a_1 \in \{0, 1\}$, and, similarly, $a_1' \in \{0, 1\}$. In view of $a_1 + a_1' = 1$, we may assume, say $a_1 = 1$, and $a_1' = 0$. From (8.27), we see that

$$\chi_i(e) = \theta_{i1} + \cdots + \theta_{in_i} = n_i,$$

and so each $\theta_{ij} = 1$. This implies that the e-action on M_i is invertible, so this action is in fact the identity. It follows that e also acts as the identity on $kG \cong \bigoplus_i n_i M_i$, so $e = 1$. QED

The theorem above first appeared in a paper of R. Swan who proved the following fact on projective modules over AG: *if P is a finitely generated (left) projective module over AG, then $k \otimes_A P$ is a finitely generated free module over kG; in particular, $\dim_k k \otimes_A P = A$-rank of P is a multiple of $|G|$.* This implies (8.26) for, if AG had a nontrivial idempotent e, then $P = AG \cdot e$ would have been a projective AG-module with A-rank strictly between 0 and $|G|$. The character-theoretic proof given above is from a paper of Takahashi. For yet another proof, see Exercise 13 below.

Exercises for §8

Ex. 8.1. Give an example of a pair of finite groups G, G' such that, for some field $k, kG \cong kG'$ as k-algebras, but $G \not\cong G'$ as groups.

Ex. 8.2. Let k be a field whose characteristic is prime to the order of a finite group G. Show that the following two statements are equivalent:
(a) each irreducible kG-module has k-dimension 1;
(b) G is abelian, and k is a splitting field for G.

Ex. 8.3. Let $G = S_3$. Show that $\mathbb{Q}G \cong \mathbb{Q} \times \mathbb{Q} \times \mathbb{M}_2(\mathbb{Q})$ and compute the central idempotents of $\mathbb{Q}G$ which give this decomposition of $\mathbb{Q}G$ into its simple components. Compute, similarly, the decompositions of $\mathbb{Q}G_1, \mathbb{Q}G_2$, where G_1 is the Klein 4-group, and G_2 is the quaternion group of order 8.

Ex. 8.4. Let $R = kG$ where k is any field and G is any group. Let I be the ideal of R generated by $ab - ba$ for all $a, b \in R$. Show that $R/I \cong k[G/G']$ as k-algebras, where G' denotes the commutator subgroup of G. Moreover, show that $I = \sum_{a \in G'} (a - 1)kG$.

Ex. 8.5. For any field k and for any normal subgroup H of a group G, show that $kH \cap \mathrm{rad}\ kG = \mathrm{rad}\ kH$.

Ex. 8.6. In the above Exercise, assume further that $[G : H]$ is finite and prime to char k. Let V be a kG-module and W be a kH-module. Show that

(1) V is a semisimple kG-module iff ${}_{kH}V$ is a semisimple kH-module.
(2) W is a semisimple kH-module iff the induced module $kG \otimes_{kH} W$ is a semisimple kG-module.
(**Hint.** Use Clifford's Theorem (8.5) and Maschke's Theorem as generalized in Ex. 6.1.)

Ex. 8.7. (Villamayor, Green-Stonehewer, Willems) Let k be a field, and H be a normal subgroup of a finite group G. Show that $rad\ kG = kG \cdot rad\ kH$ iff char $k \nmid [G : H]$. (**Hint.** For the "if" part, use the two exercises above. For the "only if" part, note that, if we view $k[G/H]$ as a left kG-module, $rad\ kH$ acts as zero on $k[G/H]$. Thus, if $rad\ kG = kG \cdot rad\ kH$, $k[G/H]$ is a semisimple kG-module and therefore a semisimple $k[G/H]$-module. Now use (6.1).)

Ex. 8.8. Let G be a finite group such that, for some field k, kG is a finite direct product of k-division algebras. Show that any subgroup $H \subseteq G$ is normal. (**Hint.** Let $m = |H|$ and $\alpha = \sum_{h \in H} h \in kH$. Show that $\frac{1}{m}\alpha$ is a central idempotent in kG.)

Ex. 8.9. Show that the First Orthogonality Relation in (8.16)(A) can be generalized to

$$\sum_{g \in G} \chi_i(g^{-1})\chi_j(hg) = \delta_{ij}|G|\chi_i(h)/n_i,$$

where h is any element in G, and $n_i = \chi_i(1)$. ((8.16)(A) was the special case of this formula for $h = 1$.)

Ex. 8.10. Under the same assumptions on kG as in Exercise 9, let $F_k(G)$ be the k-space of class functions on G, given the inner product

$$[\mu, \nu] = \frac{1}{|G|}\sum_g \mu(g^{-1})\nu(g).$$

Show that, for any class function $f \in F_k(G)$, there is a "Fourier expansion" $f = \sum_i [f, \chi_i]\chi_i$, and that, for any two class functions $f, f' \in F_k(G)$, there is a "Plancherel formula"

$$[f, f'] = \sum_i [f, \chi_i][f', \chi_i].$$

Assuming that $char\ k = 0$, show that $f = \chi_M$ for some kG-module M iff $[f, \chi_i]$ is a nonnegative integer for all i, and that M is irreducible iff $[\chi_M, \chi_M] = 1$.

Ex. 8.11. Let k be the algebraic closure of $\mathbb{F}_p$ and $K = k(t)$, where t is an indeterminate. Let G be an elementary p-group of order p^2 generated by a, b. Show that

$$(*) \qquad a \mapsto A = \begin{pmatrix} 1 & 1 \\ 0 & 1 \end{pmatrix}, \quad b \mapsto B = \begin{pmatrix} 1 & t \\ 0 & 1 \end{pmatrix}$$

defines a representation of G over K which is not equivalent to any representation of G over k.

Ex. 8.12. Let k be any field of characteristic 2, and let $G = S_4$. Let M be the kG-module given by

$$ke_1 \oplus \cdots \oplus ke_4/k(e_1 + \cdots + e_4),$$

on which G acts by permuting the e_i's. Compute the kG-composition factors of M.

Ex. 8.13. Give the following alternative proof to (8.26) (due to D. Coleman): Let $e = \sum a_g g$ and $e' = 1 - e$ be complementary idempotents in AG. Over the quotient field k of A, we have $kG = e \cdot kG \oplus e' \cdot kG$. Show that $dim_k(e \cdot kG) = \chi_{reg}(e) = a_1 \cdot |G|$, and conclude that a_1 is a nonnegative rational integer. Since $dim_k(e \cdot kG) \leq |G|$, it follows that $a_1 = 0$ or 1, and hence $e = 0$ or 1.

Ex. 8.14. (Kaplansky) Let $G = \langle x \rangle$ be a cyclic group of order 5. Show that $u = 1 - x^2 - x^3$ is a unit of infinite order in $\mathbb{Z}G$, with inverse $v = 1 - x - x^4$. Then show that

$$U(\mathbb{Z}G) = \langle u \rangle \times (\pm G) \cong \mathbb{Z} \oplus \mathbb{Z}_2 \oplus \mathbb{Z}_5.$$

For the more computationally inclined reader, show that $a = 2x^4 - x^3 - 3x^2 - x + 2$ is a unit of infinite order in $\mathbb{Z}G$, with inverse $b = 2x^4 - 3x^3 + 2x^2 - x - 1$. (**Hint.** Let ζ be a primitive 5th root of unity. Under the natural map from $\mathbb{Z}G$ to $\mathbb{Z}[\zeta]$ mapping x to ζ, xu goes to $(1 + \zeta)^2$. Now verify that $U(\mathbb{Z}G)$ maps injectively into $U(\mathbb{Z}[\zeta])$, and that $1 + \zeta$ is a fundamental unit of $\mathbb{Z}[\zeta]$, i.e., $U(\mathbb{Z}[\zeta]) = \langle 1 + \zeta \rangle \cdot \{\pm \zeta^i\}$.)

Ex. 8.15. For finite abelian groups G and H, show that $\mathbb{R}G \cong \mathbb{R}H$ as $\mathbb{R}$-algebras iff $|G| = |H|$ and $|G/G^2| = |H/H^2|$.

Ex. 8.16. Show that, for any two groups G, H, there exists a (nonzero) ring R such that $RG \cong RH$ as rings.

Ex. 8.17. Using the theory of group representations, show that for any prime p, a group G of order p^2 must be abelian.

Ex. 8.18. Let $G = \{\pm 1, \pm i, \pm j, \pm k\}$ be the quaternion group of order 8. We have shown that, over $\mathbb{C}$, G has four 1-dimensional representations, and a unique irreducible 2-dimensional representation D. Construct D explicitly, and compute the character table for G.

Ex. 8.19. Let G be the dihedral group of order $2n$ generated by two elements r, s with relations $r^n = 1$, $s^2 = 1$ and $sr\,s^{-1} = r^{-1}$. Let $\theta = 2\pi/n$.
(1) For any integer h $(0 \leq h \leq n)$, show that

$$D_h(r) = \begin{pmatrix} \cos h\theta & -\sin h\theta \\ \sin h\theta & \cos h\theta \end{pmatrix}, \quad D_h(s) = \begin{pmatrix} 1 & 0 \\ 0 & -1 \end{pmatrix}$$

defines a real representation of G.
(2) Show that, over $\mathbb{C}$, D_h is equivalent to the representation D'_h defined by

$$D'_h(r) = \begin{pmatrix} e^{-ih\theta} & 0 \\ 0 & e^{ih\theta} \end{pmatrix}, \quad D'_h(s) = \begin{pmatrix} 0 & 1 \\ 1 & 0 \end{pmatrix}.$$

(3) For $n = 2m+1$, show that $D_1, \ldots, D_m$ give all irreducible representations of G (over $\mathbb{R}$ or over $\mathbb{C}$) with dimensions > 1. For $n = 2m$, show the same for $D_1, \ldots, D_{m-1}$.
(4) Construct the character table for G.
(5) Verify that the two nonabelian groups of order 8 (the dihedral group and the quaternion group) have the same character table (upon a suitable enumeration of the characters and the conjugacy classes of the two groups).

Ex. 8.20. Let $G = S_4$, which acts irreducibly on

$$M = \mathbb{Q}e_1 \oplus \mathbb{Q}e_2 \oplus \mathbb{Q}e_3 \oplus \mathbb{Q}e_4 / \mathbb{Q}(e_1 + e_2 + e_3 + e_4).$$

Let $M' = \sigma \otimes M$, where σ denotes the sign representation of G. Show that M' is equivalent to the representation D of G as the group of rotational symmetries of the cube (or of the octahedron).

Ex. 8.21. Show that, over $\mathbb{Q}$, $G = A_5$ has four irreducible representations, of dimensions 1, 4, 5, 6 respectively.

Ex. 8.22. For any finite group G and any field k, is it true that any irreducible representation of G over k is afforded by a minimal left ideal of kG? (The answer is yes, but don't get very discouraged if you can't prove it.)

Ex. 8.23. If a finite group G has at most three irreducible complex representations, show that $G \cong \{1\}, \mathbb{Z}_2, \mathbb{Z}_3$ or S_3.

Ex. 8.24. Suppose the character table of a finite group G has the following two rows:

	g_1	g_2	g_3	g_4	g_5	g_6	g_7
μ	1	1	1	ω^2	ω	ω^2	ω
ν	2	-2	0	-1	-1	1	1

$(\omega = e^{2\pi i/3})$

Determine the rest of the character table.

Ex. 8.25. (Littlewood's Formula) Let $e = \sum_{g \in G} a_g g \in kG$ be an idempotent, where k is a field and G is a finite group. Let χ be the character of G afforded by the kG-module $kG \cdot e$. Show that for any $h \in G$,

$$\chi(h) = |C_G(h)| \cdot \sum_{g \in C} a_g,$$

where C denotes the conjugacy class of h^{-1} in G. (**Hint.** Compute $\chi(h)$ as the trace of the linear transformation $\alpha \mapsto h\alpha e$ on kG.)

Ex. 8.26. Let $G = S_3$, and k be any field of characteristic 3.
(a) Show that there are only two irreducible representations for G over k, namely, the trivial representation and the sign representation.
(b) It is known that there are exactly six (finite-dimensional) indecomposable representations for G over k (see Curtis-Reiner: Representation Theory of Finite Groups and Associative Algebras, p. 433). Construct these representations. (**Hint.** Let G act by permutation on $V = ke_1 \oplus ke_2 \oplus ke_3$,

and let $\bar{V} = V/k \cdot (e_1 + e_2 + e_3)$. Consider the kG-modules k (with trivial G-action), V, $\bar{V}$, and "twist" these by the sign representation.)

Ex. 8.27. Let G be a cyclic group of prime order $p > 2$. Show that the group of units of finite order in $\mathbb{Q}G$ decomposes into a direct product of G with $\{\pm 1\}$ and another cyclic group of order 2. (**Hint.** Look at the Wedderburn decomposition of $\mathbb{Q}G$.)

Ex. 8.28. Let G be the group of order 21 generated by two elements a, b with the relations $a^7 = 1$, $b^3 = 1$, and $bab^{-1} = a^2$.
(1) Construct the (five) irreducible complex representations of G, and compute its character table.
(2) Construct the (three) irreducible rational representations of G, and determine the Wedderburn decomposition of the group algebra $\mathbb{Q}G$. (**Hint.** Let ζ be a primitive 7th root of unity. Note that $K = \mathbb{Q}(\zeta)$ affords a six dimensional irreducible rational representation of G, with a acting as multiplication by ζ, and b acting as the Galois automorphism $\zeta \mapsto \zeta^2$. Then show that $End_{\mathbb{Q}G}(K)$ is given by the field $F = \mathbb{Q}(\sqrt{-7}) \subset K$; this gives a Wedderburn component $\mathbb{M}_3(F)$ for $\mathbb{Q}G$.)
(3) How about $\mathbb{R}G$, and the real representations of G?

§9. Linear Groups

In this section, we shall study subgroups of $GL(V)$ where V is a finite-dimensional vector space over a field k. These groups G are usually referred to as *linear groups* (or *matrix groups*). They come with a natural representation, namely that afforded by the kG-module V. Here, we no longer assume that G is finite. But since V is finite-dimensional over k, we can still apply some of the methods of representation theory to study the structure of G. There are many beautiful classical results on the structure of linear groups; we shall examine a few of these in this section. Again, our main objective is not so much to embark on a systematic study of linear groups, but rather to illustrate the relevance of the methods of ring theory to this study.

Most of the results presented in the first half of this section revolve around a famous problem which originated from the work of W. Burnside. Recall that a group G is said to be *torsion* (or *periodic*) if every $g \in G$ has a finite order. Let G be a finitely generated torsion group. If G is abelian, it is easy to show that G must be finite. In 1902, Burnside raised the following provocative question:

(9.1) General Burnside Problem (GBP). *Let G be a finitely generated torsion group. Is G necessarily finite?*

There is a weaker version of this problem, obtained by imposing a stronger hypothesis on G. A group G is said to be *of bounded exponent* if there exists a natural number $N \geq 1$ such that $g^N = 1$ for every $g \in G$. The smallest such number N is called *the exponent* of G. The weaker version of (9.1) is the following.

(9.2) Bounded Burnside Problem (BBP). *Let G be a finitely generated group of bounded exponent. Is G necessarily finite?*

One of the main goals of this section is to study these two problems for linear groups. The classical results on (GBP) and (BBP) for linear groups, in fact, provided some of the early motivation for studying these problems in general. Before we go on to study linear groups, however, it will be useful to give a quick survey of what is currently known about (GBP) and (BBP), so that the reader can have an overview of this area of study.

The answer to (GBP) is "no" in general. In 1964, Golod showed that, for any prime p, there exists an *infinite* p-group G generated by two elements. (Recall that G is a p-group if every element of G has a finite p-power order.) As for (BBP), the full answer is not completely known. If we let N be the exponent of G, the answer to (BBP) turns out to depend on N. For $N = 2$, the answer is clearly "yes" as G must be abelian. For $N = 3, 4, 6$, the answers are still "yes," by results of Burnside (1902), Sanov (1940), and M. Hall (1958). For $N \geq 72$, Novikov announced a negative answer to (BBP) in 1959; however, the details were never published. For N odd and ≥ 4381, the negative answer to (BBP) appeared in the work of Novikov and Adjan in 1968. Later, the negative answer was extended by Adjan to all odd $N \geq 665$. There was recent progress on the even N case too: in 1996, Lysënok obtained the negative answer to (BBP) for all even $N > 8000$. Thus, the finite number of cases left open for (BBP) are: odd N from 5 to 663, and even N from 8 to 8000.

There is also another version of Burnside's Problem, called the **Restricted Burnside Problem** (RBP), which asks if the "universal Burnside group" $B(r, N)$ (defined as the quotient of the free group on r generators by the normal subgroup generated by all Nth powers) has a largest finite quotient. In 1959, Kostrikin announced a positive solution to (RBP) for N prime. In the late 80s, (RBP) was affirmed by Zelmanov for all r and all N. For his spectacular solution to (RBP) in all cases, Zelmanov received a Fields Medal at the Zürich International Congress for Mathematicians in 1994.

After the brief survey above, we shall now proceed to study (GBP) and (BBP) for linear groups. The main tool is our earlier characterization of absolutely irreducible modules over algebras which was based on Burnside's Lemma (7.3). This characterization leads to the following very useful fact on traces.

(9.3) Trace Lemma. *Let k be a field and let G be a subsemigroup of the general linear group $GL_n(k)$ such that under the natural action, k^n is an absolutely irreducible module over the semigroup algebra kG. Assume that the trace function $tr\colon G \to k$ has a finite image of cardinality r. Then $|G| \leq r^{n^2}$.*

Proof. By (7.5), $kG \to \mathbb{M}_n(k)$ is onto. This means that there exist $g_1, \ldots, g_{n^2} \in G$ which form a k-basis for $\mathbb{M}_n(k)$. Consider the map

$$\varepsilon: \ \mathbb{M}_n(k) \longrightarrow k^{n^2}$$

given by

$$\varepsilon(\sigma) = (tr(\sigma g_1), \ldots, tr(\sigma g_{n^2})), \quad \sigma = (\sigma_{ij}) \in \mathbb{M}_n(k).$$

This is clearly a k-linear map. *We claim that it is a monomorphism* (and therefore an isomorphism). Indeed, suppose $\varepsilon(\sigma) = 0$. Then for any $\gamma \in \mathbb{M}_n(k)$, we have $tr(\sigma \cdot \gamma) = 0$. Letting γ be the matrix unit E_{ij}, this gives $\sigma_{ji} = 0$, so $\sigma = 0$ as claimed. It follows that $|G| = |\varepsilon(G)| \leq r^{n^2}$ since for $\sigma \in G$, each of the n^2 coordinates of $\varepsilon(\sigma)$ can take at most r different values. QED

We shall now prove the following classical result of Burnside which provides an affirmative answer to (BBP) for linear groups in certain characteristics. Note that we do not have to assume that G is finitely generated in this result.

(9.4) Burnside's First Theorem. *Let k be a field of characteristic $p \geq 0$ and let G be a subgroup of $GL_n(k)$. If G has exponent $N < \infty$ and $p \nmid N$, then $|G| \leq N^{n^3} < \infty$.*

Proof. We may clearly assume that k is algebraically closed. If $n = 1$, we have $G \subseteq k \backslash \{0\}$. Since the equation $x^N = 1$ has at most N solutions in k, the theorem is clear in this case. We now assume $n \geq 2$. For any $g \in G$, each characteristic value α of g is an Nth root of unity. Since $tr(g)$ is a sum of n such α's, it takes at most $r = N^n$ values in k as g ranges over G. If G acts irreducibly on k^n, then k^n is an *absolutely* irreducible kG-module and we have

$$|G| \leq r^{n^2} = N^{n^3}$$

by the Trace Lemma above. Now assume the G-action is reducible on k^n. After choosing a suitable basis on k^n, we may assume that the elements g of G have the form $\begin{pmatrix} g_1 & h \\ 0 & g_2 \end{pmatrix}$ where g_1, g_2 are square matrices of fixed sizes, say n_1 and n_2. Let G_i $(i = 1, 2)$ be the group of the matrices g_i which arise in this manner. Invoking an inductive hypothesis on n, we may assume that $|G_i| \leq N^{n_i^3}$ $(i = 1, 2)$. Now consider the homomorphism $G \to G_1 \times G_2$ sending $g \in G$ to (g_1, g_2). This is an injection, for if $g = \begin{pmatrix} g_1 & h \\ 0 & g_2 \end{pmatrix}$ is in the kernel, then $g_1 = I$, $g_2 = I$ and

$$I = g^N = \begin{pmatrix} I & h \\ 0 & I \end{pmatrix}^N = \begin{pmatrix} I & N \cdot h \\ 0 & I \end{pmatrix}$$

implies that $N \cdot h = 0$. Since N is not a multiple of $p = char\, k$, we have

$h = 0$. Therefore,

$$|G| \le |G_1| \cdot |G_2| \le N^{n_1^3} \cdot N^{n_2^3} \le N^{(n_1+n_2)^3} = N^{n^3}. \quad \text{QED}$$

Note that the theorem may no longer hold if we do not impose the condition that $p = char\ k$ be prime to the exponent of G. In fact, for an infinite field k of characteristic $p > 0$, the (abelian) group

$$G = \left\{ \begin{pmatrix} 1 & h \\ 0 & 1 \end{pmatrix} : h \in k \right\} \subseteq GL_2(k)$$

has exponent $N = p$, but has cardinality equal to that of k. (In fact, G is isomorphic to the additive group of k.)

By an argument very similar to that used to prove (9.4), we get the following.

(9.5) Burnside's Second Theorem. *A linear group $G \subseteq GL_n(k)$ is finite iff it has a finite number of conjugacy classes.*

Proof. ("If" part) As before, we may assume that k is algebraically closed. The hypothesis on G implies that $tr(G)$ is a finite set in k. If G acts irreducibly on k^n, we are done by the Trace Lemma as before. If G acts reducibly on k^n, we use the same notations as in the proof of (9.4). Since G_1, G_2 also have finitely many conjugacy classes, they are finite by invoking an inductive hypothesis on n. The kernel H of $G \to G_1 \times G_2$ is a normal subgroup in G, consisting of matrices $\begin{pmatrix} I & h \\ 0 & I \end{pmatrix} \in G$. This subgroup is abelian since

$$\text{(9.6)} \quad \begin{pmatrix} I & h \\ 0 & I \end{pmatrix}\begin{pmatrix} I & h' \\ 0 & I \end{pmatrix} = \begin{pmatrix} I & h+h' \\ 0 & I \end{pmatrix} = \begin{pmatrix} I & h' \\ 0 & I \end{pmatrix}\begin{pmatrix} I & h \\ 0 & I \end{pmatrix}.$$

From $[G : H] \le |G_1| \cdot |G_2| < \infty$, it follows that any $g \in H$ has only finitely many G-conjugates. (The centralizer $C_G(g) \supseteq H$ has finite index in G.) Since there are only finitely many G-conjugacy classes, H must be finite. Therefore, G is also finite. QED

Next we shall deal with the General Burnside Problem for linear groups. The main result, (9.9) below, states that (GBP) has an affirmative answer for linear groups in any characteristic. This was first proved by Schur in the case of characteristic zero, in 1911; the adaptations needed for the proof in the case of characteristic p were given by Kaplansky. Our presentation of this result is preceded by two lemmata.

(9.7) Lemma. *The* (GBP) *has an affirmative answer for any* (*finitely generated torsion*) *group G which has an abelian subgroup H of finite index.*

Proof. Let $G = g_1 H \cup \cdots \cup g_m H$ and let $\{g_1, \ldots, g_m, \ldots, g_n\}$ be a set which is closed under "inverses" and which generates the group G. Let $g_i g_j = g_r h_{ij}$ where $1 \le r \le m$ and $h_{ij} \in H$. Let H_0 be the subgroup of H generated by the (finitely many) h_{ij}'s. Since H is abelian, H_0 is finite. For any $s \leq n$ we have

$$g_s g_i g_j = g_s g_r h_{ij} = g_t h_{sr} h_{ij} \in g_t H_0,$$

for some $t \leq m$. By induction, it follows that any word in $\{g_i\colon 1 \leq i \leq n\}$ lies in $\bigcup_{i=1}^n g_i H_0$. Therefore, $G = \bigcup_{i=1}^n g_i H_0$ is finite. QED

(9.8) Lemma. *Let k be a field and G be a finitely generated torsion subgroup of $GL_n(k)$. Then G has a bounded exponent.*

Proof. We may clearly assume that k is finitely generated over its prime field P. Fix a purely transcendental extension k_0/P within k such that $[k : k_0] = r < \infty$. Now $G \subseteq GL_n(k)$ acts faithfully on k^n. Viewing k^n as k_0^{rn}, G acts faithfully on k_0^{rn}. Thus we are reduced to considering $G \subseteq GL_{rn}(k_0)$. For $g \in G$, let $m_g(t) \in k_0[t]$ be the minimal polynomial of g (as a matrix over k_0).

Case 1. $P = \mathbb{Q}$. Since $g \in G$ has finite order, the zeros of $m_g(t)$ are all roots of unity. Thus, the coefficients of $m_g(t)$ are algebraic integers. But since k_0 is purely transcendental over $\mathbb{Q}$, the only algebraic integers in k_0 are the rational integers, so $m_g(t) \in \mathbb{Z}[t]$. On the other hand, expressing the coefficients of m_g by elementary symmetric functions of its roots, we see that these coefficients are bounded in absolute value. Therefore there are only a finite number of different $m_g(t)$'s as g ranges over G. Since $m_g(t)$ uniquely determines the order of g, it follows that G has a bounded exponent.

Case 2. $P = \mathbb{F}_p$ where p is a prime. We show as in Case 1 that $m_g(t) \in \mathbb{F}_p[t]$. Since $|\mathbb{F}_p| = p$, there are again only a finite number of different $m_g(t)$'s. Therefore, we are done as before. QED

We are now ready to give the solution of (GBP) for linear groups.

(9.9) Theorem (Schur). *Let k be a field and G be a finitely generated torsion subgroup of $GL_n(k)$. Then G is finite.*

Proof. Since we already know that G must have a bounded exponent by the Lemma above, we can try to "recycle" the proof of the First Burnside Theorem (9.4). Let us, therefore, refer to the notations set up in that proof. The only difference occurs at the very end: We have (in the reducible case) a homomorphism $\varepsilon\colon G \to G_1 \times G_2$, where (by inductive hypothesis) G_1 and G_2 are finite. This implies that

$$H = ker\ \varepsilon = \left\{ \begin{pmatrix} I & h \\ 0 & I \end{pmatrix} \in G \right\}$$

has finite index in G. But H is also abelian as we have shown in (9.6). Therefore the finiteness of G follows from (9.7). QED

In view of the above results, it is useful to recall the following term introduced in Exer. (6.21): a group G is said to be *locally finite* if every finitely generated subgroup of G is finite. Clearly, such a group is torsion; the converse problem, asking if every torsion group is locally finite, is precisely (GBP). In view of this remark, we can restate (9.9) in the following equivalent form:

(9.9′) Theorem. *A linear group $G \subseteq GL_n(k)$ over a field k is torsion iff it is locally finite.*

In order to get more results on linear groups, we shall now introduce another definition:

(9.10) Definition. A linear group $G \subseteq GL(V)$ is said to be *completely reducible* if V is a completely reducible (= semisimple) module over kG, i.e., if V is a (finite, direct) sum of simple kG-submodules.

Note that in this section, the linear groups $G \subseteq GL(V)$ under consideration are usually infinite. If G is indeed infinite, the group algebra kG is *never* semisimple, by (6.3). Nevertheless, the specific kG-module V may happen to be semisimple, in which case G is by definition completely reducible. Let $Span_k(G)$ be the subspace of $End_k\ V$ spanned by the elements of G. Since G is a group, this is clearly a k-subalgebra of $End_k\ V$; it is, in fact, the image of the natural homomorphism $kG \to End_k\ V$. The following simple proposition helps to clarify the notion of a completely reducible linear group.

(9.11) Proposition. *The linear group $G \subseteq GL(V)$ is completely reducible iff the k-algebra $S := Span_k(G)$ is semisimple.*

Proof. Assume S is semisimple. Then V is a semisimple module over S and therefore over kG. Conversely, if V is semisimple over kG, then V is a *faithful* semisimple module over S. Since $(rad\ S) \cdot V = 0$, $rad\ S = 0$. As S is a finite-dimensional k-algebra, this implies that S is a semisimple ring. QED

Of course, not every linear group is completely reducible, even if $char\ k = 0$. For instance, if $dim_k\ V = 2$, it is easy to show that the group of matrices of the form $\begin{pmatrix} 1 & a \\ 0 & 1 \end{pmatrix}$ is not completely reducible.

What are some examples of completely reducible groups? If $G \subseteq GL(V)$ is finite and $p = char\ k$ does not divide $|G|$, then by Maschke's Theorem, G is always completely reducible. By a slight generalization of Maschke's Theorem (see Exercise 6.1), we have the following improved statement:

(9.12) Theorem. *Let $G \subseteq GL(V)$ and H be a subgroup of finite index m in G such that $char\ k \nmid m$. If H is completely reducible, then so is G.*

To relate Schur's Theorem (9.9) to the notion of completely reducible groups, let us make the following observation.

(9.13) Lemma. *Let $G \subseteq GL(V)$. If every finitely generated subgroup of G is completely reducible, then so is G.*

Proof. Let $g_1, \ldots, g_m \in G$ $(m < \infty)$ be chosen such that they form a k-basis of $Span_k(G)$, and let H be the subgroup of G generated by $\{g_1, \ldots, g_m\}$. Since $Span_k(G) = Span_k(H)$, the desired conclusion follows from (9.11). QED

We can now record the following nice consequence of Schur's Theorem (9.9).

(9.14) Proposition. *Let $G \subseteq GL_n(k)$ be a group such that every element $g \in G$ has finite order prime to char k. Then G is completely reducible.*

Proof. By (9.9) and Maschke's Theorem, every finitely generated subgroup H of G is finite and completely reducible. Now use (9.13). QED

In group theory, a subgroup H of a group G is said to be *subnormal* in G if there exists a finite chain of subgroups

$$G = G_0 \supseteq G_1 \supseteq \cdots \supseteq G_m = H$$

such that each G_{i+1} is normal in G_i. Clifford's Theorem (8.5) leads easily to the following sufficient condition for complete reducibility:

(9.15) Proposition. *Let H be a subnormal subgroup of a linear group $G \subseteq GL(V)$. If G is completely reducible, then so is H.*

Proof. By induction, it suffices to treat the case when H is normal in G. Write ${}_{kG}V$ as a sum of simple kG-modules V_i and apply Clifford's Theorem to each V_i. QED

Given a linear group $G \subseteq GL_n(V)$, one might ask what part of the structure of G would constitute the obstruction to G being completely reducible. This is a rather subtle problem which we shall not be able to answer fully. We can, however, account for part of the obstruction to G being completely reducible by looking at a certain radical of G, called its *unipotent radical.* This unipotent radical of G is related to the Jacobson radical of $Span_k(G)$, although, in general, the former does not determine the latter. In the balance of this section, we shall try to explain the mathematical ideas which lead to the definition of the unipotent radical. The study of this radical is interesting from the viewpoint of this chapter as it brings together various ideas from ring theory, group theory and linear algebra. Actually, some of its motivation comes from Lie's early work on solvable Lie algebras, though, in order to keep this exposition self-contained, we shall not try to develop the details of the Lie algebra connection here.

We begin by introducing the following definition.

(9.16) Definition. *Let $\lambda \in k$, and V be an n-dimensional k-vector space. A linear transformation $g \in End_k V$ is said to be λ-potent if $g = \lambda + g_0$ where g_0 is nilpotent (equivalently, if the characteristic polynomial of g is $(t - \lambda)^n \in k[t]$).*

Thus, "0-potent" means simply "nilpotent." If $\lambda \neq 0$ and g is λ-potent, then g is invertible, for, if $g = \lambda + g_0$ with g_0 nilpotent, then g^{-1} is given by

$$\lambda^{-1}(1 - \lambda^{-1}g_0 + \lambda^{-2}g_0^2 - \cdots).$$

In the important special case when $\lambda = 1$, we speak of g as a *unipotent* transformation.

Let $G \subseteq GL(V)$ be a linear group and Λ be a subset of $k^* = k\backslash\{0\}$. We shall say that G is Λ-*potent* if there is a surjection λ: $G \to \Lambda$ such that, for any $g \in G$, g is $\lambda(g)$-potent. In the special case when $\Lambda = \{1\}$, we speak of G as a *unipotent* group. If G is Λ-potent, it turns out that Λ must be a subgroup of k^* and λ must be a group homomorphism. These facts, however, are not immediate consequences of the definitions.

Note that if $g \in GL(V)$, then g is λ-potent iff g is λ times a unipotent. Thus the unipotent case is of particular importance. Before we go on, the following basic observation is in order.

(9.17) Proposition. *Let k be a field of characteristic $p > 0$. Then $g \in GL(V)$ is unipotent iff it is a p-element (i.e., its order is a power of p). Thus, a linear group $G \subseteq GL(V)$ is a unipotent group iff it is a p-group.*

Proof. If $g^{p^r} = 1$, then $(g-1)^{p^r} = 0$ so $g = 1 + (g-1)$ is unipotent. Conversely, if $g = 1 + g_0$ where $g_0^m = 0$, then for $p^r \geq m$, we have

$$g^{p^r} = 1 + g_0^{p^r} = 1. \qquad \text{QED}$$

The Proposition above suggests that the study of unipotent groups is a generalization of the study of p-subgroups of $GL(V)$ in characteristic p. In the following, we shall try to prove a few basic properties of unipotent groups (or more generally, Λ-potent groups); our considerations will be valid in all characteristics.

Suppose we fix a basis on V and identify $GL(V)$ with $GL_n(k)$. A matrix is called *unitriangular* if it is upper triangular and has 1's on the diagonal. The set of unitriangular matrices forms the *unitriangular group* $UT_n(k)$; this is clearly a unipotent group. The group $k^* \cdot UT_n(k)$ of invertible upper triangular matrices with constant diagonals is a k^*-potent group, and any subgroup of $k^* \cdot UT_n(k)$ is a Λ-potent group for some Λ. Note that the groups $UT_n(k)$, $k^* \cdot UT_n(k)$ are determined up to a conjugation in $GL(V)$. The basic result on Λ-potent groups is the following:

(9.18) Theorem (Lie, Kolchin, Suprunenko). *Let $G \subseteq GL(V)$ be a Λ-potent group (with respect to a surjection λ: $G \to \Lambda \subseteq k^*$). Then*

(1) *Any irreducible $\text{Span}_k(G)$-module M of finite k-dimension has k-dimension one.*

(2) *With respect to a suitable k-basis on V, we have $G \subseteq k^* \cdot UT_n(k)$.*

Before we proceed to the proof of this classical theorem, let us first record some of its interesting consequences. (These consequences, in fact, help clarify the meaning of the theorem itself.)

(9.19) Corollary.

(1) *The* Λ *in* (9.18) *must be a subgroup of* k^*, *and* λ *is a group homomorphism.*

(2) *Any unipotent group is conjugate in* $GL(V)$ *to a subgroup of* $UT_n(k)$.

(3) $UT_n(k)$ *is a maximal unipotent subgroup of* $GL_n(k)$, *and any maximal unipotent subgroup of* $GL_n(k)$ *is conjugate to* $UT_n(k)$.

Proof. (1), (2) are both clear from the theorem. To prove (3), note that, by Zorn's Lemma, any unipotent subgroup in $GL_n(k)$ can be enlarged into a maximal unipotent subgroup. In particular, maximal unipotent subgroups G of $GL_n(k)$ do exist. Consider any such G. By the theorem, there exists an invertible matrix ρ such that $\rho G \rho^{-1} \subseteq UT_n(k)$. Since $UT_n(k)$ is unipotent and $\rho G \rho^{-1}$ is maximal unipotent, we must have $\rho G \rho^{-1} = UT_n(k)$. This proves both statements in (3). QED

Note that, in view of (9.17), if k has characteristic $p > 0$, (3) above says precisely that $UT_n(k)$ *is a p-Sylow subgroup of* $GL_n(k)$, *and that any two p-Sylow subgroups of* $GL_n(k)$ *are conjugate*. Thus, the Sylow Theorems hold for $GL_n(k)$ for the particular prime p. This, however, does not mean that we can skip the proof in the characteristic p case, since, in general, the Sylow Theorems need not hold for infinite groups. If k is a finite field, say $k = \mathbb{F}_q$ $(q = p^r)$, then $GL_n(k)$ is a finite group of order

$$\textbf{(9.20)} \qquad (q^n - 1)(q^n - q)(q^n - q^2) \cdots (q^n - q^{n-1}),$$

while $UT_n(k)$ is a finite group of order $q^{(n^2-n)/2}$. Since $q^{(n^2-n)/2}$ is precisely the p-part of $|GL_n(k)|$ as computed in (9.20), this shows directly that $UT_n(k)$ is a p-Sylow subgroup of $GL_n(k)$ in this case. Thus, for finite fields $k = \mathbb{F}_q$ $(q = p^r)$, the statements (2), (3) of (9.19) are indeed implied by the Sylow Theorems for finite groups. In this case, incidentally, the fact that any irreducible module over a unipotent group G has k-dimension 1 (with trivial G-action) is also known to us already: see (8.4). The results (9.18), (9.19) for arbitrary fields (of any characteristic) may therefore be regarded as generalizations of the corresponding results for the special case $k = \mathbb{F}_q$.

To get another corollary out of (9.18), recall that the lower central series of a group G is defined to be the series

$$G^{(0)} \supseteq G^{(1)} \supseteq G^{(2)} \supseteq \cdots$$

where $G^{(0)} = G$, $G^{(1)} = [G, G]$ (the commutator subgroup), and, inductively, $G^{(r)} = [G, G^{(r-1)}]$. The group G is said to be *nilpotent of class r* if $G^{(r)} = \{1\}$, but $G^{(r-1)} \neq \{1\}$. We leave it as an exercise for the reader to verify that the unitriangular group $UT_n(k)$ over any field k is nilpotent of class $n - 1$ for $n \geq 2$. It follows easily from (9.18) that

(9.21) Corollary. *Any Λ-potent group in $GL_n(k)$ is nilpotent of class $< n$ (for $n \geq 2$).*

In the light of (9.17), we may regard this as a generalization of the well-known fact that any finite p-group is nilpotent.

After the above discussion of Theorem (9.18), we would now like to proceed to its proof, which will be carried out in three steps. (The notations in (9.18) will be fixed in the following.)

Step 1. *It is enough to prove* (1) *in* (9.18). For, if (1) is true, then, for any kG-composition series

$$V = V_0 \supset V_1 \supset \cdots \supset 0,$$

each V_i/V_{i+1} is 1-dimensional. With respect to a suitable basis on V, every $g \in G$ has the upper triangular form. Since g is $\lambda(g)$-potent, its diagonal entries must all be $\lambda(g)$. Therefore, $G \subseteq k^* \cdot UT_n(k)$, as desired.

Step 2. (1) (*and hence* (2)) *in* (9.18) *is true in case k is algebraically closed.* Without loss of generality, we may assume that $G \supseteq k^* = k^* \cdot I$. (If not, we can replace G by $k^* \cdot G$.) Suppose the M in (1) has k-dimension m. Let H be the image of G under $G \to End_k\, M$ and let H_0 be the (normal) subgroup of H consisting of endomorphisms of determinant 1. Since we can take mth roots in k (k is assumed to be algebraically closed here), $H = k^* \cdot H_0$. The characteristic polynomial of $h_0 \in H_0$ has the form $(t - \lambda)^m$, with $\lambda^m = \det h_0 = 1$. Therefore $tr(h_0) = m\lambda$ assumes at most m values in k. Since H_0 acts (absolutely) irreducibly on M, the Trace Lemma (9.3) implies that H_0 is finite. For $p = char\, k \geq 0$, we claim that *every p'-element $h_0 \in H_0$ is a scalar matrix*. In fact, let r be the order of h_0, with $p \nmid r$. Then h_0 satisfies the polynomials

$$t^r - 1 \quad \text{and} \quad (t - \lambda)^m,$$

and therefore h_0 also satisfies their g.c.d. But this g.c.d. is $t - \lambda$ since $t^r - 1$ has no multiple roots. Therefore, $h_0 = \lambda \cdot I$, as claimed. If $p = 0$, our claim gives $H = k^*$, so clearly $m = 1$. If $p > 0$, the claim implies that $H = k^* \cdot P$ where P is a p-Sylow subgroup of H_0. If $h_0 \in P$, then h_0 must be unipotent by (9.17) and therefore $tr(h_0) = m \cdot 1$. Since P acts (absolutely) irreducibly on M, the Trace Lemma gives $|P| = 1$. We have now $H = k^*$ so clearly m must be 1. (Alternatively, we can get $m = 1$ by using the earlier result that, for any finite p-group P, the only irreducible kG-module is k, with the trivial P-action.)

Step 3. *We shall now prove* (9.18)(1) *in the general case.* Let M and m be as above. After replacing G by its image under $G \to End_k\, M$, we may assume that G acts faithfully on M. Let E be the algebraic closure of k. By what we did in Step 1 and Step 2, there exists an E-basis in $M^E = M \otimes_k E$ with respect to which G lies in $E^* \cdot UT_n(E)$. In other words, for any $g \in G$,

$g - \lambda(g)I$ is an upper triangular matrix over E with a zero diagonal. By an easy calculation, we see that the product of any m such matrices is zero. Now choose r to be the smallest integer such that

$$(g_1 - \lambda(g_1)) \cdots (g_r - \lambda(g_r)) = 0 \quad \text{for any } g_1, \ldots, g_r \in G.$$

If $r = 1$, then $G \subseteq k^*$, so clearly $m = 1$. Now assume $r \geq 2$. Then there exist $g_2, \ldots, g_r \in G$ such that

$$w := (g_2 - \lambda(g_2)) \cdots (g_r - \lambda(g_r))v \neq 0$$

for some $v \in M$. But then $(g - \lambda(g))w = 0$ for any $g \in G$, so $k \cdot w$ is a kG-submodule of M. Since M is irreducible, we must have $M = k \cdot w$, and so $m = 1$ as desired. QED

In the case when G is a unipotent group, (9.18) is usually known as Kolchin's Theorem (proved by E.R. Kolchin in 1948 in a pioneering paper on algebraic groups). We shall now give an application of this result. (For another application, see Exercise 4.)

(9.22) Theorem. *Every linear group $G \subseteq GL(V)$ (over an arbitrary field k) has a unique maximal normal unipotent subgroup H. (We call H the unipotent radical of G.) The quotient group G/H is isomorphic to a certain completely reducible linear group over k.*

Proof. Let $\{V_1, \ldots, V_r\}$ be the composition factors of V as a kG-module. Let H be the normal subgroup of G consisting of elements $g \in G$ which act trivially on all V_i. For any such g, $(g-1)^r$ acts as zero on V and so H is unipotent. Conversely, if H_0 is any normal unipotent subgroup of G, and M is any finite dimensional irreducible $Span_k(G)$-module (e.g., any V_i), then by Clifford's Theorem (8.5), M is a semisimple $Span_k(H_0)$-module and by (9.18)(1), H_0 acts trivially on M. This shows that $H_0 \subseteq H$ and yields the following characterization of H:

(9.23) $$H = \{g \in G:\ g \text{ acts trivally on all finite-dimensional irreducible } Span_k(G)\text{-modules}\}.$$

Let $\overline{V} = V_1 \oplus \cdots \oplus V_r$. Then H is the kernel of $G \to GL(\overline{V})$ so $\bar{G} := G/H$ may be viewed as a linear group in $GL(\overline{V})$. Since $\overline{V}$ is a semisimple $k\bar{G}$-module, $\bar{G}$ is completely reducible. QED

The following consequence of this theorem (and its proof) is to be compared with (8.6) in the light of (9.17).

(9.24) Corollary. *Let $G \subseteq GL(V)$ be a linear group over k, and $S = Span_k\, G$. Then the unipotent radical H of G is given by*

$$H = \{g \in G:\ g - 1 \in rad\, S\}.$$

In particular, G is completely reducible only if $H = \{1\}$.

Note that G having a trivial unipotent radical H is only a necessary condition for $G \subseteq GL(V)$ to be completely reducible, but in general not a sufficient condition. If $H = \{1\}$, then by (9.22), G may be viewed as a completely reducible linear group acting on $\overline{V} = V_1 \oplus \cdots \oplus V_r$, but G may not be a completely reducible group acting on V itself, as the following easy example shows.

Let k be a field of characteristic $p > 0$, and let G be a finite group which has no nontrivial normal p-subgroup. Let $V = kG$ and let G act by left multiplication on V. Since G acts faithfully on V, we can view G as a subgroup of $GL(V)$. As a linear group, G has trivial unipotent radical by (9.17). However, if p divides $|G|$, V is *not* a semisimple kG-module by (6.1), and therefore G is *not* a completely reducible group acting on V.

The notion of the unipotent radical is of great significance in the theory of algebraic groups. The algebraic groups in $GL(V)$ which have trivial unipotent radicals are essentially the so-called *reductive groups.* For a more detailed study of these groups, we refer the reader to Humphreys' book "*Linear Algebraic Groups*" (Springer, 1975).

Exercises for §9

Ex. 9.1. Let $G \subseteq GL_n(k)$ be a linear group over a field k. Show that G is an f.c. group (i.e. every conjugacy class of G is finite) iff the center $Z(G)$ has finite index in G. Show that every finite group can be realized as a linear group, but not every infinite group can be realized as a linear group.

Ex. 9.2. Can every finite group be realized as an irreducible linear group? (A linear group $G \subseteq GL(V)$ is said to be irreducible if G acts irreducibly on V.)

Ex. 9.3. Let k be any field of characteristic 3, $G = S_3$ and let V be the kG-module

$$ke_1 \oplus ke_2 \oplus ke_3/k(e_1 + e_2 + e_3),$$

on which G acts by permuting the e_i's. Show that this realizes G as a linear group in $GL(V)$. Is G a completely reducible linear group? What is its unipotent radical? Determine $Span_k(G)$ and its Jacobson radical.

Ex. 9.4. Let k be a field of characteristic zero.
(1) Show that any unipotent subgroup $G \subseteq GL_n(k)$ is torsion-free.
(2) If G is a maximal unipotent subgroup of $GL_n(k)$, show that G is a divisible group.

Ex. 9.5. Let k be an algebraically closed field, and $G \subseteq GL_n(k)$ be a completely reducible linear group. Show that G is abelian iff G is conjugate to a subgroup of the group of diagonal matrices in $GL_n(k)$.

Ex. 9.6. Let k be a field of characteristic zero. Let $G \subseteq GL_n(k)$ be a linear group, and H be a subgroup of finite index in G. Show that G is completely reducible iff H is. (**Hint.** $[G : H] < \infty$ implies that H contains a normal subgroup H_0 of G such that $[G : H_0] < \infty$.)

CHAPTER 4

Prime and Primitive Rings

In commutative ring theory, three basic classes of rings are: *reduced rings*, *integral domains*, and *fields*. The defining conditions for these classes do not really make any use of commutativity, so by using exactly the same conditions on rings in general, we can define (and we have defined) the notions of *reduced rings*, *domains*, and *division rings*. However, a little careful thought will show that this is not the only way to generalize the former three classes. In fact, the defining conditions for these classes are conditions on *elements* of a ring. When we move from commutative rings to noncommutative rings, an alternative way of generalizing an "element-wise" condition should be to replace the role of elements by that of ideals. By making these changes judiciously in the basic definitions, we are led to the notions of *semiprime rings*, *prime rings*, and (*left or right*) *primitive rings*. The following chart more or less summarizes the overall situation:

commutative category		direct generalizations		alternative generalizations
{reduced comm. rings}	$\subset$	{reduced rings}	$\subset$	{semiprime rings}
$\cup$		$\cup$		$\cup$
{integral domains}	$\subset$	{domains}	$\subset$	{prime rings}
$\cup$		$\cup$		$\cup$
{fields}	$\subset$	{division rings}	$\subset$	{(1-sided) primitive rings}

This chapter is mainly concerned with the three classes of rings constituting the last column of the so-called "alternative generalizations." In §10, after introducing the notion of prime and semiprime rings, we give a brief treatment of the theory of radicals for noncommutative rings. The notions of upper, lower nilradicals and the Levitzki radical are introduced as interesting alternatives to the notion of the Jacobson radical. In general, these four radicals are mutually distinct, but for specific classes of rings, two or more of them may coincide. Some of the facts for the Jacobson radical are shown to have analogues for the other kinds of radicals. However, it is not our intention to try to extend every result on the Jacobson radical in Chapter 2 to the other kinds of radicals.

Section 11 studies the structure theory of (left) primitive rings due to N. Jacobson. Examples are given to show how this structure theory can be used to prove nontrivial results in noncommutative ring theory. In §§11–12, the importance of the class of J-semisimple rings emerges once again. These rings are the same as the semiprimitive rings, and they are characterized as subdirect products of (left) primitive rings. If a certain proposition is true for division rings, it is sometimes possible to use Jacobson's structure theory to deduce the same proposition for left primitive rings, and, via subdirect product representations, for semiprimitive rings. To deduce the same proposition for *arbitrary* rings would, however, involve "lifting through" the Jacobson radical; this is usually very difficult, and often impossible. Nevertheless, a number of remarkable theorems on rings have been proved by means of this useful procedure.

§10. The Prime Radical; Prime and Semiprime Rings

In commutative ring theory, the notions of prime ideals and radical ideals play very important roles. We begin by recalling these two basic notions. Let R be a commutative ring, and $\mathfrak{U}$ be an ideal in R. We say that $\mathfrak{U}$ is a *prime ideal* if $\mathfrak{U} \neq R$, and for $a, b \in R$,

$$ab \in \mathfrak{U} \quad \text{implies that } a \in \mathfrak{U} \text{ or } b \in \mathfrak{U}.$$

We say that $\mathfrak{U}$ is a *radical ideal* if, for $a \in R$,

$$a^n \in \mathfrak{U} \quad \text{for some } n \geq 1 \text{ implies that } a \in \mathfrak{U}.$$

In commutative algebra, it is well-known that $\mathfrak{U}$ is a radical ideal iff $\mathfrak{U}$ is an intersection of prime ideals. (If $\mathfrak{U} = R$, we regard $\mathfrak{U}$ as the intersection of an empty family of prime ideals). For any ideal $\mathfrak{U}$, there is a smallest radical ideal containing $\mathfrak{U}$, namely, the intersection of all prime ideals $\supseteq \mathfrak{U}$. This radical ideal is denoted by $\sqrt{\mathfrak{U}}$. It can also be characterized as

$$\{x \in R: \ x^n \in \mathfrak{U} \text{ for some } n \geq 1\},$$

hence the notation. As a special case, we have $\sqrt{(0)} = Nil(R)$, the ideal of nilpotent elements of R.

To begin this section, we shall try to generalize the well-known results above to the noncommutative setting. While our presentation is geared toward the noncommutative case, the results we obtain are, of course, also valid for the commutative case. Therefore, the facts mentioned in the last paragraph will all be re-proved instead of assumed. They are stated explicitly above purely as motivation for the more general treatment needed for the noncommutative case.

Let us first define the notion of a prime ideal for an arbitrary ring.

(10.1) Definition. An ideal $\mathfrak{p}$ in a ring R is said to be a *prime ideal* if $\mathfrak{p} \neq R$ and, for ideals $\mathfrak{A}, \mathfrak{B} \subseteq R$,

$$\mathfrak{A} \cdot \mathfrak{B} \subseteq \mathfrak{p} \quad \text{implies that } \mathfrak{A} \subseteq \mathfrak{p} \text{ or } \mathfrak{B} \subseteq \mathfrak{p}.$$

For an element $a \in R$, let us write $(a) = RaR$: this is the ideal generated by a in R. The following proposition offers several other characterizations of prime ideals.

(10.2) Proposition. *For an ideal $\mathfrak{p} \subsetneq R$, the following statements are equivalent*:

(1) $\mathfrak{p}$ *is prime.*

(2) *For $a, b \in R$, $(a)(b) \subseteq \mathfrak{p}$ implies that $a \in \mathfrak{p}$ or $b \in \mathfrak{p}$.*

(3) *For $a, b \in R$, $aRb \subseteq \mathfrak{p}$ implies that $a \in \mathfrak{p}$ or $b \in \mathfrak{p}$.*

(4) *For left ideals $\mathfrak{A}, \mathfrak{B}$ in R, $\mathfrak{A}\mathfrak{B} \subseteq \mathfrak{p}$ implies that $\mathfrak{A} \subseteq \mathfrak{p}$ or $\mathfrak{B} \subseteq \mathfrak{p}$.*

(4)′ *For right ideals $\mathfrak{A}, \mathfrak{B}$ in R, $\mathfrak{A}\mathfrak{B} \subseteq \mathfrak{p}$ implies that $\mathfrak{A} \subseteq \mathfrak{p}$ or $\mathfrak{B} \subseteq \mathfrak{p}$.*

Proof. It is enough to show that $(1) \Rightarrow (2) \Rightarrow (3) \Rightarrow (4) \Rightarrow (1)$. The first two implications and the last one are trivial. For $(3) \Rightarrow (4)$, assume that $\mathfrak{A}\mathfrak{B} \subseteq \mathfrak{p}$, but $\mathfrak{A} \nsubseteq \mathfrak{p}$, where $\mathfrak{A}, \mathfrak{B}$ are left ideals. Fix an element $a \in \mathfrak{A} \backslash \mathfrak{p}$. For any $b \in \mathfrak{B}$, we have $aRb \subseteq \mathfrak{A}\mathfrak{B} \subseteq \mathfrak{p}$, so by (3), $b \in \mathfrak{p}$. This shows that $\mathfrak{B} \subseteq \mathfrak{p}$. QED

As an example, note that *any maximal ideal* $\mathfrak{m}$ *in R is prime*. For, if $\mathfrak{A}, \mathfrak{B}$ are ideals not contained in $\mathfrak{m}$, then $\mathfrak{m} + \mathfrak{A} = R = \mathfrak{m} + \mathfrak{B}$, and hence

$$R = (\mathfrak{m} + \mathfrak{A})(\mathfrak{m} + \mathfrak{B}) = \mathfrak{m} + \mathfrak{A}\mathfrak{B},$$

which implies that $\mathfrak{A}\mathfrak{B} \nsubseteq \mathfrak{m}$.

In commutative algebra, prime ideals are closely tied to multiplicatively closed sets. The complement of a prime ideal is multiplicatively closed, and, given a (nonempty) multiplicatively closed set S, an ideal disjoint from S and maximal with respect to this property is always a prime ideal. We shall now prove the analogues of these facts for arbitrary rings. First we have to adapt the notion of a multiplicatively closed set to the noncommutative setting.

(10.3) Definition. A nonempty set $S \subseteq R$ is called an *m-system* if, for any $a, b \in S$, there exists $r \in R$ such that $arb \in S$.

For instance, a (nonempty) multiplicatively closed set S is an m-system. The converse is not true: for $a \in R$, $\{a, a^2, a^4, a^8, \ldots\}$ is an m-system, but not multiplicatively closed in general.

From the characterization (3) of a prime ideal, we deduce the following:

(10.4) Corollary. *An ideal $\mathfrak{p} \subseteq R$ is prime iff $R \backslash \mathfrak{p}$ is an m-system.*

(10.5) Proposition. *Let $S \subseteq R$ be an m-system, and let $\mathfrak{p}$ be an ideal maximal with respect to the property that $\mathfrak{p}$ is disjoint from S. Then $\mathfrak{p}$ is a prime ideal.*

Proof. Suppose $a \notin \mathfrak{p}$, $b \notin \mathfrak{p}$, but $(a)(b) \subseteq \mathfrak{p}$. By the maximal property of $\mathfrak{p}$, there exist $s, s' \in S$ such that $s \in \mathfrak{p} + (a)$, $s' \in \mathfrak{p} + (b)$. Take $r \in R$ with $srs' \in S$. Then

$$srs' \in (\mathfrak{p} + (a))R(\mathfrak{p} + (b)) \subseteq \mathfrak{p} + (a)(b) \subseteq \mathfrak{p},$$

a contradiction. Thus, $\mathfrak{p}$ must be a prime ideal. QED

Next we need a generalization of the notion of $\sqrt{\mathfrak{A}}$. We adopt the following:

(10.6) Definition. For an ideal $\mathfrak{A}$ in a ring R, let

$$\begin{aligned}\sqrt{\mathfrak{A}} &:= \{s \in R:\ \text{every } m\text{-system containing } s \text{ meets } \mathfrak{A}\} \\ &\subseteq \{s \in R:\ s^n \in \mathfrak{A} \text{ for some } n \geq 1\}.\end{aligned}$$

In the special case when R is a *commutative* ring, one can check that the inclusion "$\subseteq$" above is actually an equality. For, assume that some $s^n \in \mathfrak{A}$. Let S be any m-system containing s. From the definition of an m-system, there exists an $r \in R$ with $s^n r \in S$. But then S meets $\mathfrak{A}$ at $s^n r$, so $s \in \sqrt{\mathfrak{A}}$. From this, we see easily that, in the commutative case, $\sqrt{\mathfrak{A}}$ is an ideal. In the general case, whether $\sqrt{\mathfrak{A}}$ is an ideal or not is certainly not clear from the definition (10.6). We shall now prove the following result which, in particular, settles this question—affirmatively.

(10.7) Theorem. *For any ring R and any ideal $\mathfrak{A} \subseteq R$, $\sqrt{\mathfrak{A}}$ equals the intersection of all the prime ideals containing $\mathfrak{A}$. In particular, $\sqrt{\mathfrak{A}}$ is an ideal in R.*

Proof. We first prove the inclusion "$\subseteq$". Let $s \in \sqrt{\mathfrak{A}}$ and $\mathfrak{p}$ be any prime ideal $\supseteq \mathfrak{A}$. Consider the m-system $R \backslash \mathfrak{p}$. This m-system cannot contain s, for otherwise it meets $\mathfrak{A}$ and hence also $\mathfrak{p}$. Therefore, we have $s \in \mathfrak{p}$. Conversely, assume $s \notin \sqrt{\mathfrak{A}}$. Then, by definition, there exists an m-system S containing s which is disjoint from $\mathfrak{A}$. By Zorn's Lemma, there exists an ideal $\mathfrak{p} \supseteq \mathfrak{A}$ which is maximal with respect to being disjoint from S. By (10.5), $\mathfrak{p}$ is a prime ideal, and we have $s \notin \mathfrak{p}$, as desired. QED

We shall now define the notion of a semiprime ideal. It will be shown a little bit later that this is the correct generalization of the notion of a radical ideal in the commutative case.

(10.8) Definition. An ideal $\mathfrak{C}$ in a ring R is said to be a *semiprime* ideal if, for any ideal $\mathfrak{A}$ of R, $\mathfrak{A}^2 \subseteq \mathfrak{C}$ implies that $\mathfrak{A} \subseteq \mathfrak{C}$. (For instance, a prime ideal is always semiprime.)

We have the following result in parallel with (10.2).

(10.9) Proposition. *For any ideal $\mathfrak{C}$, the following statements are equivalent:*

(1) *$\mathfrak{C}$ is semiprime.*

(2) *For $a \in R$, $(a)^2 \subseteq \mathfrak{C}$ implies that $a \in \mathfrak{C}$.*

(3) *For $a \in R$, $aRa \subseteq \mathfrak{C}$ implies that $a \in \mathfrak{C}$.*

(4) *For any left ideal $\mathfrak{A}$ in R, $\mathfrak{A}^2 \subseteq \mathfrak{C}$ implies that $\mathfrak{A} \subseteq \mathfrak{C}$.*

(4)′ *For any right ideal $\mathfrak{A}$ in R, $\mathfrak{A}^2 \subseteq \mathfrak{C}$ implies that $\mathfrak{A} \subseteq \mathfrak{C}$.*

The proof of this proposition is similar to that for (10.2). It is, therefore, left as an exercise for the reader.

To parody (10.3), we define a set $S \subseteq R$ to be an *n-system* if, for any $a \in S$, there exists an $r \in R$ such that $ara \in S$. Then, it follows from (10.9) that an ideal $\mathfrak{C} \subseteq R$ is semiprime iff $R \backslash \mathfrak{C}$ is an n-system.

Next, we need the following crucial lemma relating m-systems and n-systems.

(10.10) Lemma. *Let N be an n-system in a ring R and let $a \in N$. Then there exists an m-system $M \subseteq N$ such that $a \in M$.*

Proof. We define $M = \{a_1, a_2, a_3, \ldots\}$ inductively as follows: $a_1 = a$, $a_2 = a_1 r_1 a_1 \in N$ (for some r_1), $a_3 = a_2 r_2 a_2 \in N$ (for some r_2), ..., etc. To show that M is an m-system, we must show that, for any i, j, $a_i R a_j$ contains an element of M. But if $i \le j$, $a_i R a_j$ contains $a_j R a_j$, which contains $a_{j+1} \in M$, and if $i \ge j$, $a_i R a_j$ contains $a_i R a_i$, which contains $a_{i+1} \in M$. QED

(10.11) Theorem. *For any ideal $\mathfrak{C} \subseteq R$, the following are equivalent:*

(1) *$\mathfrak{C}$ is a semiprime ideal.*

(2) *$\mathfrak{C}$ is an intersection of prime ideals.*

(3) $\mathfrak{C} = \sqrt{\mathfrak{C}}$.

(From (1) $\Leftrightarrow$ (3), we see that, in the commutative setting, semiprime ideals are precisely the radical ideals.)

Proof of (10.11). (3) $\Rightarrow$ (2) is clear since, by (10.7), $\sqrt{\mathfrak{C}}$ is an intersection of prime ideals. (2) $\Rightarrow$ (1) is also clear: in fact, from Definition (10.8), it is obvious that the intersection of any family of semiprime (or prime) ideals is always semiprime. It remains to prove (1) $\Rightarrow$ (3). For a semiprime ideal $\mathfrak{C}$, we must show that $\sqrt{\mathfrak{C}} \subseteq \mathfrak{C}$. Let $a \notin \mathfrak{C}$. Then $N := R \backslash \mathfrak{C}$ is an n-system containing a. By the Lemma above, there exists an m-system $M \subseteq N$ such that $a \in M$. But then M is disjoint from $\mathfrak{C}$, so, from Definition (10.6), $a \notin \sqrt{\mathfrak{C}}$. QED

(10.12) Corollary. *For any ideal* $\mathfrak{C} \subseteq R$, $\sqrt{\mathfrak{C}}$ *is the smallest semiprime ideal in* R *which contains* $\mathfrak{C}$.

In the special case when $\mathfrak{C} = 0$, the inclusion relation observed in Definition (10.6) shows that $\sqrt{(0)}$ is always a nil ideal. This leads us to a new notion of radical:

(10.13) Definition. For any ring R, we define $Nil_* \, R := \sqrt{(0)}$. This is called (*Baer's*) *lower nilradical* or the *Baer–McCoy radical* of R. It is the smallest semiprime ideal in R, and is equal to the intersection of all the prime ideals in R. Because of the latter, $Nil_* \, R$ is also called the *prime radical* of R in the literature. Since $Nil_* \, R$ is nil, we have by (4.11):

(10.14) $$Nil_* \, R \subseteq rad \, R.$$

(10.15) Definition. *A ring* R *is called a prime* (*resp., semiprime*) *ring if* (0) *is a prime* (*resp., semiprime*) *ideal.*

We make the following immediate observations. (a) For an ideal $\mathfrak{U} \subseteq R$, $R/\mathfrak{U}$ is prime (resp., semiprime) iff $\mathfrak{U}$ is a prime (resp., semiprime) ideal. (b) As we have stated in the Introduction, in the category of commutative rings, prime rings are the integral domains, and semiprime rings are the reduced rings. And, of course, for any commutative ring R, $Nil_* \, R$ is just $Nil \, R$, the ideal of all nilpotent elements in R.

(10.16) Proposition. *For any ring* R, *the following are equivalent*:

(1) *R is a semiprime ring.*

(2) $Nil_* \, R = 0$.

(3) *R has no nonzero nilpotent ideal.*

(4) *R has no nonzero nilpotent left ideal.*

Proof. (1) $\Longleftrightarrow$ (2) is clear from Definition (10.15). Next we shall prove

$$(4) \Rightarrow (3) \Rightarrow (1) \Rightarrow (4).$$

The first two implications are also clear. For (1) $\Rightarrow$ (4), let R be a semiprime ring and let $\mathfrak{U}$ be a nilpotent left ideal. Choose n (≥ 1) minimal such that $\mathfrak{U}^n = 0$. If $n > 1$, then $(\mathfrak{U}^{n-1})^2 = \mathfrak{U}^{2n-2} \subseteq \mathfrak{U}^n = 0$ implies that $\mathfrak{U}^{n-1} = 0$ (cf. (10.9)), contradicting the minimality of n. Thus $n = 1$ and $\mathfrak{U} = 0$. QED

The following offers a list of examples (and nonexamples) of prime and semiprime rings.

(10.17) Examples.

(a) Any domain is a prime ring.

(b) Any reduced ring is a semiprime ring.

(c) Any simple ring R is a prime ring (see the paragraph after (10.2)).

(d) For any ring R, the quotient $R/Nil_* R$ is a semiprime ring naturally associated with R. If f: $R \to S$ is a surjective homomorphism of rings, it is easy to check that $f(Nil_* R) \subseteq Nil_* S$, so f induces a surjection of semiprime rings $R/Nil_* R \to S/Nil_* S$.

(e) From (10.14), we see that $rad\ R = 0$ implies that $Nil_* R = 0$; i.e., *any semiprimitive* (= *J-semisimple*) *ring is semiprime*. In particular, semisimple rings and von Neumann regular rings are all semiprime.

(f) Any direct product of semiprime rings is semiprime. On the other hand, the direct product of two or more nonzero rings is never a prime ring.

The following Proposition and (10.20) below enable us to generate more examples of prime and semiprime rings.

(10.18) Proposition. *Let T be a set of variables which commute with one another as well as with elements of a ring R. Then the polynomial ring $A = R[T]$ is prime (resp., semiprime) iff R is prime (resp., semiprime). The same statement holds for the ring of Laurent polynomials $R[T, T^{-1}]$.*

Proof. We shall deal with the "prime" case and leave the (completely analogous) "semiprime" case to the reader. Suppose A is prime and $aRb = 0$ where $a, b \in R$. Then clearly $aAb = aR[T]b = 0$, and so $a = 0$ or $b = 0$. Conversely, suppose R is prime and $fAg = 0$, where $f, g \in R[T]$. There exists a finite set $T_0 \subseteq T$ such that $f, g \in R[T_0]$, and clearly $fR[T_0]g = 0$. Therefore, it is enough to deal with the case where T is finite. Using induction, we are finally reduced to the case $A = R[t]$ for a single variable t. In the notation above, let a, b be the leading coefficients of f and g. Then $fR[t]g = 0$ implies that $aRb = 0$, so either $a = 0$ or $b = 0$; i.e., either $f = 0$ or $g = 0$. The proof for the case of the ring of Laurent polynomials $R[T, T^{-1}]$ is similar. QED

The Proposition above leads directly to the complete determination of the prime radical of a polynomial ring $R[T]$. (It is worth noting that the arguments needed here are simpler than those needed for the determination of the Jacobson radical of $R[T]$. And also, the description obtained for $Nil_* \, R[T]$ is much more explicit.) The case of the Laurent polynomial ring $R[T, T^{-1}]$ can be handled as well.

(10.19) Theorem (Amitsur, McCoy). *For any ring R, we have $Nil_* \, R[T] = (Nil_* \, R)[T]$, and $Nil_*(R[T, T^{-1}]) = (Nil_* \, R)[T, T^{-1}]$.*

Proof. Let $I = Nil_* \, R$. Then R/I is semiprime and so, by (10.18), $R[T]/I[T] \cong (R/I)[T]$ is also semiprime. This means that $I[T]$ is a semiprime ideal in $R[T]$, so we have $I[T] \supseteq Nil_* \, R[T]$. To show the reverse inclusion, we need to show that $I[T] \subseteq \mathfrak{p}$ for any prime ideal $\mathfrak{p}$ of $R[T]$. Note that $\mathfrak{p} \cap R$ is a prime ideal in R. For, if $aRb \in \mathfrak{p} \cap R$ where $a, b \in R$, then $aR[T]b = (aRb)[T] \subseteq \mathfrak{p}$, and so we have either $a \in \mathfrak{p} \cap R$ or $b \in \mathfrak{p} \cap R$. Since $\mathfrak{p} \cap R$ is prime, we have $I \subseteq \mathfrak{p} \cap R \subseteq \mathfrak{p}$; therefore, $I[T] \subseteq \mathfrak{p}$, as desired. The same proof works for $R[T, T^{-1}]$ to give the second equation in the Theorem. QED

We also have similar results for matrix rings.

(10.20) Proposition. *A ring R is prime (resp., semiprime) iff $\mathbb{M}_n(R)$ is prime (resp., semiprime).*

(For instance, if R is any domain, then $\mathbb{M}_n(R)$ is a prime ring, although it is not a domain for $n > 1$.)

Proof. Again we shall handle only the "prime" case and leave out the "semiprime" case. Assume $R \neq 0$ is not prime. Then there are nonzero ideals A, B in R such that $A \cdot B = 0$. But then $\mathbb{M}_n(A) \cdot \mathbb{M}_n(B) = 0$, so $\mathbb{M}_n(R)$ is not prime. Conversely, if $\mathbb{M}_n(R) \neq 0$ is not prime, then it has nonzero ideals $\mathfrak{A}$, $\mathfrak{B}$ such that $\mathfrak{A} \cdot \mathfrak{B} = 0$. By (3.1), we have $\mathfrak{A} = \mathbb{M}_n(A)$ and $\mathfrak{B} = \mathbb{M}_n(B)$, where A, B are (nonzero) ideals in R. But then $\mathfrak{A} \cdot \mathfrak{B} = 0$ implies that $A \cdot B = 0$, so R is not prime. QED

(10.21) Theorem. *For any ring R, we have $Nil_* \, \mathbb{M}_n(R) = \mathbb{M}_n(Nil_* \, R)$.*

The idea of the proof is the same as that for (10.19). The details will be left to the reader (as Exercise 22).

Let us now consider the case of group rings kG. While the problem of finding a criterion for kG to be semiprimitive is unsolved, the problem of finding criteria for kG to be *prime* or *semiprime* has been completely solved by I.G. Connell and D.S. Passman, respectively. This will be presented in (A) and (B) below.

(A) Connell's Theorem. *Let k be a ring and G be a group. Then the group ring $R = kG$ is prime iff k is prime and G has no finite normal subgroup $\neq \{1\}$.*

Proof. First suppose R is prime. If A, B are ideals in k such that $A \cdot B = 0$, then $(AR) \cdot (BR) = 0$ too, and we have $AR = 0$ or $BR = 0$, so $A = 0$ or $B = 0$. This shows that k is prime. If H is a finite normal subgroup of G, then the left ideal $\mathfrak{A}$ of R generated by all $h - 1$ $(h \in H)$ is an ideal, and it is annihilated by the (left) ideal $\mathfrak{B}$ generated by $\sum_{h \in H} h$. Since R is prime, $\mathfrak{A}$ must be zero, so $H = \{1\}$. For the converse, assume that k is prime, and that G has no finite normal subgroup $\neq \{1\}$. Consider the characteristic subgroup $\Delta = \Delta(G)$ of G defined in (6.23). This group is torsion-free, for by Dietzmann's Lemma (Exercise 6.15) any torsion element in Δ and all of its G-conjugates would generate a finite normal subgroup of G. Since Δ is also an f.c. group, (6.24) implies that Δ is abelian. Therefore, any finitely generated subgroup $\Delta_0 \subseteq \Delta$ is free abelian, and by (10.18), $k\Delta_0$ is prime. An easy direct limit argument now shows that $k\Delta$ is prime. To show that kG is prime, suppose $\gamma\, kG\, \gamma' = 0$, where $\gamma, \gamma' \in kG$. Let $\pi : kG \to k\Delta$ be the $(k\Delta, k\Delta)$-bimodule homomorphism in (6.26) and (6.27). By (6.28), we have $\pi(\gamma\, kG)\pi(\gamma'\, kG) = 0$. Since $\pi(\gamma\, kG)$, $\pi(\gamma'\, kG)$ are right ideals in the prime ring $k\Delta$, one of them must be zero. It then follows from (6.27) (b) that one of $\gamma\, kG$, $\gamma'\, kG$ is zero; that is, either $\gamma = 0$ or $\gamma' = 0$. QED

(B) Passman's Theorem. *Let k be a ring and G be a group. Then the group ring $R = kG$ is semiprime iff k is semiprime and the order of any finite normal subgroup of G is not a 0-divisor in k.*

Proof. First suppose R is semiprime. Then it follows as in the above proof that k is semiprime. Let $H = \{h_1, \ldots, h_n\}$ be a finite normal subgroup of G. Then, for the ideals $\mathfrak{A}$, $\mathfrak{B}$ defined in the first part of the above proof, we have

$$(\mathfrak{A} \cap \mathfrak{B})^2 \subseteq \mathfrak{A}\mathfrak{B} = 0.$$

Since R is semiprime, $\mathfrak{A} \cap \mathfrak{B} = 0$. If $a \in k$ is such that $na = 0$, then

$$a(h_1 + \cdots + h_n) = a(h_1 - 1) + \cdots + a(h_n - 1) \in \mathfrak{A} \cap \mathfrak{B} = 0,$$

so we must have $a = 0$. This shows that $n = |H|$ is not a 0-divisor in k. *For the converse, we shall limit ourselves to the case when k is a field.* First assume *char* $k = 0$. Here we must show that R is semiprime without any condition on G. But if I is a nilpotent ideal of kG, then $I \cdot \bar{k}$ is a nilpotent ideal of $\bar{k}G$, where $\bar{k}$ is the algebraic closure of k. By (6.11)(b), $I \cdot \bar{k} = 0$, so $I = 0$. Next assume *char* $k = p$, and that G has no finite normal subgroup of order divisible by p. Then $\Delta = \Delta(G)$ is a p'-group, for, by Dietzmann's Lemma, any element in Δ of order p and its G-conjugates would generate a finite normal subgroup in G, of order divisible by p. To show that kG is semiprime, let $\gamma \in kG$ be such that $\gamma\, kG\, \gamma = 0$. Then (6.28) implies that $\pi(\gamma\, kG)\pi(\gamma\, kG) = 0$, where $\pi = \pi_\Delta$ is as defined in (6.26). Since Δ is a p'-group, (6.13) implies that the right ideal $\pi(\gamma\, kG)$ of $k\Delta$ is zero. By (6.27)(b), we conclude that $\gamma\, kG = 0$; hence $\gamma = 0$. QED

Next we shall try to relate semiprime rings to semisimple rings. We start with the following basic lemma on minimal left ideals in any ring.

(10.22) Brauer's Lemma. *Let $\mathfrak{A}$ be a minimal left ideal in a ring R. Then we have either $\mathfrak{A}^2 = 0$, or $\mathfrak{A} = Re$ for some idempotent $e \in \mathfrak{A}$.*

Proof. Assume $\mathfrak{A}^2 \neq 0$. Then $\mathfrak{A} \cdot a \neq 0$ for some $a \in \mathfrak{A}$, and therefore $\mathfrak{A} \cdot a = \mathfrak{A}$. Choose $e \in \mathfrak{A}$ such that $a = ea$. The set

$$I = \{x \in \mathfrak{A}:\ xa = 0\}$$

is a left ideal $\subsetneq \mathfrak{A}$, since $e \notin I$. Therefore $I = 0$. On the other hand, we have $e^2 - e \in \mathfrak{A}$ and $(e^2 - e)a = 0$; hence $e^2 - e = 0$. Since ${}_R\mathfrak{A}$ is minimal, we conclude that $\mathfrak{A} = Re$. QED

(10.23) Corollary. *If $\mathfrak{A}$ is a minimal left ideal in a semiprime ring R, then $\mathfrak{A} = Re$ for some idempotent $e \in \mathfrak{A}$.*

Proof. We cannot have $\mathfrak{A}^2 = 0$ by (10.16). QED

We can now prove the following analogue of our earlier result (4.14), replacing "J-semisimple" there by "semiprime."

(10.24) Theorem. *For any ring R, the following three statements are equivalent:*

(1) *R is semisimple.*

(2) *R is semiprime and left artinian.*

(3) *R is semiprime and satisfies DCC on principal left ideals.*

Proof. It suffices to prove (3) $\Rightarrow$ (1). We claim that the same arguments used in the proof of (4.14) apply here. In fact, in that proof, the only place we used the J-semisimple hypothesis on R was in proving the statement (b) there. But if R is semiprime, (b) does hold according to (10.23). Therefore, the proof for (4.14) carries over completely. QED

In the balance of this section, we shall briefly discuss two other kinds of radicals which can be associated with a ring R. These are: the *upper nil radical Nil* R*, and the *Levitzki radical L-rad R*. To define the former, we first make the following easy observation.

(10.25) Lemma. *Let $\mathfrak{A}$ be a nil left ideal, and $\mathfrak{B}$ be a nil ideal in a ring R. Then $\mathfrak{A} + \mathfrak{B}$ is a nil left ideal.*

Proof. Let $c \in \mathfrak{A} + \mathfrak{B}$. Working first in $R/\mathfrak{B}$ and lifting to R, we see that $c^n \in \mathfrak{B}$ for some $n \geq 1$. Since $\mathfrak{B}$ is nil, we have $(c^n)^m = 0$ for some $m \geq 1$. Hence $\mathfrak{A} + \mathfrak{B}$ is nil. QED

The Lemma implies, in particular, that the sum of any family of nil ideals in a ring is always nil. This leads to the following definition.

(10.26) Definition. Let $Nil^* R$ be the sum of all nil ideals in R. We call $Nil^* R$ the *upper nilradical* of R. This is the largest nil ideal of R; hence

$$Nil^* R = \{a \in R: (a) \text{ is nil}\}.$$

(10.27) Proposition. *For any ring R, we have $Nil_* R \subseteq Nil^* R \subseteq rad R$. If R is commutative, then $Nil_* R = Nil^* R = Nil R$. If R is left artinian, then $Nil_* R = Nil^* R = rad R$.*

Proof. The first inclusion follows since $Nil_* R$ is a nil ideal; the second inclusion follows (from (4.11)) since $Nil^* R$ is a nil ideal. If R is commutative, $Nil_* R$ and $Nil^* R$ both coincide with $Nil(R)$; hence they are equal. Finally, let R be left artinian. By (4.12), $rad R$ is a nilpotent ideal. Since (0) is the only nilpotent ideal in $R/Nil_* R$, it follows that $rad R \subseteq Nil_* R$. Hence, all three radicals are equal. QED

A large class of rings with a zero upper nilradical is given by group rings. Recall from (6.13) that, if k is any reduced commutative ring of prime characteristic $p > 0$, then $Nil^* kG = 0$ for any p'-group G. And in characteristic zero, if k is a ring with involution * such that

$$\sum \alpha_i^* \alpha_i = 0 \implies \text{all } \alpha_i = 0$$

(for instance k as in (6.11)(a), (b)), then $Nil^* kG = 0$ for any group G, by (6.11). In fact, using the argument in (6.11), we see the following: *If R is any ring with an involution * such that $\alpha^*\alpha = 0 \Rightarrow \alpha = 0$ in R, then R has no nonzero nil left ideals. In particular,* $Nil^* R = 0$. This applies, for instance, to any ring R of bounded operators on a complex Hilbert space which is closed with respect to $T \mapsto T^*$, where T^* denotes the *adjoint* of T. (Here, if $T^*T = 0$, then

$$\|Tv\|^2 = \langle Tv, Tv \rangle = \langle v, T^*Tv \rangle = 0 \quad \text{for all vectors } v$$

implies that $T = 0$.)

Quite generally, for any ring R and any element $x \in R$, the principal left ideal $R \cdot x$ is nil iff the principal right ideal $x \cdot R$ is nil. Therefore, R has no nonzero nil left ideal iff it has no nonzero nil right ideal. In this case, we may refer to R as a ring without nonzero nil one-sided ideals. In this connection, there is a famous unsolved problem: *If R has no nonzero nil ideals, does it* *follow that R has no nonzero nil one-sided ideals*? The truth of this was conjectured many years ago by G. Köthe.

(10.28) Köthe's Conjecture. *If* $Nil^* R = 0$, *then* R *has no nonzero nil one-sided ideals.*

It is easy to see that the following are two equivalent formulations of the same conjecture:

(10.28a) *Every nil left or right ideal of a ring* R *is contained in* $Nil^* R$.

(10.28b) *The sum of two nil left (resp., right) ideals of* R *is also nil.*

The equivalence of (10.28a) to (10.28) is clear from (10.25), and

$$(10.28a) \Longrightarrow (10.28b)$$

follows from the fact that $Nil^* R$ is a nil ideal. To see that

$$(10.28b) \Longrightarrow (10.28a),$$

note that in general, $R \cdot x$ is nil implies that $R \cdot xr$ is also nil, for any $r \in R$. From this, it follows that the sum of all nil left ideals in any ring R coincides with the sum of all nil right ideals. If (10.28b) holds, this common sum is a nil ideal, and therefore contained in $Nil^* R$. Some further interesting equivalent formulations of Köthe's Conjecture can be found in Exercise 25 of this section.

In spite of the many great advances made in ring theory in recent times, Köthe's Conjecture has remained unsolved in general. For several special classes of rings, the Conjecture has been shown to be true. One such class is, of course, the class of right artinian rings (for which $Nil^* R = rad\ R$). It turns out that the Conjecture is true even for the larger class of right noetherian rings. To prove this, we follow a remarkable argument of Y. Utumi.

(10.29) Lemma. *Assume that* R *satisfies the ACC for right annihilators* $ann_r(a) = \{x \in R:\ ax = 0\}$, *where* $a \in R$. *Then:*

(1) *Any nil one-sided ideal* $\mathfrak{U}$ *is contained in* $Nil_* R$;

(2) *Any nonzero nil right (resp., left) ideal* $\mathfrak{B}$ *contains a nonzero nilpotent right (resp., left) ideal.*

In particular, if R *is also semiprime, then every nil one-sided ideal is zero.*

Proof. (1) Assume $\mathfrak{U}$ is a nil right ideal $\not\subseteq Nil_* R$. Among the elements in $\mathfrak{U} \backslash Nil_* R$, choose a such that $ann_r(a)$ is maximal. Since $Nil_* R$ is semiprime, there exists $x \in R$ such that $axa \notin Nil_* R$. Now $ax \in \mathfrak{U}$ is nilpotent, so there exists an integer $k > 1$ such that $(ax)^k = 0 \neq (ax)^{k-1}$. Then

$$ann_r(axa) \supsetneq ann_r(a)$$

since $x(ax)^{k-2}$ belongs to $ann_r(axa)$ but not to $ann_r(a)$. Since $axa \in \mathfrak{U}\backslash Nil_* R$, this contradicts the choice of a. If $\mathfrak{U}$ is a nil *left* ideal instead, then for any $a' \in \mathfrak{U}$, $a'R$ is a nil *right* ideal, so $a'R \subseteq Nil_* R$ by the above. Therefore, we also have $\mathfrak{U} \subseteq Nil_* R$.

(2) Among the nonzero elements of $\mathfrak{B}$, choose b such that $ann_r(b)$ is maximal. It suffices to show here that $bxb = 0$ for all $x \in R$, for then we'll have $(bR)^2 = (Rb)^2 = 0$. If $\mathfrak{B}$ is a right ideal, we can repeat the argument in (1) to get $bxb = 0$. Now assume $\mathfrak{B}$ is a left ideal and $bxb \neq 0$. Then $xb \in \mathfrak{B}$ is nilpotent, so there exists an integer $k > 1$ such that $(xb)^k = 0 \neq (xb)^{k-1}$. But then $xb \in ann_r((xb)^{k-1})$ and $xb \notin ann_r(b)$, so we have

$$ann_r(b) \subsetneq ann_r((xb)^{k-1}),$$

a contradiction. QED

The last part of the lemma above leads to the quickest known proof of an earlier theorem of J. Levitzki.

(10.30) Levitzki's Theorem. *Let R be a right noetherian ring. Then every nil one-sided ideal $\mathfrak{U}$ of R is nilpotent. We have $Nil_* R = Nil^* R$, and this is the largest nilpotent right (resp., left) ideal of R.*

Proof. In view of (1) in the Lemma above, it suffices to show that $Nil_* R$ is nilpotent. Since R is right noetherian, there exists a maximal nilpotent ideal N in R. Then R/N has no nonzero nilpotent ideals, so R/N is semiprime. This shows that $N \supseteq Nil_* R$, and hence $Nil_* R = N$ is nilpotent. QED

For a little bit of history on Levitzki's Theorem, the following remark from p. 51 of N.J. Divinsky's book "*Rings and Radicals*" is of interest: "Fate seems to have had a hand in suppressing this result. Levitzki proved it in August, 1939, but because of the war and other peculiar circumstances it did not appear in print until 1950, and then in a relatively obscure journal with a minor mistake in the proof! Rumors circulated that this theorem was true but it was not noticed until Jacobson put it into his 1956 book. However, the minor flaw remained!"

Clearly, Levitzki's Theorem implies the truth of Köthe's Conjecture for right noetherian rings. There are other classes of rings for which Köthe's Conjecture has been shown to be true. For instance, let R be an algebra over a field k such that either R is algebraic over k, or $dim_k R < |k|$ (as cardinal numbers); then by (4.19) and (4.20), $Nil^* R = rad R$. In these cases, it is clear that any nil one-sided ideal is contained in $Nil^* R$. It is also known that an important class of rings called PI-algebras (over an arbitrary field k) satisfy

Köthe's Conjecture. For these algebras R, it turns out that $Nil_* R = Nil^* R$; if R is finitely generated over k, A. Braun has shown even that $Nil_* R = Nil^* R = rad R$, and that this is a *nilpotent* ideal in R.

We shall now conclude this section with some brief remarks on the Levitzki radical of a ring R. This radical is defined using the notion of "locally nilpotent" sets. We say that a set $S \subseteq R$ is *locally nilpotent* if, for any finite subset $\{s_1, \ldots, s_n\} \subseteq S$, there exists an integer N such that any product of N elements from $\{s_1, \ldots, s_n\}$ is zero. In other words, S is locally nilpotent if any subring without identity generated by a finite number of elements in S is nilpotent. As a rule, we shall use this notion for the one-sided ideals of R. Note that, if $\mathfrak{U} \subseteq R$ is a one-sided ideal, then

$$\mathfrak{U} \text{ is nilpotent} \Longrightarrow \mathfrak{U} \text{ is locally nilpotent} \Longrightarrow \mathfrak{U} \text{ is nil.}$$

One major difference between "nil" and "locally nilpotent" is the following. Let $\mathfrak{U}, \mathfrak{B}$ be one-sided nil ideals in R. If Köthe's Conjecture were true, then $\mathfrak{U}, \mathfrak{B}$ would be in $Nil^* R$ and hence $R\mathfrak{U}R$, $R\mathfrak{B}R$ and $\mathfrak{U} + \mathfrak{B}$ would all be nil. However, since Köthe's Conjecture has not been proved, these conclusions, in general, remain in doubt. The inaccessibility of these conclusions presents a major obstacle in working with nil one-sided ideals. For locally nilpotent one-sided ideals, the situation is far more pleasant, as we can see from the following result.

(10.31) Proposition. *Let $\mathfrak{U}$, $\mathfrak{B}$ be locally nilpotent one-sided ideals in R. Then $R\mathfrak{U}R$, $R\mathfrak{B}R$ and $\mathfrak{U} + \mathfrak{B}$ are locally nilpotent.*

Proof. Assume, say $\mathfrak{U}$ is a left ideal. To show that $\mathfrak{U} \cdot R$ is locally nilpotent, take a finite set $\{b_i\} \subseteq \mathfrak{U} \cdot R$ where

$$b_i = \sum_j a_{ij} r_{ij} \quad (a_{ij} \in \mathfrak{U},\ r_{ij} \in R).$$

Consider the finite set $S = \{r_{pq} a_{ij}\} \subseteq \mathfrak{U}$. Since $\mathfrak{U}$ is locally nilpotent, there exists an integer N such that the product of any N elements from S is zero. Now clearly the product of any $N + 1$ of the b_i's is zero, so $\mathfrak{U} \cdot R$ is locally nilpotent. A similar argument can be applied to show that $\mathfrak{U} + \mathfrak{B}$ is locally nilpotent. QED

Since local nilpotence is a finitary property, it follows easily from (10.31) that the sum of all locally nilpotent ideals in a ring R is locally nilpotent. We denote this sum by *L-rad* R; this is called the *Levitzki radical* of R. It is the largest locally nilpotent ideal of R, and contains every locally nilpotent one-sided ideal of R. Moreover, we have

$$\textbf{(10.32)} \qquad Nil_* R \subseteq L\text{-}rad\ R \subseteq Nil^* R \subseteq rad\ R.$$

The second inclusion is clear as *L-rad* R is a nil ideal. To see the first inclusion, it suffices to show that *L-rad* R is a semiprime ideal (since any semiprime ideal contains $Nil_* R$). Let $\mathfrak{U}$ be an ideal such that

$$\mathfrak{U}^2 \subseteq L\text{-}rad\ R.$$

We see easily that, since $\mathfrak{U}^2$ is locally nilpotent, so is $\mathfrak{U}$, and therefore $\mathfrak{U} \subseteq L\text{-}rad\ R$. This shows that $L\text{-}rad\ R$ is semiprime, as desired.

In general, the three inclusions in (10.32) are strict inclusions. First of all, it is easy to give examples of rings R for which $Nil^* R \subsetneq rad\ R$. For instance, if R is any commutative domain with a unique maximal ideal $\mathfrak{m} \neq 0$, then $rad\ R = \mathfrak{m}$, but the three other radicals are all zero. In the following, we shall give an example of a ring R for which $L\text{-}rad\ R$, and hence $Nil^* R$, are bigger than $Nil_* R$. This example is taken from a paper of J. Ram.

Let R be the ring $A[x;\sigma]$ of twisted polynomials over a commutative ring A that will be specified later. The polynomials in R are written as $\sum a_i x^i$, and are multiplied according to the twist equation $xa = \sigma(a)x$, where σ is a fixed automorphism of the ring A. We shall now define A. Fixing a field k, we take A to be the commutative k-algebra with generators t_i $(i \in \mathbb{Z})$ and relations $t_{i_1} t_{i_2} t_{i_3} = 0$, where $i_1 < i_2 < i_3$ range over all increasing arithmetic progressions of length 3 in $\mathbb{Z}$. As a k-vector space, a basis of A is given by monomials

$$\{t_{i_1}^{n_1} \cdots t_{i_r}^{n_r}\}$$

where $i_1 < \cdots < i_r$ is a sequence in $\mathbb{Z}$ containing no arithmetic progression of length 3. We shall take σ to be the k-automorphism of A (well-) defined by $\sigma(t_i) = t_{i+1}$ (for any $i \in \mathbb{Z}$). We shall show that

(1) $R = A[x;\sigma]$ *is a prime ring*, but

(2) $L\text{-}rad\ R \neq 0$.

To show (1), assume, on the contrary, that there exist nonzero polynomials f, g with $fA[x;\sigma]g = 0$. Let $a, b \in A$ be the leading coefficients of f, g. Then $fx^j g = 0$ implies that $a\sigma^{n+j}(b) = 0$ for any $j \geq 0$, where n is the degree of f. This is, however, impossible, since, by choosing j to be very large, we can guarantee that the monomials in $\sigma^{n+j}(b)$ are disjoint from those of a and that, in the expansion of $a \cdot \sigma^{n+j}(b)$, the products of the monomials in a with those in $\sigma^{n+j}(b)$ do not contain three variables whose suffixes form an arithmetic progression. This proves (1).

For (2), we shall show that *the right ideal* $t_0 x \cdot R$ *in* R *is locally nilpotent*. If so, then $t_0 x \cdot R \subseteq L\text{-}rad\ R$, and therefore $L\text{-}rad\ R \neq 0$. Consider n elements

$$t_0 x \cdot f_i \quad (1 \leq i \leq n)$$

in $t_0 x \cdot R$. We must find an integer N such that the product of any N elements among $\{t_0 x \cdot f_i\}$ is zero. Without loss of generality, we may assume that

$$f_i = a_i x^{r(i)-1}, \quad \text{where } r(i) \geq 1, \text{ and } 1 \leq i \leq n.$$

Then $t_0 x \cdot f_i = t_0 \sigma(a_i) x^{r(i)}$. For any N, a product $(t_0 x f_{i_1}) \cdots (t_0 x f_{i_{N+1}})$ has the form

$$\textbf{(10.33)}\quad a\cdot t_0\sigma^{r(i_1)}(t_0)\sigma^{r(i_1)+r(i_2)}(t_0)\cdots\sigma^{r(i_1)+\cdots+r(i_N)}(t_0)x^{r(i_1)+\cdots+r(i_{N+1})}$$
$$= at_0 t_{r(i_1)} t_{r(i_1)+r(i_2)}\cdots t_{r(i_1)+\cdots+r(i_N)} x^{r(i_1)+\cdots+r(i_{N+1})},$$

where $a \in A$. Consider the increasing sequence

$$\textbf{(10.34)}\qquad 0 < r(i_1) < r(i_1)+r(i_2) < \cdots < r(i_1)+\cdots+r(i_N).$$

Note that the gap between any two consecutive integers in this sequence is bounded by $m := \max\{r(1),\ldots,r(n)\}$. Using a classical theorem of van der Waerden, it is known that there exists an integer N, depending only on m, such that any N-term sequence as in (10.34) contains an arithmetic progression of length 3. (See the article "*Arithmetic progressions contained in sequences with bounded gaps*" by M.B. Nathanson in *Canad. Math. Bull.* **23**(1980), 491–493.) For this N, the product in (10.33) is zero, and therefore $(t_0xf_{i_1})\cdots(t_0xf_{i_{N+1}}) = 0$ for any choices of $i_1,\ldots,i_{N+1}$ from $\{1,\ldots,n\}$. This completes the proof that $t_0x\cdot R \subseteq$ *L-rad R*.

Exercises for §10

Ex. 10.0. Show that a nonzero central element of a prime ring R is not a zero-divisor in R. In particular, the center $Z(R)$ is a (commutative) domain, and char R is either 0 or a prime number.

Ex. 10.1. For any semiprime ring R, show that $Z(R)$ is reduced, and that char R is either 0 or a square-free integer.

Ex. 10.2. Let $\mathfrak{p} \subset R$ be a prime ideal, $\mathfrak{A}$ be a left ideal and $\mathfrak{B}$ be a right ideal. Does $\mathfrak{A}\mathfrak{B} \subseteq \mathfrak{p}$ imply that $\mathfrak{A} \subseteq \mathfrak{p}$ or $\mathfrak{B} \subseteq \mathfrak{p}$?

Ex. 10.3. Show that a ring R is a domain iff R is prime and reduced.

Ex. 10.4. Show that in a right artinian ring R, every prime ideal $\mathfrak{p}$ is maximal. (Equivalently, R is prime iff it is simple.)

Ex. 10.4*. For any given division ring k, list all the prime and semiprime ideals in the ring R of 3×3 upper triangular matrices over k.

Ex. 10.5. Show that the following conditions on a ring R are equivalent:
(1) All ideals $\neq R$ are prime.
(2) (a) The ideals of R are linearly ordered by inclusion, and
(b) All ideals $I \subseteq R$ are idempotent (i.e. $I^2 = I$).

Ex. 10.6. Let $R = End(V_k)$ where V is a vector space over a division ring k. Show that R satisfies the properties (1), (2) of the exercise above. In particular, every nonzero homomorphic image of R is a prime ring. (**Hint.** The ideal structure of R was determined in Exercise 3.16.)

Ex. 10.7. For any integer $n \neq 0$, show that (1) $R = \begin{pmatrix} \mathbb{Z} & n\mathbb{Z} \\ \mathbb{Z} & \mathbb{Z} \end{pmatrix}$ is a prime ring, but $R' = \begin{pmatrix} \mathbb{Z} & n\mathbb{Z} \\ 0 & \mathbb{Z} \end{pmatrix}$ is not, and (2) R is *not* isomorphic to the prime ring $\mathbb{M}_2(\mathbb{Z})$.

Ex. 10.8. (a) Show that a ring R is semiprime iff, for any two ideals $\mathfrak{U}, \mathfrak{B}$ in R, $\mathfrak{U}\mathfrak{B} = 0$ implies that $\mathfrak{U} \cap \mathfrak{B} = 0$.
(b) Let $\mathfrak{U}, \mathfrak{B}$ be left (resp. right) ideals in a semiprime ring R. Show that $\mathfrak{U}\mathfrak{B} = 0$ iff $\mathfrak{B}\mathfrak{U} = 0$. If $\mathfrak{U}$ is an ideal, show that $ann_r(\mathfrak{U}) = ann_\ell(\mathfrak{U})$.

Ex. 10.8*. Show that, with respect to inclusion, the set of semiprime ideals in any ring forms a lattice having a smallest element and a largest element. Give an example to show, however, that the sum of two semiprime ideals need not be semiprime.

Ex. 10.9. Let $I \subseteq R$ be a right ideal containing no nonzero nilpotent right ideals of R. (For instance, I may be any right ideal in a semiprime ring.) Show that the following are equivalent: (1) I_R is an artinian module; (2) I_R is a finitely generated semisimple module. In this case, show that (3) $I = eR$ for an idempotent $e \in I$.

Ex. 10.10A. Let $N_1(R)$ be the sum of all nilpotent ideals in a ring R.
(1) Show that $N_1(R)$ is a nil subideal of $Nil_* R$ which contains all nilpotent one-sided ideals of R.
(2) If $N_1(R)$ is nilpotent, show that $N_1(R) = Nil_* R$.
(3) Show that the hypothesis and conclusion in (2) both apply if ideals in R satisfy DCC.

Ex. 10.10B. Keeping the notations of Exercise 10A, give an example of a (necessarily noncommutative) ring R in which $N_1(R) \subsetneq Nil_* R$.

Ex. 10.11 (Levitzki) For any ring R and any ordinal α, define $N_\alpha(R)$ as follows. For $\alpha = 1$, $N_1(R)$ is defined as in Exercise 10A. If α is the successor of an ordinal β, define

$$N_\alpha(R) = \{r \in R:\ r + N_\beta(R) \in N_1(R/N_\beta(R))\}.$$

If α is a limit ordinal, define

$$N_\alpha(R) = \bigcup_{\beta<\alpha} N_\beta(R).$$

Show that $Nil_* R = N_\alpha(R)$ for any ordinal α with $Card\ \alpha > Card\ R$.

Ex. 10.12. Let I be a left ideal in a ring R such that, for some integer $n \geq 2$, $a^n = 0$ for all $a \in I$. Show that $a^{n-1}Ra^{n-1} = 0$ for all $a \in I$.

Ex. 10.13. (Levitzki, Herstein) Let $I \neq 0$ be a left ideal in a ring R such that, for some integer n, $a^n = 0$ for all $a \in I$.
(1) Show that I contains a nonzero nilpotent left ideal, and R has a nonzero nilpotent ideal.
(2) Show that $I \subseteq Nil_* R$.

Ex. 10.14. (Krull, McCoy) Show that any prime ideal $\mathfrak{p}$ in a ring R contains a minimal prime ideal. Using this, show that the lower nilradical $Nil_*\, R$ is the intersection of all the minimal prime ideals of R.

Ex. 10.15. Show that if the ideals in R satisfy ACC (e.g. when R is left noetherian), then R has only finitely many minimal prime ideals.

Ex. 10.16. (McCoy) For any ideal $\mathfrak{U}$ in a ring R, show that $\sqrt{\mathfrak{U}}$ consists of $s \in R$ such that every n-system containing s meets $\mathfrak{U}$.

Ex. 10.17. (Levitzki) An element a of a ring R is called *strongly nilpotent* if every sequence $a_1, a_2, a_3, \ldots$ such that $a_1 = a$ and $a_{n+1} \in a_n R a_n$ $(\forall n)$ is eventually zero. Show that $Nil_*\, R$ is precisely the set of all strongly nilpotent elements of R.

Ex. 10.18A. (1) Let $R \subseteq S$ be rings. Show that $R \cap Nil_*(S) \subseteq Nil_*(R)$.
(2) If $R \subseteq Z(S)$, show that $R \cap Nil_*(S) = Nil_*(R)$.
(3) Let R, K be algebras over a commutative ring k such that R is k-projective and $K \supseteq k$. Show that $R \cap Nil_*(R \otimes_k K) = Nil_*(R)$.

Ex. 10.18B. Let R be a k-algebra where k is a field. Let K/k be a separable algebraic field extension.
(1) Show that R is semiprime iff $R^K = R \otimes_k K$ is semiprime.
(2) Show that $Nil_*(R^K) = (Nil_*(R))^K$.

Ex. 10.19. For a ring R, consider the following conditions:
(1) Every ideal of R is semiprime.
(2) Every ideal I of R is idempotent (i.e. $I^2 = I$).
(3) R is von Neumann regular.
Show that (3) $\Rightarrow$ (2) $\Leftrightarrow$ (1), and that (1) $\Rightarrow$ (3) if R is commutative. (It is of interest to compare this exercise with Ex. 5 above.)

Ex. 10.20. Let *Rad R* denote one of the two nilradicals, or the Jacobson radical, or the Levitzki radical of R. Show that *Rad R* is a semiprime ideal. For any ideal $I \subseteq Rad\ R$, show that $Rad(R/I) = (Rad\ R)/I$. Moreover, for any ideal $J \subseteq R$ such that $Rad(R/J) = 0$, show that $J \supseteq Rad\ R$.

Ex. 10.21. Let $R[T]$ be a polynomial ring over R, and let $N = R \cap rad\ R[T]$. Show that N is a semiprime ideal of R and that $L\text{-}rad\ R \subseteq N \subseteq Nil^*\, R$, where *L-rad R* denotes the Levitzki radical of R. (**Hint.** Show that $(L\text{-}rad\ R)[T]$ is a nil ideal in $R[T]$.)

Ex. 10.22. Supply the details for the proof of (10.21).

Ex. 10.23. Let I be a nil left ideal of a ring R.
(1) Show that the set of matrices in $\mathbb{M}_n(R)$ whose kth column consists of elements of I and whose other columns are zero is a nil left ideal of $\mathbb{M}_n(R)$.
(2) If $T_n(R)$ is the ring of $n \times n$ upper triangular matrices over R, show that $T_n(I)$ is a nil left ideal in $T_n(R)$.

Ex. 10.24. (Krempa-Amitsur) Let I be an ideal of a ring R such that, for all n, $\mathbb{M}_n(I)$ is a nil ideal in $\mathbb{M}_n(R)$. Show that $I[t] \subseteq rad\ R[t]$. (**Hint.** Reduce to showing that any polynomial $f(t) = 1 + a_1 t + \cdots + a_n t^n$ is a unit in $R[t]$ for any $a_1, \ldots, a_n \in I$. Let $g(t) = 1 + \sum_{i=1}^{\infty} b_i t^i$ be the inverse of f in $R[[t]]$, and let A be the companion matrix

$$\begin{pmatrix} 0 & & & -a_n \\ 1 & 0 & & -a_{n-1} \\ & \ddots & \ddots & \vdots \\ & & 1 & -a_1 \end{pmatrix}.$$

Show that $A^n \in \mathbb{M}_n(I)$ and hence A is nilpotent. Conclude from this that $b_i = 0$ for sufficiently large i.)

Ex. 10.25. Using (23) and (24), show that the following are equivalent:

(1) Köthe's Conjecture ("the sum of two nil left ideals in any ring is nil").
(1)′ The sum of two nil 1-sided ideals in any ring is nil.
(2) If I is a nil ideal in any ring R, then $\mathbb{M}_n(I)$ is nil for any n.
(2)′ If I is a nil ideal in any ring R, then $\mathbb{M}_2(I)$ is nil.
(3) $Nil^*(\mathbb{M}_n(R)) = \mathbb{M}_n(Nil^*(R))$ for any ring R and any n.
(4) If I is a nil ideal in any ring R, then $I[t] \subseteq rad\ R[t]$.
(5) $rad\ (R[t]) = (Nil^*\ R)[t]$ for any ring R.

(Note that, if true, (5) would give a much sharper form of Amitsur's Theorem in (5.10). But of course (5) may very well be false. As to (4), Amitsur had once conjectured that, *if $I \subseteq R$ is a nil ideal, then $I[t]$ is also nil.* This would certainly have implied the truth of (4) (and hence of Köthe's Conjecture!). However, A. Smoktunowicz [00] has recently produced a counterexample to Amitsur's conjecture. While this counterexample did not disprove Köthe's Conjecture, it certainly seemed to have lent credence to the (long-held) suspicion that the Conjecture is *false*.)

§11. Structure of Primitive Rings; the Density Theorem

In this section, we introduce a new class of rings called *left primitive* rings. The central result in the section is the Density Theorem of Jacobson and Chevalley. For the class of left artinian rings, this theorem gives another approach to the Wedderburn–Artin Theorem on the structure of (artinian) simple rings. For rings possibly without chain conditions, the Density Theorem sheds light on the structure of left primitive rings, characterizing them as "dense" rings of linear transformations on right vector spaces over division rings. This structure theorem on left primitive rings may therefore be viewed as a generalization of the Wedderburn–Artin Theorem on artinian simple rings.

To lead up to the definition of a left primitive ring, we first call attention to the following characterization of a semiprimitive (= J-semisimple) ring, which is an easy consequence of (4.1).

(11.1) Proposition. *A ring R is semiprimitive iff R has a faithful semisimple left module M.*

Proof. Suppose M exists. Since (by 4.1)) *rad* R acts as zero on all left simple R-modules, we have $(rad\ R) \cdot M = 0$. Then the faithfulness of M implies that $rad\ R = 0$, so R is semiprimitive. Conversely, assume $rad\ R = 0$. Let $\{M_i\}$ be a complete set of mutually nonisomorphic simple left R-modules. Then $M = \bigoplus_i M_i$ is semisimple, and

$$ann(M) = \bigcap_i ann(M_i) = rad\ R$$

by (4.1). Since $rad\ R = 0$, M is a faithful R-module. QED

Motivated by the Proposition above, we now give the definition of a left primitive ring.

(11.2) Definition. A ring R is said to be *left* (resp., *right*) *primitive* if R has a faithful simple left (resp., right) module. (Note that such R is necessarily $\neq 0$.)

While the notion of semiprimitivity is left-right symmetric, the notion of primitivity is not. An example of a left primitive ring which is not right primitive was constructed by G. Bergman in 1965. Other such examples were found later by A.V. Jategaonkar.

Before we study left primitive rings in more detail, it is useful to extend the notion of left primitivity from rings to ideals.

(11.3) Definition. An ideal $\mathfrak{A} \subseteq R$ is said to be *left* (resp., *right*) *primitive* if the quotient ring $R/\mathfrak{A}$ is left (resp., right) primitive.

We have the following easy characterization of left primitive ideals.

(11.4) Proposition. *An ideal $\mathfrak{A}$ in R is left primitive iff $\mathfrak{A}$ is the annihilator of a simple left R-module.*

Proof. First suppose $\mathfrak{A} = ann\ M$, where M is a simple left R-module. Then M may be viewed as a simple $R/\mathfrak{A}$-module, and as such, it is faithful. Therefore $R/\mathfrak{A}$ is a left primitive ring. Conversely, suppose $R/\mathfrak{A}$ is a left primitive ring, and let M be a faithful simple left $R/\mathfrak{A}$-module. Then, viewed as an R-module, ${}_RM$ remains simple, and its annihilator in R is $\mathfrak{A}$. QED

From (4.2) and (11.4), we have the following

(11.5) Corollary. *The Jacobson radical rad R is the intersection of all the left (resp., right) primitive ideals in R.*

Next we try to relate left primitive rings to other classes of rings we studied before.

(11.6) Proposition. *A simple ring is left (and right) primitive. A left primitive ring is both semiprimitive and prime.*

Proof. The first statement is obvious since, if R is simple, R must act faithfully on *any* nonzero module. The fact that a left primitive ring is semiprimitive is clear from (11.1) and (11.2). Finally, let R be a left primitive ring, and let M be a faithful simple left R-module. Consider any nonzero ideal $\mathfrak{U}$ in R. Clearly $\mathfrak{U} \cdot M$ is an R-submodule of M, and the faithfulness of ${}_RM$ implies that $\mathfrak{U} \cdot M \neq 0$. Therefore $\mathfrak{U} \cdot M = M$. If $\mathfrak{B}$ is another nonzero ideal in R, we then have

$$(\mathfrak{B}\mathfrak{U})M = \mathfrak{B}(\mathfrak{U}M) = \mathfrak{B} \cdot M = M,$$

so $\mathfrak{B}\mathfrak{U} \neq 0$. This verifies that R is a prime ring. QED

The Proposition we just proved completes the following chart of implications, which is to be compared with the chart in the Introduction to Chapter 4:

$$\begin{array}{ccccc} \text{semisimple} & \Longrightarrow & \text{semiprimitive} & \Longrightarrow & \text{semiprime} \\ \Uparrow\text{(if } DCC) & & \Uparrow & & \Uparrow \\ \text{simple} & \Longrightarrow & \text{left primitive} & \Longrightarrow & \text{prime} \end{array}$$

In general, none of the implications is reversible. However, for left artinian rings, the horizontal implications can be replaced by equivalences, as we shall now show.

(11.7) Proposition. *Let R be a left artinian ring. Then*

(1) *R is semisimple $\Longleftrightarrow$ R is semiprimitive $\Longleftrightarrow$ R is semiprime.*

(2) *R is simple $\Longleftrightarrow$ R is left (resp., right) primitive $\Longleftrightarrow$ R is prime.*

Proof. The two equivalences in (1) follow respectively from (4.14) and (10.24). To prove (2), it is enough to show that, if R is prime (and left artinian), then R is simple. Since R is semiprime, it follows from (1) that R is semisimple. If there is more than one simple component, R would fail to be prime. Therefore there is only one simple component, i.e., R is simple. QED

In the category of commutative rings, the notion of (left) primitive rings also does not add anything new, in view of the following observation.

(11.8) Proposition. *A commutative ring R is a (left) primitive ring iff it is a field.*

Proof. The "if" part is clear. For the converse, let R be primitive and let M be a faithful simple left R-module. Then $M \cong R/\mathfrak{m}$ for some maximal ideal

$\mathfrak{m}$ in R. Since $\mathfrak{m} \cdot M = 0$, it follows that $\mathfrak{m} = 0$. This clearly implies that R is a field. QED

In principle, left primitive rings are ubiquitous. For, if R is any nonzero ring, and M is any simple left R-module, then $R/ann(M)$ is a left primitive ring. To give a more explicit example, we proceed as follows. Let k be any division ring, V_k be a right k-vector space, and let $E = End(V_k)$, operating on the left of V. Clearly, V is a faithful simple left E-module, so E is a left primitive ring. If $dim_k V = n < \infty$, then of course $E \cong \mathbb{M}_n(k)$ is an artinian simple ring. But if $dim_k V$ is infinite, then E gives an example of a nonsimple, noncommutative, and nonartinian left primitive ring. The class of left primitive rings constructed above is important because we shall see later in this section that a general left primitive ring R "resembles" $E = End(V_k)$ in a certain sense.

We have pointed out before that a left primitive ring need not be right primitive. However, for the class of rings which possess minimal one-sided ideals, it turns out that left primitivity and right primitivity are both equivalent to primeness, and furthermore, these properties imply the uniqueness of the isomorphism type of the faithful simple left (resp., right) modules. This result will be proved in (11.11) below. To prepare ourselves for this proof, we first point out an interesting fact concerning one-sided minimal ideals in a semiprime ring.

(11.9) Lemma. *Let R be a semiprime ring, and $a \in R$. If Ra is a minimal left ideal, then aR is a minimal right ideal.*

Proof. It suffices to show that, for any nonzero element $ar \in aR$, we have $a \in arR$. Since R is semiprime, $arsar \neq 0$ for some $s \in R$. Let φ: $Ra \to Ra$ be the R-homomorphism defined by

$$\varphi(x) = xrsa, \quad \text{for any } x \in Ra.$$

Since $\varphi(a) = arsa \neq 0$ and Ra is simple, φ is an isomorphism. Let ψ be the inverse of φ. Then

$$a = \psi\varphi(a) = \psi(arsa) = ar\psi(sa) \in arR. \qquad \text{QED}$$

(11.10) Remark. If R is not semiprime, the conclusion of the lemma may no longer hold. For instance, if k is any division ring and R is the ring of 2×2 upper triangular matrices over k, then the matrix unit $a = E_{11}$ generates a minimal left ideal $Ra = kE_{11}$, but $aR = E_{11}k + E_{12}k$ is not a minimal right ideal as it contains the ideal $I = E_{12}k$. The ring R here is *not* semiprime since $I^2 = 0$.

With the aid of (11.9), we can now prove the following.

(11.11) Theorem. *Let R be a ring with a minimal left ideal $\mathfrak{A}$. The following properties are equivalent*:

(1) *R is prime.*

(2) *R is left primitive.*

(3) *R is right primitive.*

If these properties hold, then R also has a minimal right ideal $\mathfrak{B}$. Any faithful simple left (resp., right) R-module is isomorphic to ${}_R\mathfrak{A}$ (resp., $\mathfrak{B}_R$).

Proof. In general, (2) or (3) implies (1). Now assume (1). We claim that ${}_R\mathfrak{A}$ is a faithful R-module. Indeed, if $r \in R$ is such that $r\mathfrak{A} = 0$, then $(Rr)\mathfrak{A} = 0$, so by (10.2), $Rr = 0$, i.e., $r = 0$. Since ${}_R\mathfrak{A}$ is also a simple module, (2) follows. Now consider *any* faithful simple left R-module M. Then $\mathfrak{A} \cdot m \neq 0$ for some $m \in M$, so by the irreducibility of M, we have $M = \mathfrak{A} \cdot m$. The map

$$\mathfrak{A} \longrightarrow \mathfrak{A} \cdot m = M$$

sending $a \in \mathfrak{A}$ to $am \in M$ is clearly an R-module isomorphism from $\mathfrak{A}$ to M. By (11.9), R has also a minimal right ideal, so the remaining conclusions follow from left-right symmetry. QED

The above theorem applies nicely to the ring $E = End(V_k)$, where V is a nonzero right vector space over the division ring k. In fact, let $V = vk \oplus V_0$ where $v \in V$ is a fixed nonzero vector, and let $e \in E$ be the projection of V onto vk with kernel V_0. The map φ: $Ee \to {}_EV$ defined by $\varphi(re) = r(v)$ (for every $r \in E$) is clearly a surjective E-homomorphism. If $\varphi(re) = 0$, then

$$(re)(v) = r(v) = 0 \quad \text{and} \quad (re)(V_0) = 0$$

imply that $re = 0$. Thus, φ: $Ee \to V$ is an E-isomorphism, so Ee is a minimal left ideal of E. Since E is left primitive, (11.11) implies that it is also right primitive. Moreover, $Ee \cong {}_EV$ is the unique faithful simple left E-module, and eE is the unique faithful simple right E-module (up to isomorphism).

Recall that, by Theorem 3.10, if a simple ring has a minimal left ideal, then it is already left and right artinian. The remarks in the last paragraph show however that, for left primitive rings, this situation does not prevail. In fact, the left (and right) primitive ring $E = End(V_k)$ in the last paragraph has a minimal left (resp., right) ideal, but if $dim_k\ V$ is infinite, E is neither left nor right artinian.

For any ring R, let $soc({}_RR)$ be the socle of R as a left R-module, i.e., $soc({}_RR)$ is the sum of all minimal left ideals of R. (If there are no minimal left ideals, this sum is defined to be zero.) We call $soc({}_RR)$ the *left socle* of R, and define the *right socle* $soc(R_R)$ similarly. It is easy to see that both socles are ideals of R. In general, these may be different ideals. But *for semiprime rings R, they are equal* in view of (11.9). In this case we may write $soc(R)$ for either socle. This notation can be used in particular for 1-sided primitive rings. For instance, the left (and right) primitive ring $E = End(V_k)$ considered above has a nonzero socle. In Exercise 18 below, several criteria are given for a left primitive ring to have a nonzero socle, and a method for computing this socle is described.

Next we shall give a few more examples of left primitive rings. In particular, we would like to construct left primitive rings which admit nonisomorphic faithful simple left modules. (Such rings have necessarily zero socles.) The first example below is, in fact, a simple domain.

Let k be a division ring of characteristic zero, and let δ be a non-inner derivation on k. Then, by Amitsur's Theorem (3.16), the differential polynomial ring $R = k[x;\delta]$ is a simple domain. For any $a \in k$, $M_a = R/R(x-a)$ is easily seen to be a simple left R-module. To work with M_a more explicitly, we use the decomposition

$$R = R(x-a) \oplus k$$

to identify M_a with k. Under this identification, the R-action on $k = M_a$ is determined by $x * b = ba + \delta b$ for $b \in k$, in view of

$$xb = bx + \delta b = b(x-a) + ba + \delta b.$$

Since R is simple, any M_a is faithful. *When is $M_a \cong M_{a'}$, as R-modules?* Suppose $f\colon M_a \to M_{a'}$ is an R-isomorphism. Then there exists $c \in k\backslash\{0\}$ such that $f(b) = bc$ for all $b \in k$, and we have $f(x * b) = x * f(b)$, that is,

$$\begin{aligned}(ba + \delta b)c &= bca' + \delta(bc)\\ &= bca' + (\delta b)c + b\delta c,\end{aligned}$$

or equivalently, $a = ca'c^{-1} + \delta c \cdot c^{-1}$. Conversely, if this equation holds for some $c \neq 0$, then, by reversing the above argument, we have $M_a \cong M_{a'}$. We say that a is *δ-conjugate to a'* if

$$a = ca'c^{-1} + \delta c \cdot c^{-1} \quad \text{for some } c \neq 0.$$

By what we said above (or by a direct check), δ-conjugacy is an equivalence relation on k, and *the isomorphism classes of simple left R-modules of the type M_a are in one-one correspondence with the δ-conjugacy classes of k.* Note that the class of 0 consists of all "logarithmic derivatives" $\delta c \cdot c^{-1}$, where $c \in k\backslash\{0\}$. Therefore, there exist two nonisomorphic (faithful) simple left R-modules if k has an element which is *not* a logarithmic derivative with respect to δ. Here, R has zero socle, since it is a domain but not a division ring.

In the above example, R was simple. To get some nonsimple examples, we shall look at skew polynomial rings of the type $R = k[x;\sigma]$, where σ is an endomorphism of the division ring k. We begin by noting that Euclidean algorithm does hold in R for one-sided division: If $f(x)$, $g(x)$ are left polynomials in R with $f \neq 0$, then there exist unique $q(x)$ and $r(x)$ such that

$$g(x) = q(x)f(x) + r(x),$$

with $r(x) = 0$ or $deg\ r(x) < deg\ f(x)$. In particular, if $\mathfrak{A}$ is a left ideal in R, the usual Euclidean algorithm argument shows that $\mathfrak{A}$ is a principal left ideal generated by any polynomial in $\mathfrak{A}$ with the least degree. As for the ideals of R, we have the following result.

(11.12) Proposition. *Assume that σ is not an automorphism of finite inner order on k. (This includes the case when σ is not onto.) Then the nonzero ideals of $R = k[x;\sigma]$ are $R \cdot x^m$, $m \geq 0$.*

Proof. Using the twist equation $xa = \sigma(a)x$, it is easy to see that each $R \cdot x^m$ is an ideal. Conversely, let $\mathfrak{A}$ be any nonzero ideal of R. Then $\mathfrak{A} = R \cdot f$ where

$$f = x^m + a_{m-1}x^{m-1} + \cdots + a_n x^n$$

with $m \geq n \geq 0$ and $a_n \neq 0$. We are done if we can show that $m = n$. First, note that

$$fx - xf = (a_{m-1} - \sigma(a_{m-1}))x^m + \cdots + (a_n - \sigma(a_n))x^{n+1}$$

belongs to $\mathfrak{A} = R \cdot f$, so it must be of the form cf for some $c \in k$. Comparing the coefficients of x^n, we see that $c = 0$, so $\sigma(a_i) = a_i$ for every i. Next, for any $a \in k$, $fa - \sigma^m(a)f \in \mathfrak{A}$ has degree $< m$, so it must be zero. Looking at its coefficient of x^n, we get $a_n\sigma^n(a) - \sigma^m(a)a_n = 0$. Therefore, $\sigma^n(a_n a) = \sigma^m(a a_n)$, and so

$$a_n a = \sigma^{m-n}(a a_n) = \sigma^{m-n}(a)a_n,$$

since σ is always injective. If $m > n$, this would imply that σ is an automorphism of finite inner order. Since we assumed this is not the case, we must have $m = n$ and $f = x^m$. QED

(11.13) Proposition. *Under the hypothesis of* (11.12), *$M_a = R/R(x-a)$ is a faithful simple left R-module for every $a \in k\backslash\{0\}$, so R is a left primitive ring. We have $M_a \cong M_{a'}$ as R-modules iff $a = \sigma(c)a'c^{-1}$ for some $c \in k\backslash\{0\}$.*

Proof. Clearly, M_a is a simple module. Also, it is easy to check that

$$x^m \notin R(x-a) \quad \text{for } a \neq 0 \text{ and } m \geq 0,$$

so by (11.12), $R(x-a)$ does not contain any nonzero ideal of R. Therefore, $ann(M_a) = 0$, so M_a is a faithful simple left R-module for every $a \neq 0$. As before, we shall *identify* M_a with k, using the decomposition $R = R(x-a) \oplus k$. Here, the x-action on $k = M_a$ is given by $x * b = \sigma(b)a$ for $b \in k$, since

$$xb = \sigma(b)x = \sigma(b)(x-a) + \sigma(b)a \in R.$$

Now suppose we have an isomorphism $f\colon M_a \to M_{a'}$. Then there exists $c \in k\backslash\{0\}$ such that $f(b) = bc$ for all $b \in k = M_a$. The homomorphism condition $f(x * b) = x * f(b)$ here amounts to $\sigma(b)ac = \sigma(bc)a'$, or equivalently $a = \sigma(c)a'c^{-1}$. Conversely, if this holds for some $c \neq 0$, then by reversing the above argument we have $M_a \cong M_{a'}$. QED

We say that $a \in k\backslash\{0\}$ is *σ-conjugate to* $a' \in k\backslash\{0\}$ if $a = \sigma(c)a'c^{-1}$ for some $c \in k\backslash\{0\}$. Under the hypothesis of (11.12), we see that *the isomorphism classes of the faithful simple left R-modules of the form M_a are in one-one correspondence with the σ-conjugacy classes of $k\backslash\{0\}$*. Using this, we can construct a left primitive ring R with infinitely many distinct faithful simple left modules. For instance, let $k = \mathbb{R}(t)$ and let $R = k[x;\sigma]$, where σ is the

$\mathbb{R}$-automorphism of k sending t to $t+1$. In k, define

$$deg(f/g) = deg\, f - deg\, g \quad \text{for } f, g \in \mathbb{R}[t]\backslash\{0\}.$$

Then, for any nonzero $c = c(t) \in \mathbb{R}(t)$,

$$\sigma(c)c^{-1} = c(t+1)/c(t)$$

has degree 0. Therefore $a(t)$ is σ-conjugate to $a'(t)$ only if they have the same degree. In particular, $M_1, M_t, M_{t^2}, \ldots$ are mutually nonisomorphic faithful simple left R-modules.

In the last two examples, we exploited the presence of a nontrivial twist on the skew polynomial rings $k[x;\delta]$ and $k[x;\sigma]$ to give the proof of their left primitivity. However, even in the untwisted case, the class of rings $k[x]$ affords some nice examples of left (and right) primitive rings, as the following result shows.

(11.14) Proposition. *Let k be a division ring which is not algebraic over its center C, and let $R = k[x]$. Then for any $a \in k$ which is not algebraic over C, $M_a = R/R(x-a)$ (resp., $R/(x-a)R$) is a faithful simple left (resp., right) R-module. In particular, R is left and right primitive. The isomorphism classes of simple modules of the type M_a ($a \in k$) are in one-one correspondence with the conjugacy classes of k.*

Proof. We shall work with left modules in this proof, the case of right modules being similar. Since $R(x-a)$ is a maximal left ideal, $M_a = R/R(x-a)$ is a simple left R-module. As before, we identify M_a with k via the decomposition $R = R(x-a) \oplus k$. Here, as in the proof of (11.13), the x-action on $k = M_a$ is $x * b = ba$ for any $b \in k$. Assume that $ann(M_a) \neq 0$. Since $ann(M_a)$ is an ideal, Exercise 1.16 shows that it contains a nonzero $g(x) = \sum_i c_i x^i \in C[x]$. But then

$$0 = g(x) * 1 = \sum_i (c_i x^i) * 1 = \sum_i c_i a^i$$

shows that a is algebraic over C, a contradiction. We can show as in the two earlier examples that

$$M_a \cong M_{a'} \quad \text{iff} \quad a = ca'c^{-1} \quad \text{for some } c \in k\backslash\{0\}.$$

This gives the last conclusion in the Proposition. QED

Having given the above examples of left primitive rings, we shall now return to the general theory. Our next goal is to prove the Density Theorem of Jacobson and Chevalley (11.16). The structure theorem on left primitive rings ((11.19) below) will be seen to be a consequence of this Density Theorem.

First let us define the notion of density. Let R, k be two rings, and $V = {}_RV_k$ be an (R,k)-bimodule. We write $E = End(V_k)$, which operates on V from the left. We say that R *acts densely* on V_k if, for any $f \in E$ and any $v_1, \ldots, v_n \in V$, there exists $r \in R$ such that $rv_i = f(v_i)$ for $i = 1, 2, \ldots, n$. To explain why this property is referred to as "density," we make the following observation. Let $\mathscr{T}$ be the topology on E defined by taking as a basis all sets of the form

$$\{g \in E: \ g(v_i) = v_i' \quad (1 \le i \le n)\},$$

where n is any natural number, and v_i, v_i' are arbitrary elements of V. It is easy to see that R acts densely on V_k (in the sense defined above) iff, under the natural map from R to E, the image of R is a dense subring of E with respect to the topology $\mathscr{T}$.

Before stating the Density Theorem, we need the following preliminary result.

(11.15) Lemma. *In the notation above, assume that ${}_RV$ is a semisimple R-module, and that $k = End({}_RV)$. Then any R-submodule W of V is an E-submodule (and, of course, conversely).*

Proof. Take a suitable R-submodule W' of V such that $V = W \oplus W'$, and let $e \in k$ be the projection of V on W with respect to this decomposition. Then, for any $f \in E$, we have

$$f(W) = f(We) = (fW)e \subseteq W,$$

so W is an E-submodule of V. QED

The Lemma above prepared us for the proof of the following fundamental result.

(11.16) Density Theorem (Jacobson, Chevalley). *Let R be a ring and V be a semisimple left R-module. Then, for $k = End({}_RV)$, R acts densely on V_k.*

Proof. As before, let $E = End(V_k)$. For $f \in E$ and $v_1, \ldots, v_n \in V$, we seek an element $r \in R$ such that $rv_i = f(v_i)$ for all i. The idea of the proof, after N. Bourbaki, is to apply the Lemma above to the *semisimple* R-module $\tilde{V} = V^n$ (direct sum of n copies of V). First we compute that

$$\begin{aligned} \tilde{k} &:= End({}_R\tilde{V}) = End({}_RV^n) \\ &\cong \mathbb{M}_n(End({}_RV)) = \mathbb{M}_n(k). \end{aligned}$$

Now define $\tilde{f}: \tilde{V} \to \tilde{V}$ by taking $\tilde{f} = (f, f, \ldots, f)$. *We claim* that $\tilde{f} \in End(\tilde{V}_{\tilde{k}})$. To see this, let $\tilde{e} \in \tilde{k}$, and represent $\tilde{e}$ as a matrix (e_{ij}), where $e_{ij} \in k$. Then, for any $(w_1, \ldots, w_n) \in \tilde{V}$:

$$\begin{aligned} \tilde{f}((w_1, \ldots, w_n)\tilde{e}) &= \tilde{f}\Big(\sum w_i e_{i1}, \ldots, \sum w_i e_{in}\Big) \\ &= \Big(f\Big(\sum w_i e_{i1}\Big), \ldots, f\Big(\sum w_i e_{in}\Big)\Big) \\ &= \Big(\sum f(w_i)e_{i1}, \ldots, \sum f(w_i)e_{in}\Big) \\ &= (f(w_1), \ldots, f(w_n))\tilde{e} \\ &= (\tilde{f}(w_1, \ldots, w_n))\tilde{e}, \end{aligned}$$

as desired. Now consider the cyclic R-submodule $\tilde{W}$ of $\tilde{V}$ generated by $(v_1, \ldots, v_n) \in \tilde{V}$. By the preceding Lemma (applied to $\tilde{W} \subseteq \tilde{V}$), $\tilde{W}$ is stabilized by $End(\tilde{V}_{\tilde{k}})$, in particular by $\tilde{f}$. Thus,

$$\tilde{f}(v_1, \ldots, v_n) = (f(v_1), \ldots, f(v_n))$$

remains in $\tilde{W} = R \cdot (v_1, \ldots, v_n)$, so there exists an $r \in R$ such that $f(v_i) = rv_i$ for $i = 1, 2, \ldots, n$. QED

(11.17) Corollary. *Let R, V, k and E be as in the Density Theorem. If V_k is finitely generated as a (right) k-module, then the natural map ρ: $R \to E$ is onto.*

Proof. Let $v_1, \ldots, v_n \in V$ be a finite set of generators of V as a right k-module. Let $f \in E$. By the Density Theorem, there exists an $r \in R$ such that $rv_i = f(v_i)$ for all i. For any $v \in V$, write $v = \sum v_i a_i$ where $a_i \in k$. Then

$$rv = \sum r(v_i a_i) = \sum (rv_i) a_i = \sum (f(v_i)) a_i$$
$$= f\left(\sum v_i a_i\right) = f(v),$$

and so $f = \rho(r)$. QED

The main case of interest in the Density Theorem (11.16) is when V is a *simple* R-module. However, for the above formulation to work, it is necessary to deal more generally with *semisimple* modules, for, even if ${}_RV$ itself is simple, the auxilliary module $\tilde{V}$ (to which we applied Lemma (11.15)) is only semisimple. Therefore, we may as well let ${}_RV$ be a semisimple module (instead of a simple module) in the statement of (11.16).

In the important case when V is a simple left R-module, the endomorphism ring $k = End({}_RV)$ is a division ring by Schur's Lemma, so V is a right vector space over k. This suggests that linear algebra should play a role in studying the action of R on the simple module ${}_RV$.

To proceed more formally, let us recall a useful definition in linear algebra. Let V_k be a right vector space over a division ring k, and let $E = End(V_k)$. A subset $S \subseteq E$ is said to be *m-transitive* on V if, for any set of $n \leq m$ linearly independent vectors $v_1, \ldots, v_n$ and any other set of n vectors $v'_1, \ldots, v'_n$ in V, there exists $s \in S$ such that $s(v_i) = v'_i$ for all i. We say that S is a *dense set of linear transformations* on V_k if S is m-transitive for all (finite) m. The following Proposition ensures that this terminology is consistent with the one we have adopted before.

(11.18) Proposition. *Let V be an (R, k)-bimodule where k is a division ring. Let $E = End(V_k)$ and let ρ: $R \to E$ be the natural map. Then R acts densely on V_k iff $\rho(R)$ is a dense ring of linear transformations on V.*

Proof. The "only if" part is clear. For the "if" part, assume $\rho(R)$ is a dense ring of linear transformations on V. Let $f \in E$ and let $v_1, \ldots, v_n \in V$. After a reindexing, we may assume that $v_1, \ldots, v_m$ are k-linearly independent, and that each v_i is a k-linear combination of $v_1, \ldots, v_m$. Since $\rho(R)$ is m-transitive on V, there exists $r \in R$ such that $rv_i = f(v_i)$ for $i \leq m$. By linearity, it follows that this equation holds for all $i \leq n$. QED

Combining (11.16), (11.17) and (11.18), we obtain the following important result which is sometimes also referred to as the "Density Theorem."

(11.19) Structure Theorem for Left Primitive Rings. *Let R be a left primitive ring and V be a faithful simple left R-module. Let k be the division ring $End({}_RV)$. Then R is isomorphic to a dense ring of linear transformations on V_k. Moreover:*

(1) *If R is left artinian, then $n := dim_k\ V$ is finite, and $R \cong \mathbb{M}_n(k)$.*

(2) *If R is not left artinian, then $dim_k\ V$ is infinite, and for any integer $n > 0$, there exists a subring R_n of R which admits a ring homomorphism onto $\mathbb{M}_n(k)$.*

(Note that (1) here recovers the Wedderburn–Artin Theorem for left artinian simple rings, since these are exactly the left artinian left primitive rings, by (11.7). The proof below is independent of, and more general than, our earlier proof in §3. Thus, (11.19) may be thought of as a generalization of the Wedderburn–Artin Theorem.)

Proof. Since ${}_RV$ is faithful, the natural map

$$\rho\colon R \longrightarrow E := End(V_k)$$

is injective. By (11.16) and (11.18), $\rho(R)$ is a dense ring of linear transformations on V_k. This proves the first conclusion. Now assume $dim_k\ V = n < \infty$. By (11.17), we have $\rho(R) = E$, so $R \cong \mathbb{M}_n(k)$ and R is left artinian. Next assume $dim_k\ V$ is infinite. Fix a sequence of linearly independent vectors $v_1, v_2, \ldots$ in V, and let

$$V_n = \sum_{i=1}^{n} v_i k \quad (1 \leq n < \infty).$$

Finally, let

$$R_n = \{r \in R\colon\ r(V_n) \subseteq V_n\} \quad \text{(a subring of } R\text{);}$$

$$\mathfrak{A}_n = \{r \in R\colon\ r(V_n) = 0\} \quad \text{(an ideal of } R_n \text{ and a left ideal of } R\text{).}$$

Then $R_n/\mathfrak{A}_n$ acts faithfully on the k-space V_n. By the n-transitivity of R (on V), any k-linear endomorphism of V_n can be realized as the action of some $r \in R_n$. Therefore the natural map $R_n/\mathfrak{A}_n \to End((V_n)_k)$ is an isomorphism, giving $R_n/\mathfrak{A}_n \cong \mathbb{M}_n(k)$. Moreover, by $(n+1)$-transitivity, there exists $r \in R$ such that

$$rv_1 = \cdots = rv_n = 0 \quad \text{but} \quad rv_{n+1} \neq 0,$$

so $\mathfrak{A}_n \supsetneqq \mathfrak{A}_{n+1}$ for all n. Therefore, $\mathfrak{A}_1 \supseteq \mathfrak{A}_2 \supseteq \cdots$ is a strictly decreasing chain of left ideals in R, so R is not left artinian. A moment's reflection shows that we have now proved both (1) and (2) in the Theorem. QED

Instead of starting with a left primitive ring R and looking at a faithful simple R-module ${}_RV$ and its endomorphism ring k, we can also start with an arbitrary division ring k and an arbitrary right vector space V_k. From this perspective, we look at $E = End(V_k)$ and consider rings of linear transformations $R \subseteq E$. We have defined earlier the notion of m-transitivity for R ($m = 1, 2, \ldots$) and R is dense iff it is m-transitive for all m. As a further consequence of the Density Theorem, we shall now show that, to check the density of R, it is enough to check 2-transitivity.

(11.20) Theorem. *Let R be a ring of linear transformations on a nonzero right vector space V over a division ring k. Then:*

(1) *R is 1-transitive iff ${}_RV$ is a simple R-module. If this is the case, R is a left primitive ring.*

(2) *The following are equivalent:* (a) *R is 2-transitive;* (b) *R is 1-transitive and $End({}_RV) = k$;* (c) *R is dense in $E := End(V_k)$.*

Proof. (1) is obvious. For (2), we have (b) $\Rightarrow$ (c) by (1) and the Density Theorem, and (c) $\Rightarrow$ (a) is clear. Therefore, we only need (a) $\Rightarrow$ (b). Assume R is 2-transitive (in particular 1-transitive). Consider any $\lambda \in End({}_RV)$. Fix a nonzero vector $v \in V$; we claim that v and $v\lambda$ are k-linearly dependent. Indeed, if not, there would exist (by 2-transitivity) an $r \in R$ such that $rv = 0$ but $r(v\lambda) \neq 0$. However, since $\lambda \in End({}_RV)$, we have $r(v\lambda) = (rv)\lambda = 0$, a contradiction. Now write $v\lambda = va$, where $a \in k$. For any $w \in V$, the 1-transitivity of R implies that $w = sv$ for some $s \in R$. But then

$$w\lambda = (sv)\lambda = s(v\lambda) = s(va) = (sv)a = wa.$$

Therefore $\lambda = a \in k$. This shows that $End({}_RV) = k$. QED

In certain special cases, the 1-transitivity of $R \subseteq E$ may already imply 2-transitivity (and hence density). For instance, if k is an algebraically closed field, and $dim_k\, V < \infty$, then for any k-subalgebra

$$R \subseteq E = End(V_k),$$

1-transitivity of R will already force R to be E. In general, however, 1-transitivity is weaker than density. To see this, we can construct an example as follows. Let k be a field which is not algebraically closed, and let $R \supset k$ be a field extension of finite degree > 1. We view R as a right k-vector space V and embed R in $E = End(V_k)$ by identifying $r \in R$ with the left multiplication by r on R. Since R is a field, ${}_RV$ is simple so R is 1-transitive on V. On the other hand, an easy check shows that R is not 2-transitive on V. Note that here, $End({}_RV) \cong R \supsetneqq k$.

As an illustration, we shall construct some explicit examples of dense rings of linear transformations (thereby obtaining more examples of left primitive rings). The first example, due to I. Kaplansky, is prompted by the question: what can we say about the center $Z(R)$ of a left primitive ring R? Since a left primitive ring R is also a prime ring, $Z(R)$ is a (commutative) domain by Exercise 10.0. The following well-known example of Kaplansky shows that $Z(R)$ can be *any* prescribed commutative domain.

Let A be a commutative domain and let k be its quotient field. Let

$$V_k = \bigoplus_{i=1}^{\infty} e_i k,$$

and let R be the subring of $E = End(V_k)$ consisting of endomorphisms of V with matrices of the form

$$\textbf{(11.21)} \qquad r = \begin{pmatrix} M & & & \\ & a & & \\ & & a & \\ & & & \ddots \end{pmatrix},$$

where M is any finite matrix over k, and $a \in A$. This ring R is clearly dense in E so it is a left primitive ring. What is its center $Z(R)$? Say the matrix r above lies in $Z(R)$, with $M \in \mathbb{M}_n(k)$. Then certainly $M \in Z(\mathbb{M}_n(k))$ so M is a scalar matrix, say $b \cdot I_n$ $(b \in k)$. By the same analysis, $\begin{pmatrix} M & 0 \\ 0 & a \end{pmatrix}$ is also a scalar matrix in $\mathbb{M}_{n+1}(k)$, so we must have $b = a$. Therefore,

$$Z(R) = \{a \cdot I : \ a \in A\}$$

is isomorphic to A. (For another construction, see Exercise 11.)

Our second example of a dense ring of linear transformations follows ideas of Jacobson and Samuel. Let k be a field, V and E be as above, and let $f \in E$ be defined by

$$f(e_1) = 0, \quad f(e_i) = e_{i-1} \quad \text{for} \quad i \geq 2.$$

Further, let $g \in E$ be *any* endomorphism with the property that

(11.22) *For any* $m \geq 1, g^m e_1 = e_{r(m)}$ *where* $\lim_{m \to \infty} r(m) = \infty$.

Let R be the k-subalgebra of E generated by f and g. (Note that we have $k \subseteq Z(E)$ since k is a field.) We claim that *R acts irreducibly on V* (so R is a left primitive ring). For, if $W \subseteq V$ is a nonzero R-submodule, there exists a nonzero vector $w \in W$ which has the shortest representation as a linear combination of the e_i's, say

$$w = e_{i_1} a_1 + e_{i_2} a_2 + \cdots + e_{i_n} a_n,$$

where $i_1 < \cdots < i_n$, and each $a_j \neq 0$. Applying f^{i_1}, we get a nonzero element $f^{i_1}(w)$ with a shorter representation unless $n = 1$. Therefore W contains e_{i_1}, and also

$$f^{i_1-1}(e_{i_1}) = e_1.$$

Applying to e_1 a large power of g followed by a suitable power of f, we see that W contains e_j for all $j \geq 1$, so $W = V$. Next we claim that R *is dense in* E. For this, it suffices (by the Density Theorem) to show that $End({}_R V) = k$. Let $\lambda \in End({}_R V)$. Since $f(e_1\lambda) = (f(e_1))\lambda = 0$ and $ker(f) = e_1 k$, we can write $e_1\lambda = e_1 a$ for some $a \in k$. For any j, we can write $e_j = f^s g^t e_1$ for suitable integers s, t. But then

$$e_j\lambda = (f^s g^t e_1)\lambda = f^s g^t (e_1 a) = e_j a,$$

hence $\lambda = a \in k$.

By choosing g differently satisfying (11.22), we get different examples of left primitive rings R. For instance, if we define g by $g(e_i) = e_{i+1}$ for all $i \geq 1$, it may be checked that all relations between f and g are consequences of the relation $fg = 1$; i.e., the k-homomorphism $D\colon k\langle x, y\rangle \to R$ given by $D(x) = f$ and $D(y) = g$ has kernel $(xy - 1)$. Assuming this fact (which is the content of Exercise 9), it follows that

$$k\langle x, y\rangle/(xy - 1)$$

is a left primitive ring. On the other hand, by choosing g in another way, it is possible to arrange that there be *no* relations between f and g, so that R is k-isomorphic to $k\langle x, y\rangle$.

(11.23) Proposition (Samuel). *Define g by $g(e_i) = e_{i^2+1}$ for all i. Then the k-homomorphism D: $k\langle x, y\rangle \to R$ defined by $D(x) = f$, $D(y) = g$ is an isomorphism. In particular, it follows that the free algebra $A := k\langle x, y\rangle$ is left primitive.*

Proof. The definition of g certainly fulfills the condition specified in (11.22). Thus, once we have shown that D is an isomorphism, the left primitivity of A follows. Via the representation D, V becomes a left A-module, with x and y acting respectively as f and g. Our job is to show that A acts *faithfully* on V. Let us say that an element $z \in A$ is "eventually zero" on V if $ze_i = 0$ for all sufficiently large i. We will show that, *if z is eventually zero on V, then* $z = 0 \in A$. In particular, this will establish the faithfulness of ${}_A V$.

Let H be any monomial in x and y. For sufficiently large i, He_i has the form $e_{h(i)}$, where h is a (uniquely determined) monic polynomial in $\mathbb{Z}[t]$ with $deg\ h = 2^d$, where $d = deg_y\ H$. (For instance, if $H = yx^2y$, then for $i \geq 2$,

$$H \cdot e_i = yx^2 e_{i^2+1} = ye_{i^2-1} = e_{h(i)},$$

where $h(t) = (t^2 - 1)^2 + 1$.) We claim that

(11.24) *If H, H' are different monomials in x and y, then $h(t) \neq h'(t)$ in $\mathbb{Z}[t]$.*

First assume this claim. If $z = \sum a_j H_j \in A$ where $a_j \in k$ and $\{H_j\}$ are different monomials, then for sufficiently large i,

(11.25) $$ze_i = \left(\sum a_j H_j\right) e_i = \sum a_j e_{h_j(i)}.$$

By (11.24), the h_j's are all different, so for large i, the $h_j(i)$'s are all different. Therefore, if $z \neq 0$ in A, then in view of (11.25), z cannot be "eventually zero" on V.

We now prove the claim (11.24). If H and H' end with the same letter (x or y), we are done by invoking an inductive hypothesis. Therefore, we need only consider the case where $H = H_1 x$ and $H' = H_1' y$. The polynomial $h'(t)$ associated with H' has only even powers of t. However, since the polynomial associated with x is $t - 1$, the $h(t)$ associated with H is seen (inductively) to be of the form

$$t^{2^m} - nt^{2^m - 1} + \cdots,$$

where $m = deg_y\, H_1$ and n is a *positive* integer. Thus $h(t) \neq h'(t)$. QED

(11.26) Corollary. *For any field k, any free k-algebra in finitely or countably many (and at least two) indeterminates is a left primitive ring.*

Proof. Since the algebra R above is freely generated by f and g, it follows from (1.2) that the algebra R_∞ generated by $\{fg^i\colon i \geq 0\}$ is also a free algebra on these generators. We have $R_\infty = \bigcup_{n=1}^{\infty} R_n$, where

$$R_n = k\langle\{fg^i:\ 0 \leq i \leq n\}\rangle$$

is a free algebra on $n+1$ generators. We claim that R_2 (and hence R_n for $2 \leq n \leq \infty$) *acts irreducibly on* V. Once we have proved this, it follows that R_n is left primitive for $2 \leq n \leq \infty$. To prove the claim, we proceed as before: Let $W \subseteq V$ be a nonzero R_2-submodule of V. The same argument used in the proof of the irreducibility of R shows that $e_1 \in W$. But then W also contains

$$fg^2(e_1) = e_4, \quad (fg^2)^2(e_1) = (fg^2)(e_4) = e_{289}, \quad \ldots \quad ,$$

and hence W contains all e_i. QED

Note that, in the above, $R_1 = k\langle f, fg\rangle$ *does not* act irreducibly on V, since $e_1 \cdot k$ is an R_1-submodule. However, we need not work with R_1 as the case of the free algebra on two generators has already been dealt with.

At this point, we should point out that E. Formanek has proved that, *over any field k, any free algebra* $k\langle X\rangle$ *with* $|X| \geq 2$ *is always left (and right) primitive.* We shall not prove this result for an arbitrary field k here. However, using a different method, we can prove this result for any *countable* field k. In fact, we can even relax the assumption that k be a field! The result we shall prove is the following:

(11.27) Theorem (E. Formanek). *Let k be any (not necessarily commutative) countable domain, and let* $\{x_i\colon i \in I\}$ *be any set of independent indeterminates* *over k. If $|I| \geq 2$, then $k\langle\{x_i\colon i \in I\}\rangle$ (the ring generated freely over k by $\{x_i\}$) is a left primitive ring.*

It is remarkable that, in this Theorem, we do not need any hypothesis on the domain k, other than that it be countable. The balance of this section will be devoted to the proof of this beautiful result. To begin the proof, let us observe the following slight reformulation of the notion of a left primitive ring.

(11.28) Lemma. *A ring R is left primitive iff there exists a left ideal $\mathfrak{A} \subsetneq R$ which is comaximal with any ideal $\mathfrak{A}' \neq 0$. ($\mathfrak{A}$ and $\mathfrak{A}'$ comaximal means that $\mathfrak{A} + \mathfrak{A}' = R$.)*

Proof. If such an $\mathfrak{A}$ exists, we may assume (after an application of Zorn's Lemma) that it is a maximal left ideal. The annihilator of the simple left R-module $R/\mathfrak{A}$ is an ideal in $\mathfrak{A}$, and so it must be zero. This shows that R is left primitive. Conversely, if R is left primitive, there exists a faithful simple left R-module, which we may take to be $R/\mathfrak{A}$ for some (maximal) left ideal $\mathfrak{A} \subsetneq R$. A nonzero ideal $\mathfrak{A}'$ cannot lie in $\mathfrak{A}$ (for otherwise $\mathfrak{A}'$ annihilates $R/\mathfrak{A}$) and so must be comaximal with $\mathfrak{A}$. QED

In order to prove (11.27), we state a more general result of Formanek on semigroup rings.

(11.29) Theorem. *Let A, B be two semigroups $\neq \{1\}$, and let $G = A * B$ be their free product. Let k be any domain with $|k| \leq |G|$. Then the semigroup ring $R = kG$ is left primitive unless $|A| = |B| = 2$.*

Assuming this theorem, let us first complete the proof of (11.27). We refer to the notations there. Since $|I| \geq 2$, we can write the free semigroup generated by $\{x_i\colon i \in I\}$ as a free product $G = A * B$, where A, B are free semigroups $\neq \{1\}$. If k is countable, (11.29) applies to show that

$$kG = k\langle x_i\colon\ i \in I\rangle$$

is a left primitive ring. Incidentally, this kind of argument also leads to another proof of the fact that any commutative domain k is the center of a left primitive ring R, since we can take R to be $k\langle x_i\colon i \in I\rangle$ for a sufficiently large indexing set I.

We shall now begin the proof of (11.29). The proof will be presented in the case when A, B are both infinite. This is the case needed for the application in the paragraph above. The case when one (or both) of A, B is finite uses similar ideas and will be left to the reader.

Recall that the nonidentity elements of $G = A * B$ are reduced words whose letters belong alternately to $A^* = A\backslash\{1\}$ and to $B^* = B\backslash\{1\}$. We say that such a word has type AB if it begins with a letter in A^* and ends with a letter in B^*. Similarly we can define words of type AA, BA and BB. The *length* of a reduced word is defined to be the total number of its letters; the length of the identity element is taken to be zero.

By interchanging A and B if necessary, we may assume that $|A| \geq |B|$. Since $|k| \leq |G|$, we have $|R| = |A|$. For the rest of the proof, we fix a 1–1 correspondence $a \mapsto \alpha(a)$ between A^* and $R \backslash \{0\}$. For each $a \in A^*$, we shall define another element $\beta(a) \in R$ with the following properties:

(1) *For each* $a \in A^*$, $\beta(a) \in 1 + R \cdot \alpha(a) \cdot R$.

(2) $\{\beta(a) : \ a \in A^*\}$ *generates a left ideal* $\mathfrak{A} \subsetneq R$.

If we can construct such $\beta(a)$'s, then the left ideal $\mathfrak{A}$ in (2) will be comaximal with any nonzero ideal $\mathfrak{A}'$ in R (since $\mathfrak{A}'$ contains some $\alpha(a)$). The left primitivity of R then follows from (11.28).

For $r \in R \backslash \{0\}$, let *max-supp*(r) denote the set of elements in G of maximal length in the support of r. For each $a \in A^*$, fix an element $\alpha_0(a) \in$ *max-supp*$(\alpha(a))$. We now fix an element $b \in B^*$, and define $\beta(a)$ to be

(11.30)
$$\begin{cases} 1 + b\alpha(a)a + \alpha(a)ab & \text{if } \alpha_0(a) \text{ has type } AB \text{ or } \alpha_0(a) = 1, \\ 1 + b\alpha(a)ba + \alpha(a)bab & \text{if } \alpha_0(a) \text{ has type } AA, \\ 1 + \alpha(a)bab + a\alpha(a)ba & \text{if } \alpha_0(a) \text{ has type } BA, \\ 1 + \alpha(a)ab + a\alpha(a)a & \text{if } \alpha_0(a) \text{ has type } BB. \end{cases}$$

Clearly the property (1) is satisfied, so we need only verify (2). In the course of proving (2), we will also be able to see why the definition of $\beta(a)$ is arranged as in (11.30).

First let us analyze the case when $\alpha_0(a)$ is of type AB. Say

$$\alpha_0(a) \in \textit{max-supp}(\alpha(a))$$

has length n. Since it has type AB, $b\alpha_0(a)a$ and $\alpha_0(a)ab$ are both reduced words. They have length $n+2$ and therefore belong to *max-supp*$(\beta(a))$. Thus, any element in *max-supp*$(\beta(a))$ has length $n+2$, and ends with a or ab. Note also that $\alpha_0(a)ab$ begins in A and $b\alpha_0(a)a$ begins in B.

After similarly analyzing the other three cases, we see that the definitions in (11.30) have been arranged so that (for each $a \in A^*$) $\beta(a)$ has the following properties:

(3) *Each element in max-supp*$(\beta(a))$ *ends with either a or ab.*

(4) *There exist elements in max-supp*$(\beta(a))$ *which begin in A and which begin in B.*

Using these properties, we can now verify (2). Consider a left multiple $r \cdot \beta(a)$ $(r \in R \backslash \{0\})$ and let $r_0 \in$ *max-supp*(r). In view of (4), we can choose an element

$$\beta_0(a) \in \textit{max-supp}(\beta(a))$$

such that $r_0 \cdot \beta_0(a)$ is already a reduced word in G and hence belongs to *max-supp*$(r \cdot \beta(a))$. Furthermore, any element in *max-supp*$(r \cdot \beta(a))$ arises in this way; in particular, by (3), such an element must end with a or with ab.

Now consider a finite sum $\sum_{i=1}^m r_i\beta(a_i)$, where $r_i \in R\backslash\{0\}$ and $a_1, \ldots, a_m$ are different elements in A^*. From the foregoing analysis, we see that the *max-supp*$(r_i\beta(a_i))$'s are disjoint for $1 \le i \le m$. In particular, $\sum_{i=1}^m r_i\beta(a_i)$ can never be 1, and so $\{\beta(a): a \in A^*\}$ generates a left ideal $\subsetneq R$, as desired.

QED

Exercises for §11

Ex. 11.1. Show that a homomorphic image of a left primitive ring need not be left primitive.

Ex. 11.2. Show that a ring R can be embedded into a left primitive ring iff either *char* R is a prime number $p > 0$, or $(R, +)$ is a torsion-free abelian group.

Ex. 11.3. Let R be a left primitive ring. Show that for any nonzero idempotent $e \in R$, the ring $A = eRe$ is also left primitive.

Ex. 11.4. Which of the following implications are true?
(a) R left primitive $\Longleftrightarrow$ $\mathbb{M}_n(R)$ left primitive.
(b) R left primitive $\Longleftrightarrow$ $R[t]$ left primitive.

Ex. 11.5. Let R be a ring which acts faithfully and irreducibly on a left module V. Let $v \in V$ and $\mathfrak{A}$ be a nonzero right ideal in R. Show that $\mathfrak{A} \cdot v = 0 \Longrightarrow v = 0$.

Ex. 11.5*. For any left ideal I in a ring R, define the *core* of I to be the sum of all ideals in I. Thus, *core*(I) is the (unique) largest ideal of R contained in I.
(1) Show that *core*$(I) =$ *ann*(V) where V is the left R-module R/I. (In particular, V is faithful iff *core*$(I) = 0$.)
(2) Show that R/I is faithful only if $I \cap Z(R) = 0$, where $Z(R)$ is the center of R.

Ex. 11.6. (Artin-Whaples) Let R be a simple ring with center k (which is a field by Exercise 3.4). Let $x_1, \ldots, x_n \in R$ be linearly independent over k. Show that, for any $y_1, \ldots, y_n \in R$, there exist $a_1, \ldots, a_m$ and $b_1, \ldots, b_m$ in R such that $y_i = \sum_{j=1}^m a_j x_i b_j$ for every i. (**Hint.** Let R^{op} be the opposite ring of R. Let $A = R \otimes_k R^{\text{op}}$ act on R by the rule $(a \otimes_k b^{\text{op}})x = axb$. Show that R is a simple left A-module with $End({}_AR) = k$. Then apply the Density Theorem.)

Ex. 11.7. Let $E = End(V_k)$ where V is a right vector space over the division ring k. Let R be a subring of E and $\mathfrak{A}$ be a nonzero ideal in R. Show that R is dense in E iff $\mathfrak{A}$ is dense in E.

Ex. 11.7*. Let $E = End(V_k)$ be as in Exercise 7, and let $R \subseteq E$ be a dense subring.

(1) For any $a \in R$ with finite rank, show that $a = ara$ for some $r \in R$.
(2) Deduce that the set

$$S = \{a \in R: \ rank(a) < \infty\}$$

is a von Neumann regular ring (possibly without identity).

Ex. 11.8. Let $V = \bigoplus_{i=1}^{\infty} e_i k$ where k is a field. For any n, let S_n be the set of endomorphisms $\lambda \in E = \text{End}(V_k)$ such that λ stabilizes $\sum_{i=1}^{n} e_i k$ and $\lambda(e_i) = 0$ for $i \geq n+1$. Show that

$$S = \bigcup_{n=1}^{\infty} S_n \subseteq E$$

is a dense set of linear transformations. For any i, j, let $E_{ij} \in E$ be the linear transformation which sends e_j to e_i and all $e_{j'}$ $(j' \neq j)$ to zero. Show that any k-subalgebra R of E containing all the E_{ij}'s is dense in E and hence left primitive.

Ex. 11.9. (Jacobson) Keep the notations above and define $f, g \in E$ by $g(e_i) = e_{i+1}$, $f(e_i) = e_{i-1}$ (with the convention that $e_0 = 0$). Let R be the k-subalgebra of E generated by f and g.
(1) Use Exercise 8 to show that R acts densely on V_k.
(2) Show that R is isomorphic to $S := k\langle x, y\rangle/(xy-1)$, with a k-isomorphism matching f with $\bar{x}$ and g with $\bar{y}$. (**Hint.** Show that for any i, j, $g^{i-1}f^{j-1} - g^i f^j = E_{ij}$, in the notation of Exercise 8.)

Ex. 11.10. For a field k, construct two left modules V, V' over the free algebra $R = k\langle x, y\rangle$ as follows. Let $V = V' = \sum_{i=1}^{\infty} e_i k$. Let R act on V by:

$$xe_i = e_{i-1}, \quad ye_i = e_{i^2+1},$$

and let R act on V' by

$$xe_i = e_{i-1}, \quad ye_i = e_{i^2+2}$$

(with the convention that $e_0 = 0$). Show that V, V' are nonisomorphic faithful simple left R-modules. (**Hint.** To show that $V \not\cong V'$ as R-modules, note that $x^2 y$ annihilates e_1 in V, but does not annihilate any nonzero vector in V'.)

Ex. 11.11. Let A be a subring of a field K. Show that the subring S of $K\langle x_1, \ldots, x_n\rangle$ $(n \geq 2)$ consisting of polynomials with constant terms in A is a left primitive ring with center A.

Ex. 11.12. Let k be a field of characteristic zero. Represent the Weyl algebra $R = k\langle x, y\rangle/(xy - yx - 1)$ as a dense ring of linear transformations on an infinite-dimensional k-vector space V, and restate the Density Theorem in this context as a theorem on differential operators. (**Hint.** Let ${}_R V = k[y]$ with $y \in R$ acting as left multiplication by y and $x \in R$ acting as $D = d/dy$ (formal differentiation). Choosing the basis $\{e_n = y^n: \ n \geq 0\}$ on V, we have

$ye_n = e_{n+1}$ and $xe_n = ne_{n-1}$ (with $e_{-1} = 0$). The R-module ${}_RV$ is simple, with $End({}_RV) = k$ (more generally, see Exercise 3.18), so R acts densely on V_k. Elements $\sum a_{ij}y^ix^j \in R$ act as differential operators $\sum a_{ij}y^iD^j$ on $k[y]$.)

Ex. 11.13. Let R be a left primitive ring such that $a(ab - ba) = (ab - ba)a$ for all $a, b \in R$. Show that R is a division ring. (**Hint.** Let ${}_RV$ be a faithful simple R-module with $k = End({}_RV)$. It suffices to show that $dim_k V = 1$. Assume instead there exist k-linearly independent vectors u, $v \in V$. By the Density Theorem, there exist a, $b \in R$ such that $au = u$, $av = 0$, and $bu = v$, $bv = 0$. The equation $a(ab - ba) = (ab - ba)a$ applied to u shows that $v = 0$, a contradiction.)

Ex. 11.14. Let R be a left primitive ring such that $1 + r^2$ is a unit for any $r \in R$. Show that R is a division ring.

In the next four exercises (15 through 18), let R be a left primitive ring, ${}_RV$ be a faithful simple R-module, and k be the division ring $End({}_RV)$. Recall that, by the Density Theorem, R acts densely on V_k.

Ex. 11.15. For any k-subspace $W \subseteq V$, let

$$ann(W) = \{r \in R : \ rW = 0\},$$

and, for any left ideal $\mathfrak{A} \subseteq R$, let

$$ann(\mathfrak{A}) = \{v \in V : \ \mathfrak{A}v = 0\}.$$

Suppose $n = dim_k\ W < \infty$. *Without assuming the Density Theorem*, show by induction on n that $ann(ann(W)) = W$. From this equation, deduce that R acts densely on V_k. If, in addition, R is left artinian, show that $dim_k\ V < \infty$, $R = End(V_k)$, and $ann(ann(\mathfrak{A})) = \mathfrak{A}$ for any left ideal $\mathfrak{A} \subseteq R$. In this case, $W \mapsto ann(W)$ gives an inclusion-reversing one-one correspondence between the subspaces of V_k and the left ideals of the (simple artinian) ring R.

Ex. 11.16. For any $r \in R$ of rank m (i.e. $dim_k\ rV = m$), show that there exist $r_1, \ldots, r_m \in Rr$ of rank 1 such that $r = r_1 + \cdots + r_m$. (**Hint.** Write $rV = e_1k \oplus \cdots \oplus e_mk$ and apply the Density Theorem.)

Ex. 11.17. (1) Show that $\mathfrak{A} \subseteq R$ is a minimal left ideal of R iff $\mathfrak{A} = Re$ where $e \in R$ has rank 1, and that (2) $\mathfrak{B} \subseteq R$ is a minimal right ideal of R iff $\mathfrak{B} = eR$ where $e \in R$ has rank 1.

Ex. 11.18. Show that the following statements are equivalent:
(1) $soc(R) \neq 0$,
(2) R contains a projection of V onto a line,
(3) there exists a nonzero $r \in R$ of finite rank, and
(4) for any finite-dimensional k-subspace $W \subseteq V$, R contains a projection of V onto W.
Finally, show that $soc(R) = \{r \in R : \ rank(r) < \infty\}$.

Ex. 11.19. Show that a ring R is left primitive iff R is prime and R has a faithful left module of finite length.

Ex. 11.20. For any division ring k with center C, show that the following are equivalent:
(1) $k[x]$ is left primitive,
(2) $k[x]$ is right primitive,
(3) there exists $0 \neq f \in k[x]$ such that $k[x]f \cap C[x] = 0$.
(**Hint.** Show that, if $0 \neq gf \in C[x]$, then $gf = fg$.)

Ex. 11.21. Show that the Density Theorem (11.16) need not hold if the $_RV$ there is *not* a semisimple module. (**Hint.** Let $R = \begin{pmatrix} k & k \\ 0 & k \end{pmatrix}$ act on $V = \begin{pmatrix} k \\ k \end{pmatrix}$ by matrix multiplication. We have shown before (7.4) that $End(_RV) = k$, but no element of R can take $\begin{pmatrix} 1 \\ 0 \end{pmatrix}$ to $\begin{pmatrix} 0 \\ 1 \end{pmatrix}$, so R is not even 1-transitive. For another counterexample, take $R = \mathbb{Z}$, $V = \mathbb{Q}$.)

§12. Subdirect Products and Commutativity Theorems

In this section, we introduce the notion of subdirect products and try to explain how to use such subdirect product representations in the general study of the structure of rings. We start by formally defining a subdirect product representation.

(12.1) Definition. Let R and $\{R_i: i \in I\}$ be rings, and $\varepsilon: R \to \prod_{i \in I} R_i$ be an injective ring homomorphism. We say that ε *represents* R *as a subdirect product of the* R_i's if each of the maps $R \to R_i$ (obtained by composing ε with the coordinate projections) is *onto*. (More informally, we say that R is a subdirect product of the R_i's.) We say that the subdirect product representation above is a *trivial* representation if one of the maps $R \to R_i$ is an isomorphism.

We observe that R can be represented as a subdirect product of

$$\{R_i: i \in I\}$$

iff there exists, for every i, a surjective ring homomorphism $\varphi_i: R \to R_i$ such that $\bigcap_{i \in I} ker\ \varphi_i = 0$. The case of a trivial representation occurs when one of the φ_i's is already an isomorphism. This case is not of much interest since the presence of the other φ_i's will no longer play a significant role.

(12.2) Proposition and Definition. *For a nonzero ring R, the following statements are equivalent:*

(1) *Every representation of R as a subdirect product of other rings is trivial.*

(2) *The intersection of all nonzero ideals of R is nonzero.*

(3) *R has a nonzero ideal which is contained in all other nonzero ideals.*

If R satisfies these properties, we shall say that R is subdirectly irreducible; otherwise, we say that R is subdirectly reducible.

Proof. We have clearly (2) $\Leftrightarrow$ (3), so what we need is (1) $\Leftrightarrow$ (2).

(1) $\Rightarrow$ (2). Let $\{\mathfrak{A}_i: i \in I\}$ be the nonzero ideals of R and let $R_i = R/\mathfrak{A}_i$. If $\bigcap \mathfrak{A}_i = 0$, we have an obvious subdirect product representation $R \to \prod R_i$ with each $R \to R_i$ *not* an isomorphism. This contradicts (1).

(2) $\Rightarrow$ (1). Consider any subdirect product representation $\varepsilon\colon R \to \prod R_i$ and let $\mathfrak{A}_i = ker(R \to R_i)$. If each $\mathfrak{A}_i \neq 0$, then $\bigcap \mathfrak{A}_i \neq 0$ by (2), contradicting the fact that ε is injective. Therefore, some $\mathfrak{A}_i = 0$; i.e., ε is a trivial representation. QED

Let us say that an ideal $L \subseteq R$ is *little* if $L \neq 0$ and L is contained in every nonzero ideal of R. By (12.2), *R has a little ideal L iff R is subdirectly irreducible*; in this case, clearly, L is unique and is equal to the intersection of all nonzero ideals in R. In the following, we shall give some examples of subdirectly reducible and subdirectly irreducible rings.

(1) Clearly, *any simple ring R is subdirectly irreducible.* (Its little ideal is given by R itself.)

(2) A semisimple ring is subdirectly irreducible iff it has only one simple component.

(3) $\mathbb{Z}/(p^n)$ (p a prime) *and* $k[t]/(p(t)^n)$ (k a field, $p(t)$ an irreducible polynomial) *are subdirectly irreducible,* for all $n > 0$. Their little ideals are, respectively, $(p^{n-1})/(p^n)$ and $(p(t)^{n-1})/(p(t)^n)$.

(4) *$\mathbb{Z}$ is not subdirectly irreducible,* since it cannot have a little ideal. In fact, for any infinite set of primes $\{p_i\}$, the natural map

$$\mathbb{Z} \longrightarrow \prod \mathbb{Z}/(p_i^{n_i}) \quad (n_i > 0)$$

is a nontrivial representation of $\mathbb{Z}$ as a subdirect product of subdirectly irreducible rings. (A similar statement can be made about the polynomial ring $k[t]$ in (3), noting that $k[t]$ has always infinitely many primes.)

(5) *Let $R = \mathbb{Z}G$ where G is a finite group. Then R is not subdirectly irreducible.* By (4), we may assume that $G \neq \{1\}$. Since $\mathbb{Q}G$ is semisimple, but not simple, we have an isomorphism

$$\varepsilon\colon \mathbb{Q}G \longrightarrow C_1 \times \cdots \times C_r$$

where $r \geq 2$ and the C_i's are simple rings. Then each projection $\mathbb{Q}G \to C_i$ has a nonzero kernel, so $\mathbb{Z}G \to C_i$ has also a nonzero kernel. Letting A_i be the image of $\mathbb{Z}G$ in C_i, we have then a nontrivial subdirect product representation

$$\varepsilon: \ \mathbb{Z}G \longrightarrow A_1 \times \cdots \times A_r.$$

For instance, if G is the cyclic group of order n, then ε represents $\mathbb{Z}G$ as a subdirect product of the rings $\mathbb{Z}[\zeta_d]$, where d ranges over all positive divisors of n.

(6) *Let R be a prime ring. If $soc(R) \neq 0$, then R is subdirectly irreducible, with $soc(R)$ as its little ideal.* To see this, it suffices to see that any minimal left ideal $\mathfrak{A}$ is contained in every nonzero ideal I. Since R is prime, we have $0 \neq I \cdot \mathfrak{A} \subseteq \mathfrak{A}$. The minimality of $\mathfrak{A}$ then implies that $\mathfrak{A} = I \cdot \mathfrak{A} \subseteq I$.

(7) The converse of (6) is false. Indeed, let R be a nonartinian simple ring. Then R is prime and subdirectly irreducible, but $soc(R) = 0$ by (3.10).

(8) *A left primitive ring need not be subdirectly irreducible.* For instance, the left primitive ring R constructed in (11.13) has clearly no little ideal (by (11.12)), so R is subdirectly reducible.

The role of subdirectly irreducible rings is seen from the following result.

(12.3) Birkhoff's Theorem. *Any nonzero ring R can be represented as a subdirect product of subdirectly irreducible rings.*

Proof. For any $a \neq 0$ in R, let $\mathfrak{m}_a$ be an ideal maximal with respect to the property that $a \notin \mathfrak{m}_a$. (Such an ideal exists by Zorn's Lemma.) In $R/\mathfrak{m}_a$, any nonzero ideal must contain $a + \mathfrak{m}_a$ so $R/\mathfrak{m}_a$ is subdirectly irreducible. Since $\bigcap_{a \neq 0} \mathfrak{m}_a = 0$, the natural map

$$R \longrightarrow \prod_{a \neq 0} R/\mathfrak{m}_a$$

represents R as a subdirect product of the subdirectly irreducible rings $R/\mathfrak{m}_a$. QED

This theorem suggests that, in a way, we may view the subdirectly irreducible rings as the building blocks for arbitrary rings. However, the structure of subdirectly irreducible rings can still be very complicated, and one cannot realistically hope to describe them completely. In the category of *commutative* rings, some work has been done by McCoy and Divinsky toward the determination of the subdirectly irreducible rings. We shall carry out this determination only in the easy case of commutative reduced rings.

(12.4) Proposition. *Let R be a commutative reduced ring. Then R is subdirectly irreducible iff R is a field.*

Proof. ("Only if"). Let R be subdirectly irreducible. Clearly R has only trivial idempotents. (If e is an idempotent other than 0 and 1, then

$$R \cong Re \times R \cdot (1 - e)$$

is a nontrivial (sub)direct product representation.) Fix an element $a \neq 0$ in the little ideal of R. Since $a^2 \neq 0$, we have $a \in (a^2)$ so $a = a^2 b$ for some $b \in R$. But then $ab = a^2 b^2 = (ab)^2$ implies that $ab = 1$ (since $ab \neq 0$). For any $c \neq 0$, we have $a \in Rc$. Since a is a unit, c is also a unit. Therefore, R is a field. QED

Returning now to the study of arbitrary rings, we have the following result.

(12.5) Theorem. *A nonzero ring R is semiprime (resp., semiprimitive) iff R is a subdirect product of prime (resp., left primitive) rings.*

Proof. Assume R is semiprime (resp., semiprimitive). Let $\{\mathfrak{A}_i\}$ be the family of prime (resp., left primitive) ideals in R. Then $\bigcap \mathfrak{A}_i = 0$ so R is a subdirect product of the prime (resp., left primitive) rings $\{R/\mathfrak{A}_i\}$. Conversely, assume there is a subdirect product representation $R \to \prod R_i$ where the R_i's are prime (resp., left primitive) rings. Then, for any i,

$$\mathfrak{A}_i := ker(R \longrightarrow R_i)$$

is a prime (resp., left primitive) ideal, and $\bigcap \mathfrak{A}_i = 0$. Since $Nil_* R$ (resp., rad R) is contained in $\bigcap \mathfrak{A}_i = 0$, it must be zero. QED

There also exists a parallel result which gives a subdirect product characterization of a (possibly noncommutative) reduced ring. For this result, we need a lemma. Let us say that an ideal $\mathfrak{p}$ in a ring R is *completely prime* if $R/\mathfrak{p}$ is a domain. Clearly a completely prime ideal is always prime.

(12.6) Lemma. *Let R be a reduced ring. If $\mathfrak{p}$ is a minimal prime in R, then $\mathfrak{p}$ is completely prime.*

Proof (following Rowen [88]). Let $S = R \backslash \mathfrak{p}$, and let $S' \supseteq S$ be the (multiplicative) monoid generated by S. *We claim that* $0 \notin S'$. For otherwise, we'll have an equation $s_1 \cdots s_n = 0$, with all $s_i \in S$ and with n minimal. Clearly, $n \geq 2$. Since R is reduced and $(s_n R s_1 \cdots s_{n-1})^2 = 0$, we have $s_n R s_1 \cdots s_{n-1} = 0$. But $\mathfrak{p}$ is prime, so there exists an element $s := s_n r s_1 \in S$, for some $r \in R$. We have then $s s_2 \cdots s_{n-1} = 0$, in contradiction to the minimal choice of n. With the knowledge that $0 \notin S'$, we can "enlarge" (0) to a prime ideal $\mathfrak{p}'$ disjoint from S' (using (10.5)). But $\mathfrak{p}$ is a minimal prime, so we must have $\mathfrak{p}' = \mathfrak{p}$; that is, $S' = S$. Thus, S' is closed under multiplication, so $R/\mathfrak{p}$ is a domain. QED

(12.7) Theorem (Andrunakievich-Ryabukhin). *A nonzero ring R is reduced iff R is a subdirect product of domains.*

Proof. First assume there is a subdirect product representation $R \to \prod R_i$ where the R_i's are domains. If $a \in R$ is nilpotent, then a maps to zero in each R_i, so $a = 0$. Conversely, assume R is reduced. Since $Nil_* R$ is a nil ideal, we have $Nil_* R = 0$. Let $\{\mathfrak{p}_i\}$ be the family of minimal prime ideals in R. By Exercise 10.14,

$$\bigcap \mathfrak{p}_i = Nil_* R = 0,$$

so R is a subdirect product of $\{R/\mathfrak{p}_i\}$. By the Lemma above, each $R/\mathfrak{p}_i$ is a domain. QED

In passing, we observe the following. Call an ideal $\mathfrak{A} \subseteq R$ *reduced* if $R/\mathfrak{A}$ is a reduced ring. It follows easily from the results above that an ideal $\mathfrak{A} \subseteq R$ *is reduced iff* $\mathfrak{A}$ *is an intersection of completely prime ideals.* This may be regarded as an analogue of the result that an ideal $\mathfrak{A} \subseteq R$ is semiprime iff $\mathfrak{A}$ is an intersection of prime ideals.

It is also worthwhile to point out that, using (12.6), we can actually derive a *necessary and sufficient* condition for all minimal primes in a ring R to be completely prime. This result should be viewed as a self-strengthening of (12.6).

(12.6)′ Theorem (G. Shin). *Every minimal prime in a ring R is completely prime iff every nilpotent element of R is in $Nil_*(R)$.*

Proof. Note that the latter condition amounts to the factor ring $\bar{R} := R/Nil_*(R)$ being reduced. Suppose this holds, and let $\mathfrak{p}$ be any minimal prime of R. Then its image $\bar{\mathfrak{p}}$ in $\bar{R}$ is also a minimal prime. By (12.6), $R/\mathfrak{p} \cong \bar{R}/\bar{\mathfrak{p}}$ is a domain, so $\mathfrak{p}$ is completely prime in R. Conversely, let $\{\mathfrak{p}_i\}$ be the family of minimal primes in R, and suppose each $\mathfrak{p}_i$ is completely prime. Then, each $R/\mathfrak{p}_i$ is a domain. Using Exercise 10.14 again, we have $Nil_*(R) = \bigcap_i \mathfrak{p}_i$. Thus, $\bar{R}$ embeds in $\prod_i R/\mathfrak{p}_i$, which implies that $\bar{R}$ is reduced. QED

In the literature, there is a somewhat strange name given to the rings R satisfying the second condition in (12.6)′: they are called "2-primal rings". *These are the rings in which $Nil_*(R)$ consists precisely of all the nilpotent elements of R.* (For instance, reduced rings and commutative rings are both 2-primal.) In such rings R, we have obviously $Nil_*(R) = Nil^*(R)$, and any nil 1-sided ideal lies in $Nil_*(R)$; in particular, Köthe's Conjecture holds for R.

We will not go into more details on 2-primal rings here, but in the exercise set for this section, we shall describe a couple of other conditions on rings that are (strictly) between "reduced" and "2-primal": see Ex. (12.18).

In studying the structure of rings, it is often useful to keep in mind the following chart of basic objects and their relationships:

$$\text{(12.8)}\qquad \begin{array}{ccc} \text{Rings} \xrightarrow{\text{mod radical}} & \text{Semiprimitive Rings} & \\ & \Big\downarrow {\scriptstyle \text{subdir. prod. representation}} & \\ & \text{Left Primitive Rings} \xrightarrow[\text{Density Theorem}]{} & \text{Division Rings and their matrix rings} \end{array}$$

When we try to prove certain theorems about rings, it is sometimes possible to make use of the chart above to reduce the proofs from one class of rings to another, simpler, class. To begin with, we would first test the desired theorems on division rings and their matrix rings. If these theorems hold up, then, using the Density Theorem, we would test them for left primitive rings. If they still hold up, we would next use subdirect product representations to test them for semiprimitive rings. In the literature, a number of results on semiprimitive rings have been obtained in this manner. For any ring R, these results are then valid modulo the Jacobson radical *rad* R. Lifting these results to R itself is usually more difficult, and sometimes impossible; however, whenever it can be done, one will get results for a general ring R.

Needless to say, the procedure sketched above is no panacea for proving all theorems, even only for semiprimitive rings. Nevertheless it is an important procedure, and it has proved to be effective in a number of instances. As an illustration, we shall show how this procedure works in proving several interesting commutativity theorems due to Jacobson, Herstein, and Kaplansky.

(12.9) Jacobson–Herstein Theorem. *A ring R is commutative iff:*

(*) *For any $a, b \in R$, there exists an integer $n(a,b) > 1$ such that $(ab - ba)^{n(a,b)} = ab - ba$.*

Of course, the thrust of the theorem is in its "if" part. Note that this result includes as a special case the following theorem of Jacobson, which was its predecessor.

(12.10) Jacobson's Theorem. *Let R be a ring such that, for any $a \in R$, $a^{n(a)} = a$ for some integer $n(a) > 1$. Then R is commutative.*

Jacobson's Theorem was in part motivated by Wedderburn's classical result that any finite division ring is commutative. Clearly (12.10) applied to finite division rings would yield Wedderburn's result. However, Wedderburn's result is *not* to be viewed as a corollary of (12.10) since it is usually assumed in the proof of (12.10) (and (12.9)).

The truth of (12.9) for division rings will be proved in the next chapter (see (13.9)) when we study division rings in more detail. Here we shall assume (12.9) for division rings and show how to deduce it for arbitrary rings by using the general procedure of (12.8). As we have mentioned before, we need only deal with the "if" part.

Step 1. (12.9) *is true for any left primitive ring* R. By the structure theorem for left primitive rings, there exists a division ring k such that either (1) $R \cong \mathbb{M}_m(k)$ for some m, or (2) for any m, there exists a subring $R_m \subseteq R$ which has a ring homomorphism onto $\mathbb{M}_m(k)$. However, for $m \geq 2$, $\mathbb{M}_m(k)$ can *never* satisfy $(*)$. For if we take a, b to be the matrix units E_{11} and E_{12}, then $ab - ba = b$, so

$$(ab - ba)^n = b^n = 0 \neq ab - ba \quad \text{for any } n \geq 2.$$

Since R and hence its subrings and their homomorphic images satisfy $(*)$, we must have $R \cong k$, so we are back to the case of division rings.

Step 2. (12.9) *is true for any semiprimitive ring* R. We have a subdirect product representation $R \to \prod R_i$, where the R_i's are left primitive rings. Each R_i, being a homomorphic image of R, satisfies $(*)$, and is therefore commutative. Since R is isomorphic to a subring of $\prod R_i$, it is also commutative.

Step 3. (12.9) *is true for any ring* R. By Step 2, we know that $R/rad\ R$ is commutative. Therefore, for any $a, b \in R$, the additive commutator $d = ab - ba$ lies in $rad\ R$. For $n = n(a, b) > 1$, we have $d^n = d$, so $d(1 - d^{n-1}) = 0$. Since

$$1 - d^{n-1} \in 1 + rad\ R$$

is a unit, it follows that $d = 0$, so R is commutative. QED

In the literature, there is a host of other commutativity theorems that can be proved by similar techniques. We shall restrict ourselves to only one more such result, as follows.

(12.11) Theorem (Herstein, Kaplansky). *For any semiprimitive ring* R, *the following statements are equivalent*:

(1) R *is commutative*;

(2) *For any* $a, b \in R$, *the additive commutator* $ab - ba$ *lies in the center* $Z(R)$ *of* R;

(3) *For any* $a \in R$, $a^{n(a)} \in Z(R)$ *for some* $n(a) \geq 1$.

Proof. As before, it suffices to prove (2) $\Rightarrow$ (1) and (3) $\Rightarrow$ (1). The proofs of these implications for division rings will be given later in the next chapter (see (13.5) and (15.15)). Assuming these later results, the proofs of

$$(2) \Longrightarrow (1) \quad \text{and} \quad (3) \Longrightarrow (1)$$

for semiprimitive rings can be given along the same lines of Step 1 and Step 2 above. It is clear that (2) and (3) are each inherited by the subrings of R and their quotient rings, so the only thing left to be checked is that (2) and (3) cannot be satisfied by $\mathbb{M}_m(k)$ $(m \geq 2)$ for any division ring k. In fact, for any nonzero ring k, if we let $a = E_{11}$ and $b = E_{12}$ in $\mathbb{M}_m(k)$ as before, then $ab - ba = b \notin Z(\mathbb{M}_m(k))$, and $a^n = a \notin Z(\mathbb{M}_m(k))$ for any $n \geq 1$ and $m \geq 2$.[1] QED

The theorem above gives an example of a result which can be proved by the reduction techniques of (12.8) for semiprimitive rings, but not for arbitrary rings. In fact, counterexamples for (2) $\Rightarrow$ (1) and (3) $\Rightarrow$ (1) for general rings can be obtained as follows. Let k be a field, and let

$$R = \left\{ \begin{pmatrix} x & y & z \\ 0 & x & w \\ 0 & 0 & x \end{pmatrix} : x, y, z, w \in k \right\}.$$

This ring is easily checked to be noncommutative, but for any $a, b \in R$, $ab - ba$ is a scalar multiple of the matrix unit E_{13} which is in the center of R. Thus, (2) holds but (1) does not. Here, $rad\ R = \left\{ \begin{pmatrix} 0 & y & z \\ 0 & 0 & w \\ 0 & 0 & 0 \end{pmatrix} \right\}$ has dimension 3 over k, and $R/rad\ R \cong k$. To construct a counterexample for (3) $\Rightarrow$ (1), let $R = \mathbb{F}_p G$ where G is a noncommutative finite p-group. Then R is noncommutative. The Jacobson radical J of R is given by the augmentation ideal, and satisfies $J^{|G|} = 0$ (see (8.8)). Now consider any $a \in R$. If $a \in J$, then $a^{|G|} = 0 \in Z(R)$. If $a \notin J$, then, since $R/J \cong \mathbb{F}_p$, there exists $r \in \mathbb{F}_p$ such that $a - r \in J$; but then $(a - r)^{|G|} = 0$ and so

$$a^{|G|} = r^{|G|} = r \in \mathbb{F}_p \subseteq Z(R).$$

Therefore, R satisfies (3), but not (1).

Exercises for §12

Ex. 12.0. (1) Characterize rings R which are subdirect products of fields. (2) Characterize rings S which can be embedded into direct products of fields.

---

[1] In connection with (12.11), we should point out that Herstein has shown that (3) $\Rightarrow$ (1) holds already for *semiprime* rings.

(3) Characterize rings T which can be embedded into a direct product of $\mathbb{F}_2$'s and $\mathbb{F}_3$'s.

Ex. 12.1. Let R be a subdirectly irreducible ring. Show that if R is semiprimitive (resp. semiprime, reduced), then R is left primitive (resp. prime, a domain). In particular, show that R is left primitive iff R is right primitive.

Ex. 12.2. (1) Show that a commutative domain cannot have exactly three ideals. (2) Show that there exist (noncommutative) domains A which have exactly three ideals. (3) Give an example of a nonsimple subdirectly irreducible domain.

Ex. 12.3A. Let k be a field of characteristic p and G be a finite p-group, where p is a prime. Show that the ring $A = kG$ is subdirectly irreducible, with its little ideal given by $k\sigma$ where $\sigma = \sum_{g \in G} g \in A$.

Ex. 12.3B. The following exercise appeared in an algebra text: "The zero-divisors in a subdirectly irreducible ring (together with zero) form an ideal." Give a counterexample! Then suggest a remedy.

Ex. 12.3C. (McCoy) Let R be a commutative subdirectly irreducible ring which is not a field. Let L be the little ideal of R. Prove the following statements:
(1) $L = aR$ for some a with $a^2 = 0$.
(2) $\mathfrak{m} := ann(a)$ is a maximal ideal of R.
(3) $\mathfrak{m} = I$ (the set of all 0-divisors of R together with 0).
(4) $ann(\mathfrak{m}) = aR$, and $a \in Nil(R) \subseteq \mathfrak{m}$.
(5) If R is noetherian, then $Nil(R) = \mathfrak{m}$, and R is an artinian local ring.

Ex. 12.3D. Let R be a commutative artinian ring. Show that R is subdirectly irreducible iff $soc(R)$ is a minimal ideal.

Ex. 12.4. The following exercise appeared in an algebra text: "Let $n \geq 2$ and a, b be elements in a ring D. If

$$(*) \qquad a^n - b^n = (a-b)(a^{n-1} + a^{n-2}b + \cdots + ab^{n-2} + b^{n-1}),$$

then $ab = ba$." Show that
(1) this is true for $n = 2$;
(2) this is false for $n = 3$, even in a division ring D;
(3) if (*) holds for $n = 3$ for *all* a, b in a ring D, then indeed $ab = ba$ for all a, b.

Ex. 12.5. Call a ring R *strongly* (von Neumann) *regular* (or *abelian regular*) if, for any $a \in R$, there exists $x \in R$ such that $a = a^2x$. If R is strongly regular, show that R is semiprimitive, and that R is left primitive iff it is a division ring. Using this, show that any strongly regular ring is a subdirect product of division rings.

Ex. 12.6A. (Arens-Kaplansky, Forsythe-McCoy) Show that the following conditions on a ring R are equivalent:

(1) R is strongly regular;
(2) R is von Neumann regular and reduced;
(3) R is von Neumann regular and every idempotent in R is central;
(4) Every principal right ideal of R is generated by a central idempotent.
(**Hint.** Do the cycle of implication (1) $\Rightarrow$ (2) $\Rightarrow$ (3) $\Rightarrow$ (4) $\Rightarrow$ (1). Here, (2) $\Rightarrow$ (3) is true for any ring R.)

Ex. 12.6B. (Jacobson, Arens-Kaplansky) Let R be a reduced algebraic algebra over a field k. Let $r \in R$, and $\varphi(x) \in k[x]$ be its minimal polynomial over k.
(1) Show that any root $a \in k$ of $\varphi(x)$ is a simple root.
(2) Show that R is strongly regular.

Ex. 12.6C. (Ehrlich) Show that any strongly regular ring R is unit-regular in the sense of Exercise 4.14B, that is, for any $a \in R$, there exists $u \in U(R)$ such that $a = aua$.

Ex. 12.7. In the Hint to Exercise 6A, we have mentioned the fact that any idempotent e in a reduced ring is central. Now prove the following more general fact: in any ring R, an idempotent $e \in R$ is central if (and only if) it commutes with all nilpotent elements in R.

Ex. 12.8A. Prove the following commutativity theorem without using the machinery of subdirect products: Let R be a ring such that $(ab)^2 = a^2b^2$ for any elements $a, b \in R$. Then R is commutative.

Ex. 12.8B. Let R be a ring possibly without an identity. If $a^2 - a \in Z(R)$ for every $a \in R$, show that R is a commutative ring.

Ex. 12.8C. Let R be a ring such that, for any $a \in R$, $a^{n(a)} = a$ for some integer $n(a) > 1$. By Jacobson's Theorem (12.10), R is a commutative ring. Show that R has Krull dimension 0.

Ex. 12.9. Give elementary proofs for the following special cases of Jacobson's Theorem (12.10): Let $n \in \{2, 3, 4, 5\}$ and let R be a ring such that $a^n = a$ for all $a \in R$. then R is commutative. (**Hint.** Assume $n = 3$. Use the Hint for Ex. 12.6A to show that $a^2 \in Z(R)$. Now expand $(a+1)^2$, $(a+1)^3$ to show that $2a$, $3a \in Z(R)$. Next assume $n = 4$. Show as above that $a^2 + a \in Z(R)$, whence $ab + ba \in Z(R)$. Finally, $a(ab + ba) = (ab + ba)a$ gives $a^2b = ba^2$. The case $n = 5$ is more daunting, but still doable along the same lines.)

Ex. 12.10. Let R be a ring such that $a^6 = a$ for any $a \in R$. Show that $a^2 = a$ for any $a \in R$, i.e. R is a (commutative) Boolean ring. (**Hint.** Note that $char(R) = 2$ and expand $a + 1 = (a+1)^6$.)

Ex. 12.11. Let p be a fixed prime. Following McCoy, define a nonzero ring R to be a *p-ring* if $a^p = a$ and $pa = 0$ for all $a \in R$. Assuming Jacobson's Theorem (12.10), show that:

(1) A ring R is a p-ring iff it is a subdirect product of $\mathbb{F}_p$'s;
(2) A finite ring R is a p-ring iff it is a finite direct product of $\mathbb{F}_p$'s.

Ex. 12.12. For any field k, show that there exists a k-division algebra D with two elements $a \neq c$ such that $a^2 - 2ac + c^2 = 0$.

Ex. 12.13. Let R be a nonzero ring such that every subring of R is a division ring. Show that R must be an algebraic field extension of some $\mathbb{F}_p$.

Ex. 12.14. Show that a ring R is reduced iff, for any elements $a_1, \ldots, a_r \in R$ and any positive integers $n_1, \ldots, n_r$:

$$a_1 \cdots a_r \neq 0 \Longrightarrow a_1^{n_1} \cdots a_r^{n_r} \neq 0.$$

Ex. 12.15. Let a, b be elements in a reduced ring R, and r, s be two positive integers that are relatively prime. If $a^r = b^r$ and $a^s = b^s$, show that $a = b$ in R. (**Hint.** Use (12.7).)

Ex. 12.16. (Cohn) A ring R is said to be *reversible* if $ab = 0 \in R \Rightarrow ba = 0$. Show that:
(1) A ring R is a domain iff R is prime and reversible.
(2) R is a reduced ring iff R is semiprime and reversible.

Ex. 12.17. A ring R is said to be *symmetric* if $abc = 0 \in R \Rightarrow bac = 0$. Show that R is symmetric iff, for any n,

$$a_1 \cdots a_n = 0 \in R \Longrightarrow a_{\pi(1)} \cdots a_{\pi(n)} = 0 \quad \text{for any permutation } \pi.$$

Ex. 12.18. Show that, for any ring R:

$$R \text{ is reduced} \Rightarrow R \text{ is symmetric} \Rightarrow R \text{ is reversible} \Rightarrow R \text{ is 2-primal},$$

but each of these implications is irreversible!

Ex. 12.19. (1) If R is a left artinian ring for which $R/\mathit{rad}\, R$ is reduced, show that R is 2-primal.
(2) Show that a finite direct product of 2-primal rings is 2-primal.
(3) (Birkenmeier-Heatherly-Lee) If R is 2-primal, show that so is the polynomial ring $R[T]$ (for any set of commuting variables T).

CHAPTER 5

Introduction to Division Rings

In the category of rings, the most "perfect" objects are the rings in which we can not only add, subtract, and multiply, but also divide (by nonzero elements). These rings are called *division rings*, or *skew fields*, or *sfields*. No matter how we call them, it is clear that a careful study of their properties would be vital for the development of ring theory in general. In this introductory chapter, we shall give an exposition on the basic theory of division rings, starting with Wedderburn's beautiful theorem that any finite division ring is commutative. This landmark result, proved by Wedderburn in 1905, has fascinated generations of algebraists and inspired a long sequence of more general commutativity theorems, by Jacobson, Kaplansky, Herstein, and others. We shall study some of these results in §13, and go on to study maximal subfields in division rings, polynomial equations over division rings, and ordered division rings in §§15, 16, and 18. In §14, we present several types of elementary constructions of division rings, thus providing some basic examples with which to illustrate the general theory. This section is written independently of §13, so it is possible for the reader to start this chapter by first reading §14 to see the basic examples before reading §13.

Fields are special examples of division rings, and, of course, there is a rich and very extensive theory of fields (and field extensions). This theory, however, belongs more properly to the domain of commutative algebra. Since in this book our main interest is in noncommutative rings, our presentation on division rings will focus more on the *noncommutative* aspects of the theory. Thus the usual theory of fields and field extensions will almost be completely ignored in this chapter.

All division rings may be broadly classified into two types, according to whether they are finite dimensional (as vector spaces) over their centers. To distinguish these types, we speak of *centrally finite* division rings and *centrally infinite* division rings. The theory of centrally finite division rings is very well developed, and certainly deserves a separate book for its study. For this, we refer the reader to Jacobson's treatise "Finite-dimensional Division Algebras over Fields," Springer-Verlag, 1996. In this chapter, our general policy is to try to avoid imposing the "centrally finite" hypothesis on division rings. Thus, the results we obtain in §§13–14 will be largely applicable to *all* division rings.

§13. Division Rings

In this beginning section, we shall study some of the most basic results concerning division rings. The following notations will be used consistently throughout: D denotes a division ring and D^* denotes its multiplicative group; for $a, b \in D$ and $x, y \in D^*$, $ab - ba$ is called an *additive commutator*, and $x^{-1}y^{-1}xy$ is called a *multiplicative commutator*. For any subset $S \subseteq D$, $C(S) = C_D(S)$ denotes

$$\{d \in D:\ ds = sd \text{ for all } s \in S\}$$

(the centralizer of S in D). Note that $C(S)$ is a division subring of D containing $Z(D)$, the center of D.

The first result we shall present here is Wedderburn's classic theorem which states that all finite division rings are commutative. This beautiful result was discovered by Wedderburn in 1905. It is remarkable that this discovery was made only within a couple of years of E.H. Moore's classification of finite commutative fields (1903).

(13.1) Wedderburn's "Little" Theorem. *Let D be a finite division ring. Then D is a (finite) field.*

Proof. The center F of D is a finite field, say $|F| = q$ (a prime power ≥ 2). We want to prove that $n := dim_F\, D$ is 1. *Assume that* $n > 1$ and write down the "class equation" for the finite group D^*:

$$|D^*| = q^n - 1 = q - 1 + \sum [D^* : C(a)^*].$$

Here, a ranges over a (nonempty) set of representatives of non-singleton conjugacy classes of D^*. Write $r = r(a) = dim_F\, C(a)$. Then $1 \leq r < n$, and the transitivity formula for dimensions shows that $r \,|\, n$. Rewriting the class equation, we have

$$(*) \qquad q^n - 1 = q - 1 + \sum \frac{q^n - 1}{q^r - 1}.$$

Since $r \,|\, n$, we have the following factorization in $\mathbb{Z}[x]$:

$$x^n - 1 = \Phi_n(x)(x^r - 1)h(x) \quad (h(x) \in \mathbb{Z}[x]),$$

where $\Phi_n(x)$ is the nth cyclotomic polynomial. This equation implies that each $(q^n - 1)/(q^r - 1)$ is an integer divisible by $\Phi_n(q)$. From $(*)$, it follows that $\Phi_n(q) \mid (q - 1)$. In particular,

$$q - 1 \geq |\Phi_n(q)| = \prod |q - \zeta|,$$

where ζ ranges over all primitive nth roots of unity. This is absurd since $n > 1$ and $q \geq 2$ clearly imply that $|q - \zeta| > q - 1 \geq 1$ for each ζ. QED

The class equation in the form $(*)$ above was given in Wedderburn's original proof of his "Little" Theorem in 1905. From this equation, Wedderburn derived a contradiction by appealing to a number-theoretic theorem of Birkhoff and Vandiver, which is a rather deep result. The above "surprise finish," which replaces the use of the Birkhoff–Vandiver Theorem by a simple application of cyclotomic polynomials was given in a paper of Witt in 1931.

We shall record two easy corollaries.

(13.2) Corollary. *Any finite subring R of a division ring D is a field.*

Proof. R is easily seen to be a division ring, so (13.1) applies. QED

Another remarkable consequence of Wedderburn's Little Theorem concerns the structure of finite subgroups of the multiplicative group of a division ring D. In the case when D is commutative, it is a well-known result in elementary field theory that any finite subgroup of D^* is cyclic. It turns out that this result remains valid for arbitrary division rings D, as long as *char* $D \neq 0$.

(13.3) Corollary. *Let D be a division ring of characteristic $p > 0$, and G be a finite subgroup of D^*. Then G is cyclic.*

Proof. Let $F = \mathbb{F}_p$ be the prime field of D, and let

$$K = \left\{\sum \alpha_i g_i : \ \alpha_i \in F, \ g_i \in G\right\}.$$

This is a finite subring of D, and so by (13.2) it is a field. Since G is a subgroup of K^*, G is cyclic. QED

If *char* $D = 0$, the result above, of course, does not hold. For instance, the division ring D of the real quaternions contains the quaternion group $\{\pm 1, \pm i, \pm j, \pm k\}$ which is not cyclic. Another, bigger, finite group contained in D^* is the binary tetrahedral group

$$\{\pm 1, \pm i, \pm j, \pm k, (\pm 1 \pm i \pm j \pm k)/2\}$$

of order 24 (see (1.1)). This leads to the following interesting question: *what finite groups can occur as subgroups of the multiplicative groups of division rings* (*of characteristic zero*)? The complete answer to this question has been given by Amitsur in 1955, but it will not be presented here.

Next we shall make some elementary observations on additive commutators in a division ring.

(13.4) Proposition. *Let D be a division ring. If an element $y \in D$ commutes with all additive commutators in D, then $y \in Z(D)$.*

Proof. If $y \notin Z(D)$, we have $xy \neq yx$ for some $x \in D$. Consider the equation $x(xy) - (xy)x = x(xy - yx)$. Since y commutes with the additive commutators $x(xy) - (xy)x$ and $xy - yx$ ($\neq 0$), it must commute with x, a contradiction. QED

(13.5) Corollary. *If all additive commutators are central in a division ring D, then D is a field.*

If S is a subset in a division ring D, the division ring generated by S is, by definition, the intersection of all division subrings of D containing S. This is the smallest division subring of D containing S. We have the following second consequence of (13.4).

(13.6) Corollary. *Let D be a noncommutative division ring. Then D is generated as a division ring by all of its additive commutators together with $Z(D)$. (In other words, D is generated as a $Z(D)$-division algebra by all of its additive commutators.)*

Proof. If $x \notin Z(D)$, we have $xy \neq yx$ for some $y \in D$. The $Z(D)$-division algebra generated by additive commutators of D contains $x(xy) - (xy)x$ and $xy - yx$ ($\neq 0$), and hence it contains x. Therefore, it is D. QED

For an element a in a ring D, write $\delta_a \colon D \to D$ for the map defined by $x \mapsto ax - xa$. This map is a *derivation* in the sense that $\delta_a(x + y) = \delta_a(x) + \delta_a(y)$ and $\delta_a(xy) = x\delta_a(y) + \delta_a(x)y$ for all $x, y \in D$. We say that δ_a is the *inner derivation* of D associated with a. An additive subgroup of D is said to be a *Lie ideal* if it is invariant under all δ_a ($a \in D$). Our last result on additive commutators concerns Lie ideals in division rings.

(13.7) Proposition. *Let $K \subsetneq D$ be division rings such that K is a Lie ideal in D. If char $K \neq 2$, then $K \subseteq Z(D)$.*

Proof. Consider any element $a \in D \backslash K$, and any $c \in K$. *We claim that they must commute.* Indeed, from $\delta_a^2(c) = ca^2 - 2aca + a^2c \in K$ and

$$\delta_{a^2}(c) = a^2c - ca^2 \in K,$$

we can add to get

$$2(a^2c - aca) = 2a\delta_a(c) \in K.$$

If $\delta_a(c) \neq 0$, we have $2\delta_a(c) \in K^*$, and hence $a \in K^*$, a contradiction. Therefore $\delta_a(c) = ac - ca = 0$, as claimed. Now consider any element $c' \in K^*$. Then a and ac' are both in $D \backslash K$, so, by the foregoing, they commute with c. But then $c' = a^{-1} \cdot ac'$ also commutes with c. This shows that $c \in Z(D)$, and hence $K \subseteq Z(D)$. QED

Later, we shall see that the four results above for additive commutators also have valid analogues for multiplicative commutators. Right now, we move on to prove a basic lemma on division rings from Herstein [68].

(13.8) Herstein's Lemma. *Let D be a division ring of characteristic $p > 0$. Suppose a is a noncentral, torsion element of D^*. Then there exists $y \in D^*$ such that $yay^{-1} = a^i \neq a$, for some $i > 0$. Moreover, y can be chosen to be an additive commutator in D.*

Proof. Adjoining a to the prime field $\mathbb{F}_p$, we get a finite field $K = \mathbb{F}_p[a]$. Writing $|K| = p^n$, we have, in particular, $a^{p^n} = a$. Let $\delta = \delta_a$, which is not the zero derivation, since a is not central. For $z \in K$, $\delta(z) = 0$, so δ is K-linear on ${}_KD$. *The main step in the proof is to show that δ has an eigenvector* in ${}_KD$.

Think of $\delta \in E := End({}_KD)$ as $\lambda - \rho$, where $\lambda, \rho \in E$ are defined by $\lambda(x) = ax$ and $\rho(x) = xa$ (for any $x \in D$). Since λ and ρ commute, and E has characteristic p, we have $\delta^{p^n} = (\lambda - \rho)^{p^n} = \lambda^{p^n} - \rho^{p^n}$, so[1]

$$\delta^{p^n}(x) = a^{p^n}x - xa^{p^n} = ax - xa = \delta(x).$$

Thus, $\delta^{p^n} = \delta \in E$. Using the factorization

$$t^{p^n} - t = \prod_{b \in K} (t - b) \in K[t]$$

and computing in the K-algebra E, we have

$$0 = \delta^{p^n} - \delta = \left(\prod_{b \in K^*} (\delta - b) \right) \cdot \delta.$$

Since $\delta \neq 0$, this implies that, for some $b_0 \in K^*$, $\delta - b_0 \in E$ is *not* a monomorphism. (Recall that monomorphisms are left-cancellable.) This means that $(\delta - b_0)x = 0$ for some $x \in D^*$: this x is then an eigenvector for δ with eigenvalue $b_0 \in K^*$.

From $\delta(x) = ax - xa = b_0x$, we get $xax^{-1} = a - b_0 \in K \backslash \{a\}$. In the cyclic group K^*, xax^{-1} and a have the same order and so they generate the same cyclic subgroup. Hence $xax^{-1} = a^i$ $(\neq a)$ for some $i > 0$. If we further replace

---

[1] Actually, the method used here yields good information on powers of δ in any characteristic. By the Binomial Theorem in commutative algebra, we have $(\lambda - \rho)^k = \sum_{i=0}^{k} (-1)^i \binom{k}{i} \lambda^{k-i} \rho^i$ for any k. This leads to an explicit formula $\delta^k(x) = \sum_{i=0}^{k} (-1)^i \binom{k}{i} a^{k-i} x a^i$ for any x.

x by the additive commutator $y = \delta(x) = ax - xa \neq 0$, then

$$ya = (ax - xa)a = aa^i x - a^i xa = a^i y,$$

so $yay^{-1} = a^i \neq a$. QED

As a matter of fact, the Lemma above remains true if $\operatorname{char} D = 0$. A characteristic-free proof of (13.8) can be given for *all* division rings D with the help of Dickson's Theorem (16.8): see Exercise (16.17) below.

At the end of the last chapter, we studied the Jacobson–Herstein Theorem (12.9) on commutativity, and showed that if the theorem holds true for division rings, then it holds true for arbitrary rings. However, the proof in the division ring case was not yet given. With the help of Herstein's Lemma, we can now return to tie up this loose end.

(13.9) Theorem. *Let D be a division ring such that, for any $a, b \in D$, there exists an integer $n = n(a, b) > 1$ such that $(ab - ba)^n = ab - ba$. Then D is a field.*

Proof. By the given hypothesis, any nonzero additive commutator has finite order in D^*. Let us assume that $D \neq F = Z(D)$. By (13.5), there exists an additive commutator $a = bb' - b'b \notin F$. For any $c \in F^*$,

$$ca = (cb)b' - b'(cb)$$

is also a nonzero additive commutator. Since a and ca both have finite order, there exists an integer $k > 0$ such that

$$1 = a^k = (ca)^k = c^k a^k,$$

so $c^k = 1$. From this, it follows that *char* F = *char* $D > 0$. Since the element a above is noncentral and torsion, Herstein's Lemma yields an additive commutator $y \in D^*$ such that $yay^{-1} = a^i \neq a$, where $i > 0$. By the given hypothesis, y is also torsion in D^*. Since y normalizes the cyclic group $\langle a \rangle$, the product $\langle a \rangle \cdot \langle y \rangle$ is a *finite subgroup* of D^*. By (13.3), this subgroup is commutative: this contradicts the fact that $yay^{-1} \neq a$. QED

For later reference, we shall give another application of Herstein's Lemma. Here we shall use the following notation: for a subfield F in a division ring D, and a subset $S \subseteq D$, we write $F(S)$ to denote the division subring of D generated by F and S. Note that if the elements of S commute with themselves and with the elements of F, then $F(S)$ is a subfield of D.

(13.10) Theorem. *Let D be an infinite division ring with center F. Then for any $a \in D$, $F(a)$ is contained in an infinite subfield K of D. In particular, the centralizer $C_D(a)$ ($\supseteq K$) is infinite.*

Proof. Clearly we may assume that $F \subsetneq D$ and that $F(a)$ is finite. In particular, F is finite. We may also assume that $a \notin F$, for, if otherwise, we can simply replace a by an element outside of F. Now a is noncentral and torsion in D^*, so by Herstein's Lemma, there exists $y \in D^*$ such that $yay^{-1} = a^i \neq a$, where $i > 0$. Letting $\langle y \rangle$ act on the group $\langle a \rangle$ by conjugation, we see that some y^n ($n > 0$) must act trivially on $\langle a \rangle$ (since $\langle a \rangle$ is finite). This means that y^n commutes with a, so $K := F(a, y^n)$ is a field (containing $F(a)$). But by the last part of the proof of (13.9), y must have infinite order. Hence $K \supseteq \langle y^n \rangle$ is an *infinite* field, as desired. QED

Our next batch of results concern algebraic algebras over fields. Recall that an *algebraic algebra* over a field F is an algebra each of whose elements is algebraic over F. Note that if such an algebra D is a domain, then it is in fact a division algebra. (For $d \in D$, $F[d]$ is a field.) In the following, we shall determine all algebraic division algebras over finite fields and the real field $\mathbb{R}$.

(13.11) Theorem (Jacobson). *Let D be an algebraic division algebra over a finite field F. Then D is commutative (and is therefore an algebraic field extension of F).*

Proof. Let $p = char\ F$. For any $d \in D$, $F[d]$ is a finite algebraic extension of F, so it is a finite field. If $p^n = |F[d]|$, then $d^{p^n} = d$. In particular, the hypothesis of (13.9) holds for D, so (13.9) implies that D is a field. QED

The determination of algebraic division algebras over the real field $\mathbb{R}$ goes back to the 19th century: it was accomplished by Frobenius in a paper published in 1877. Note that in the following statement of Frobenius' Theorem, the algebraic algebra is *not* assumed to be finite-dimensional over $\mathbb{R}$ to begin with.

(13.12) Frobenius' Theorem. *Let D be an algebraic division algebra over $\mathbb{R}$. Then, as an $\mathbb{R}$-algebra, D is isomorphic to $\mathbb{R}$, $\mathbb{C}$, or $\mathbb{H}$ (the division algebra of real quaternions).*

Proof. We may assume that $dim_{\mathbb{R}}\ D \geq 2$ (for otherwise $D = \mathbb{R}$). Take an element $\alpha \in D \backslash \mathbb{R}$. Then $\mathbb{R}[\alpha]$ is a proper algebraic extension of $\mathbb{R}$, so $\mathbb{R}[\alpha] \cong \mathbb{C}$. In the following, *we shall fix a copy of $\mathbb{C}$ in D, and view D as a left vector space* over $\mathbb{C}$. The symbol i shall denote the complex number $\sqrt{-1} \in \mathbb{C}$.

Let $D^+ = \{d \in D\colon\ di = id\} \supseteq \mathbb{C}$, and $D^- = \{d \in D\colon\ di = -id\}$. These are $\mathbb{C}$-subspaces of ${}_{\mathbb{C}}D$ with $D^+ \cap D^- = 0$. We claim that $D^+ \oplus D^- = D$. Indeed, if $a \in D$, it is easy to see that

$$d^+ := ia + ai \in D^+, \quad \text{and} \quad d^- := ia - ai \in D^-.$$

Since $d^+ + d^- = 2ia$, we have

$$a = (2i)^{-1}(d^+ + d^-) \in D^+ + D^-.$$

(**Note.** If we look at the $\mathbb{C}$-linear map $\lambda\colon D \to D$ sending $a \in D$ to ai, the equation $D = D^+ \oplus D^-$ amounts exactly to the eigenspace decomposition of λ. However, the argument used above is straightforward and even more elementary.)

How big are the $\mathbb{C}$-subspaces D^+ and D^-? For any $d^+ \in D^+$, $\mathbb{C}[d^+]$ is an algebraic field extension of $\mathbb{C}$, so it must be $\mathbb{C}$ itself, which shows that $D^+ = \mathbb{C}$. If $D^- = 0$ we are done, so assume $D^- \neq 0$. Fix an element $z \in D^- \backslash \{0\}$. Then the $\mathbb{C}$-linear map $\mu : D^- \to D^+$ sending $x \in D^-$ to xz is injective. Since $dim_{\mathbb{C}}\, D^+ = 1$, it follows that $dim_{\mathbb{C}}\, D^- = 1$, and so

$$dim_{\mathbb{R}}\, D = 2\, dim_{\mathbb{C}}\, D = 4.$$

The element z is algebraic over $\mathbb{R}$ so $z^2 \in \mathbb{R} + \mathbb{R}z$. On the other hand, $z^2 = \mu(z) \in D^+ = \mathbb{C}$, so

$$z^2 \in \mathbb{C} \cap (\mathbb{R} + \mathbb{R}z) = \mathbb{R}.$$

If $z^2 > 0$ in $\mathbb{R}$, we can write $z^2 = r^2$ for some $r \in \mathbb{R}$; this leads to $z = \pm r \in \mathbb{R}$, a contradiction. Therefore, $z^2 < 0$ in $\mathbb{R}$, and we can write $z^2 = -r^2$ instead, for some $r \in \mathbb{R}^*$. Letting $j = z/r$, we have $j^2 = -1 = i^2$, $ji = -ij$, and

$$D = \mathbb{C} \oplus \mathbb{C}j = \mathbb{R} \oplus \mathbb{R}i \oplus \mathbb{R}j \oplus \mathbb{R}ij,$$

so D is a copy of the real quaternions. QED

The above completely elementary proof of Frobenius' Theorem is, I believe, due to R. Palais. From this proof, it would almost seem that, even if one had *not* encountered Hamilton's quaternions before, the analysis of the structure of D given above would have led one unmistakably to the discovery of the division ring of real quaternions. We should also remark that Frobenius' Theorem remains valid if $\mathbb{R}$ is replaced by any real-closed field, that is, a field k such that $\sqrt{-1} \notin k$ and $k(\sqrt{-1})$ is algebraically closed. For such a field, it can be shown that $k = k^2 \cup (-k^2)$, so the proof above can be carried over *verbatim.*

Frobenius' Theorem on real division algebras was almost certainly the very first substantial result obtained in the classification theory of algebras (then known as hypercomplex systems). Theorem (13.12) has been imitated many times over in the 20th century in different contexts of classification. Most notable examples are: the Gel'fand-Mazur Theorem for commutative Banach division algebras, Hopf's Theorem on commutative nonassociative real division algebras, and more generally, the Kervaire-Milnor Theorem for nonassociative real division algebras, etc. The first of these used techniques from functional analysis, while the second and the third used techniques from topology.

If the ground field F is $\mathbb{Q}$ instead of a finite field or the real field, there will by *many* algebraic division algebras D over F. The ones which are finite-dimensional over $\mathbb{Q}$ are classified by a deep theorem of Albert, Brauer, Hasse, and Noether, which says that D must be a "cyclic algebra" over its center. (Cyclic algebras will be defined in §14.) This result is beyond the scope of our book. On the other hand, the *infinite*-dimensional algebraic division algebras over $\mathbb{Q}$ are not completely classified. We shall not say anything more in this direction. However, let us at least sketch the construction of an example of a *noncommutative, infinite-dimensional, algebraic division algebra over* $\mathbb{Q}$. Let $p_1 < p_2 < \cdots$ be any sequence of primes. It is known that for each n, there exists a $\mathbb{Q}$-division algebra A_n of dimension p_n^2, with $Z(A_n) = \mathbb{Q}$. By a standard theorem which we'll assume here without proof,

$$D_n = A_1 \otimes_{\mathbb{Q}} \cdots \otimes_{\mathbb{Q}} A_n$$

is also a division algebra, with center $\mathbb{Q}$. Viewing D_n as a subalgebra of $D_{n+1} = D_n \otimes_{\mathbb{Q}} A_{n+1}$, we can form $D = \bigcup_{n \geq 1} D_n$, which is clearly an infinite-dimensional algebraic division algebra over $Z(D) = \mathbb{Q}$.

Next we shall turn our attention to the study of multiplicative commutators in a division ring. We shall occasionally drop the word "multiplicative" and just speak of commutators, as is the common practice. It turns out that some of the results we proved earlier ((13.4) through (13.7)) for additive commutators have analogues for multiplicative commutators. In order to get these new results, we first derive a couple of identities. Let a, c be two *noncommuting* elements in a division ring D. Let $b = a - 1 \in D^*$. Then

$$\begin{aligned} a(a^{-1}ca - b^{-1}cb) &= ca - ab^{-1}cb \\ &= c(b+1) - (b+1)b^{-1}cb \\ &= c - b^{-1}cb \neq 0, \end{aligned} \tag{13.13}$$

and so

$$a(a^{-1}cac^{-1} - b^{-1}cbc^{-1}) = 1 - b^{-1}cbc^{-1} \neq 0. \tag{13.14}$$

Note that the *RHS* of (13.13) and (13.14) are nonzero since $b = a - 1$ does not commute with c.

(13.15) Proposition. *Let D be a division ring. If an element $c \in D$ commutes with all multiplicative commutators, then $c \in Z(D)$.*

Proof. Assume, instead, that $ca \neq ac$ for some $a \in D$. Let $b = a - 1 \in D^*$ as above and use (13.14). By hypothesis, c commutes with $a^{-1}cac^{-1}$ and $b^{-1}cbc^{-1}$, so by (13.14), c commutes with a, a contradiction. QED

(13.16) Corollary. *If all multiplicative commutators are central in a division ring D, then D is a field.*

At this point, it is of interest to mention the following conjecture of Herstein:

> *Assume that, for any two nonzero elements a, b in a division ring D, there exists a positive integer $n(a, b)$ such that*
>
> $$(aba^{-1}b^{-1})^{n(a,b)} \in Z(D),$$
>
> *then D is a field.*

Herstein has proved this conjecture in the case when D is centrally finite, or $Z(D)$ is uncountable, but the general case seems to be still open.

We shall next prove a famous theorem due independently to Cartan, Brauer, and Hua. This is a multiplicative analogue of the result (13.7), except that here no assumption on the characteristic is needed. It is convenient to adopt the following terminology from group theory: For a pair of division rings $K \subseteq D$, we say that K is normal in D if, for any $x \in D^*$, $xKx^{-1} \subseteq K$ (in other words, if K^* is a normal subgroup of D^*).

(13.17) Cartan–Brauer–Hua Theorem. *Suppose K is normal in D as above and $K \neq D$. Then $K \subseteq Z(D)$.*

Proof. Consider any element $a \in D \backslash K$ and any $c \in K$. *We claim that they must commute.* Once we have proved this, the same argument used in the second half of the proof of (13.7) shows that $c \in Z(D)$. To prove our claim, assume that a, c do not commute, and write $b = a - 1 \in D^*$. In the identity (13.13), $a^{-1}ca$, $b^{-1}cb$ as well as c are in K^*, so we get $a \in K^*$, a contradiction. QED

(13.18) Corollary. *Let D be a division ring and $d \in D \backslash Z(D)$. Then D is generated as a division ring by all the conjugates of d.*

Proof. Let K be the division subring of D generated by all the conjugates of d. For any $x \in D^*$, $x^{-1}Kx$ contains all conjugates of d, so $x^{-1}Kx \supseteq K$. This gives $xKx^{-1} \subseteq K$, so K is normal in D. Since $d \in K$ is not central in D, the Theorem implies that $K = D$. QED

We record one more corollary of (13.17) which is to be contrasted with (13.6).

(13.19) Corollary. *A noncommutative division ring D is generated as a division ring by all of its multiplicative commutators.*

Proof. Let K be the division subring of D generated by all of its commutators. Clearly K is invariant under all automorphisms of D, in particular under all inner automorphisms. Therefore, K is normal in D. But D is not commutative, so by (13.16), some multiplicative commutator is not central. Thus $K \nsubseteq Z(D)$, and (13.17) implies that $K = D$. QED

In the literature, there are many different generalizations of the Cartan–Brauer–Hua Theorem. Let us mention, for instance, a generalization due to C. Faith. Let K be a division subring of a division ring D such that $K \neq D$ and $K \nsubseteq Z(D)$. The Cartan–Brauer–Hua Theorem says that K^* cannot be a normal subgroup of D^*. Faith has generalized this by showing that the group-theoretic normalizer $N_{D^*}(K^*)$ must have infinite index in D^*. There are also various other generalizations of the Cartan–Brauer–Hua Theorem to simple rings, semisimple rings, matrix rings, and to "twisted settings," etc. However, see Exercise 9.

Finally, we end this section by studying briefly the multiplicative group D^* of a division ring D. Again, there are very rich results in the literature in this area of study, but we shall limit ourselves to only a couple of results which concern the subgroup structure of D^*.

Recall that, for any group G, the *upper central series* of G is defined to be the series

$$\{1\} \subseteq G_1 \subseteq G_2 \subseteq \cdots \subseteq G,$$

where $G_1 = Z(G)$, $G_2/G_1 = Z(G/G_1), \ldots$, etc. The group G is said to be *nilpotent* if $G_n = G$ for some integer n.

(13.20) Theorem. *Let D be a division ring, and $\{1\} \subseteq G_1 \subseteq G_2 \subseteq \cdots$ be the upper central series of the group $G = D^*$. Then $G_1 = G_2 = G_3 = \cdots$.*

Proof. We may assume that D is not commutative. Suppose there exists an element $c \in G_2 \backslash G_1$. Then $c \notin Z(D)$, so we have $ca \neq ac$ for some $a \in D$. Let $b = a - 1 \in D^*$ and try to use the tricky identity (13.14). Since $\bar{c} \in Z(G/G_1)$, we have $x^{-1}cxc^{-1} \in G_1$ for every $x \in D^*$. From (13.14), we have an equation

$$a(\alpha - \beta) = 1 - \beta \neq 0, \quad \text{where } \alpha, \beta \in G_1.$$

Since $G_1 \cup \{0\} = Z(D)$ is a field, this implies that $a \in G_1$, a contradiction.

QED

We deduce easily from (13.20) the following pleasant consequence.

(13.21) Corollary. *The multiplicative group D^* of a division ring D is nilpotent iff D is a field.*

We remark that this result remains true if the word "nilpotent" is replaced by "solvable." However, this stronger version (due to L.K. Hua) is considerably more difficult to prove.

Our last goal in this section is to give some information on the group-theoretic index $[D^* : K^*]$ for a pair of division rings $K \subseteq D$. As it turns out, if $K \subsetneq D$, the index $[D^* : K^*]$ is almost never finite. To formulate this result more generally, we proceed as follows.

Let V be any right vector space over a division ring K. Then K^* acts on $V^* := V \backslash \{0\}$ by right multiplication, and we can form the orbit space V^*/K^*. We shall denote this space by $\mathbb{P}(V)$ and call it the *projective space* associated with V. In the case when K is commutative and *dim* $V_K \geq 2$, it is "geometrically obvious" that the projective space $\mathbb{P}(V)$ is infinite except when V itself is finite. The usual proof of this fact is based on the possibility of finding a "line" in $\mathbb{P}(V)$. This proof, in fact, does not depend on the commutativity of K, so we can use it to get a similar result in the general case.

(13.22) Theorem. *Let V_K be a right vector space over a division ring K with $dim(V_K) \geq 2$. Then $\mathbb{P}(V)$ is finite iff V (and hence K) is finite.*

Proof. ("only if") Let v_1, v_2 be two K-independent vectors in V. To find a "line" in $\mathbb{P}(V)$, we define a map λ: $K \to \mathbb{P}(V)$ by

$$\lambda(k) = (v_1 + v_2 k)K^*.$$

We claim that this map is injective. In fact, if $k, k' \in K$ are such that $(v_1 + v_2 k)K^* = (v_1 + v_2 k')K^*$, then

$$v_1 + v_2 k = (v_1 + v_2 k')k'' \quad \text{for some } k'' \in K^*.$$

Comparing the coefficients, we have $k'' = 1$ and $k = k'k''$, so $k = k'$. Assume now that $\mathbb{P}(V)$ is finite. Then by the injectivity of λ, K^* is finite and so V is also finite. QED

(In the proof above, λ represents an "affine line" in $\mathbb{P}(V)$. If we had chosen to use homogeneous coordinates, the same construction would have given a "projective line" in $\mathbb{P}(V)$. To some readers, this may be a bit more satisfactory.)

Applying (13.22) to rings, we deduce immediately the following ring-theoretic result which is essentially due to R. Brauer and C. Faith.

(13.23) Corollary. *Let K be a division subring of a ring R and let $V \supsetneq K$ be a subspace of the vector space R_K. Let $V^* = V \backslash \{0\}$ and let V^*/K^* be the orbit space of the right K^*-action on V^*. Then V^*/K^* is finite iff V is finite.*

Actually, the hypothesis that V be a subspace of R_K is stronger than necessary. By analyzing the proof of (13.22), we see that the conclusion of (13.23) is valid for any set $V \supsetneq K$ in R such that $V \cdot K \subseteq V$ and $V + K \subseteq V$. (In the proof, we work with $v_1 = 1$ and any $v_2 \in V \backslash K$.)

(13.24) Corollary. *For a pair of division rings $K \subsetneq D$, we have*

$$[D^* : K^*] < \infty \iff D \textit{ is finite.}$$

Consider a division subring D of a division ring E, and an element a in E. For any $d \in D^*$, we say that dad^{-1} is a *D-conjugate* of a. By conjugation, D^* acts on the set of D-conjugates of a; the isotropy subgroup of a under this action is K^*, where K is the division ring $D \cap C_E(a)$. Thus, the set of D-conjugates of a is in a one-one correspondence with the coset space D^*/K^*. Applying (13.24), we deduce the following result.

(13.25) Corollary. *In the notation above, assume that the division ring D is infinite. Then either there is only one D-conjugate of a (i.e., $D \subseteq C_E(a)$) or there are infinitely many D-conjugates of a.*

Our final result in this section is a combination of Wedderburn's Little Theorem and (13.25).

(13.26) Herstein's Theorem. *If a is a noncentral element in a division ring D, then a has infinitely many conjugates in D.*

Proof. Since a is a noncentral element, Wedderburn's Little Theorem implies that D is infinite. The result now follows by applying (13.25) with $E = D$.

QED

In closing, we remark that, in any division ring D, the cardinality of the conjugacy class of any noncentral element $a \in D$ is in fact equal to the cardinality of D. This is a result of W. Scott.

Exercises for §13

Ex. 13.1. Show that a nonzero ring D is a division ring iff, for any $a \neq 1$ in D, there exists an element $b \in D$ such that $a + b = ab$. (Cf. Ex. 4.2.)

Ex. 13.2. Let L be a domain and K be a division subring of L. If L is finite-dimensional as a right K-vector space, show that L is also a division ring.

Ex. 13.3. Show that any finite prime ring R is a matrix ring over a finite field.

Ex. 13.4. For any division ring D, show that any finite abelian subgroup of D^* is cyclic.

Ex. 13.5. Show that an element in a division ring D commutes with all its conjugates iff it is central.

Ex. 13.6. Let D be an algebraic division algebra over a field k. Let $a, b \in D^*$ be such that $bab^{-1} = a^n$, where $n \geq 1$. Show that a, b generate a finite-dimensional division k-subalgebra of D.

Ex. 13.7. (Brauer) Let $K \subseteq D$ be two division rings, and let $N = N_{D^*}(K^*)$, $C = C_{D^*}(K^*)$ be the group-theoretic normalizer and centralizer of the subgroup $K^* \subseteq D^*$. For any $h \in D \backslash \{0, -1\}$, show that $h, 1 + h \in N$ iff $h \in K \cup C$. Using this, give another proof for the Cartan-Brauer-Hua Theorem (13.17). (**Hint.** Assume that h, $1 + h \in N$ but $h \notin C$. Find $\delta \in K$ such that $\delta_0 = h\delta h^{-1} \neq \delta$, and let $\delta_1 = (h+1)\delta(h+1)^{-1}$. Subtract to show that $\delta - \delta_1 = (\delta_1 - \delta_0)h$. Since $\delta \neq \delta_0$, we also have $\delta_1 \neq \delta_0$, and so $h = (\delta_1 - \delta_0)^{-1}(\delta - \delta_1) \in K$.)

Ex. 13.8. (Cf. Herstein's Lemma (13.8)) Let D be a division ring of characteristic $p > 0$ with center F. Let $a \in D \backslash F$ be such that $a^{p^n} \in F$ for some $n \geq 1$. Show that there exists $b \in D$ such that $ab - ba = 1$, and there exists $c \in D$ such that $aca^{-1} = 1 + c$. (**Hint.** As in the proof of (13.8), $\delta = \delta_a$ is nilpotent. Let k be the smallest integer such that $\delta^{k+1} \equiv 0$, and let $x \in D$ be such that $\delta^k(x) \neq 0$. For $c := \delta^{k-1}(x) \neq 0$, show that $b := c\delta(c)^{-1}a$ satisfies $\delta(b) = a$.)

Ex. 13.9. The following example of Amitsur shows that the Cartan-Brauer-Hua Theorem (13.17) does not extend directly to simple rings. Let $F = \mathbb{Q}(x)$ and $A = F[t;\delta]$ be the ring of polynomials $\{\sum g_i(x)t^i\}$ over F with multiplication defined by the twist $tg(x) = g(x)t + \delta(g(x))$, where $\delta(g(x))$ denotes the formal derivative of $g(x)$. Show that

(1) A is a simple domain,
(2) $U(A) = F^*$, and
(3) F is invariant under all automorphisms of A, but $F \nsubseteq Z(A)$.

Ex. 13.10. Let k be an algebraically closed field and D be a division k-algebra. Assume that either D is an algebraic algebra over k, or $\dim_k D <$ Card k (as cardinal numbers). Show that $D = k$. (**Hint.** For the second part, use the argument in the proof of (4.20) to show that D must be an algebraic k-algebra.)

Ex. 13.11. (Jacobson) Let A be a reduced algebraic algebra over a finite field $\mathbb{F}_q$. Show that A is commutative. (**Hint.** Show that, if $\psi(x) \in \mathbb{F}_q[x]$ is such that $\psi(0) \neq 0$, then $\psi(x) | (x^N - 1)$ for some integer N. Then use Ex. 12.6B(1) and Jacobson's Theorem (12.10).)

Ex. 13.12. Let D be a division ring, and $x, y \in D^*$ be such that $\omega = xyx^{-1}y^{-1}$ lies in the center F of D.

(1) For any integers m, n, show that $x^n y^m x^{-n} y^{-m} = \omega^{mn}$.
(2) If x is algebraic over F, show that ω is a root of unity.
(3) If ω is a primitive kth root of unity, show that $(y + x)^k = y^k + x^k$.

(**Hint for** (2). Write down a polynomial equation for x over F of the smallest degree and conjugate it by y.)

Ex. 13.13. Let D be a division ring and $A = (a_{ij})$ be an $m \times n$ matrix over D. Define the row rank r (resp. column rank c) of A to be the *left* (resp. *right*) dimension of the row (resp. column) space of A as a subspace of ${}_D(D^n)$ (resp. $(D^m)_D$). Show that $r = c$. (**Hint.** Let B be an $r \times n$ matrix whose r rows form a basis of the left row space of A. Expressing the rows of A as *left* combinations of those of B, we get $A = B'B$ for a suitable $m \times r$ matrix B'. But the equation $A = B'B$ also expresses the columns of A as *right* combinations of the r columns of B'; hence $c \leq r$. This proof appeared in Lam [86].)

Ex. 13.14. Keep the notations in Exercise 13. The common value $r = c$ is called the *rank* of the matrix A. Show that the following statements are equivalent:
(1) *rank* $A = s$.
(2) s is the largest integer such that A has an $s \times s$ invertible submatrix.
(3) A has an $s \times s$ invertible submatrix M such that any $(s+1) \times (s+1)$ submatrix of A containing M is not invertible.
(This is called "Kronecker's Rank Theorem.")

Ex. 13.15. Show that $rank(A) = rank(A^t)$ for all matrices A over a division ring D iff D is a field. (**Hint.** For $a, b \in D$, $A = \begin{pmatrix} 1 & b \\ a & ab \end{pmatrix}$ has rank 1, but $A^t = \begin{pmatrix} 1 & a \\ b & ab \end{pmatrix}$ has rank 2 unless $ab = ba$.)

Ex. 13.16. Let G be the group of order 21 generated by two elements a, b with the relations $a^7 = 1$, $b^3 = 1$, and $bab^{-1} = a^2$. Using the Wedderburn decomposition of $\mathbb{Q}G$ obtained in Ex. (8.28)(2), show that G cannot be embedded in the multiplicative group of *any* division ring D. (**Hint.** If $G \subseteq U(D)$, then $char(D) = 0$, and the $\mathbb{Q}$-span of G in D is a division ring which must be a Wedderburn component of $\mathbb{Q}G$.)

Ex. 13.17. Give a ring-theoretic proof for the conclusion of the above exercise (without using the representation theory of groups). (**Hint.** Suppose $G \subseteq U(D)$, where D is a division ring (or just a domain). From $ba = a^2b$ and $b^2a = a^4b^2$, show that

$$0 = (b^2 + b + 1)a = a^4b^2 + a^2b + a = a^2(1 - a^2)b - a^4 + a \in D,$$

which implies that b commutes with a. For generalizations of this argument, see Lam [01].)

§14. Some Classical Constructions

In the last section, we discussed some of the basic properties of division rings, but did not provide enough examples. To remedy this, we shall now devote the present section to the explicit construction of examples of division rings. Of course, we have no lack of examples of fields, but, from the viewpoint of this book, fields are "trivial" examples of division rings. Our primary interest here is in constructing nice examples of *noncommutative* division rings.

Since the center F of a division ring D is a field, we can regard D as an algebra over F, so $dim_F\, D$ makes sense. Depending on whether $dim_F\, D$ is finite or infinite, we can classify all division rings into two broad categories. For convenience, we introduce the following terminology.

(14.1) Definition. *A division ring D is called centrally finite if D is finite-dimensional over its center. Otherwise, D is called centrally infinite.*

In a manner of speaking, centrally finite division rings are not very far from being commutative, while centrally infinite ones are highly noncommutative. In this section, we shall construct examples of both centrally finite and centrally infinite division rings. The examples we shall present are all classical in origin: they are due to D. Hilbert (1899), L. E. Dickson (1906), A. Mal'cev (1948), and B.H. Neumann (1949). Hilbert's example, which arose from his study of the independence of axioms in geometry, was the first example known of a centrally infinite division ring. A careful study of Hilbert's example apparently led Dickson to the construction of cyclic algebras, an all-important class of centrally finite algebras. The Mal'cev–Neumann construction of "Laurent series" division rings is also related to Hilbert's example, with additional motivation coming from the earlier work of H. Hahn (1907) on the embedding of ordered abelian groups into groups of Laurent series. The Mal'cev–Neumann construction has also applications to the problem of embedding domains into division rings and to valuation theory.

First let us discuss Hilbert's example. This was, in fact, already described in §1. Let k be a field (for simplicity), and σ be a fixed automorphism of k. We write $D = k((x, \sigma))$ for the ring of formal Laurent series $\sum_{i=n}^{\infty} a_i x^i$, where $n \in \mathbb{Z}$ and $a_i \in k$, with multiplication defined by the twist equation $xa = \sigma(a)x$ (for all $a \in k$). We have seen in §1 that D is in fact a division ring. The following proposition computes the center $Z(D)$ of D and determines, in particular, when D is centrally finite.

(14.2) Proposition. *For $D = k((x, \sigma))$ as above, let $k_0 \subseteq k$ be the fixed field of σ; i.e., $k_0 = \{a \in k : \sigma(a) = a\}$. Then*

$$Z(D) = \begin{cases} k_0 & \text{if } \sigma \text{ has infinite order,} \\ k_0((x^s)) & \text{if } \sigma \text{ has a finite order } s. \end{cases}$$

In particular, the division ring D is centrally finite iff σ has a finite order.

Proof. Consider a series $f = \sum_{i=n}^{\infty} a_i x^i \in Z(D)$, and let j be an index such that $a_j \neq 0$. For any scalar $a \in k$, we have $(\sum a_i x^i)a = a(\sum a_i x^i)$, so, comparing the coefficients for x^j, we get $a_j \sigma^j(a) = a a_j$; hence $\sigma^j(a) = a$.

Case 1. *σ has infinite order.* In this case, the above argument implies that $a_j \neq 0$ is only possible for $j = 0$. Hence $f = a_0$. Since we also have $a_0 x = x a_0 = \sigma(a_0)x$, it follows that $\sigma(a_0) = a_0$; i.e., $a_0 \in k_0$. Conversely, any $a_0 \in k_0$ clearly commutes with any Laurent series, so $Z(D) = k_0$. Since $dim_{k_0} D$ is clearly infinite, D is not centrally finite in this case.

Case 2. *σ has a finite order s.* The first paragraph of the proof shows that if $a_j \neq 0$ in a series $f = \sum_{i=n}^{\infty} a_i x^i \in Z(D)$, then $\sigma^j = \mathrm{Id}_k$, so $s \mid j$. But we also have $fx = xf$, which implies that each $a_i \in k_0$. Thus $f \in k_0((x^s))$ (the ordinary Laurent series field in x^s over k_0). Conversely, it is easy to verify that any monomial ax^{sj} with $a \in k_0$ commutes with any Laurent series in D, so $Z(D) = k_0((x^s))$. Let $F = k_0((x^s))$ and $K = k((x^s))$. Since x^s commutes with all elements of k, K is again the ordinary Laurent series field in x^s over k. We have $dim_F\, K = dim_{k_0}\, k = s$ by Galois Theory, and, since

(14.3) $$D = K \cdot 1 \oplus K \cdot x \oplus \cdots \oplus K \cdot x^{s-1},$$

the dimension of D as a left K-vector space is also s. By the transitivity formula for dimensions, it follows that $dim_{Z(D)}\, D = dim_F\, D = s^2$. In particular, D is a centrally finite division ring. QED

In Hilbert's original example, the field k was taken to be $\mathbb{Q}(t)$ and σ was taken to be the $\mathbb{Q}$-automorphism on $\mathbb{Q}(t)$ which sends t to $2t$. The elements in D have then the form

$$\sum_{i=n}^{\infty} a_i(t)x^i \quad (a_i(t) \in \mathbb{Q}(t))$$

with multiplication dictated by $x \cdot a(t) = a(2t)x$. Since σ here clearly has infinite order, the division ring of all Laurent series $\sum_{i=n}^{\infty} a_i(t)x^i$ is not centrally finite. Historically, this was the first example of a centrally infinite division ring.

Our next goal is to present Dickson's construction of cyclic algebras. To see where the ideas came from, we go back to the analysis of the *second* case in the proof of (14.2). Referring to the notations used there, we note that F is the fixed field of K under the automorphism $\tilde{\sigma}$ which takes x^s to x^s and acts as σ on k. Thus K/F is a Galois extension with $Gal(K/F) = \langle \tilde{\sigma} \rangle$, a cyclic group of order s. The division ring D is represented as a left K-vector space by (14.3). Under this representation, the multiplication in D is determined by the rule

(14.4) $$x \cdot \left(\sum_{i=n}^{\infty} b_i x^{is} \right) = \left(\sum_{i=n}^{\infty} \sigma(b_i) x^{is} \right) x = \tilde{\sigma} \left(\sum_{i=n}^{\infty} b_i x^{is} \right) \cdot x,$$

and the fact that $x^s \in F \subseteq K$. In 1906, Dickson extracted the key features of D as described above to formulate the important definition of a *cyclic algebra*. We shall now explain Dickson's construction.

Let K/F be a cyclic (Galois) extension, i.e., a finite, separable, normal field extension with a cyclic Galois group, say generated by an automorphism σ of order $s = dim_F\, K$. Fixing a nonzero element $a \in F$ and a symbol x, we let

$$D = K \cdot 1 \oplus K \cdot x \oplus \cdots \oplus K \cdot x^{s-1},$$

and multiply elements in D by using the distributive law, and the two rules

(14.5) $$x^s = a, \quad x \cdot b = \sigma(b)x \quad (\text{for any } b \in K).$$

It is easy to see that $F \subseteq Z(D)$, so D is an F-algebra, of dimension s^2. This algebra is denoted by $(K/F, \sigma, a)$, and is called the *cyclic algebra associated with* $(K/F, \sigma)$ *and* $a \in F \backslash \{0\}$. For instance, the division algebra D in Case 2 of the proof of (14.2) is such a cyclic algebra (though the σ here was denoted by $\tilde{\sigma}$ in (14.4)). The division algebra $\mathbb{H}$ of real quaternions is also a cyclic algebra, as we can take $F = \mathbb{R}$, $K = \mathbb{C}$, $\sigma =$ complex conjugation on $\mathbb{C}$, $a = -1$, and $x = j$. (We have $\mathbb{H} = \mathbb{R} \oplus \mathbb{R}i \oplus \mathbb{R}j \oplus \mathbb{R}ij = \mathbb{C} \oplus \mathbb{C}j$ and

$$j(\alpha + \beta i) = (\alpha - \beta i)j = \sigma(\alpha + \beta i)j$$

for $\alpha, \beta \in \mathbb{R}$.)

A more efficient way of constructing $(K/F, \sigma, a)$ is by using the skew polynomial ring $B = K[t; \sigma]$. Recall from §1 that this ring consists of left polynomials $\sum b_i t^i$ ($b_i \in K$) which are multiplied by using the rule $tb = \sigma(b)t$ (for any $b \in K$). The cyclic algebra $(K/F, \sigma, a)$ is simply (isomorphic to) the quotient algebra $B/(t^s - a)$, where $(t^s - a)$ denotes the ideal in B generated by the central polynomial $t^s - a$.

For a general cyclic algebra $D = (K/F, \sigma, a)$ as defined above, we have the following elementary properties.

(14.6) Theorem.

(1) *D is a simple F-algebra with $Z(D) = F$.*

(2) *$C_D(K)$ (the centralizer of K in D) is K itself.*

(3) *K is a maximal subfield of D.*

Proof. For (1), let $\mathfrak{A}$ be a nonzero ideal in D. Choose a nonzero element

$$z = b_{i_1} x^{i_1} + \cdots + b_{i_r} x^{i_r} \in \mathfrak{A} \quad (b_{i_j} \in K,\ 0 \le i_1 < \cdots < i_r \le s - 1)$$

with r as small as possible. Clearly, each $b_{i_j} \neq 0$, so b_{i_j} is a unit in K (and hence in D). *We claim that* $r = 1$. Once we prove this, then, since x is a unit in D (with inverse $a^{-1}x^{s-1}$), so is $z = b_{i_1} x^{i_1}$. This implies that $\mathfrak{A} = D$, as desired. To prove our claim, assume, instead, that $r \ge 2$. Since $\sigma^{i_1} \neq \sigma^{i_r}$, there exists $b \in K$ such that $\sigma^{i_1}(b) \neq \sigma^{i_r}(b)$. The ideal $\mathfrak{A}$ contains the following two elements:

$$zb = b_{i_1} \sigma^{i_1}(b) x^{i_1} + \cdots + b_{i_r} \sigma^{i_r}(b) x^{i_r},$$
$$\sigma^{i_1}(b) z = b_{i_1} \sigma^{i_1}(b) x^{i_1} + \cdots + b_{i_r} \sigma^{i_1}(b) x^{i_r}.$$

Hence $\mathfrak{A}$ also contains their difference

$$b_{i_2}(\sigma^{i_2}(b) - \sigma^{i_1}(b)) x^{i_2} + \cdots + b_{i_r}(\sigma^{i_r}(b) - \sigma^{i_1}(b)) x^{i_r}.$$

This element is nonzero since x^{i_r} appears with a nonzero coefficient. This contradicts the minimal choice of r, thus proving the simplicity of D. For (2), we need only prove that $C_D(K) \subseteq K$. Let

$$d = \sum_{i=0}^{s-1} b_i x^i \in C_D(K)$$

where $b_i \in K$. For any $b \in K$, $bd = db$ shows that $bb_i = b_i \sigma^i(b)$ for $i \le s - 1$. If $b_i \neq 0$ for some *positive* $i \le s - 1$, this would imply that σ^i is the identity on K, a contradiction. Thus $d = b_0 \in K$, showing that $C_D(K) = K$. From this, we can easily deduce (3), for, if L is a subfield of D containing K, then $L \subseteq C_D(K) = K$. Finally, to finish the proof, we need to show that $F \subseteq Z(D)$ is an equality. Let $b \in Z(D)$. Then $b \in C_D(K) = K$. From $bx = xb = \sigma(b)x$, we see that $b = \sigma(b)$. Since σ generates the Galois group of K/F, Galois Theory implies that $b \in F$. QED

In general, the cyclic algebra $D = (K/F, \sigma, a)$ need not be a division algebra. For instance, if $a = 1$, we have $x^s = 1$, and so

$$(1 - x)(1 + x + \cdots + x^{s-1}) = 0.$$

Thus, if $s > 1$, $1 - x$ and $1 + x + \cdots + x^{s-1}$ are zero-divisors in D. More precisely, we have the following explicit criterion for the simple F-algebra D to be split. Here, $N = N_{K/F}$ denotes the field norm from K to F, and $\dot{K} = K \backslash \{0\}$.

(14.7) Theorem. *We have $D \cong \mathbb{M}_s(F)$ as an F-algebra iff $a \in N_{K/F}(\dot{K})$.*

Proof. First assume that $a \in N(\dot{K})$. Then $N(d)a = 1$ for some $d \in \dot{K}$. For $y := dx \in D = (K/F, \sigma, a)$, we have

$$y^s = \sigma^{s-1}(d) \cdots \sigma(d) d x^s = N(d)a = 1,$$

and for any $b \in K$, $yb = dxb = d\sigma(b)x = \sigma(b)y$. From this, we see easily that $D \cong (K/F, \sigma, 1)$, so it suffices to show that

$$D' := (K/F, \sigma, 1) \cong \mathbb{M}_s(F).$$

Let $B = K[t; \sigma]$ as before, and view D' as $B/(t^s - 1)$. Since

$$t^s - 1 = (t^{s-1} + \cdots + 1)(t - 1) \in B,$$

$(t^s - 1)$ is contained in the maximal left ideal $B \cdot (t - 1)$. Therefore D' has a simple left module $M = B/B \cdot (t - 1) \cong K$ which has F-dimension s. The module action gives an F-algebra homomorphism $D' \to End_F(M)$. Since both algebras have F-dimension s^2 and D' is simple, we have $D' \cong End_F(M) \cong \mathbb{M}_s(F)$. (For another proof of this, see Exercise 4.) Conversely, if $D \cong \mathbb{M}_s(F)$, then $D \cong B/(t^s - a)$ has a simple module of F-dimension s. Such a module must be isomorphic to B/Bf for some principal left ideal $Bf \supseteq (t^s - a)$. (Recall that B is a principal left ideal domain: see (1.25).) Since

$$s = dim_F\, B/Bf = (dim_F\, K)\, deg\, f,$$

we see that $deg\, f = 1$. After a scaling (from the left), we may assume that $f = t - c$, where $c \in K$. We have

$$t^s - a = (b_{s-1}t^{s-1} + \cdots + b_1 t + b_0)(t - c),$$

where $b_i \in K$. Multiplying out the *RHS* and comparing coefficients, we get successively

$$\begin{aligned} b_{s-1} &= 1, \\ b_{s-2} &= \sigma^{s-1}(c), \\ b_{s-3} &= \sigma^{s-1}(c)\sigma^{s-2}(c), \\ \cdots & \quad \cdots \\ b_0 &= \sigma^{s-1}(c)\cdots\sigma(c), \end{aligned}$$

and finally $a = b_0 c = \sigma^{s-1}(c)\cdots\sigma(c)c = N(c)$. QED

(14.8) Corollary. *Suppose s is a prime number. Then $D = (K/F, \sigma, a)$ is a division algebra iff $a \notin N_{K/F}(\dot{K})$.*

Proof. Again let us write $N = N_{K/F}$. If $a \in N(\dot{K})$, then $D \cong \mathbb{M}_s(F)$ by (14.7), so D is not a division algebra. Conversely, assume that D is not a division algebra. By the Wedderburn–Artin Theorem, the simple F-algebra D is isomorphic to $\mathbb{M}_r(E)$ for some division F-algebra E, where r is necessarily >1. Comparing F-dimensions, we get $s^2 = r^2 \, dim_F \, E$. Since s is prime, we must have $r = s$ and $dim_F \, E = 1$, so $D \cong \mathbb{M}_s(F)$. From (14.7) again, we conclude that $a \in N(\dot{K})$. QED

Example. Let $F = E(t)$ and $K = F(\sqrt{t})$, where E is a field of characteristic not 2, and t is transcendental over E. Let σ be the F-automorphism on K taking $\sqrt{t}$ to $-\sqrt{t}$. Then, for $a \in \dot{E}$,

$$D := (K/F, \sigma, a) \text{ is a division algebra iff } a \notin E^2.$$

To see this, it suffices (by (14.8)) to show that $a \notin N(\dot{K})$ iff $a \notin E^2$. The "only if" part is clear. For the "if" part, assume that $a = N(f(t) + \sqrt{t}\, g(t))$, where $f, g \in E(t)$. Writing $f = f_0/h$ and $g = g_0/h$ where $f_0, g_0, h \in E[t]$ and $h \neq 0$, we have $ah(t)^2 = f_0(t)^2 - tg_0(t)^2$. A comparison of leading coefficients on the two sides shows that $a \in E^2$.

If s is not a prime, it is not easy to decide in general when a cyclic algebra $D = (K/F, \sigma, a)$ is a division algebra. However, there is a well-known *sufficient* condition, again in terms of $N = N_{K/F}$, which will guarantee that D is a division algebra. This sufficient condition was found by Wedderburn in 1914. Consider the quotient group $\dot{F}/N(\dot{K})$, which is an abelian group of exponent dividing s (since $d^s = N(d)$ for any $d \in \dot{F}$). Wedderburn's sufficient condition is expressed in terms of the order of the image of a in this group. Note that, in the special case when s is a prime, Wedderburn's result below reduces to the sufficiency part of (14.8).

(14.9) Wedderburn's Theorem. *Suppose the image of a in $\dot{F}/N(\dot{K})$ has order s (i.e., $a, a^2, \ldots, a^{s-1} \notin N(\dot{K})$). Then $D = (K/F, \sigma, a)$ is a division F-algebra.*

Nowadays this is viewed as a cohomological theorem, and is proved in most textbooks using 2-cocycle calculations. In keeping with the classical nature of the present section, however, we would like to offer a non-cohomological proof, following the one given in Appendix 1 in Dickson's book [23], but incorporating certain simplifications suggested by J.-P. Tignol. I thank Tignol for his role in helping me understand this nice classical proof.

To begin the proof of (14.9), we work again with the skew polynomial ring $B = K[t;\sigma]$. Let $z = t^s$, which lies in the center of B. Then $K[z] \subseteq B$, and we can view B as a *left* $K[z]$-module. As such, B has a free basis $\{1, t, \dots, t^{s-1}\}$. Each polynomial $\beta \in B$ acts by *right* multiplication on B as a $K[z]$-endomorphism of B. Let $n(\beta) \in K[z]$ denote the determinant of this $K[z]$-endomorphism. Then we have the usual multiplicative property

$$n(\beta\beta') = n(\beta)n(\beta') \quad \text{for } \beta, \beta' \in B.$$

(Note that "n" here is like an algebra norm, except that we cannot call B a $K[z]$-algebra since $K[z]$ is not in the center of B.)

The σ-action on K extends to a σ-action on $B = K[t;\sigma]$ by letting $\sigma(t) = t$. With this extended action, we have $t\beta = \sigma(\beta)t$ for every $\beta \in B$. Write

$$\beta = \beta_0 + \beta_1 t + \cdots + \beta_{s-1}t^{s-1},$$

where $\beta_i \in K[z]$. With respect to the $K[z]$-basis $\{1, t, \dots, t^{s-1}\}$ on B, the right multiplication by β has the matrix

$$(*) \qquad \begin{pmatrix} \beta_0 & \beta_1 & \beta_2 & \cdots & \beta_{s-1} \\ z\sigma(\beta_{s-1}) & \sigma(\beta_0) & \sigma(\beta_1) & \cdots & \sigma(\beta_{s-2}) \\ z\sigma^2(\beta_{s-2}) & z\sigma^2(\beta_{s-1}) & \sigma^2(\beta_0) & \cdots & \sigma^2(\beta_{s-3}) \\ \vdots & \vdots & \vdots & & \vdots \\ z\sigma^{s-1}(\beta_1) & z\sigma^{s-1}(\beta_2) & z\sigma^{s-1}(\beta_3) & \cdots & \sigma^{s-1}(\beta_0) \end{pmatrix}$$

and so $n(\beta)$ is the determinant of this matrix. We make the following observations on $n(\beta)$.

(14.10) Lemma.

(1) *For $\beta = \beta_0 + \beta_1 t + \cdots + \beta_{s-1}t^{s-1}$ as above, we have $n(\beta) \in F[z]$. The constant term of $n(\beta)$ is $N(b_0)$ where b_0 is the constant term of β_0. (In particular, the restriction of n to K is $N = N_{K/F}$.)*

(2) *Assume that all $\beta_i \in K$. Then $deg_z\, n(\beta) = deg_t\, \beta$, and if β is monic in t, $n(\beta)$ has leading coefficient $(-1)^{d(s-1)}$, where $d = deg_t\, \beta$.*

Proof. From $t\beta = \sigma(\beta)t$, we have $n(t)n(\beta) = n(\sigma(\beta))n(t)$. Since $n(t) = \pm z$, this gives $n(\beta) = n(\sigma(\beta))$. On the other hand, by $(*)$, $n(\sigma(\beta)) = \sigma(n(\beta))$. Thus $n(\beta) \in K[z]$ is invariant under σ, so $n(\beta) \in F[z]$. Setting $z = 0$ in $(*)$, we find that the constant term of $n(\beta)$ is

$$b_0\sigma(b_0)\cdots\sigma^{s-1}(b_0) = N(b_0),$$

where b_0 is the constant term of β_0 (as a polynomial in z). The conclusions in (2) follow easily by working with the determinant of $(*)$ by setting $\beta_d = 1$ and $\beta_i = 0$ for $i > d := deg_t\, \beta$. QED

Now we are ready to give the proof of Wedderburn's Theorem.

Proof of (14.9). Assume that $(K/F, \sigma, a) \cong B/(t^s - a)$ is *not* a division algebra. Then $(t^s - a) \subsetneq \mathfrak{B}$ for some left ideal $\mathfrak{B} \subsetneq B$. Again using the fact that B is a principal left ideal domain, we can write $\mathfrak{B} = B \cdot \beta$ for some

$$\beta = t^d + \cdots + b_1 t + b_0$$

with $b_i \in K$ and $1 \leq d \leq s - 1$. Then $t^s - a = \beta'\beta$ for some $\beta' \in B$. Applying the norm map, we have

$$n(\beta')n(\beta) = n(z - a) = (z - a)^s.$$

By (14.10)(1), $n(\beta), n(\beta') \in F[z]$, and, therefore, by (14.10)(2),

$$n(\beta) = (-1)^{d(s-1)}(z - a)^d.$$

Comparing constant terms and using (14.10)(1) again, it follows that

$$N(b_0) = (-1)^{d(s-1)}(-a)^d,$$

and so

$$a^d = (-1)^{ds} N(b_0) = N((-1)^d b_0),$$

contradicting the hypothesis of (14.9). QED

Note that we have a commutative diagram

$$\begin{array}{ccccc} F[z] & \subset & K[z] & \subset & B = K[t;\sigma] \\ \downarrow & & \downarrow & & \downarrow \\ F & \subset & K & \subset & D = \bigoplus_{i=1}^{s-1} Kx^i \end{array}$$

where the vertical maps send t to x and $z = t^s$ to $x^s = a$. The norm map n: $B \to F[z]$ can be "specialized" to D as follows. View D as a left K-vector space as shown above and for any $\alpha \in D$, let $n(\alpha)$ be the determinant of the K-endomorphism on D given by right multiplication of α. Then for

$$\alpha = b_0 + \cdots + b_{s-1}x^{s-1} \quad (b_i \in K),$$

the matrix $M(\alpha)$ of the α-action with respect to the left K-basis $\{1, x, \ldots, x^{s-1}\}$ is obtained from $(*)$ by replacing z there by a and the β_i's by b_i's. Thus,

$$\textbf{(14.11)} \quad n(\alpha) = \det M(\alpha) = \det \begin{pmatrix} b_0 & b_1 & \cdots & b_{s-1} \\ a\sigma(b_{s-1}) & \sigma(b_0) & \cdots & \sigma(b_{s-2}) \\ a\sigma^2(b_{s-2}) & a\sigma^2(b_{s-1}) & \cdots & \sigma^2(b_{s-3}) \\ \vdots & \vdots & & \vdots \\ a\sigma^{s-1}(b_1) & a\sigma^{s-1}(b_2) & \cdots & \sigma^{s-1}(b_0) \end{pmatrix}.$$

As in (14.10), we can show that $n(\alpha) \in F$. The map $\alpha \mapsto M(\alpha)$ defines an F-algebra homomorphism

$$M\colon D \longrightarrow End({}_K D) \cong \mathbb{M}_s(K)$$

(where the K-endomorphisms on D are written on the right). In terms of the F-algebra generators of D, this M is described uniquely (in the K-basis $\{1, x, \ldots, x^{s-1}\}$) by

(14.12)

$$M(x) = \begin{pmatrix} 0 & 1 & 0 & \cdots & 0 \\ 0 & 0 & 1 & \cdots & 0 \\ \vdots & \vdots & \ddots & \ddots & \vdots \\ & & & & 1 \\ a & 0 & 0 & \cdots & 0 \end{pmatrix}, \quad M(b) = \begin{pmatrix} b & & & 0 \\ & \sigma(b) & & \\ & & \ddots & \\ 0 & & & \sigma^{s-1}(b) \end{pmatrix},$$

where $b \in K$. Note that M represents D faithfully as an F-algebra of K-matrices. Since $M(D)$ commutes elementwise with the scalar matrices $K \subseteq \mathbb{M}_s(K)$, we have a K-algebra homomorphism

$$M \otimes 1\colon D \otimes_F K \longrightarrow M(D) \cdot K \subseteq \mathbb{M}_s(K).$$

Here, $D \otimes_F K$ and $\mathbb{M}_s(K)$ both have dimension s^2 over K. By (14.6)(1) and a result about scalar extensions in the next section (see (15.1)(3)), $D \otimes_F K$ is a simple K-algebra. Thus, we have an isomorphism

(14.13)
$$M \otimes 1\colon D \otimes_F K \xrightarrow{\cong} \mathbb{M}_s(K)$$

which, on the factor $D = D \otimes 1$, is given explicitly by (14.12). This says that K is a splitting field for the simple F-algebra D, which provides a nice illustration of a general property of s-dimensional subfields in s^2-dimensional simple algebras with center F. The map $n\colon D \to F$ defined by $n(\alpha) = det\, M(\alpha)$ (cf. (14.11)) is called the *reduced norm* on D in the standard terminology of the theory of simple algebras.

At this point, we should point out that, although the norm condition given in Wedderburn's Theorem (14.9) is a sufficient condition for the cyclic algebra $D = (K/F, \sigma, a)$ to be a division algebra, it is in general *not* a necessary condition, in case $s = [K : F]$ fails to be a prime. In the following, we shall construct a cyclic *division* algebra of the form $D = (K/F, \sigma, -1)$ with $s = 4$. Since -1 has order ≤ 2 in $\dot{F}/N(\dot{K})$, we see that Wedderburn's norm condition is not a necessary condition for D to be a division algebra.

The example we shall present is essentially due to Brauer; our exposition below follows closely the suggestions of Tignol. Let $K = \mathbb{Q}(y, z)$, where y, z are commuting indeterminates, and let σ be the $\mathbb{Q}$-automorphism of order 4 on K defined by $\sigma(y) = z$, $\sigma(z) = -y$. We write F and L respectively for the fixed fields K^{σ} and K^{σ^2}, so K/F is a quartic cyclic extension containing the quadratic subextension L/F, with $Gal(K/F) = \langle \sigma \rangle$. *Our goal is to show that*

$D := (K/F, \sigma, -1)$ *is a division algebra*. We write D as usual in the form

$$K \oplus Kx \oplus Kx^2 \oplus Kx^3, \quad \text{with } x^4 = -1.$$

We shall first show that the centralizer $C_D(x^2)$ is a division algebra.

An easy computation shows that

$$C_D(x^2) = L \oplus Lx \oplus Lx^2 \oplus Lx^3.$$

Note that $j := x^2$ commutes with elements of L, with $j^2 = x^4 = -1$. Now for $i = \sqrt{-1} \in \mathbb{C}$, let

$$\mathbb{Q}' = \mathbb{Q}(i), \quad K' = K(i) = \mathbb{Q}'(y, z), \quad L' = L(i), \quad F' = F(i),$$

and extend the action of σ to K' by defining $\sigma(i) = i$. Then

$$K'^{\sigma} = K(i)^{\sigma} = K^{\sigma}(i) = F',$$

and similarly $K'^{\sigma^2} = L'$, with $Gal(K'/F') = \langle\sigma\rangle$. We now have

$$\begin{aligned} C_D(x^2) &= L \oplus Lx \oplus Lj \oplus Ljx \\ &= (L \oplus Lj) \oplus (L \oplus Lj)\,x \\ &\cong (L'/F', \sigma, i), \end{aligned}$$

where, of course, the σ here means $\sigma|_{L'}$. To show that this centralizer is a division algebra, we try to identify the quadratic extension L'/F' more explicitly. Let

$$y' = (y + iz)(y - iz) = y^2 + z^2,$$

and $z' = y - iz$, in K'. Then $K' = \mathbb{Q}'(y', z')$ and, in terms of these new variables, the σ-action is simply

$$\sigma(y') = y', \quad \sigma(z') = iz'.$$

This enables us to compute quickly the fixed fields of σ and σ^2 on K', namely,

$$\begin{aligned} F' &= K'^{\sigma} = \mathbb{Q}'(y', z'^4), \quad \text{and} \\ L' &= K'^{\sigma^2} = \mathbb{Q}'(y', z'^2). \end{aligned}$$

Applying the example given after (14.8) (with $E = \mathbb{Q}'(y')$ and $t = z'^4$), we see that $C_D(x^2) = (L'/F', \sigma, i)$ is indeed a division algebra. (The fact that i is a nonsquare in $\mathbb{Q}' = \mathbb{Q}(i)$ implies readily that i is also a nonsquare in $\mathbb{Q}'(y')$.)

Having done the above work, it is now not difficult to show that D itself is a division algebra. Let

$$R = \{f_0 + f_1 x + f_2 x^2 + f_3 x^3 : \ f_j \in \mathbb{Q}[y, z]\};$$

this is a subring of D. If D contains zero-divisors, then R also does, by clearing denominators. Therefore, it suffices to show that R is a domain. Let

R_d be the $\mathbb{Q}$-vector space spanned by

$$\{y^m z^n x^k : m+n=d,\ k \geq 0\}$$

in D. We see easily that

$$R = \bigoplus_{d=0}^{\infty} R_d \quad \text{with } R_d R_{d'} \subseteq R_{d+d'},$$

so R is a graded ring. If R has zero-divisors, then, taking their highest homogeneous components, we will have some nonzero $\alpha \in R_d$, $\beta \in R_{d'}$ with $\alpha\beta = 0$. By left-multiplying α and/or right-multiplying β by y if necessary, we may assume that d and d' are both *even*. But then α and β are in $C_D(x^2)$, since

$$x^2 y^m z^n = (-1)^{m+n} y^m z^n x^2$$

implies that $x^2\gamma = (-1)^e \gamma x^2$ for any $\gamma \in R_e$. This contradicts the fact that $C_D(x^2)$ is a division ring, so we have completed the proof that $D = (K/F, \sigma, -1)$ is a division ring.

Although there are fields F for which Wedderburn's norm condition on a is not a necessary condition for a cyclic algebra $(K/F, \sigma, a)$ to be a division algebra, there are also fields for which it is. Most notably, over any algebraic number field F, the converse of (14.9) also holds, so $(K/F, \sigma, a)$ is a division algebra *if and only if* the image of a has order $[K : F]$ in $\dot{F}/N(\dot{K})$. The proof of this fact, however, requires deep results from algebraic number theory.

Wedderburn's Theorem (14.9) enables us to construct explicitly many examples of centrally finite cyclic division algebras. For $s = 2$, we get essentially the so-called *generalized quaternion algebras*. For instance, the centralizer $C_D(x^2) = (L'/F', \sigma, i)$ we encountered in the above construction is such an algebra. A detailed treatment for generalized quaternion algebras can be found in Lam [73], so we shall not dwell on this case here. Instead, we present below some examples of cyclic division algebras of dimension s^2 where $s = [K : F] \geq 3$. We start with a lovely example of Dickson in the case $s = 3$ over the rational field.

Let $F = \mathbb{Q}$ and $E = \mathbb{Q}(\zeta)$, where $\zeta = e^{2\pi i/7}$ is a primitive 7th root of unity. Then $[E : \mathbb{Q}] = 6$ and $Gal(E/\mathbb{Q}) = \langle\tau\rangle$ where τ is the automorphism on E (of order 6) defined by $\tau(\zeta) = \zeta^3$. Let K be the unique subfield of E such that $[K : \mathbb{Q}] = 3$. Then $K/\mathbb{Q}$ is a cyclic extension of degree 3. Using this, we shall construct explicitly a cyclic $\mathbb{Q}$-division algebra of dimension nine.

First let us try to understand K a little better. We think of E as a subfield of $\mathbb{C}$, and consider

$$v = \zeta + \zeta^{-1} = 2\cos(2\pi/7) \in E \cap \mathbb{R}.$$

A short computation shows that $v^3 + v^2 - 2v = 1$. Since

$$f(t) := t^3 + t^2 - 2t - 1$$

is irreducible over $\mathbb{Q}$, it is the minimal polynomial of v, and so we have $K = \mathbb{Q}(v) = E \cap \mathbb{R}$. The Galois group $G = Gal(K/\mathbb{Q})$ is $\{1, \sigma, \sigma^2\}$, where σ is the restriction of τ^2 to K. The three conjugates of $v \in K$ are

$$v = 2\cos(2\pi/7),$$
$$\sigma(v) = v^2 - 2 = 2\cos(4\pi/7), \quad \text{and}$$
$$\sigma^2(v) = 1 - v - v^2 = 2\cos(8\pi/7) = 2\cos(6\pi/7).$$

In order to construct a cyclic division algebra from $K/\mathbb{Q}$, we need an element $a \in \dot{\mathbb{Q}} \backslash N(\dot{K})$ where $N = N_{K/\mathbb{Q}}$. To this end, we first compute explicitly the norm form N. For $\alpha = p + qv + rv^2$ where $p, q, r \in \mathbb{Q}$, left multiplication by α on K has matrix

$$\begin{pmatrix} p & r & q - r \\ q & p + 2r & 2q - r \\ r & q - r & p - q + 3r \end{pmatrix}$$

with respect to the $\mathbb{Q}$-basis $\{1, v, v^2\}$. A determinant computation yields

$$N(\alpha) = p^3 + q^3 + r^3 - p^2q - 2pq^2 + 5p^2r + 6pr^2 - q^2r - 2qr^2 - pqr.$$

Let n be any even integer. *We claim that, if $n \in N(\dot{K})$, then $8 | n$.* In fact, write $n = N((p + qv + rv^2)/m)$ where $p, q, r, m \in \mathbb{Z}$, and $m > 0$ is chosen as small as possible. Then $m^3 n = N(p + qv + rv^2)$. Computing *mod* 2, we have $s \equiv s^2 \equiv s^3$ for any $s \in \mathbb{Z}$, and so

$$m^3 n \equiv p + q + r + pq + pr + qr + pqr$$
$$\equiv 1 + (p + 1)(q + 1)(r + 1) \pmod 2.$$

Since n is even, p, q, r must *all* be even. But then m must be odd (by its minimal choice). Writing $p = 2p_0$, $q = 2q_0$, and $r = 2r_0$, we have

$$m^3 n = 8N(p_0 + q_0 v + r_0 v^2) \in 8\mathbb{Z},$$

so $n \in 8\mathbb{Z}$ as claimed.

From the above, we see, in particular, that $2, 4 \notin N(\dot{K})$. Therefore, by (14.8), the cyclic algebra $D = (K/\mathbb{Q}, \sigma, 2)$ is a (9-dimensional) $\mathbb{Q}$-division algebra, with $Z(D) = \mathbb{Q}$. Explicitly, we have

(14.14a) $\quad K = \mathbb{Q}(v), \quad D = K \oplus K \cdot x \oplus K \cdot x^2 = \mathbb{Q}\langle v, x \rangle,$

with the relations

(14.14b) $\quad v^3 + v^2 - 2v - 1 = 0, \quad x^3 = 2, \quad \text{and} \quad xv = (v^2 - 2)x.$

Remark. The connoisseur in number theory would no doubt have noticed that the proof for

$$n \in (2\mathbb{Z}) \cap N(\dot{K}) \Longrightarrow 8 | n$$

can be given quite a bit more efficiently by using local number theory.

In fact, since $t^3 + t^2 - 2t - 1$ is irreducible over $\mathbb{Z}/2\mathbb{Z}$, it is also irreducible over $\mathbb{Q}_2$, the field of 2-adic numbers. Thus, $\mathbb{Q}_2(v)$ is *unramified* over $\mathbb{Q}_2$. Let v_2 denote the normalized (additive) valuation on $\mathbb{Q}_2(v)$. If $n \in 2\mathbb{Z}$ and $n = N(\alpha)$ ($\alpha \in \mathbb{Q}$), then $v_2(\alpha)$ must be >0 and so

$$v_2(n) = v_2(N(\alpha)) = v_2(\alpha \cdot \sigma(\alpha) \cdot \sigma^2(\alpha)) \geq 3.$$

Since v_2 is just the usual 2-adic valuation on $\mathbb{Q}$, this means that $8 \mid n$.

Of course, in the case when $s = [K : F]$ is a prime (as in Dickson's example), one can refer directly to (14.8), so one does not need Wedderburn's Theorem (14.9). In order to show a true application of Wedderburn's result, we shall now give an example for which $s = [K : F]$ is completely arbitrary. This example was pointed out in Jacobson [75], p. 83.

Let K_0/F_0 be a cyclic extension of degree s, with $G_0 = Gal(K_0/F_0) = \langle \sigma_0 \rangle$. Let t be an indeterminate over K_0, and let $K = K_0(t)$, $F = F_0(t)$. We can extend σ_0 to an automorphism σ on K by letting $\sigma(t) = t$. Then the fixed field of σ is $F = F_0(t)$, and K/F is Galois with $Gal(K/F) = \langle \sigma \rangle$. We claim that *the (multiplicative) order of t in $\dot{F}/N(\dot{K})$ is precisely* s, so $(K/F, \sigma, t)$ is a cyclic *division* algebra over F. To prove our claim, consider an arbitrary element of K, say

$$f = (a_0 + \cdots + a_m t^m)/(b_0 + \cdots + b_n t^n),$$

where $a_i, b_i \in K_0$ and $a_m \neq 0 \neq b_n$. Computing the norm, we have

$$\begin{aligned} N_{K/F}(f) &= \frac{(\sum a_i t^i)(\sum \sigma_0(a_i) t^i) \cdots (\sum \sigma_0^{s-1}(a_i) t^i)}{(\sum b_j t^j)(\sum \sigma_0(b_j) t^j) \cdots (\sum \sigma_0^{s-1}(b_j) t^j)} \\ &= \frac{N_0(a_0) + \cdots + N_0(a_m) t^{sm}}{N_0(b_0) + \cdots + N_0(b_n) t^{sn}}, \end{aligned}$$

where $N_0 = N_{K_0/F_0}$. If we have $t^k = N_{K/F}(f)$ where $k > 0$ and f is as above, then

$$N_0(a_0) + \cdots + N_0(a_m) t^{sm} = t^k [N_0(b_0) + \cdots + N_0(b_n) t^{sn}].$$

Since $N_0(a_m) \neq 0 \neq N_0(b_n)$, a comparison of the degrees on the two sides gives $sm = k + sn$, and so $s \mid k$, as claimed. This completes the proof that $D = (K/F, \sigma, t)$ is a cyclic division F-algebra.

To write down a more concrete example, let $K_0 = \mathbb{Q}(t_1, \ldots, t_s)$ and let F_0 be the fixed subfield of the automorphism σ_0 which cyclically permutes $\{t_1, \ldots, t_s\}$. Then K_0/F_0 is Galois of degree s, with $Gal(K_0/F_0) = \langle \sigma_0 \rangle$. Introducing a new indeterminate t as above, we get a cyclic division algebra $D = (K/F, \sigma, t)$ over $F = F_0(t)$. In this F-algebra, the cyclic algebra relations simplify to

(14.15) $\quad x^s = t \quad \text{and} \quad x \cdot f(t_1, \ldots, t_s) = f(t_2, \ldots, t_s, t_1) \cdot x$

for all $f \in \mathbb{Q}(t_1, \ldots, t_s)$.

We now come to the last topic in this section, which is the Mal'cev–Neumann construction of Laurent series division rings. Our presentation here follows closely Neumann [49]. The main idea of this new construction is that one can combine Hilbert's twisted Laurent series construction with the usual construction of group rings to get a much bigger class of division rings. In Hilbert's examples, the noncommutative feature essentially arises from the Hilbert twist, but in the Mal'cev–Neumann examples, there is an additional noncommutative feature arising from the use of possibly noncommutative (ordered) groups. In the case when only commutative (ordered) groups are used and the Hilbert twist is taken to be trivial, the idea of the construction goes back much earlier to H. Hahn (1907). However, the generalization to the noncommutative case is highly nontrivial, and was successfully completed only upon the appearance of the papers of Mal'cev and Neumann in 1948–49.

In order to present the Mal'cev–Neumann construction, it is convenient to record first a few facts on subsets of a totally ordered set G. Recall that a subset $S \subseteq G$ is *well-ordered* (or *WO* for short) if every nonempty subset of S has a least element.

(14.16) Lemma. *Let $(G, <)$ be a totally ordered set. For any subset $S \subseteq G$, the following statements are equivalent:*

(1) *S is WO.*

(2) *S satisfies DCC (i.e., any sequence $s_1 \geq s_2 \geq s_3 \geq \cdots$ in S is eventually constant).*

(3) *Any sequence $\{s_1, s_2, s_3, \ldots\}$ in S contains a subsequence $\{s_{n(1)}, s_{n(2)}, s_{n(3)}, \ldots\}$ (where $n(1) < n(2) < n(3) < \cdots$) such that $s_{n(1)} \leq s_{n(2)} \leq s_{n(3)} \leq \cdots$.*

Proof. Since (3) $\Rightarrow$ (2) and (2) $\Rightarrow$ (1) are both obvious, it is enough to show that (1) $\Rightarrow$ (3). Let $\{s_1, s_2, s_3, \ldots\}$ be a sequence in a *WO* subset S of G. Choose $n(1)$ so that $s_{n(1)} = \min\{s_i \colon i \geq 1\}$. Then choose $n(2) > n(1)$ so that $s_{n(2)} = \min\{s_i \colon i > n(1)\}, \ldots$, etc. This produces a nondecreasing subsequence $s_{n(1)} \leq s_{n(2)} \leq \cdots$, as desired. QED

(14.17) Lemma. *Let S, T be WO subsets of a totally ordered set $(G, <)$. Then $S \cup T$ is WO. If $(G, <)$ is an ordered group, then*

$$U := S \cdot T = \{st \colon\ s \in S,\ t \in T\}$$

is also WO. Moreover, for any $u \in U$, there exist only a finite number of ordered pairs (s, t) $(s \in S,\ t \in T)$ such that $u = st$.

Proof. We omit the trivial proof of the first conclusion. For the second conclusion, assume, instead, that U is *not WO*. By (14.16), there would exist a strictly decreasing sequence

$$s_1t_1 > s_2t_2 > \cdots$$

where $s_i \in S$, $t_i \in T$. After replacing $\{s_1, s_2, \ldots\}$ by a subsequence, we may assume (since S is *WO*) that $s_1 \leq s_2 \leq \cdots$. If $t_i \leq t_{i+1}$ for some i, we would have

$$s_it_i \leq s_{i+1}t_i \leq s_{i+1}t_{i+1},$$

a contradiction. Thus we must have $t_1 > t_2 > t_3 > \cdots$. But this contradicts the fact that T is *WO*. This proves the second conclusion in the Lemma, and the third conclusion follows from a similar argument. QED

We are now ready to present the general Mal'cev–Neumann construction of Laurent series rings. For this construction, we fix a base ring R and an ordered group $(G, <)$. We assume that G is multiplicatively written, and write

$$P = \{x \in G\colon\ x > 1\}$$

for the positive cone of the ordering on G. Furthermore, we fix a group homomorphism ω from G to $Aut(R)$, the group of automorphisms of the ring R; the image of $g \in G$ under ω will be denoted by ω_g.

As a set, the Mal'cev–Neumann ring $A = R((G, \omega))$ consists of certain formal, but not necessarily finite, sums

$$\alpha = \sum_{g \in G} \alpha_g g \quad (\text{"Laurent series"})$$

where the α_g's are elements of R. (We think of such a formal series $\alpha = \sum_{g \in G} \alpha_g g$ as a function $\alpha\colon G \to R$ defined by $\alpha(g) = \alpha_g$ for all $g \in G$.) For each such α, we define the *support* of α by $supp(\alpha) := \{g \in G\colon \alpha_g \neq 0\}$. Now define

(14.18) $$A = R((G, \omega)) = \{\alpha = \sum \alpha_g g:\ supp(\alpha) \subseteq G \text{ is } WO\}.$$

In A, we add and multiply elements according to the following formal rules:

(14.19) $$\sum_{g \in G} \alpha_g g + \sum_{g \in G} \beta_g g = \sum_{g \in G} (\alpha_g + \beta_g) g,$$

(14.20) $$\left(\sum_{g \in G} \alpha_g g\right)\left(\sum_{h \in G} \beta_h h\right) = \sum_{u \in G} \left(\sum \alpha_g \omega_g(\beta_h)\right) u,$$

where the last sum is over all (g, h) such that $gh = u$. Since we may restrict g and h respectively to $supp(\alpha)$, $supp(\beta)$, and these supports are *WO* sets in G, the last sum in (14.20) is finite by (14.17). Also, since

$$supp(\alpha + \beta) \subseteq supp(\alpha) \cup supp(\beta),$$
$$supp(\alpha\beta) \subseteq supp(\alpha) \cdot supp(\beta),$$

the supports on the *LHS* are both *WO* by (14.17). Therefore, addition and multiplication are well-defined in A. Having made this observation, it is straightforward to check that $(A, +, \cdot)$ is a ring. The subring of A consisting of all *finite* sums $\alpha = \sum \alpha_g g$ (i.e., sums of finite support) is just the twisted group ring $R * G$ defined in §1, which may be denoted here by $R[G, \omega]$. As usual, we shall identify R with the subring $R \cdot 1 \subseteq A$, and identify G with the subgroup $1 \cdot G$ of invertible elements in A. If ω happens to be the trivial homomorphism, the resulting untwisted ring of Laurent series will be denoted by $R((G))$.

Of course, the idea of multiplying two "series" α and β by (14.20) stems from the distributive law and the twist law $g \cdot r = \omega_g(r)g$, where $r \in R$ and $g \in G$. In the special case when G is an infinite cyclic group $\{x^n\colon n \in \mathbb{Z}\}$ ordered by the positive cone $P = \{x^n\colon n > 0\}$, the homomorphism

$$\omega\colon\ G \longrightarrow Aut(R)$$

is specified by a single automorphism $\sigma := \omega_x$. In this case, the twist law boils down to $x \cdot r = \sigma(r)x$ (for $r \in R$), and

$$A = R((\langle x \rangle, \omega)) = \left\{ \sum_{i=n}^{\infty} \alpha_i x^i\colon\ \alpha_i \in R,\ n \in \mathbb{Z} \right\}$$

is just Hilbert's twisted Laurent series ring $R((x, \sigma))$, noting that *WO*-subsets of $\mathbb{Z}$ are just nonempty subsets which are bounded below.

The reason we are interested in $R((G, \omega))$ in this section is given in the next theorem. Note that in this result, no additional assumption on the homomorphism $\omega\colon G \to Aut(R)$ is needed.

(14.21) Theorem. *Assume R is a division ring, and $(G, <)$ and ω are as above. Then $A = R((G, \omega))$ is also a division ring.*

The proof of this interesting result is based on the following crucial lemma on ordered groups (G, P):

(14.22) Lemma. *Let S be a WO subset of P in the ordered group (G, P). Let $S^n = \{s_1 \cdots s_n\colon\ s_i \in S\}$ for $n \geq 1$, and let $S^\infty = \bigcup_{n \geq 1} S^n \subseteq P$. Then*

(1) *S^∞ is WO, and*

(2) *any $u \in S^\infty$ lies in only finitely many S^n's.*

Of course, by (14.17) and induction, we know that each S^n $(n \geq 1)$ is *WO*. However, an infinite union of *WO* subsets of G need not be *WO*! Therefore, the conclusion (1) in the Lemma is not immediate. This part (1) is, in fact, the crux of the Lemma; its proof is long and rather technical. For this reason, it is convenient to first assume the truth of (1) in the Lemma. Using this, we shall deduce its part (2), and give a proof for Theorem 14.21. We shall then return to give the tricky proof for part (1) of (14.22).

Proof of (1) $\Rightarrow$ (2) in (14.22). Assume that there is a counterexample u to (2). Since S^∞ is WO by (1), there exists a least counterexample $u \in S^\infty$. For $1 \le i < \infty$, write $u = s_{i1}s_{i2}\cdots s_{in_i}$ where $2 \le n_1 < n_2 < \cdots$, and $s_{ij} \in S$. Since

$$u = s_{i1} \cdot (s_{i2} \cdots s_{in_i}) \in S \cdot S^\infty,$$

and both S and S^∞ are WO, (14.17) implies that there is an element $v \in G$ such that

$$s_{i2} \cdots s_{in_i} = v$$

for infinitely many i's. This v clearly lies in infinitely many S^n's, but $s_{i1} > 1$ for all i implies that $v < u$. This contradicts the choice of u as the least counterexample. QED

(14.23) Corollary. (*No assumption on R here.*) *Let*

$$\alpha = \sum \alpha_g g \in A = R((G, \omega))$$

be such that $S := supp(\alpha)$ *lies in* P. *Then for any* $a_0, a_1, \cdots \in R$, *the sum* $a_0 + a_1\alpha + a_2\alpha^2 + \cdots$ *gives a well-defined element of* A.

Proof. Since $supp(\alpha^n) \subseteq S^n$, each $g \in G$ can lie in $supp(\alpha^n)$ only for finitely many n's, according to (14.22)(2). Therefore, the sum

$$\gamma = a_0 + a_1\alpha + a_2\alpha^2 + \cdots$$

makes sense. Moreover, $supp(\gamma)$ is WO since it lies in

$$\{1\} \cup \bigcup_{n \ge 1} S^n = \{1\} \cup S^\infty.$$

Therefore, γ is an element of A. QED

We are now suitably equipped to give the

Proof of (14.21). Assume R is a division ring and consider a nonzero element $\beta = \sum \beta_g g \in A$. Let g_0 be the least element in $supp(\beta)$. Then $\beta_{g_0}^{-1}\beta g_0^{-1} = 1 - \alpha$ where $\alpha \in A$ has $supp(\alpha) \subseteq P$. By (14.23),

$$\gamma = 1 + \alpha + \alpha^2 + \cdots$$

is a well-defined element of A, and a routine formal check shows that γ is an inverse of $1 - \alpha$. Therefore, $1 - \alpha$ is a unit in A, and so $\beta = \beta_{g_0}(1 - \alpha)g_0$ is also a unit in A. QED

To tie the loose end, it remains to give a proof for (1) of Lemma (14.22). We proceed as follows.

In the ordered group (G, P), we say that two elements s and t in P are *relatively archimedean* (written $s \sim t$) if $s \le t^m$ and $t \le s^n$ for some positive integers m, n. It is easy to check that "$\sim$" is an equivalence relation on P.

The equivalence class of $s \in P$ will be denoted by $[s]$, which is called the *archimedean class* of s. Given two archimedean classes $[r]$ and $[s]$, we define $[r] < [s]$ if $r^n < s$ for all $n \geq 1$. We see easily that "$<$" is well-defined (independently of the choice of the class representatives), and gives a *total* ordering on the set of all the archimedean classes of G. As usual, $[r] \leq [s]$ shall mean either $[r] < [s]$ or $[r] = [s]$.

For any elements $s_1, \ldots, s_n \in P$, we always have

$$[s_1 \cdots s_n] = [\max\{s_1, \ldots, s_n\}].$$

In fact, if, say $s_i = \max\{s_1, \ldots, s_n\}$, then $s_1 \cdots s_n \leq s_i^n$ and $s_i \leq s_1 \cdots s_n$ (since all $s_j > 1$). This shows that $s_i \sim s_1 \cdots s_n$, and so $[s_1 \cdots s_n] = [s_i]$. We shall now proceed to

Proof of (1) in (14.22). For S as in (14.22), assume that S^∞ is *not WO*. Then there exists a strictly decreasing sequence $u_1 > u_2 > \cdots$ in S^∞, say $u_i = s_{i1}s_{i2}\cdots s_{in_i}$, where $s_{ij} \in S$. We claim that the sequence of archimedean classes $[u_1] \geq [u_2] \geq \cdots$ is eventually constant. To see this, let

$$s_i = \max\{s_{i1}, \ldots, s_{in_i}\} \in S.$$

By the foregoing observation, $[u_i] = [s_i]$ so we have $[s_1] \geq [s_2] \geq \cdots$. Since $\{s_1, s_2, \ldots\} \subseteq S$ has a smallest element, say s_{i_0}, the sequence $[s_1] \geq [s_2] \geq \cdots$ must stabilize after i_0 terms, as claimed.

Let $U = \min\{[u_i]\colon i \geq 1\} = [s_{i_0}]$. A different choice of a strictly decreasing sequence in S^∞, say $u_1' > u_2' > \cdots$, would lead to another archimedean class U'. Since any such class is the class of an element in S, we may assume that our initial $u_1 > u_2 > \cdots$ has been chosen such that U is as small as possible. After discarding a finite number of u_i's, we may assume that $U = [u_i] = [s_i]$ for all $i \geq 1$.

Consider the nonempty set $\{s \in S\colon\ [s] = U\}$. This has a least element, say s_U. Since $[s_U] = [u_1]$, there exists an integer $m \geq 1$ such that $u_1 \leq s_U^m$. We may further assume that our sequence $u_1 > u_2 > \cdots$ (subject to all foregoing conditions) has been so chosen that the m we took above is as small as possible. We represent each u_i in one of the following four forms:

$$u_i = \begin{cases} s_i \\ v_i s_i \\ s_i w_i \\ v_i s_i w_i \end{cases}$$

where $v_i, w_i \in S^\infty$. Only a finite number of the u_i's can be of the first type, for otherwise we would have a strictly decreasing sequence in S, which is impossible. Therefore, there must exist a sequence of the u_i's of *one* of the other three types, say, the fourth type. (The other two types are, in fact, simpler and can be similarly handled.) After passing to a subsequence, we may assume that $u_i = v_i s_i w_i$ for all i. Let $B = \{v_i\colon i \geq 1\}$, $C = \{w_i\colon i \geq 1\}$

and let $D = \{s_i: i \geq 1\} \subseteq S$. If B and C are both WO, then by (14.17) (applied twice), BDC is also WO, and $u_1 > u_2 > \cdots$ in BDC gives a contradiction. Thus, we may assume, say, B is *not* WO. After replacing the v_i's by a subsequence, we may therefore assume that $v_1 > v_2 > \cdots$ in $B \subseteq S^\infty$. We have seen earlier that

$$V := \min\{[v_i]:\ i \geq 1\}$$

exists, and, since $v_i \leq u_i$, we have $V \leq U$. By the minimal choice of U, we get $V = U$ (and hence $s_V = s_U$). As before, we may assume that $[v_1] = [v_2] = \cdots$. From

$$v_1 s_U \leq v_1 s_1 \leq v_1 s_1 w_1 = u_1 \leq s_U^m,$$

we see that $m \geq 2$ (for otherwise $v_1 \leq 1$). But then cancellation of s_U implies that $v_1 \leq s_U^{m-1} = s_V^{m-1}$: this contradicts the minimal choice of m. QED

Now the proof of Theorem (14.21) is complete. We have the following important consequence:

(14.24) Corollary. *Let R be any division ring, and $(G, <)$ and ω be as above. Then the twisted group ring $R[G, \omega]$ can be embedded in a division ring, namely $R((G, \omega))$.*

Let $\{x_i:\ i \in I\}$ be a set of independent indeterminates each of which commutes with R. The free ring $R\langle x_i: i \in I\rangle$ generated by $\{x_i\}$ over R is a subring of the group ring $R[G]$, where G is the free group generated by $\{x_i\}$. Since (by (6.31)) any free group can be ordered, we have the following consequence of (14.24) by taking ω to be the trivial homomorphism:

(14.25) Corollary (Mal'cev, Neumann, Moufang). *For any division ring R, the free ring $R\langle x_i:\ i \in I\rangle$ can be embedded in a division ring.*

The $R((G, \omega))$ construction gives many examples of centrally infinite division rings. We shall not compute the center of $A = R((G, \omega))$ in general, but shall content ourselves with the remark that, using the argument in the proof of Case 1 in (14.2), we get easily the following conclusion:

(14.26) Corollary. *If R is a field, $(G, <)$ is a nontrivial ordered group and ω: $G \to Aut(R)$ is an injective homomorphism, then*

$$Z(A) = R^G := \{r \in R:\ \omega_g(r) = r \ \textit{ for all } g \in G\},$$

and the division ring A is centrally infinite.

In this section, we have only touched lightly upon the problem of constructing division rings. There are many other methods available in the literature, e.g., Noether's method of constructing centrally finite division algebras by crossed products (1929), Köthe's method of constructing centrally infinite division algebras by using infinite tensor products (1931), and Ore's method of constructing division algebras by using one-sided rings of quotients (1930). Some of these methods will be studied in more detail in *Lectures*. In this section, we have presented what we regard as the most elementary constructions of division rings. On the one hand, these constructions are of great historical importance; on the other hand, they can be presented in an entirely elementary way, without assuming any knowledge of the general theory of division rings. These two factors make the examples presented here particularly suitable for this introductory section.

For more extensive surveys on the construction of division rings, we refer the reader to Cohn [77], [95], and Dauns [82].

Exercises for §14

Ex. 14.1. Let D be a ring containing a division ring K. Let $[D:K]_\ell$ (resp. $[D:K]_r$) denote the dimension (as a cardinal number) of D as a left (resp. right) K-vector space. The following construction shows that $[D:K]_\ell \neq [D:K]_r$ in general. Let k be a field and $K = k(t)$ where t is an indeterminate. Let $D = K \oplus K \cdot x$, made into a ring with the multiplication rules $x^2 = 0$ and $x \cdot b(t) = b(t^2)x$ for any $b(t) \in K$. Show that $K \cdot x = xK \oplus txK$, and so $[D:K]_\ell = 2$, $[D:K]_r = 3$. If we set $K_n = k(t^n)$, then the transitivity formulas for left and right dimensions show that $[D:K_n]_\ell = 2n$, $[D:K_n]_r = 3n$.

Ex. 14.2. The following question was raised by E. Artin: *if $D \supseteq K$ are a pair of division rings, is $[D:K]_\ell = [D:K]_r$?* Show that this holds if D is centrally finite. (**Hint.** Let A be the subring of D generated by K and $Z(D)$. Show that A is a division ring and that $[A:K]_\ell = [A:K]_r$. Then use the transitivity formulas for left and right dimensions to show that $[D:A]_\ell = [D:A]_r$ and $[D:K]_\ell = [D:K]_r$.)
Remark. In 1961, P.M. Cohn found examples of division rings $D \supseteq K$ such that $[D:K]_r = 2$ but $[D:K]_\ell = \infty$. Later, A. Schofield has even found examples $D \supseteq K$ where $[D:K]_\ell$ and $[D:K]_r$ are both finite but not equal.

Ex. 14.3. For K/F a cyclic extension with $Gal(K/F) = \langle\sigma\rangle$, let $D = (K/F, \sigma, a)$ be a cyclic algebra, where $a \in F$. Show that, if we allow a to be zero, the resulting algebra D is no longer simple (unless $K = F$).

Ex. 14.4. Let $D = (K/F, \sigma, a)$ be a cyclic algebra for which there exists $c \in K^*$ with $N_{K/F}(c) = a$. Give an alternative proof for the splitting of D by constructing an explicit isomorphism L from D to $End_F(K)$. (**Hint.** For $b \in K$, let $\lambda(b) \in End_F(K)$ denote the left multiplication by b on K. Define L by $L(b) = \lambda(b)$ and $L(x) = \lambda(c)\sigma$ and show that L respects the defining relations in the cyclic algebra. Then use the simplicity of D.)

Ex. 14.5. In Dickson's example of a 9-dimensional $\mathbb{Q}$-algebra D, as defined in (14.14)(a) and (b), show that, for any nonzero $\alpha \in K$, we have

$$(*) \qquad \mathbb{Q}(\alpha x) \cong \mathbb{Q}(\sqrt[3]{2N(\alpha)}) \quad \text{and} \quad \mathbb{Q}(ax^2) \cong \mathbb{Q}(\sqrt[3]{4N(\alpha)})$$

(where $N = N_{K/\mathbb{Q}}$), and that these are maximal subfields of D. Note that $K = \mathbb{Q}(v)$ is Galois over $\mathbb{Q}$, but these new maximal subfields are not.

Ex. 14.6. Refer to the notations in (14.2) and assume that σ has finite order s. In this case, $D = k((x, \sigma)) = (K/F, \sigma, x^s)$, where $F = k_0((x^s))$ and $K = k((x^s))$. Show that $K' = k_0((x))$ as well as K are maximal subfields of D. If $s > 1$, show that K and K' are not isomorphic over F.

Ex. 14.7. Let $\zeta \in \mathbb{C}\backslash\{1\}$ be an sth root of unity, where s is a prime. Let $k = \mathbb{Q}(\zeta)$, $K = k(y)$, and $F = k(y^s)$, where y is an indeterminate commuting with k.
(1) Show that K/F is a cyclic (Kummer) extension with $Gal(K/F) = \langle\sigma\rangle$, where σ is defined by $\sigma(y) = \zeta y$.
(2) Show that $\zeta \notin N_{K/F}(K^*)$. Therefore, $D = (K/F, \sigma, \zeta)$ is an s^2-dimensional division algebra over its center F. (The cyclic algebra relations boil down to $x^s = \zeta$, $xy = \zeta yx$. Note that the former relation already implies that $x\zeta = \zeta x$.)
(3) Show that K and $K' = F(x)$ are both maximal subfields of D. (Note that $K = \mathbb{Q}(\zeta, y)$ and $K' = k(x)(y^s) \cong \mathbb{Q}(\zeta', y)$, where ζ' is a primitive s^2th root of unity. In particular, K and K' are nonisomorphic fields.)

Ex. 14.8. Let a be an integer which is not a norm from the extension $E = \mathbb{Q}(\sqrt[3]{2})$ of $\mathbb{Q}$. Show that the $\mathbb{Q}$-algebra with generators ω, α, x and relations

$$(*) \quad \omega^2 + \omega + 1 = 0, \quad \alpha^3 = 2, \quad x^3 = a, \quad \omega\alpha = \alpha\omega, \quad \omega x = x\omega, \quad x\alpha = \omega\alpha x$$

is an 18-dimensional $\mathbb{Q}$-division algebra with center $\mathbb{Q}(\omega)$.

Ex. 14.9. Let $A = R((G, \omega))$ be as in (14.18), where R is an arbitrary ring. For any $\alpha \in A$ such that

$$supp(\alpha) \subseteq P := \{g \in G \colon\ g > 1\}$$

and any positive integer n which is invertible in R, show that $1 + \alpha$ has an nth root in A. (**Hint.** The key point is to show that the binomial coefficients $\binom{1/n}{k}$ make sense in $\mathbb{Z}[1/n]$, and hence in R.)

Ex. 14.10. Let $A = R((G))$ (with trivial ω), where R is a division ring and $(G, <)$ is a (not necessarily abelian) ordered group. Define a map $\varphi\colon A^* \to G$ by $\varphi(\alpha) = \min(supp\ \alpha)$ for any $\alpha \in A^*$. Show that φ is a Krull valuation of the division ring A with value group G in the sense that it satisfies the following properties:
(1) $\varphi(\alpha\beta) = \varphi(\alpha)\varphi(\beta)$ for any $\alpha, \beta \in A^*$, and
(2) $\varphi(\alpha + \beta) \geq \min\{\varphi(\alpha), \varphi(\beta)\}$ for any $\alpha, \beta, \alpha + \beta \in A^*$.

Ex. 14.11. Let $A = R((G, \omega))$ be as in (14.18), where R is a division ring. In case G is an additively written ordered abelian group, it is convenient to introduce a symbol x and write the elements of G "exponentially" as $\{x^g : g \in G\}$. The elements of A are then written as $a = \sum_{g \in G} a_g x^g$. Now let G be the additive group $\mathbb{R}$ with the usual ordering. Let A_1 be the subset of A consisting of $\sum_{n=1}^{\infty} a_n x^{g_n}$ where $a_n \in R$ and $\{g_1, g_2, \ldots\}$ is any strictly increasing sequence in $\mathbb{R}$ with $\lim_{n \to \infty} g_n = +\infty$. Show that A_1 is a proper division subring of A. If

$$A_2 = \left\{ \sum_{n=1}^{\infty} a_n x^{g_n} : \ a_n \in R \ \text{ and } \ g_1 < g_2 < \cdots \ \text{in } \mathbb{R} \right\},$$

is A_2 a subring of A?

Ex. 14.12. Let K be a division ring and σ be an automorphism of K such that $\sigma^2(b) = aba^{-1}$ for all $b \in K$, where a is an element of K fixed by σ. Let $D = K \oplus Kx$ where $x^2 = a$ and $xb = \sigma(b)x$ for all $b \in K$.
(1) Show that D is a division ring iff there does not exist $c \in K$ such that $a = \sigma(c)c$.
(2) Compute $Z(D)$.
(3) Show that D is centrally finite iff K is.

Ex. 14.13A. Let G be a group of automorphisms of a field K, and let F be the fixed field of G. Let $A = K * G$ be the skew group ring of G over K, with respect to the natural action of G on K, as defined in (1.11). By adapting the arguments used in the proof of (14.6), show that
(1) A is a simple ring with center F.
(2) $K (= K \cdot 1)$ is a maximal subfield of A.

Ex. 14.13B. (W. Sinnott) Keep the notations in Exercise 13A.
(1) Show that K is a left A-module under the action

$$\left(\sum a_\sigma \sigma \right) \cdot c = \sum a_\sigma \sigma(c).$$

(2) Show that ${}_AK$ is faithful and simple, with $End({}_AK) \cong F$.
(3) Using the Wedderburn-Artin Theorem (but without assuming *any* facts from Galois Theory), show that $|G| < \infty$ iff $[K : F] < \infty$, and that, in this case, $|G| = [K : F]$.

Ex. 14.13C. (Berger-Reiner) Keep the notations in Exercise 13A, and assume that $n = |G| < \infty$. The field K has the natural structure of a left FG-module. Noether's *Normal Basis Theorem* in Galois Theory states that the FG-module K is free of rank 1, i.e. there exists an element $c \in K$ such that $K = \bigoplus_{\sigma \in G} F \cdot \sigma(c)$.
(1) Deduce this theorem from the Krull-Schmidt Theorem for finite-dimensional FG-modules (see (19.22)).
(2) Deduce from (1) that, for any subgroup $H \subseteq G$, the fixed field $L = K^H$ has a primitive element over F.

Ex. 14.14. In the $\mathbb{Q}$-division algebra D defined in (14.14), compute the inverse of the element $v+x$. (The answer is $[x^2+(v^2+v+1)x+(v^2+v-2)]/3$.)

Ex. 14.15. Let $K=\mathbb{Q}(v)$ be the cubic field defined by the minimal equation $v^3+av+b=0$ where a, b are odd integers. Show that for any element $\alpha = p+qv+rv^2 \in K$ where $p, q, r \in \mathbb{Z}$, we have

$$N_{K/\mathbb{Q}}(\alpha) \equiv 1+(p+1)(q+1)(r+1) \pmod 2.$$

Using this, show that, for any even integer n, if $n \in N_{K/\mathbb{Q}}(\dot{K})$, then $8\,|\,n$.

Ex. 14.16. Show that $K=\mathbb{Q}(v)$ with $v^3-3v+1=0$ is a cyclic cubic field, and find the conjugates of v. Then, proceeding as in Dickson's example (14.14), show that $D=\mathbb{Q}\langle v, x\rangle$ defined by the relations

$$v^3-3v+1=0, \quad x^3=2, \quad \text{and} \quad xv=(v^2-2)x$$

is a 9-dimensional division algebra with center $\mathbb{Q}$.

Ex. 14.17. Do the same for $K=\mathbb{Q}(v)$ with $v^3-7v+7=0$. (**Hint.** The conjugates of v are $3v^2+4v-14$, and $-3v^2-5v+14$.)

§15. Tensor Products and Maximal Subfields

In this section, we shall investigate the behavior of division rings under the tensor product operation. This investigation turns out to be important for the study of subfields and maximal subfields of division rings which is the main topic for this section. Most of the results in the first half of this section can be developed more generally for simple rings instead of division rings. However, since our attention in this chapter is focused on division rings, we shall not try to give the more general treatment here. In this section, the division rings under investigation are not always assumed to be centrally finite; in particular, about half of the results here are meaningful for arbitrary division rings.

Recall that for algebras D, D' over a field F, we can form the tensor product algebra $D \otimes_F D'$, in which $D = D \otimes 1$ and $D' = 1 \otimes D'$ are commuting subalgebras. Since we are studying division rings in this section, a natural question to ask would be: what is the structure of $D \otimes_F D'$ when D, D' are both division F-algebras? In the case when F is exactly the center of D, this question is partly answered in the following theorem.

(15.1) Theorem. *Let D, D' be F-algebras, where $F=Z(D)$ is a field, and let $R := D \otimes_F D'$.*

(1) *The centralizer $C_R(D)$ of $D = D \otimes 1$ in R is $D' = 1 \otimes D'$.*

(2) $Z(R) = Z(D')$.

(3) *Assume D, D' are both division algebras. Then R is a simple F-algebra. It is an artinian algebra if $\dim_F D' < \infty$ (but not conversely).*

Proof. Fix an F-basis $\{d_i' : i \in I\}$ in D' and write

$$R = D \otimes_F \left(\bigoplus F \cdot d_i'\right) = \bigoplus (D \otimes_F d_i'). \tag{15.2}$$

Let $y = \sum c_i \otimes d_i'$ (a finite sum) be in $C_R(D)$, where $c_i \in D$. For any $d \in D$, $dy = yd$ implies that $\sum dc_i \otimes d_i' = \sum c_i d \otimes d_i'$. From (15.2), it follows that $c_i \in Z(D) = F$, and therefore $y = \sum 1 \otimes c_i d_i' \in D'$. Since $D' \subseteq C_R(D)$, this shows that $C_R(D) = D'$. Using this, it follows readily that $Z(R) = Z(D')$. For (3), assume that D, D' are division algebras and let $\mathfrak{A} \neq 0$ be an ideal in R. Let

$$z = \sum_{j=1}^{m} d_j \otimes d_{n_j}'$$

be a nonzero element in $\mathfrak{A}$, with m chosen minimal. Since D is a division ring, we may clearly assume that $d_1 = 1$. For any $d \in D$,

$$dz - zd = \sum_{j=2}^{m} (dd_j - d_j d) \otimes d_{n_j}'$$

lies in $\mathfrak{A}$. By the minimal choice of m, this element must be zero. Therefore, for $j \geq 2$, $dd_j = d_j d$, and so $d_j \in Z(D) = F$. We have now

$$z = \sum_{j=1}^{m} 1 \otimes d_j d_{n_j}' \in D' \backslash \{0\}.$$

Since D' is a division ring, z has an inverse in D' and hence in R. This shows that R is a simple ring. Assume now $\dim_F D' < \infty$. Then I above is a finite set. As a left D-vector space, R is finite-dimensional with basis $\{1 \otimes d_i' : i \in I\}$. Since any left ideal of R is a D-subspace of ${}_D R$, it follows that R is left (and hence also right) artinian. QED

Note that the simple ring $D \otimes_F D'$ may be artinian even without $\dim_F D'$ being finite. For instance, if D is the division ring of rational quaternions, with $Z(D) = \mathbb{Q}$, then for $D' = \mathbb{R}$, $D \otimes_{\mathbb{Q}} D'$ is the division ring of real quaternions, which is, of course, artinian.

Next we shall specialize to the study of the tensor product of a division ring D with the opposite ring K^{op} of a division subring K of D containing the center of D. Recall that the opposite ring K^{op} consists of "formal" elements $\{a^{\text{op}} : a \in K\}$ with addition and multiplication given by

$$a^{\text{op}} + b^{\text{op}} = (a+b)^{\text{op}}, \quad a^{\text{op}} \cdot b^{\text{op}} = (ba)^{\text{op}}.$$

In K^{op}, the multiplication of elements is "turned around"; more precisely, the opposite ring K^{op} is canonically anti-isomorphic to K. Note that if K is a division ring, then so is K^{op}.

(15.3) Theorem. *Let D be a division ring with center F. Let K be a division subring of D containing F, and let $L = C_D(K)$ (also a division subring $\supseteq F$). Then D can be made into a faithful simple left module over $R := D \otimes_F K^{\text{op}}$ in such a way that $End({}_R D) \cong L$. From the Density Theorem* (11.16), *it follows that R acts as a dense ring of linear transformations on D_L.*

Proof. The left action of R on D is given by

$$(d \otimes a^{\text{op}})(v) = dva \quad \text{for} \quad d, v \in D \quad \text{and} \quad a \in K.$$

Since left multiplication by d commutes with right multiplication by a by the associative law, it is easy to check that the above action is well-defined, making D into a left R-module. The ring of endomorphisms of ${}_R D$ is computed as follows. Let $f \in End({}_R D)$ (written on the right of D). Since f commutes with left multiplication by elements of D, f must be the right multiplication on D by an element $c \in D$. Since this multiplication also commutes with right multiplication by elements of K, we have $c \in C_D(K) = L$. Thus $End({}_R D) \cong L$. Clearly, the R-module ${}_R D$ is simple (since ${}_D D$ is already simple) and faithful (since R is a simple ring, by (15.1)). QED

Using (15.3), we can prove the following remarkable result.

(15.4) Theorem. *With the notation in* (15.3), *the following statements are equivalent*:

(1) $dim_F\, K < \infty$.

(2) $dim(D_L) < \infty$.

(3) *The simple ring $R = D \otimes_F K^{\text{op}}$ is artinian.*

If any of these conditions holds, say $r = dim(D_L) < \infty$, then $dim_F\, K$ is also equal to r, and $R = D \otimes_F K^{\text{op}} \cong \mathbb{M}_r(L)$. Moreover, $C_D(L) = K$, and $dim_F\, D = (dim_F\, K) \cdot (dim_F\, L)$ (as cardinal numbers).

Proof. (1) $\Rightarrow$ (3) follows from (15.1), and (2) $\Leftrightarrow$ (3) follows from (11.17). *For the rest of the proof, assume $r = dim(D_L) < \infty$.* We must prove (1) and the other conclusions. By (11.19), the natural map

$$R \longrightarrow End(D_L) \cong \mathbb{M}_r(L)$$

is onto, so $R \cong \mathbb{M}_r(L)$. As left R-modules, ${}_R R \cong r \cdot ({}_R D)$. In particular, as left D-modules, ${}_D R \cong r \cdot ({}_D D)$; therefore, $dim({}_D R) = r$. Since $dim_F\, K = dim({}_D R)$, it follows that $dim_F\, K = r < \infty$, proving (1). By the transitivity formula for dimensions, we have

$$dim_F\, D = dim(D_L) \cdot dim_F\, L = (dim_F\, K)(dim_F\, L).$$

(Note that in the case when D is centrally infinite, this equation amounts to $dim_F\, D = dim_F\, L$, since $dim_F\, K < \infty$ here.)

Finally, to show that $C_D(L) = K$, it suffices to see that any $b \in C_D(L)$ must lie in K. The right multiplication by b on D, denoted by ρ_b, belongs to $End(D_L)$. Therefore, ρ_b corresponds to left multiplication by an element $y \in D \otimes_F K^{\text{op}}$ on D. Since ρ_b also commutes with left multiplication by any element of D, we have $y \in C_R(D)$. Therefore, by (15.1), $y = 1 \otimes a^{\text{op}}$ for some $a \in K$. Letting y and ρ_b operate on $1 \in D$, we conclude that $b = a \in K$.

QED

According to this Theorem, if K is any division subring of D with infinite dimension over F, then $D \otimes_F K^{\text{op}}$ is neither left nor right artinian. Thus, the tensor product of two very nice artinian algebras over F may fail to be either left or right artinian. Letting $K = D$ (and $L = F$), we have, in particular, the following conclusion:

(15.5) Corollary. *A division ring D with center F is centrally finite iff $D \otimes_F D^{\text{op}}$ is a (simple) artinian ring. If $n = dim_F\, D < \infty$, then*

$$D \otimes_F D^{\text{op}} \cong End(D_F) \cong \mathbb{M}_n(F).$$

From left-right symmetry, we can also draw the following conclusion from (15.4):

(15.6) Corollary. *In (15.3), assume that $r = dim_F\, K < \infty$. Then $dim(D_L) = dim({}_L D)$ $(= r)$.*

In (15.4), the conclusions that

$$C_D(C_D(K)) = K \quad \text{and} \quad dim_F\, D = (dim_F\, K)(dim_F\, C_D(K))$$

(under the assumption that $dim_F\, K < \infty$) are often referred to as the *Double Centralizer Theorem*. We note in passing that, if K is *not* assumed to be finite-dimensional over F, the equation $C_D(C_D(K)) = K$ need not hold in general. For a counterexample, see Exercise 4 below.

A powerful method for studying division rings is to investigate the properties of their maximal subfields. The tensor product results given in the discussion above were developed in part to facilitate the study of maximal subfields. By definition, a subfield K in a ring D is said to be a *maximal subfield* of D if K is not properly contained in another field lying in D. If D itself is a division ring, then to say that K is a maximal subfield is equivalent to saying that K is a maximal commutative subring, since any commutative subring of D is contained in a subfield of D. We have the following easy characterization of maximal subfields of a division ring which is independent of the tensor product results.

(15.7) Proposition. *A subfield K of a division ring D is a maximal subfield iff $C_D(K) = K$. If this is the case, then $K \supseteq Z(D)$.*

Proof. Let $L = C_D(K) \supseteq K$. If $L = K$, then for any subfield K' of D containing K, we have $K' \subseteq L = K$, so K is a maximal subfield. Conversely, assume K is a maximal subfield. For any $c \in L$, we can adjoin c to K to form a field $K(c)$. By the maximality of K, we must have $K(c) = K$ and so $c \in K$. Applying this argument to $c \in Z(D)$, we see that $Z(D) \subseteq K$. QED

If we now bring the tensor product results to bear, we can prove the following important fact about maximal subfields in a division ring. (The case of cyclic algebras provides a good illustration of this result; see (14.12) and (14.13).)

(15.8) Theorem. *Let D be a division ring with center F and let K be a maximal subfield of D. Then the scalar extension algebra $D \otimes_F K$ is a simple algebra which acts as a dense ring of linear transformations on D_K. The following statements are equivalent*:

(1) *$D \otimes_F K$ is artinian.*

(2) *$\dim(D_K) < \infty$.*

(3) *$\dim({}_K D) < \infty$.*

(4) *$\dim_F K < \infty$.*

(5) *D is centrally finite.*

Assume $r = \dim_F K < \infty$. Then $\dim(D_K) = \dim({}_K D) = r$,

$$D \otimes_F K \cong End(D_K) \cong \mathbb{M}_r(K),$$

and $\dim_F D = r^2$ (a perfect square). Moreover, a subfield $E \supseteq F$ is a maximal subfield of D iff $\dim_F E = \sqrt{\dim_F D}$.

Proof. Here $L := C_D(K)$ is just K by (15.7), and K^{op} may be identified with K since K is commutative. Therefore, all conclusions follow from (15.3), (15.4) and (15.6), except the "if" part in the last statement of the theorem. Finally, this "if" part follows easily from the "only if" part since any subfield $E \subseteq D$ can be enlarged to a maximal subfield of D. QED

This theorem can be combined with a classical theorem of Artin and Schreier to give some nice results. The Artin–Schreier Theorem we shall use here states the following: *If K is an algebraically closed field and F is a subfield in K with $1 < \dim_F K < \infty$, then F must be a real-closed field with $F(\sqrt{-1}) = K$.* If K is assumed to be of characteristic zero, this is a relatively easy result. The more difficult part of the argument is devoted to showing that K *must* have characteristic zero. A full proof of the Artin–Schreier Theorem can be found on p. 674 of Jacobson [89].

(15.9) Theorem. *Let D be a noncommutative division ring containing an algebraically closed field K such that $r := dim(D_K) < \infty$. Then $F := Z(D)$ is a real-closed field, and D is the division ring of quaternions over F.*

Proof. Clearly K must be a maximal subfield of D. By (15.8), $dim_F\, D = r^2$ and $dim_F\, K = r$. Since $D \neq F$, we have $r > 1$, so the Artin–Schreier Theorem implies that F is real-closed with $F(\sqrt{-1}) = K$; in particular, $r = 2$. By Frobenius' Theorem (the finite-dimensional version of (13.12), for real closed fields), D is the division ring of quaternions over F. QED

By using similar techniques, we can also obtain an extension of Frobenius' Theorem which is due to Gerstenhaber and Yang [60]. Our proof here is quicker and much more conceptual than the one which appeared in their original paper.

(15.10) Theorem (Gerstenhaber–Yang). *Let D be a noncommutative division ring containing a real-closed field R such that $s := dim(D_R) < \infty$. Then $F := Z(D)$ is a real-closed field, and D is the division ring of quaternions over F. There exists an element $i \in C_D(R)$ with $i^2 = -1$ such that $R(i) = F(i)$ in D (but R and F need not be isomorphic).*

Proof. Let K be a maximal subfield of D containing R. *We claim* that $K \neq R$. For, if $K = R$, then $dim(D_K) = s > 1$ since D is not commutative. By (15.8) we have $s = dim_F\, K = dim_F\, R$, and so F has codimension $2s > 2$ in the algebraically closed field $R(\sqrt{-1})$. This is impossible by the Artin–Schreier Theorem.[2] Therefore $K \supsetneq R$, and, since K is algebraic over R, K must be algebraically closed with $K = R(i)$ where $i \in K \subseteq C_D(R)$ and $i^2 = -1$. Now (15.9) gives the desired result. QED

To see that F and R above need not be isomorphic, use the fact that the complex number field $\mathbb{C}$ contains real-closed subfields R with $\mathbb{C} = R(\sqrt{-1})$ but $R \not\cong \mathbb{R}$. Thus, in the division ring $\mathbb{H}$ of quaternions over $\mathbb{R}$, $\mathbb{H}$ has left and right dimensions $= 4$ over R, but R is noncentral and not isomorphic to $\mathbb{R} = Z(\mathbb{H})$.

In the second half of this section, we shall be concerned with the existence problem of separable algebraic elements over the center (or a subfield of the center) of a division ring. The presence of such separable elements has proved to be useful in understanding the structure of division algebras in general. The following existence theorem is due to Noether in the case of centrally finite division algebras, and to Jacobson in the general case of algebraic algebras.

---

[2] Note that we only need the characteristic zero case of the Artin–Schreier Theorem here.

(15.11) Noether–Jacobson Theorem. *Let D be a noncommutative division ring which is an algebraic algebra over a field F (which need not be $Z(D)$). Then there exists an element in $D \backslash F$ which is separable over F.*

Proof. (Herstein) We may assume that *char* $F = p > 0$. If we deny the conclusion of the theorem, every element in $D \backslash F$ would be purely inseparable over F. Fix any element $a \in D \backslash Z(D)$, and fix an integer n such that $a^{p^n} \in F$. Let $\delta = \delta_a \colon D \to D$ be defined by $\delta(x) = ax - xa$. As in the proof of (13.8),

$$\delta^{p^n}(x) = a^{p^n}x - xa^{p^n} = 0$$

since $a^{p^n} \in F \subseteq Z(D)$. Fix an $x \in D$ such that $\delta(x) \neq 0$, and let r be the largest integer such that $y := \delta^r(x) \neq 0$. Then $\delta(y) = 0$ implies that y commutes with a, so $z := y^{-1}a$ also commutes with a. Since $r \geq 1$, we can write $y = \delta(u) = au - ua$ for some $u \in D$. Multiplying by z from the right, we get

$$a = a(uz) - (uz)a = av - va,$$

where $v = uz$. Thus $v = 1 + a^{-1}va$. Take an integer $r \geq 0$ such that $v^{p^r} \in F$. Then we have

$$\begin{aligned} v^{p^r} &= (1 + a^{-1}va)^{p^r} \\ &= 1 + (a^{-1}va)^{p^r} \\ &= 1 + a^{-1}v^{p^r}a \\ &= 1 + v^{p^r}, \end{aligned}$$

a blatant contradiction. QED

In the case of centrally finite division rings, we have the following important application of (15.11).

(15.12) Theorem. *Let D be a centrally finite division ring with center F. Then D has a maximal subfield that is separable over F. In fact, every subfield $E \subseteq D$ separable over F can be enlarged into a maximal subfield $K \subseteq D$ that is separable over F.*

Proof. Given E, let $K \supseteq E$ be a maximal separable field extension of F in D. (Such a field certainly exists since $dim_F\ D < \infty$.) We are done if we can show that K is a maximal subfield of D. Assume otherwise, and let $L := C_D(K)$. By (15.7), $L \supsetneq K$ and by (15.4),

$$Z(L) = L \cap C_D(L) = L \cap K = K.$$

In particular, L is not commutative. Applying (15.9) to the finite-dimensional K-division algebra L, there exists an element $z \in L \backslash K$ that is separable over K. Then $K(z)$ is a separable field extension of K and hence of F. Since $K(z) \supsetneq K$, this contradicts the choice of K. QED

Note that the proof above depends on the Double Centralizer Theorem on K, which requires the hypothesis that $dim_F\, K < \infty$. Thus, it is not clear how this proof could extend to, say, the case of algebraic algebras. In fact, if D is an algebraic division algebra with center F, I do not know if D must contain a maximal subfield that is separable over F.

In Section 12, we stated a commutativity theorem due to Kaplansky, which says that *if R is a semiprimitive ring such that, for any $a \in R$, $a^{n(a)} \in Z(R)$ for some positive integer $n(a)$, then R is commutative* (cf. (12.11)). We showed already in §12 that it would be enough to prove this result for division rings R. We shall now try to present this proof using the results in this section.

For convenience, let us adopt the following terminology: Given a pair of rings $S \subseteq R$, we say that *R is radical over S* if, for any $a \in R$, $a^{n(a)} \in S$ for some $n(a) \geq 1$. In order to prove Kaplansky's Theorem for division rings, we first need to understand field extensions $K \subseteq L$ where L is radical over K (i.e., L^*/K^* is a torsion group). Fortunately, such field extensions can be completely characterized.

(15.13) Proposition. *Let $F \subsetneq K$ be a field extension where K is radical over F, and let P be the prime field of F. Then char $P = p > 0$, and either K is purely inseparable over F, or K is algebraic over P. (Conversely, if these conclusions hold, then clearly K is radical over F.)*

Proof. We may assume that the algebraic extension K/F is *not* purely inseparable (for otherwise we are done). Hence there is an element $a \in K \backslash F$ that is separable over F. Let E be a finite normal extension of F containing a. Since $a \notin F$, there is an automorphism φ of E over F such that $b := \varphi(a) \neq a$. Fix an integer $n > 0$ such that $a^n \in F$. Then

$$b^n = \varphi(a)^n = \varphi(a^n) = a^n$$

implies that $b = \omega a$ where $\omega \neq 1$ is an nth root of unity in E. Similarly, since $\varphi(a+1) = b+1$ and $(a+1)^m \in F$ for some $m > 0$, there is an mth root of unity $\omega' \in E$ such that $b+1 = \omega'(a+1)$. Now $\omega \neq \omega'$, otherwise

$$b+1 = \omega(a+1) = b+\omega,$$

contrary to $\omega \neq 1$. Eliminating b, we have $a = (\omega'-1)/(\omega-\omega')$. Since ω, ω' are roots of unity, this shows that *a is algebraic over the prime field P*. Now consider any element $r \in F$. Repeating the foregoing argument with $a+r$ instead of a, we see that $a+r$, and hence r, is algebraic over P. In short, F, and hence K, is algebraic over P. It remains only to show that *char* $P > 0$. For any integer r, our argument applied to $a+r$ shows that there is an equation

$$a + r = (\omega_r' - 1)/(\omega_r - \omega_r') \tag{15.14}$$

where $\omega_r \neq \omega_r'$ are suitable roots of unity. Upon reexamining our argument, we see that all the ω_r, ω_r' are found in the field $E_0 := P(a,b)$. Since

$[E_0 : P] < \infty$, E_0 contains only a finite number of roots of unity. Thus *char* P must be nonzero, for otherwise we would not be able to account for the infinite number of elements in (15.14). QED

With the preparation above, we can now offer a proof for Kaplansky's Theorem in the division ring case.

(15.15) Theorem. *Let D be a division ring that is radical over its center F. Then $D = F$.*

Proof. Assume $D \neq F$. Since D is an algebraic algebra over F, the Noether–Jacobson Theorem (15.11) implies that there exists an element $c \in D \backslash F$ that is separable over F. The field $K := F(c)$ is radical over F, so we can apply (15.13) to $K \supsetneqq F$. Since K is separable over F, the conclusion is that K is algebraic over its prime field, which is some finite field $\mathbb{F}_p$. Now any $d \in D$ is algebraic over F so d is also algebraic over $\mathbb{F}_p$. This shows that the division ring D is an algebraic algebra over $\mathbb{F}_p$. But then by Jacobson's Theorem (13.11), D is commutative. This is a contradiction. QED

To conclude this section, we shall give another application of the Noether–Jacobson Theorem to the study of the structure of division rings. The following two results may be viewed as analogues of the Theorem on Primitive Elements in field theory. Our presentation here follows Jacobson [56].

(15.16) Theorem (Brauer, Albert). *Let D be a division ring of dimension r^2 over its center F. Then there exist elements $\alpha, \beta \in D$ such that the r^2 elements $\{\alpha^i \beta \alpha^j : 0 \leq i, j < r\}$ form an F-basis for D.*

Proof. By (15.12), there exists a maximal subfield K of D that is separable over F. Then, by the Theorem on Primitive Elements, $K = F(\alpha)$ for some α which satisfies a minimal polynomial $f(t) \in F[t]$ of degree r. View D as a left $D \otimes_F K^{\text{op}}$-module by the action $(d \otimes a^{\text{op}})(v) = dva$, where $d, v \in D$ and $a \in K$. Since K is commutative, we shall identify K^{op} canonically with K. In the proof of (15.3), we have observed that D is a faithful $(D \otimes_F K)$-module; thus, D is also a faithful $(K \otimes_F K)$-module. Let $T \in End({}_K D)$ be defined by $T(v) = v\alpha$, for $v \in D$. Then $f(T) = 0$. We claim that $1, T, \ldots, T^{r-1}$ are K-independent in $End({}_K D)$. For, if $\sum_{i=0}^{r-1} a_i T^i = 0$ where $a_i \in K$, then

$$\left(\sum_{i=0}^{r-1} a_i \otimes \alpha^i \right)(v) = \sum_{i=0}^{r-1} a_i v \alpha^i = \sum_{i=0}^{r-1} a_i T^i v = 0$$

for all $v \in D$, and hence $\sum_{i=0}^{r-1} a_i \otimes \alpha^i = 0$ in $K \otimes_F K$. Since

$$K \otimes_F K = \bigoplus_{i=0}^{r-1} K \otimes \alpha^i,$$

it follows that all $a_i = 0$. Therefore, as a K-linear transformation on ${}_KD$, T has minimal polynomial $f(t)$. Since $dim({}_KD) = r$, we conclude from standard facts in linear algebra that the space ${}_KD$ is T-cyclic, i.e., there exists an element $\beta \in D$ such that D is spanned over K by all $T^j\beta$ ($j \geq 0$). Therefore

$$D = \sum_{j=0}^{r-1} K \cdot T^j\beta = \sum_{j=0}^{r-1} \left(\sum_{i=0}^{r-1} F\alpha^i \right) \beta\alpha^j = \sum_{0 \leq i,j < r} F \cdot \alpha^i\beta\alpha^j.$$

Since $dim_F\, D = r^2$, it follows that $\{\alpha^i\beta\alpha^j : 0 \leq i,j < r\}$ forms an F-basis for D. QED

(15.17) Corollary. *Let D be a division ring of dimension r^2 over its center F. Then there exist two conjugate elements α, α' in D such that $\{\alpha^i\alpha'^j : 0 \leq i,j < r\}$ forms an F-basis for D. In particular, D is generated as an F-algebra by α and α'.*

Proof. Right multiplying the basis $\{\alpha^i\beta\alpha^j : 0 \leq i,j < r\}$ in (15.16) by β^{-1}, we get a new basis $\{\alpha^i(\beta\alpha\beta^{-1})^j : 0 \leq i,j < r\}$ for ${}_FD$. Now let $\alpha' = \beta\alpha\beta^{-1}$. QED

Exercises for §15

Ex. 15.1. (Hua, Kaplansky) Suppose, in a noncommutative division ring D, an integer $n(x) > 0$ is assigned to every element $x \in D$ in such a way that $n(axa^{-1}) = n(x)$ for every $a \in D^*$. Show that $\{x^{n(x)} : x \in D^*\}$ generates D as a division ring. (**Hint.** The division ring K generated by $\{x^{n(x)} : x \in D^*\}$ is invariant under all inner automorphisms of D. If $K \neq D$, then $K \subseteq Z(D)$ by (13.17). But then D is commutative by (15.15).)

Ex. 15.2. (Jacobson) Let D be an algebraic division algebra of infinite dimension over its center F. Show that there exist elements in D of arbitrarily high degree over F. (**Hint.** Assume otherwise and let $K \supseteq F$ be a finite separable field extension of F in D with the largest degree. By (15.4), K is the center of $L = C_D(K)$. If $K \subsetneq L$, there would exist an element $\alpha \in L \backslash K$ separably algebraic over K, by (15.11). Therefore $K = L$ so K is a maximal subfield of D. But then (15.8) implies that $dim_F\, D < \infty$.)

Ex. 15.3. Let R be a field, $(G, <)$ be a nontrivial ordered group, and ω: $G \to Aut(R)$ be an injective group homomorphism. Show that R is a maximal subfield of the Mal'cev-Neumann division ring $D = R((G, \omega))$.

Ex. 15.4. Let $R = \mathbb{Q}(y)$ and let σ be the $\mathbb{Q}$-automorphism of R defined by $\sigma(y) = 2y$. Let D be the division ring $R((x, \sigma))$. Show that $K_1 = \mathbb{Q}(y)$ and $K_2 = \mathbb{Q}((x))$ are both maximal subfields of D. Let $L_1 = \mathbb{Q}(y^2)$ and $L_2 = \mathbb{Q}((x^2))$, both of which contain $Z(D) = \mathbb{Q}$. Show that $C_D(L_i) = K_i$ and so $C_D(C_D(L_i)) = K_i \supsetneq L_i$ for $i = 1, 2$. (This provides counterexamples to the Double Centralizer Theorem for subfields of D infinite-dimensional over the center.)

Ex. 15.5. Let D be a centrally finite division ring with center k. Let $[D, D]$ be the additive subgroup generated by all additive commutators $ab - ba$, where $a, b \in D$. Show that $[D, D] \subsetneq D$. (**Hint.** Let K be a maximal subfield of D. Extending scalars to K, $[D, D]^K \subseteq [D^K, D^K]$. Note that D^K is a matrix algebra over K and $[D^K, D^K]$ consists of matrices of trace zero.)

Ex. 15.6. Let D be a centrally finite division algebra with center k, and let K be a subfield of D containing k. Show that K is a splitting field for D iff K is a maximal subfield of D.

§16. Polynomials over Division Rings

In this section, we shall be interested in polynomials over a division ring D and roots of these polynomials in rings R containing D. Since the coefficients of these polynomials may not commute with elements of R, the notion of roots must be defined carefully. We proceed as follows.

For any ring R, let $R[t]$ denote the polynomial ring in one variable t over R, where t commutes elementwise with R. For a polynomial

$$f(t) = \sum_{i=0}^{n} a_i t^i \in R[t],$$

and an element $r \in R$, we define $f(r)$ (the evaluation of f at r) to be the element $\sum_{i=0}^{n} a_i r^i \in R$. Note that although

$$\sum_{i=0}^{n} a_i t^i = \sum_{i=0}^{n} t^i a_i$$

in the polynomial ring $R[t]$, the two elements $\sum_{i=0}^{n} a_i r^i, \sum_{i=0}^{n} r^i a_i$ of R may be different. To evaluate $f(r)$, we have to first express f in the form $\sum_{i=0}^{n} a_i t^i$, and then substitute r for t. Note that from $f(t) = g(t)h(t) \in R[t]$, it does not follow that $f(r) = g(r)h(r)$, i.e., evaluation at r is in general *not* a ring homomorphism from $R[t]$ to R. For a simple example, consider $g(t) = t - a$, $h(t) = t - b$, where a, b are two noncommuting elements of R. Then $f(t) := g(t)h(t) = t^2 - (a + b)t + ab$, and so

$$f(a) = a^2 - (a + b)a + ab = ab - ba \neq 0 = g(a)h(a).$$

(16.1) Definition. An element $r \in R$ is said to be a *right root* of $f(t) \in R[t]$ if $f(r) = 0$.

Since (mostly) only right roots will be considered in the text, we shall often omit the adjective "right" and simply call them roots. It is easy to establish the following noncommutative form of the Remainder Theorem.

(16.2) Proposition. *An element $r \in R$ is a root of a nonzero polynomial $f(t) \in R[t]$ iff $t - r$ is a right divisor of $f(t)$ in $R[t]$. The set of polynomials in $R[t]$ having r as a root is the left ideal $R[t] \cdot (t - r)$.*

Proof. It suffices to prove the first statement. If $f(t)$ is of the form

$$\left(\sum c_i t^i\right)(t - r) = \sum c_i t^{i+1} - \sum c_i r t^i,$$

then

$$f(r) = \sum c_i r^{i+1} - \sum c_i r \cdot r^i = 0.$$

Conversely, assume $f(r) = 0$. By the usual Euclidean Algorithm, $f(t) = g(t)(t - r) + s$ for some $g(t) \in R[t]$ and some $s \in R$. The first part shows that r is a root of $g(t)(t - r)$. Thus $0 = f(r) = s$; i.e., $f(t) = g(t)(t - r)$. QED

We shall now specialize to division rings. The fact that any nonzero element has an inverse in a division ring makes possible the following result.

(16.3) Proposition. *Let D be a division ring and let $f(t) = g(t)h(t) \in D[t]$. Let $d \in D$ be such that $a := h(d) \neq 0$. Then*

$$f(d) = g(ada^{-1})h(d).$$

In particular, if d is a root of f but not of h, then ada^{-1} is a root of g.

Proof. Let $g(t) = \sum b_i t^i$. Then $f(t) = \sum b_i h(t) t^i$, so

$$f(d) = \sum b_i h(d) d^i = \sum b_i a d^i a^{-1} a$$

$$= \sum b_i (ada^{-1})^i a = g(ada^{-1})h(d).$$

The last conclusion follows since D has no zero-divisors. QED

Over a field, a polynomial of degree n has at most n distinct roots. Over a division ring, this is no longer true. For instance, in the division ring $\mathbb{H}$ of the real quaternions, i, j, k are all roots of $t^2 + 1$. In fact, since any conjugate of i is a root of $t^2 + 1$ and since i has infinitely many conjugates (e.g., by (13.26), or by a direct check), $t^2 + 1$ has infinitely many roots in $\mathbb{H}$. Nevertheless the above-mentioned fact on polynomials over fields does have a reasonable analogue for division rings, as follows.

(16.4) Theorem (Gordon–Motzkin). *Let D be a division ring and let f be a polynomial of degree n in $D[t]$. Then the roots of f lie in at most n conjugacy classes of D. If $f(t) = (t - a_1) \cdots (t - a_n)$ where $a_1, \ldots, a_n \in D$, then any root of f is conjugate to some a_i.*

Proof. We proceed by induction on n, the case $n = 1$ being clear. For $n \geq 2$, let $c \in D$ be a root of f and write $f(t) = g(t)(t - c)$ by (16.2). Suppose $d \neq c$ is another root of f. Then by (16.3), d is conjugate to a root of $g(t)$. Invoking an inductive hypothesis, d lies in a union of at most $n - 1$ conjugacy classes of D. Therefore, the roots of f lie in at most n conjugacy classes of D. The second conclusion of the theorem follows similarly. QED

Next we shall prove some results on polynomials whose coefficients are from the center F of a division ring D. Note that if $a \in D$ is a root of a polynomial $f(t) \in F[t]$, then all conjugates of a are also roots of f. (This follows by conjugating the equation $f(a) = 0$ by the nonzero elements of D.) We shall say that a conjugacy class A is algebraic over F if one (and hence all) of its elements is algebraic over F. In this case, the elements of A have the same minimal polynomial over F, which we shall call the minimal polynomial of A.

(16.5) Lemma. *Let D be a division ring with center F and A a conjugacy class of D which is algebraic over F with minimal polynomial $f(t) \in F[t]$. If a polynomial $h(t) \in D[t] \backslash \{0\}$ vanishes identically on A (i.e., $h(A) = 0$), then $\deg h \geq \deg f$.*

Proof. Assuming the conclusion is false, we can pick a polynomial

$$h(t) = t^m + d_1 t^{m-1} + \cdots + d_m \in D[t]$$

such that $h(A) = 0$ and $m < \deg f$ is as small as possible. Since $h(t) \notin F[t]$, there exists some $d_i \notin F$ so we can pick an element $e \in D^*$ not commuting with d_i. For any $b \in D$, let us write b' for the conjugate ebe^{-1}. For any $a \in A$, we can conjugate the equation

$$a^m + d_1 a^{m-1} + \cdots + d_m = 0$$

to get

$$(a')^m + d_1'(a')^{m-1} + \cdots + d_m' = 0.$$

On the other hand, we also have

$$(a')^m + d_1(a')^{m-1} + \cdots + d_m = 0.$$

Thus, the polynomial $H(t) = \sum_{j=1}^{m} (d_j - d_j')t^{m-j}$ vanishes on $eAe^{-1} = A$. This polynomial is not the zero polynomial (since $d_i \neq d_i'$) and its degree is $< m$. This contradicts the choice of m. QED

(16.6) Theorem. *In the notation above, a polynomial $h(t) \in D[t]$ vanishes on A iff $h(t) \in D[t] \cdot f(t)$.*

Proof. For any $a \in A$, $f(a) = 0$ implies that $f(t) \in D[t] \cdot (t - a)$. If $h(t) \in D[t] \cdot f(t)$, then we also have $h(t) \in D[t] \cdot (t - a)$, so $h(A) = 0$. Conversely, assume $h(A) = 0$ where $0 \neq h \in D[t]$. By the division algorithm, we can write

$h(t) = q(t)f(t) + h_1(t)$, where $h_1 = 0$ or $deg\, h_1 < deg\, f$. But then $h(A) = 0$ implies that $h_1(A) = 0$, so by the Lemma, $h_1 = 0$; i.e., $h_1(t) = q(t)f(t)$.

QED

Before we move on to give the applications of the theorem above, it is worthwhile to mention that the argument used in the proof of (16.5) can be used also to show the following interesting fact, which is well-known in the commutative case:

(16.7) Theorem. *Let D be an infinite division ring. Then no nonzero $h(t) \in D[t]$ can vanish identically on D.*

Proof. Assuming the contrary, we can pick a monic polynomial

$$h(t) = t^m + d_1 t^{m-1} + \cdots + d_{m-1} t \in D[t]$$

of the least degree such that $h(D) = 0$. Arguing as in the proof of (16.5), we can show that all d_i must belong to $F := Z(D)$; i.e., $h(t) \in F[t]$. Since $h(F) = 0$, F is a finite field. But now $h(D) = 0$ implies that D is an algebraic algebra over F. By Jacobson's Theorem (13.11), D is commutative, so $D = F$ is finite, a contradiction. QED

Coming back now to the applications of (16.3)–(16.6), we shall prove two classical results about polynomials over division rings, the first one due to Dickson, and the second one to Wedderburn.

(16.8) Dickson's Theorem. *Let a, b be two elements in a division ring D both of which are algebraic over $F = Z(D)$. Then a, b are conjugate in D iff they have the same minimal polynomial over F.*

Proof. We have already observed the "only if" part before. For the converse, let A be the conjugacy class determined by a, and assume a, b have the same minimal polynomial $f(t) \in F[t]$. Since $f(b) = 0$, $f(t)$ has a factor $t - b$ in the polynomial ring over the *field* $F(b)$, say

$$f(t) = h(t)(t - b) = (t - b)h(t),$$

where $h(t) \in F(b)[t]$. By (16.5), there exists $a' \in A$ such that $h(a') \neq 0$. Since $f(a') = 0$, (16.3) implies that a conjugate of a' is a root of $t - b$. This means that b is a conjugate of a', and hence of a. QED

The next result, due to Wedderburn, states that if an irreducible polynomial over the center of a division ring D has a root in D, then the polynomial "splits completely" in $D[t]$. This result has no analogue in the commutative theory, and thus represents a new phenomenon in the noncommutative theory.

(16.9) Wedderburn's Theorem. *Let D be a division ring with center F, and let A be a conjugacy class of D which is algebraic over F with minimal polynomial*

$f(t) \in F[t]$ of degree n. Then there exist $a_1, \ldots, a_n \in A$ such that

$$f(t) = (t - a_n) \cdots (t - a_1) \in D[t].$$

Also, $f(t)$ is the product of the same linear factors, permuted cyclically. The element $a_1 \in A$ here can be arbitrarily prescribed.

Proof. Fix an element $a_1 \in A$, so $f(t) \in D[t] \cdot (t - a_1)$. Take a factorization

$$f(t) = g(t)(t - a_r) \cdots (t - a_1)$$

with $g(t) \in D[t]$, $a_1, \ldots, a_r \in A$, where r is chosen as large as possible. *We claim that $h(t) := (t - a_r) \cdots (t - a_1)$ vanishes identically on A.* In fact, let $a \in A$, so $f(a) = 0$. If $h(a) \neq 0$, (16.3) implies that $g(a_{r+1}) = 0$ for a conjugate a_{r+1} of a. But then we can write $g(t) = g_1(t)(t - a_{r+1})$ for some $g_1(t) \in D[t]$, and so $f(t)$ has a right factor

$$(t - a_{r+1})(t - a_r) \cdots (t - a_1),$$

contradicting the choice of r. Therefore $h(A) = 0$, and (16.5) implies that $r = n := \deg f$; i.e., $f(t) = (t - a_n) \cdots (t - a_1)$.

To prove the conclusion about the cyclic permutations of the factors, we simply note that whenever $f(t) \in F[t] \setminus \{0\}$ factors as $f_1(t) f_2(t)$ in $D[t]$, then f_1 and f_2 must commute. In fact, from $f = f_1 f_2$, we get

$$f_1 f = f f_1 = f_1 (f_2 f_1)$$

so cancellation of f_1 yields $f = f_2 f_1$. QED

(16.10) Corollary. *In the notation of the theorem, if*

$$f(t) = t^n + d_1 t^{n-1} + \cdots + d_n \in F[t],$$

then $-d_1$ is a sum of elements of A and $(-1)^n d_n$ is a product of elements of A.

The factorization of the minimal polynomial $f(t)$ in $D[t]$ in Wedderburn's Theorem is, of course, very far from being unique, since a_1 can be chosen to be any element of A. If the element a_1 is not central (so $n \neq 1$), we know from Herstein's Theorem (13.26) that A is infinite, so $f(t)$ has infinitely many distinct factorizations into linear factors in $D[t]$. In the context of the theory of polynomial equations over division rings, Herstein's Theorem has the following remarkable generalization:

(16.11) Theorem (Gordon–Motzkin). *Let D be a division ring, and let $g(t) = \sum_{i=0}^n c_i t^i \in D[t]$. Let Γ be the set of roots of g in D, and let A be a conjugacy class of D. If $|\Gamma \cap A| \geq 2$, then $|\Gamma \cap A|$ is infinite.*

Proof. Let us fix an element $a \in A$. Which conjugates dad^{-1} of a are in Γ? Since

$$g(dad^{-1}) = \sum c_i (dad^{-1})^i = \left(\sum c_i d a^i \right) d^{-1},$$

we must look for $d \in D^*$ for which $\sum c_i da^i = 0$. Consider the additive homomorphism $L: D \to D$ defined by $L(d) = \sum c_i da^i$, for $d \in D$. Let $K = C_D(a)$, which is a division subring of D. Then L is right K-linear, since, for any $k \in K$,

$$L(d \cdot k) = \sum c_i(dk)a^i = \sum c_i da^i \cdot k = L(d)k.$$

Let $V = ker(L)$ which is a right K-subspace of D_K. We have a surjective mapping

$$\lambda: \; V^* = V \backslash \{0\} \longrightarrow \Gamma \cap A$$

defined by $\lambda(d) = dad^{-1}$ for $d \in V^*$. For this map λ, we have

$$\begin{aligned} \lambda(d) = \lambda(d') &\Longleftrightarrow dad^{-1} = d'ad'^{-1} \\ &\Longleftrightarrow d^{-1}d' \in C_D(a)^* = K^* \\ &\Longleftrightarrow d' \in dK^*. \end{aligned}$$

Therefore, λ *induces a one-one correspondence between* $\Gamma \cap A$ *and the projective space* $\mathbb{P}(V_K)$ *associated with the vector space* V_K. Now assume $|\Gamma \cap A| \geq 2$. Then V_K has at least two "lines" through the origin; i.e., $dim(V_K) \geq 2$. Since D is noncommutative (in view of $|\Gamma \cap A| \geq 2$), our earlier result (13.10) implies that $K = C_D(a)$ is infinite. Then, by (13.22), $\mathbb{P}(V_K)$ is also infinite. Using the one-one correspondence set up above, we conclude that $\Gamma \cap A$ is infinite. QED

Note that Herstein's Theorem (13.26) corresponds to the special case of (16.11) when $g(t)$ is the zero polynomial. In this case $L = 0$ and $V = D$ in the notation above. The projective space $\mathbb{P}(V_K)$ here is just the coset space $D^*/C_D(a)^*$, and the one-one correspondence defined by λ is the usual one between the conjugates of a and the cosets of D^* modulo $C_D(a)^*$.

(16.12) Corollary. *If* $g(t) \in D[t]$ *has degree* n *and* Γ *is its set of roots in* D, *then either* $|\Gamma| \leq n$ *or* $|\Gamma|$ *is infinite.*

Proof. Assume $|\Gamma| > n$, say $r_1, \ldots, r_{n+1}$ are distinct roots of g in D. Since by (16.4) the elements of Γ lie in at most n conjugacy classes of D, two of the r_i's must lie in the same conjugacy class, say A. Then $|\Gamma \cap A| \geq 2$ and the theorem implies that $|\Gamma \cap A|$ is infinite. Therefore, $|\Gamma|$ is infinite. QED

In the case when D is a field, if $c_1, \ldots, c_n$ are different elements of D, then there is a unique monic polynomial in $D[t]$ of degree n vanishing on $c_1, \ldots, c_n$, namely, $(t - c_1) \cdots (t - c_n)$. This fact can be generalized to the noncommutative case as follows.

(16.13) Theorem (Bray–Whaples). *Let* D *be a division ring and* $c_1, \ldots, c_n$ *be* n *pairwise nonconjugate elements in* D. *Then there is a unique polynomial*

$g(t) \in D[t]$, monic of degree n, such that $g(c_1) = \cdots = g(c_n) = 0$. Moreover, $g(t)$ has the following properties:

(a) *$c_1, \ldots, c_n$ are all the roots of g in D.*

(b) *If $h(t) \in D[t]$ vanishes on all c_i $(1 \leq i \leq n)$, then $h(t) \in D[t] \cdot g(t)$.*

Proof. We proceed by induction on n. For $n = 1$, $t - c_1$ is the polynomial we want, and it clearly has the properties (a), (b). Now assume $n \geq 2$. As inductive hypothesis, we assume that there exists a monic polynomial $f(t)$ of degree $n-1$ with $c_2, \ldots, c_n$ as *all* its roots and such that any polynomial vanishing on $c_2, \ldots, c_n$ has $f(t)$ as a right factor. In particular, the polynomial $g(t)$ we are looking for must have the form

$$g(t) = (t - d)f(t)$$

for some $d \in D$. Since $f(c_1) \neq 0$, (16.3) implies that $g(c_1) = 0$ iff $d = f(c_1)c_1 f(c_1)^{-1}$. This shows the *existence* and *uniqueness* of g. Now it only remains for us to prove the properties (a) and (b) for the $g(t)$ thus constructed. To prove (b), let $h(t) \in D[t]$ be such that

$$h(c_i) = 0 \quad (1 \leq i \leq n).$$

Write $h(t) = p(t)g(t) + r(t)$ where either $r(t) = 0$ or $deg\, r(t) < n$. Since $g(c_i) = h(c_i) = 0$, we have $r(c_i) = 0$ too $(1 \leq i \leq n)$, and so $r(t) = 0$ (for otherwise $r(t)$ can have roots in at most $n-1$ conjugacy classes, by (16.4)). Thus, $h(t) = p(t)g(t)$, as desired. Finally, to prove (a), let $c \in D$ be a root of $g(t) = (t-d)f(t)$, where $d = f(c_1)c_1 f(c_1)^{-1}$ as before. We distinguish the following two cases:

Case 1. *c is a root of $f(t)$.* In this case, $c \in \{c_2, \ldots, c_n\}$ by the inductive hypothesis, so we are done.

Case 2. *c is not a root of $f(t)$.* Then by (16.3), c is conjugate to d and hence to c_1. *We claim that $c = c_1$.* Indeed, assume otherwise. Take a quadratic polynomial

$$\lambda(t) = (t - e)(t - c_1)$$

which vanishes on c_1 and c. (Such a polynomial can be constructed uniquely by choosing e to be

$$(c - c_1)c(c - c_1)^{-1},$$

as in the first part of the proof.) Again by (16.3), the roots of λ are all conjugate to c_1. Write $g(t) = q(t)\lambda(t) + s(t)$ where either $s(t) = 0$ or $deg\, s(t) \leq 1$. Then $s(t)$ vanishes on c_1 and c, and so $s(t) = 0$; i.e., $g(t) = q(t)\lambda(t)$. For $i \geq 2$ we have $\lambda(c_i) \neq 0$, so, by (16.3), c_i is conjugate to a root of $q(t)$. But then $q(t)$ has roots in $n-1$ different conjugacy classes, which is impossible by (16.4) since $deg\, q(t) = n - 2$. Therefore we have shown that $c = c_1$. QED

In field theory, a field F is said to be algebraically closed if every nonconstant polynomial in one variable over F has a root in F. In analogy to this, it is reasonable to define that a *division ring D is right algebraically closed if every nonconstant polynomial in one variable over D has a right root in D*. In view of (16.2), this is equivalent to saying that every $f(t) \in D[t]$ splits completely into a product of linear factors in $D[t]$. The notion of a *left algebraically closed* division ring is similarly defined.

For *centrally finite* division rings, it turns out that "left algebraically closed" and "right algebraically closed" are equivalent properties. Moreover, the algebraically closed centrally finite division rings can be completely classified. In the following, this classification will be carried out. In the commutative case, it is well-known that algebraically closed fields are classified by their characteristics, and by their transcendence degrees over their prime field. Therefore, in the work below, it is sufficient to focus our attention on the noncommutative case. The first half of the classification is achieved in the following result. The proof we present below is a simplification of the one which appeared in Niven [41].

(16.14) Theorem (Niven, Jacobson). *Let R be a real-closed field and D be the division ring of quaternions over R. Then D is right (and left) algebraically closed.*

Proof. For any quaternion $q = a + bi + cj + dk \in D$, we define $\bar{q}$ to be the quaternionic conjugate $a - bi - cj - dk$. It is easy to see that $q \mapsto \bar{q}$ is an R-linear anti-automorphism of D with fixed field R. For any polynomial $f(t) = \sum q_r t^r \in D[t]$, define $\bar{f}(t)$ to be $\sum \bar{q}_r t^r$. Then we can check that, for any pair of polynomials $f, g \in D[t]$, $\overline{fg} = \bar{g}\bar{f}$. In particular,

$$\overline{\bar{f}f} = \bar{f}\bar{\bar{f}} = \bar{f}f$$

implies that $\bar{f}f \in R[t]$, for any polynomial $f \in D[t]$. By induction on $n = deg\, f$, we shall show that f has a (right) root in D. For $n = 1$, this is obvious, so assume $n \geq 2$. Since $R(i) \subseteq D$ is an algebraic closure of R, $\bar{f}f$ has a root α in $R(i)$. By (16.3), either α is a (right) root of f, or a conjugate β of α is a (right) root of $\bar{f}$. In the former case we are done. In the latter case, if $f(t) = \sum q_r t^r$, we have $\sum \bar{q}_r \beta^r = 0$, so $\sum \bar{\beta}^r q_r = 0$, i.e., $\bar{\beta}$ is a *left* root of $f(t)$. By (16.2) (applied to left roots), we can write $f(t) = (t - \bar{\beta})g(t)$ where $g(t) \in D[t]$ has degree $n - 1$. By the induction hypothesis, $g(t)$ has a right root $\gamma \in D$. But then γ is also a right root for $f(t)$, as desired. (A similar argument shows that every $f \in D[t]$ has a *left* root in D.) QED

To complete the classification of (noncommutative) centrally finite division rings which are algebraically closed, we now proceed to the following strong converse of (16.14):

(16.15) Theorem (Baer). *Let D be a noncommutative centrally finite division ring with center R such that any nonconstant polynomial in $R[t]$ has a root in*

D (e.g., D is right algebraically closed). Then R is a real-closed field, and D is the division ring of quaternions over R.

Proof. Let $n = dim_R\, D$. For any irreducible polynomial $f(t) \in R[t]$, there is a root α of f in D. Therefore

$$deg\, f = dim_R\, R(\alpha) \le dim_R\, D = n.$$

Using this, we shall prove that *the field R is perfect*. If *char* R is zero, there is nothing to prove, so assume *char* R is a prime p. For any $\alpha \in R$, we need to show that α has a pth root in R. Let C be a fixed algebraic closure of R, and, for any m, let α^{1/p^m} be the unique p^m-th root of α in C. Since the R-dimensions of the $R(\alpha^{1/p^m})$'s are bounded, the sequence $R(\alpha^{1/p}) \subseteq R(\alpha^{1/p^2}) \subseteq \cdots$ must stabilize. Therefore, there exists an m such that

$$\alpha^{1/p^{m+1}} \in R(\alpha^{1/p^m}).$$

Raising this equation to the p^m-th power, we get $\alpha^{1/p} \in R^{p^m}(\alpha) \subseteq R$, as desired.

Among the simple extensions of R in C, pick K to be of the largest R-dimension. (K exists since the degrees of simple extensions of R are bounded.) If there exists an element $\beta \in C \backslash K$, then $K(\beta) \supseteq R$ is a (finite) *separable* extension of R since R is perfect, and so $K(\beta)$ is a simple extension of R with dimension $> dim_R\, K$, contradicting the choice of K. Therefore, we must have $K = C$ and so $dim_R\, C < \infty$. On the other hand, $R \ne C$ for otherwise $D = C$ is commutative. Hence, by the Artin–Schreier Theorem (cf. §15), R is a real-closed field. By Frobenius' Theorem (the finite-dimensional version of (13.12), for real-closed fields), D is the division ring of quaternions over R. QED

Combining (16.14) and (16.15), we have the following classification theorem.

(16.16) Theorem. *The centrally finite noncommutative division rings which are right algebraically closed are precisely the division rings of quaternions over real-closed fields. These division rings are also left algebraically closed.*

The classification of these division rings is now complete in the sense that we have reduced it to the classification of real-closed fields. On the other hand, the structure of algebraically closed division rings which are centrally infinite does not seem to be well understood.

We now finish this section by giving more information on the solution of polynomial equations over quaternion algebras. These results go back to Niven [41], but our proofs here are considerably simpler than Niven's original proofs. We first make the following observation about "quadratic" conjugacy classes in a division ring.

(16.17) Lemma. *Let D be a division ring with center F, and let A be a conjugacy class of D which has a quadratic minimal polynomial $\lambda(t)$ over F. If $f(t) \in D[t]$ has two roots in A, then $f(t) \in D[t] \cdot \lambda(t)$ and $f(A) = 0$.*

Proof. Write $f(t) = q(t)\lambda(t) + (at + b)$, where $q(t) \in D[t]$ and $a, b \in D$. Then $at + b$ has two roots in A, and so $a = b = 0$. Thus $f(t) = q(t)\lambda(t)$, which vanishes identically on A. QED

In the following, we shall take R again to be a real-closed field, and $D = R \oplus Ri \oplus Rj \oplus Rk$ to be the division ring of quaternions over R. The next result gives some criteria for a polynomial $f(t) \in D[t]$ to have infinitely many roots in D.

(16.18) Proposition (Niven). *In the notation above, the following conditions on a nonzero polynomial $f(t) \in D[t]$ are equivalent:*

(1) *$f(t)$ has infinitely many roots in D.*

(2) *There exist $a, b \in R$ with $b \neq 0$ such that $f(a + bi) = f(a - bi) = 0$.*

(3) *$f(t)$ has a right factor $\lambda(t)$ which is an irreducible quadratic in $R[t]$.*

If these conditions hold, then f vanishes on the conjugacy class of $a + bi$.

Proof. Assume (1). Then by (16.4), $f(t)$ has two roots in a certain conjugacy class A. The minimal polynomial $\lambda(t)$ of A over R is an irreducible quadratic in $R[t]$. By (16.17), we have $f(t) \in D[t] \cdot \lambda(t)$ (and $f(A) = 0$). This shows that (1) $\Rightarrow$ (3). Conversely, assume (3) and let c be a root of $\lambda(t)$ in $R[i]$. By (13.26), c has infinitely many conjugates in D. All of these are roots of $\lambda(t)$ and hence of $f(t)$, so we have (3) $\Rightarrow$ (1). Finally, (3) $\Rightarrow$ (2) is obvious, and (2) $\Rightarrow$ (3) follows easily from (16.2) and (16.3). QED

In case we already know one root α of a polynomial $f(t) \in D[t]$, the result above may be used as follows. If $\alpha \in R$, we write $f(t) = g(t)(t - \alpha)$ and find the remaining roots of f by solving $g(t) = 0$. If $\alpha \notin R$, take a conjugate $\alpha' \neq \alpha$ of α and test whether $f(\alpha') = 0$. If $f(\alpha') = 0$, then $f(t) = h(t)\lambda(t)$ where $\lambda(t)$ is the minimal polynomial of α over R. In this case, *all* conjugates of α are roots of f, and we find the remaining roots of f by solving $h(t) = 0$. If $f(\alpha') \neq 0$, then $f(t)$ has only one root in the conjugacy class of α. In this case, we continue to look for roots of f in other conjugacy classes.

In solving polynomial equations over quaternion algebras, there is one case worth mentioning in which we get only a finite number of roots. This is the case when the polynomial has all but its constant coefficient in the center.

(16.19) Proposition. *Let D, R be as above and $f(t) = \sum_{r=0}^{n} a_r t^r$ where $a_0 \in D \backslash R$ and $a_1, \ldots, a_n \in R$. Then $f(t)$ has at most n roots in D.*

Proof. Let $\alpha \in D$ be any root of $f(t)$. Then α commutes with $\sum_{r=1}^{n} a_r \alpha^r = -a_0$, so $\alpha \in C_D(R(a_0))$. Since $R(a_0)$ is a maximal subfield, we have $C_D(R(a_0)) = R(a_0)$ and so the roots of $f(t)$ are all in $R(a_0)$. The desired conclusion now follows since $f(t)$ has at most n roots in the field $R(a_0)$. QED

(16.20) Corollary (Niven). *For $a \in D \backslash R$, the equation $t^n = a$ has exactly n solutions in D, all of which lie in $R(a)$.*

Exercises for §16

In the following exercises, D denotes a division ring with center F.

Ex. 16.1. Let K be a division subring of D and let $f(t) = h(t)k(t) \neq 0$ in $D[t]$. If $f(t), k(t) \in K[t]$, show that $h(t) \in K[t]$.

Ex. 16.2. Let K be a division subring of D. An element $d \in D$ is said to be *(right) algebraic over K* if $f(d) = 0$ for some polynomial $f(t) \in K[t] \backslash \{0\}$. Among these polynomials, there is a unique monic $f_0(t) \in K[t]$ of the least degree, which is called the *minimal polynomial of d over K*. Is $f_0(t)$ always irreducible in $K[t]$? In case $f_0(t)$ is indeed irreducible in $K[t]$, are two roots of $f_0(t)$ in D always conjugate by an element of $C_D(K)$?

Ex. 16.3. In the division ring $\mathbb{H}$ of real quaternions, find all roots of the quadratic polynomials

$$f_1(t) = (t - j)(t - i), \quad f_2(t) = (t - (j - i))(t - i),$$

and of the cubic polynomial $f_3(t) = (t - k)(t - j)(t - i)$.

Ex. 16.4. Let $c_0, \ldots, c_n \in D$ be pairwise nonconjugate elements. Show that the $(n + 1) \times (n + 1)$ Vandermonde matrix

$$V = V(c_0, \ldots, c_n) = \begin{pmatrix} 1 & \cdots & 1 \\ c_0 & \cdots & c_n \\ c_0^2 & \cdots & c_n^2 \\ \vdots & & \vdots \\ c_0^n & \cdots & c_n^n \end{pmatrix}$$

is invertible over D. If $d_0, \ldots, d_n$ are any given elements in D, show that there exists a unique polynomial

$$f(t) = a_n t^n + \cdots + a_0 \in D[t]$$

such that $f(c_i) = d_i$ for $0 \leq i \leq n$.

Ex. 16.5. Let a, b, c be three *distinct* elements in D.
(1) Show that the 3×3 Vandermonde matrix $V = V(a, b, c)$ is not invertible iff

$$(b - a)b(b - a)^{-1} = (c - a)c(c - a)^{-1}$$

(2) Show that V is invertible if a, b, c do not lie in a single conjugacy class of D. (The general theory of Vandermonde matrices over division rings was developed in Lam [86].)

Ex. 16.6. If a conjugacy class A of D is not algebraic over F, show that no $h(t) \in D[t] \backslash \{0\}$ can vanish identically on A.

Ex. 16.7. Let $f(t) = g(t)h(t)$ where $g \in F[t]$ and $f, h \in D[t]$. Show that $d \in D$ is a root of f iff d is a root of g or a root of h.

Ex. 16.8. (Cf. Exercise 1.16) For a polynomial $f(t) \in D[t]$, write (f) for the ideal generated by f. By a central factor of f, we mean a polynomial $h \in F[t]$ which divides f. Assume $f \neq 0$.
(a) Let $f(t) = g(t)f_0(t)$, where $f_0(t)$ is a central factor of $f(t)$ of the largest possible degree. Show that $(f) = (f_0)$.
(b) Show that $(f) = D[t]$ iff f has no nonconstant central factor.
(c) If f vanishes on some conjugacy class A of D, then $(f) \subsetneq D[t]$.

Ex. 16.9. (Jacobson) Assume that $d = dim_F\, D < \infty$. For any polynomial $f(t) = \sum_{i=0}^n a_i t^i \in R = D[t]$ of degree n, show that there exists a nonzero polynomial $g(t) \in R$ of degree $n(d-1)$ such that $f(t)g(t) = g(t)f(t) \in F[t]$. As a consequence, show that R is neither left primitive nor right primitive. [**Hint.** Assume $a_n = 1$. Let

$$V = De_1 \oplus \cdots \oplus De_n$$

be a left D-vector space with basis $e_1, \ldots, e_n$, and let A be the D-endomorphism defined by: $e_i A = e_{i+1}$ for $i < n$, and

$$e_n A = -a_0 e_1 - a_1 e_2 - \cdots - a_{n-1} e_n.$$

(The matrix of A with respect to $\{e_i\colon\ 1 \le i \le n\}$ is the companion matrix of f, with last row $(-a_0, -a_1, \ldots, -a_{n-1})$.) Identify $E = End({}_DV)$ with $\mathbb{M}_n(D)$ via the basis $\{e_i\colon\ 1 \le i \le n\}$, and view $\mathbb{M}_n(D)$ as a (D, D)-bimodule. Show that $e_1(A^i a) = a(e_1 A^i)$ for any $i < n$, and, using this, show that $f(A) = \sum_{i=0}^n a_i A^i = 0$ and that, if $r(t) \in D[t]$ has degree $< n$, then $r(A) = 0$ implies that $r(t) \equiv 0$. Let $m(t) \in F[t]$ be the minimal polynomial of A as an endomorphism of the nd-dimensional vector space ${}_FV$ and deduce from $m(A) = 0$ that $f(t)$ is a right factor of $m(t)$ in $D[t]$.]

Ex. 16.10. Let a, b be elements of D which satisfy the same minimal polynomial $f(t)$ over the center F of D. For $c \in D^*$, P.M. Cohn has shown that the equation $ax - xb = c$ has a solution $x \in D$ iff $f(t)$ has a right factor $(t - cbc^{-1})(t - a)$ in $D[t]$. Give a proof for the "only if" part. (**Hint.** Let $y = xbx^{-1} - a = -cx^{-1} \neq 0$ and write $f(t) = g(t)(t - a)$. Then

$$0 = f(xbx^{-1}) = g(yxbx^{-1}y^{-1})y$$

by (16.3), and so $yxbx^{-1}y^{-1} = cbc^{-1}$ is a root of g.)

Ex. 16.11. Let $a \in D$ be algebraic over F. Show that the "metro equation" $ax - xa = 1$ has a solution $x \in D$ iff a is not separable over F. (You may use Cohn's result mentioned in Exercise 10.)

Ex. 16.12. (Cohn) Let $a, b \in A$, where A is an algebra over a commutative ring k. Suppose there exists $f(t) \in k[t]$ with $f(b) = 0$ and $f(a) \in U(A)$. For any $c \in A$, show that the equation $ax - xb = c$ has a unique solution $x \in A$. (**Hint.** Let L denote left multiplication by a and R denote right multiplication by b, on A. Then $LR = RL$, $f(R) = 0$, while $f(L)$ is a bijection. To solve the equation $(L - R)(x) = c$, note that

$$f(L) = f(L) - f(R) = (L - R) \cdot g(L, R) = g(L, R)(L - R)$$

for some $g \in k[t, s]$. This implies that $L - R$ is a bijection.)

Ex. 16.13. Let A be a k-algebra where k is a field, and let $a, b \in A$ be algebraic elements over k with minimal polynomials $m_a(t), m_b(t) \in k[t]$. Consider the following conditions:
(1) $m_a(t), m_b(t)$ are relatively prime in $k[t]$;
(2) for any $c \in A$, the equation $ax - xb = c$ has a unique solution $x \in A$;
(3) for $x \in A$, $ax = xb \Longrightarrow x = 0$.
Show that (1) $\Rightarrow$ (2) $\Rightarrow$ (3), and that (3) $\Rightarrow$ (1) if A is a simple algebra with $\dim_k A < \infty$ and $Z(A) = k$. Does (3) $\Rightarrow$ (1) hold in general for any finite-dimensional algebra with center k?

Ex. 16.14. Let $a \in D$ where D is a division ring. If a is the only root of the polynomial $(t - a)^2$ in D, show that a is the only root of $(t - a)^n$ in D for any $n \geq 2$.

Ex. 16.15. Let $a \in D$ where D is a division algebra over a field F. Suppose $F(a)$ is a separable quadratic extension of F. Show that the polynomial

$$f(t) = (t - a)^n = \sum_{i=0}^{n} (-1)^i \binom{n}{i} a^i t^{n-i}$$

has a unique root (namely a) in D. (In particular, this is always the case for every $a \in D$ if D is a generalized quaternion division algebra over a field F of characteristic not 2.)

Ex. 16.16. Let $\mathbb{H}$ be the division ring of real quaternions. For $\alpha = w + xi + yj + zk \in \mathbb{H}$, write $\bar{\alpha} = w - xi - yj - zk$, and define the trace and the norm of α by

$$T(\alpha) = \alpha + \bar{\alpha} = 2w, \quad N(\alpha) = \alpha\bar{\alpha} = w^2 + x^2 + y^2 + z^2.$$

Using Dickson's Theorem (16.8), show that two quaternions $\alpha, \beta \in \mathbb{H}$ are conjugate iff $T(\alpha) = T(\beta)$ and $N(\alpha) = N(\beta)$. Does this conclusion also hold in the division ring of rational quaternions?

Ex. 16.17. Use Dickson's Theorem (16.8) to show that Herstein's Lemma (13.8) holds in a division ring D of *any* characteristic. (**Hint.** Say $a \in D^*$ is noncentral of order n. Factor $t^n - 1$ over the center into monic irreducibles, and show that one of the factors must vanish on a and some $a^i \neq a$.)

Ex. 16.18. (Lam-Leroy) Let $f(t) = (t - d)g(t) \in D[t]$. If D is a centrally finite division ring, show that $f(d') = 0$ for a suitable conjugate d' of d.

CHAPTER 6

Ordered Structures in Rings

The ring of integers $\mathbb{Z}$ has a natural ordering structure which behaves compatibly with the operations of addition and multiplication in $\mathbb{Z}$. If we axiomatize the basic properties of the ordering structure on $\mathbb{Z}$ with respect to addition and multiplication, we arrive at the notion of an *ordered ring*. In this short chapter, we shall give an introduction to some of the basic facts concerning ordered rings.

Historically, the study of orderings on fields is of great importance. The fundamental work of Artin and Schreier in the 1920's relating orderings to the notion of "formal reality" of fields was a landmark in the development of modern algebra. In the early 1950's, the Artin–Schreier criterion for the existence of orderings on fields was successfully generalized to rings. In the first section of this chapter, we shall describe fully this generalization, following the work of J.-P. Serre, T. Szele, G. Pickert, and R.E. Johnson. After this, the chapter concludes with a section on ordered division rings.

Before we go on, we should remark that the orderings on rings we study in this chapter are by no means of the most general kind. The orderings as defined in this chapter can only exist for rings without zero-divisors. To make the notion of orderings meaningful for more general rings, our definition of ordering in §17 would need to be modified. However, as one might expect, the theory of orderings for rings admitting zero-divisors is a bit more complicated. Since it is not our intention to go deeply into this subject, we find it more convenient in this chapter to work with a more restrictive notion of orderings. For a comprehensive treatment of the different kinds of orderings (total orderings, partial orderings, lattice orderings, etc.) on rings which may have zero-divisors, we refer the reader to Fuchs [63].

§17. Orderings and Preorderings in Rings

First we define formally the notion of an ordered ring. By saying that a ring R is ordered, we mean there is a (transitive) total ordering "$<$" given on R such that, for all elements $a, b, c \in R$, we have

$$a < b \Longrightarrow a + c < b + c,$$

$$0 < a,\ 0 < b \Longrightarrow 0 < ab.$$

The *positive cone* of the ordering "$<$" is defined to be $P := \{c \in R: 0 < c\}$. Clearly, P has the following three properties:

(17.1) $P + P \subseteq P$.

(17.2) $P \cdot P \subseteq P$.

(17.3) $P \cup (-P) = R \backslash \{0\}$.

Conversely, if we are given a set P satisfying these three axioms, then, defining a total ordering on R by: $a < b \Longleftrightarrow b - a \in P$, it is easy to check that R becomes an ordered ring under "$<$". For this reason, we shall refer to a set P satisfying (17.1), (17.2), (17.3) above as an *ordering* on R. The ring of integers $\mathbb{Z}$, the field of rational numbers $\mathbb{Q}$, and the field of real numbers $\mathbb{R}$, with their usual orderings, are the quintessential examples. For some examples of noncommutative ordered rings, see Exercises 1, 2 and 3 below.

(17.4) Proposition. *Let P be an ordering on a ring $R \neq 0$. Then*

$$P \cap (-P) = \varnothing \quad \text{(null set)},$$

$1 \in P$, and R is a domain with characteristic zero.

Proof. If $a \in P \cap -P$, then $0 = a + (-a) \in P + P \subseteq P$, contradicting the property (17.3) above. Next, note that one of ± 1 belongs to P, so $1 = 1^2 = (-1)^2 \in P \cdot P \subseteq P$. From this, it follows further that, for any natural number n,

$$n \cdot 1 = 1 + \cdots + 1 \in P.$$

Therefore, *char* $R = 0$. Finally, if $b, c \in R \backslash \{0\}$, then for suitable choices of the signs, $(\pm b)(\pm c) \in P \cdot P \subseteq P$, and so $bc \neq 0$. This shows that R is a domain. QED

The Proposition above gives certain necessary conditions on a ring R in order that an ordering exists in R. In general, these conditions are not sufficient to guarantee the existence of an ordering. The first major question concerning ordered rings is therefore: What is a *necessary and sufficient* condition for a ring R to be "orderable"?

In the case when R is a field, this question was satisfactorily answered by Artin and Schreier: In 1927, they showed that *a field R is orderable iff R is "formally real" in the sense that -1 is not a sum of squares in R*. The latter is clearly a necessary condition for the existence of an ordering on R; the nontrivial part of the Artin–Schreier Theorem is that it is also sufficient. Using this important theorem as the basis, Artin and Schreier launched a very fruitful study of the arithmetic properties of formally real fields.

In this section, we shall present a generalization of the Artin–Schreier Theorem to possibly noncommutative rings. It is not clear at this point how we should define "formal reality" in a ring, but we would hope to arrive at a suitable definition which will turn out to be equivalent to the existence of an ordering on the ring. Here we follow J.-P. Serre's idea of first extending the notion of orderings to that of preorderings.

By definition, a *preordering* in a ring R is a subset $T \subseteq R\backslash\{0\}$ satisfying the following two properties:

(17.5) $T + T \subseteq T$.

(17.6) *For $a_1, \ldots, a_m \in R\backslash\{0\}$ and $t_1, \ldots, t_n \in T$, the product of $a_1, a_1, \ldots, a_m, a_m, t_1, \ldots, t_n$, taken in any order, lies in T.*

Since we shall be referring to the property (17.6) rather frequently, it is convenient to introduce the following notation: For arbitrary elements $a_1, \ldots, a_m \in R$ and nonnegative integers $i_1, \ldots, i_m$, we shall write $per(a_1^{i_1} \cdots a_m^{i_m})$ to mean a product of the following $i_1 + \cdots + i_m$ factors

$$\underbrace{a_1, \ldots, a_1}_{i_1 \text{ times}}, \ldots, \underbrace{a_m, \ldots, a_m}_{i_m \text{ times}}$$

permuted in any way. In this notation, the property (17.6) above may be expressed in the form: $per(a_1^2 \cdots a_m^2 t_1 \cdots t_n) \in T$, for any $a_1, \ldots, a_m \in R\backslash\{0\}$ and $t_1, \ldots, t_n \in T$.

Using the same type of arguments as before, we can easily establish the following analogue of (17.4):

(17.7) Proposition. *Let T be a preordering on a ring $R \neq 0$. Then $T \cap -T = \varnothing$, $1 \in T$, and R is a domain with characteristic zero.*

The fact that $axa = (-a)x(-a)$ implies that any ordering in R is always a preordering. More generally, *the intersection of an arbitrary (non-empty) family of orderings is also a preordering*. Later in this section, we shall be able to characterize the preorderings in R which arise in this way.

For any preordering $T \subseteq R\backslash\{0\}$ and any nonzero element $b \in R$, we shall write T_b for the set of all sums of elements of the form $per(b^i a_1^2 \cdots a_m^2 t_1 \cdots t_n)$, where $a_1, \ldots, a_m \in R\backslash\{0\}$, $t_1, \ldots, t_n \in T$, and $i, m, n \geq 0$.

(17.8) Lemma. *The following statements are equivalent:*

(1) T_b *is not a preordering in* R.

(2) *There exists an equation* $t' + bt = 0$ *where* $t, t' \in T$.

(3) *There exists an equation* $t'' + t'b = 0$ *where* $t', t'' \in T$.

Proof. If (2) holds, then $t'b + btb = 0$, so (3) holds with $t'' := btb \in T$. Similarly, we have (3) $\Rightarrow$ (2). Since (2) $\Rightarrow$ (1) is clear, we are left only with the proof of (1) $\Rightarrow$ (2). It is easy to see that the set T_b satisfies the two properties (17.5) and (17.6). Therefore, T_b fails to be a preordering iff $0 \in T_b$, i.e., iff there exists an equation

$$\text{(17.9)} \qquad 0 = \sum per(b^i a_1^2 \cdots a_m^2 t_1 \cdots t_n),$$

where, in each term, $a_1, \ldots, a_m \in R \backslash \{0\}$ and $t_1, \ldots, t_n \in T$. Note that:

for even i, $\quad per(b^i a_1^2 \cdots a_m^2 t_1 \cdots t_n) \in T$, and

for odd i, $\quad b \cdot per(b^i a_1^2 \cdots a_m^2 t_1 \cdots t_n) \in T$,

since T is a preordering. Grouping terms in (17.9) according to the parity of i, we have therefore an equation $0 = t + r$ where $t \in T$ and $br \in T$. (Note that there must exist a term with odd i in (17.9) lest $t = 0$.) Left multiplying by b, we get $bt + t' = 0$, where $t' = br \in T$. QED

The Lemma we just proved enables us to characterize the orderings among the preorderings.

(17.10) Theorem. *A preordering* $T \subseteq R \backslash \{0\}$ *is an ordering iff* T *is maximal as a preordering.*

Proof. First assume T is an ordering. If there exists a preordering $T' \supsetneq T$, then, for $a \in T' \backslash T$, we have $-a \in T \subseteq T'$ and so

$$0 = a + (-a) \in T' + T' \subseteq T',$$

a contradiction. Thus, T is a maximal preordering. Conversely, assume T is a maximal preordering. If T is not an ordering, then there exists an element b such that $b, -b \notin T$. Since T_b satisfies (17.5), (17.6) and $T_b \supsetneq T$, the Lemma above implies that there exists an equation $t_1 + bt_2 = 0$, where $t_1, t_2 \in T$. Applying the same argument to $-b$, we get a similar equation $t_3 - bt_4 = 0$, where $t_3, t_4 \in T$. But then for $t_5 := (bt_2)(bt_4) \in T$, we get $t_1 t_3 + t_5 = 0$, a contradiction. QED

For any nonzero ring R, define $T(R)$ to be the set of all sums of terms of the form $per(a_1^2 \cdots a_m^2)$ where $a_i \in R \backslash \{0\}$. As before, it is easy to check that

$T(R)$ satisfies the two properties (17.5), (17.6). We define the ring R to be *formally real* if $0 \notin T(R)$. If this is the case, then $T(R)$ is a preordering in R: we shall call it the *weak preordering* of R since it is contained in every preordering of R. We have now the following complete generalization of the Artin–Schreier Theorem.

(17.11) Theorem (R.E. Johnson). *For any ring $R \neq 0$, the following statements are equivalent:*

(1) *R is formally real.*

(2) *R has a preordering.*

(3) *R has an ordering.*

Proof. (3) $\Rightarrow$ (1) and (1) $\Rightarrow$ (2) are clear. For (2) $\Rightarrow$ (3), fix a preordering T in R. By Zorn's Lemma, T can be enlarged into a maximal preordering T_1. By (17.10), T_1 is an ordering for R. QED

In the theory of formally real fields, it is well-known that any preordering T in a field F is the intersection of all the orderings containing T. In the case when $T = T(F)$, this says that an element $a \in F\backslash\{0\}$ is a sum of squares in F iff a is positive in each ordering of the (formally real) field F. In this form, the result is due to Artin, who used it as one of the tools in his famous solution of Hilbert's 17th Problem (concerning the structure of positive semidefinite rational functions). In the following, we shall try to see how these results can be generalized from fields to arbitrary rings.

For any preordering T in a ring, we have always the following equalities:

$$
\textbf{(17.12)}\qquad
\begin{aligned}
&\{a \in R:\ at \in T \text{ for some } t \in T\}\\
&= \{a \in R:\ t'a \in T \text{ for some } t' \in T\}\\
&= \{a \in R:\ ab^2 \in T \text{ for some } b \neq 0\}\\
&= \{a \in R:\ b'^2 a \in T \text{ for some } b' \neq 0\}.
\end{aligned}
$$

In fact, if $at = t'$ where $t, t' \in T$, then $t'a = ata \in T$, and also $at^2 = t't \in T$. We shall denote the set defined in (17.12) by $\tilde{T}$, and call it the *division closure* of T. Clearly, $0 \notin \tilde{T}$ and $T \subseteq \tilde{T}$. It is not difficult to check directly that $\tilde{T}$ is a preordering of R (see Exercise 7 below). We need not give the details of this check here since the fact that $\tilde{T}$ is a preordering is already clear from the following characterization of $\tilde{T}$.

(17.13) Theorem. *For any preordering $T \subseteq R\backslash\{0\}$, the division closure $\tilde{T}$ of T is equal to the intersection T' of all the orderings of R containing T.*

Proof. For any ordering $P \supseteq T$, we have $\tilde{P} \supseteq \tilde{T}$. Since clearly $\tilde{P} = P$, this implies that $T' \supseteq \tilde{T}$. To complete the proof, we shall show that, for any

$a \neq 0$, $a \notin \tilde{T}$ implies that $a \notin P$ for some ordering $P \supseteq T$ (for then $a \notin T'$). Consider T_{-a} which is defined in the paragraph preceding (17.8). Since $a \notin \tilde{T}$, (17.8) implies that T_{-a} is a preordering in R. As we saw before, T_{-a} can be enlarged into an ordering P of R. But then $P \supseteq T$, and $-a \in T_{-a} \subseteq P$ implies that $a \notin P$. QED

We have the following two immediate consequences of (17.13).

(17.14) Corollary. *A preordering $T \subseteq R \backslash \{0\}$ is an intersection of orderings iff T is "division-closed" in the sense that, for $a \in R$ and $t \in T$, $at \in T$ implies that $a \in T$.*

(17.15) Corollary. *In a formally real ring R, a nonzero element $a \in R$ is totally positive (i.e., positive in all orderings of R) iff there exists $b \in R \backslash \{0\}$ such that ab^2 belongs to the weak preordering $T(R)$.*

In general, the weak preordering $T(R)$ need not be division-closed, so the totally positive element a above need not belong to $T(R)$. For an explicit example, see Exercise 8 below.

Let R' be a ring containing the ring R. Then, by intersecting with R, any ordering (resp., preordering) T' of R' "restricts" to an ordering (resp., preordering) of R. If T' restricts to T, we shall also say that T "extends" to T'. Using the techniques we have introduced in this section, we also obtain a nice criterion for the extendibility of an ordering of R to one of R'.

(17.16) Theorem. *Let $R \subseteq R'$ be rings. Then an ordering P of R can be extended to an ordering of R' iff, in R', 0 is not a sum of elements of the form $per(a_1^2 \cdots a_m^2 t_1 \cdots t_n)$, where $a_1, \ldots, a_m \in R' \backslash \{0\}$ and $t_1, \ldots, t_n \in P$.*

Proof. The "only if" part is clear. For the "if" part, assume that the set T' of all sums of elements of the form $per(a_1^2 \cdots a_m^2 t_1 \cdots t_n)$ above does not contain zero. Then T' is a preordering of R'. Let P' be any ordering of R' containing T'. Then $P \subseteq T' \cap R \subseteq P' \cap R$ clearly implies that $P = P' \cap R$, since P is an ordering. QED

If R is any *commutative* domain, it is well-known that any ordering on R extends uniquely to an ordering on its quotient field. In the noncommutative case, we shall see later (in *Lectures*) that a domain need not be embeddable in a division ring. Nevertheless, one can show that any ordering of a domain R can always be extended, uniquely, to an ordering of any "ring of quotients" of R. By a ring of quotients of R, we mean (in this particular context) a domain $R' \supseteq R$ such that, for any $x \in R'$, there exist $a, b \in R \backslash \{0\}$ such that $ax \in R$ and $xb \in R$.

(17.17) Theorem (Albert, Neumann, Fuchs). *Let $R \subseteq R'$ be domains such that R' is a ring of quotients of R. Then any ordering P of R extends uniquely to an ordering P' of R'.*

Proof. Let $P' = \{x \in R' : \exists a, b \in P \text{ such that } axb \in P\}$. We claim that

(17.18)(a) $$P' = \{x \in R' : \exists a \in P \text{ such that } ax \in P\}.$$

For this, it suffices to prove the inclusion "$\subseteq$". Let $x \in P'$, and let $a, b \in P$ be such that $axb \in P$. Fix an element $c \in R\backslash\{0\}$ such that $c \cdot ax \in R$. We may clearly assume that $c \in P$; then $c(axb) \in P \cdot P \subseteq P$. This implies that $cax \in P$, where $ca \in P$, as desired. Similarly, we can prove that

(17.18)(b) $$P' = \{x \in R' : \exists b \in P \text{ such that } xb \in P\}.$$

Also, it is easy to see that $R \cap P' = P$, and $P' \cup (-P') = R'\backslash\{0\}$. To show that P' is an ordering of R', it remains only to show that

$$x, y \in P' \Longrightarrow x + y, \ \ xy \in P'.$$

By (17.18) ((a) and (b)), there exist $a, b \in P$ such that $ax \in P$ and $yb \in P$. But then

$$a(x+y)b = (ax)b + a(yb) \in P \quad \text{and}$$
$$a(xy)b = (ax)(yb) \in P,$$

so by definition, $x + y \in P'$, $xy \in P'$. Therefore, P' is an ordering of R' extending P. The uniqueness of P' is clear. QED

Note that the definition of an ordering on a ring does not depend on the existence of an identity element. Thus, we can talk about orderings on rings which are possibly without identity. The proof of the theorem above did not make any use of an identity element on R. Therefore, the theorem proved in (17.17) holds also for any pair of rings $R \subseteq R'$ possibly without identity, where R' is a ring of quotients of R. This observation is worthwhile because it leads quickly to the following nice result of G. Grätzer and E. Schmidt.

(17.19) Corollary. *Let R be a domain and I be a nonzero ideal of R. Then any ordering on I (as a ring possibly without identity) extends uniquely to an ordering on R.*

Proof. Fix an element $a \neq 0$ in I. For any $x \in R$, we have $ax \in I$ and $xa \in I$, so R is a ring of quotients of I. Now apply (17.17). QED

In general, of course, the class of ordered rings is too large and varied to admit any good classification theorem. The subclass of *archimedean* ordered rings, however, is small enough to be completely described. In the rest of this section, we shall prove the classification theorem for archimedean ordered rings, which essentially goes back to Hilbert. First let us define what is meant by an archimedean ordered ring. Let a be a positive element in an ordered ring $(R, <)$. We say that a is *infinitely large* if $a > n$ $(= n \cdot 1)$ for any integer $n \geq 1$, and that a is *infinitely small* if $na < 1$ for any integer $n \geq 1$.

(17.20) Lemma. *For any ordered ring $(R, <)$, the following two properties are equivalent:*

(1) *For any $a, b > 0$ in R, there exists an integer $n \geq 1$ such that $na > b$.*

(2) *R has neither infinitely large nor infinitely small elements.*

If (1) *or* (2) *holds, $(R, <)$ is said to be an archimedean ordered ring.*

Proof. (1) $\Rightarrow$ (2) is clear. Now assume (2), and consider $a, b > 0$. By (2), we have $b < n$ and $ma > 1$ for suitable integers $m, n \geq 1$. But then $mna > n > b$, as desired. QED

Note that if $(R, <)$ is an ordered division ring, then, for $a > 0$, a is infinitely large iff a^{-1} is infinitely small. Thus, in this case, $(R, <)$ is archimedean iff R has no infinitely large elements, iff R has no infinitely small elements.

(17.21) Theorem. *Let $(R, <)$ be an archimedean ordered ring. Then*

(1) *R is a commutative ring.*

(2) *$(R, <)$ is order-isomorphic to a unique subring of $\mathbb{R}$ (with the induced ordering).*

(3) *The only order-preserving ring automorphism of R is the identity map.*

Proof. For any $a \in R$, let

$$U_a = \{m/n: \ m, n \in \mathbb{Z}, \ n > 0, \ na < m\},$$
$$L_a = \{m/n: \ m, n \in \mathbb{Z}, \ n > 0, \ m \leq na\}.$$

Since $(R, <)$ is archimedean, each of these sets contains an integer; in particular, they are nonempty subsets of $\mathbb{Q}$. It is easy to see that $\{L_a, U_a\}$ is a Dedekind cut on $\mathbb{Q}$, so $\{L_a, U_a\}$ defines a real number $f(a)$. We have now a mapping $f \colon R \to \mathbb{R}$. *We claim that $a < b \Rightarrow f(a) < f(b)$.* In fact, pick an integer $n \geq 1$ such that $n(b - a) > 2$, and let m be the smallest integer $> na$. Then

$$na \geq m - 1 > m + 1 - n(b - a)$$

leads to $nb > m + 1 > m > na$, which in turn implies that

$$f(b) \geq \frac{m+1}{n} > \frac{m}{n} \geq f(a).$$

In particular, f is injective. Next, consider two arbitrary elements $a, b \in R$. It is immediate that $U_a + U_b \subseteq U_{a+b}$ and $L_a + L_b \subseteq L_{a+b}$. In view of these, we deduce from the properties of Dedekind cuts that $f(a+b) = f(a) + f(b)$. Similarly, we can show that $f(ab) = f(a)f(b)$. Thus, $(R, <)$ is order-isomorphic to $f(R)$ with the induced ordering from $\mathbb{R}$. In particular, R is commutative. Since f is clearly the only order-embedding of R into $\mathbb{R}$, the conclusion (3) and the uniqueness part of the conclusion (2) in the theorem both follow. QED

Exercises for §17

Ex. 17.1. Let $(R, <)$ be an ordered ring and $(G, <)$ be a multiplicative ordered group. In the group ring $A = RG$, define P to be the set of $\sum_{i=1}^{n} r_i g_i$ where $r_1 > 0$ in R and $g_1 < g_2 < \cdots < g_n$ in G. Show that (A, P) is an ordered ring. (The same conclusion holds already if $(G, <)$ is an ordered semigroup.)

Ex. 17.2. Using (6.31), show that any free algebra F over a formally real field k can be ordered.

Ex. 17.3. Let R by the Weyl algebra generated over $\mathbb{R}$ by x and y, with the relation $xy - yx = 1$. Elements of R have the canonical form

$$r = r_0(x) + r_1(x)y + \cdots + r_n(x)y^n,$$

where each $r_i(x) \in \mathbb{R}[x]$, $r_n(x) \neq 0$ (if $r \neq 0$). Let $P \subset R$ be the set of all nonzero elements $r \in R$ above for which $r_n(x)$ has a positive leading coefficient. Show that P defines an ordering "$<$" on R on which

$$\mathbb{R} < x < x^2 < \cdots < y < xy < x^2y < \cdots < y^2 < xy^2 < x^2y^2 < \cdots.$$

Ex. 17.4. In $R = \mathbb{Q}(t)$, define P to be the set of $f(t)/g(t)$ where f, g are polynomials with positive leading coefficients. Show that (R, P) is a non-archimedean ordered field, and that all the order-automorphisms of (R, P) are induced by $t \mapsto at + b$ where $a, b \in \mathbb{Q}$, $a > 0$.

Ex. 17.5. Let $k \subseteq \mathbb{R}$ be a field of real algebraic numbers, and let π be any transcendental real number. Let $P \subseteq k[t]$ be the set of polynomials $f(t)$ such that $f(\pi) > 0$. Show that P is an archimedean ordering in $A = k[t]$.

Ex. 17.6. If a commutative domain R is not formally real, must -1 be a sum of squares in R?

Ex. 17.7. Show by a direct calculation that, in any ring R, the division closure $\tilde{T}$ of any preordering T is also a preordering.

Ex. 17.8. Show that

$$R = \mathbb{R}[x_1, \ldots, x_n, y, z]/(x_1^2 + \cdots + x_n^2 - y^2 z)$$

is formally real, and that its weak preordering $T = T(R)$ is not division-closed, i.e. $T \neq \tilde{T}$. (**Hint.** Using the homomorphism $f: R \to \mathbb{R}$ defined by $f(\bar{x}_1) = \cdots = f(\bar{x}_n) = f(\bar{y}) = 0$, $f(\bar{z}) = -1$, show that $\bar{z} \in \tilde{T} \setminus T$.)

Ex. 17.9A. (E. Artin) For a formally real field k, show that the weak preordering T of $R = k[t]$ is division-closed.

Ex. 17.9B. Let k be any formally real field and let $r \geq 2$. Show that there exist polynomials in $R = k[x_1, \ldots, x_r]$ which are sums of squares in $k(x_1, \ldots, x_r)$, but not in R.

Ex. 17.10. Without using Dedekind cuts, show that an archimedean ordered ring $(R, <)$ is commutative, and that the only order-automorphism of $(R, <)$ is the identity. (**Hint.** Let $a, b > 0$ in R, and $m \geq 1$ be any integer. Choose an integer n such that $(n-1)a \leq mb < na$. Then $m(ab - ba) < a^2$, and so $ab - ba \leq 0$.)

Ex. 17.11. Give an example of an ordered ring R which has infinitely large elements but no infinitely small elements. Give an example of an ordered ring other than $\mathbb{Z}$ for which there exists no element between 0 and 1.

Ex. 17.12. Let $(R, <)$ be any ordered ring which is algebraic over a subfield $F \subseteq Z(R)$. If F is archimedean with respect to the induced ordering, show that $(R, <)$ is an archimedean ordered field.

Ex. 17.13. Let (R, P) be an ordered ring, and $\sigma: R \to R$ by any endomorphism of the additive group of R such that $\sigma(P) \subseteq P$. Assume that, for any $r \in R$, there exists an integer $n \geq 1$ such that $\sigma^n(r) = r$. Show that σ is the identity map.

Ex. 17.14. Let (R, P) be an ordered ring for which P is a well-ordered set, i.e. every nonempty subset of P has a smallest element. Show that (R, P) is order-isomorphic to $\mathbb{Z}$ with its usual ordering.

§18. Ordered Division Rings

In this section, we shall specialize our study of ordering structures to division rings. In this case, many of the results proved in the last section take on a simpler form. We shall begin by noting these simplifications.

Throughout this section, D denotes a division ring, and D^* denotes its multiplicative group of nonzero elements. If P is an ordering in D, then clearly P is a subgroup of D^*. In fact, with the total ordering induced from D, *P is itself a multiplicative ordered group*, with the ordering cone $\{a \in P: a > 1\}$. Since P has index 2 in D^*, P is a normal subgroup of D^*. With a little work, we show below that a similar result also holds for preorderings.

(18.1) Proposition. *A set $T \subseteq D^*$ in a division ring D is a preordering iff $T + T \subseteq T$, $T \cdot T \subseteq T$ and $D^{*2} \subseteq T$. If T is a preordering then $t \in T \Rightarrow t^{-1} \in T$, and T contains the commutator group $[D^*, D^*]$. In particular, T is a normal subgroup of D^*, and D^*/T is an abelian group of exponent* 2.

Proof. For the first conclusion, we need only show that the axiom (17.6) for T is equivalent to the two properties $T \cdot T \subseteq T$ and $D^{*2} \subseteq T$. Clearly, (17.6) implies these properties. Conversely, assume these properties hold, and consider

$$x = per(a_1^2 \cdots a_m^2 t_1 \cdots t_n)$$

where $a_1, \ldots, a_m \in D^*$ and $t_1, \ldots, t_n \in T$. Using the relation $aba = (ab)^2(b^{-1})^2 b$, we can rewrite x as a product $c_1 \cdots c_r$, where each c_i is either a (nonzero) square or is an element of T. Hence, $x \in T$. Next, we note that

$$aba^{-1}b^{-1} = a^2(a^{-1}b)^2(b^{-1})^2,$$

so $[D^*, D^*] \subseteq T$. The other conclusions in the Proposition now follow easily. QED

For convenience, we shall call any element of the form $a_1^2 \cdots a_m^2$ a *square-product*. By the observations made in the proof of the Lemma, *any commutator in D is a square-product, and any element $per(a_1^2 \cdots a_m^2)$ is also a square-product*. Recall that, in the last section, we wrote $T(D)$ for the set of sums of elements of the form $per(a_1^2 \cdots a_m^2)$ where m is arbitrary, and the a_i's are nonzero. *For a division ring D, $T(D)$ is then the set of sums of nonzero square-products*. By definition, D is formally real iff $0 \notin T(D)$. Since all the nonzero square-products form a subgroup of D^*, we see that *D is formally real iff $-1 \notin T(D)$*. Combining this observation with (17.11), we get

(18.2) Theorem (Szele, Pickert). *A division ring D can be ordered iff -1 is not a sum of square-products in D.*

If D is a field, any square-product is, of course, just a square. In this case, (18.2) recovers the classical theorem of Artin and Schreier on the orderability of a field. If D is a division ring but not a field, then a square-product need not be a perfect square. If D is not formally real, we can write -1 as a sum of square-products, but not necessarily as a sum of squares. An explicit example to this effect will be constructed later in this section (see (18.7)).

Recall that, for a preordering $T \subseteq D^*$ and any element $b \in D$, we have defined T_b in the paragraph preceding (17.8). In view of the observations made before (18.2), T_b boils down to $T + bT$ $(= T + Tb)$ in the case of division rings. Also, since T is a subgroup of D^*, it is clearly division-closed in the sense of (17.14). Thus, combining (17.14) and (17.8), we have

(18.3) Theorem. *Let T be any preordering in a division ring D. Then T is the intersection of all the orderings containing T. For $b \in D$, $T_b = T + bT$ is a preordering of D iff $b \notin -T$.*

(18.4) Corollary. *Assume char $D \neq 2$. Then an element $a \in D^*$ is totally positive (i.e., positive with respect to all orderings of D) iff a is a sum of square-products.*

Proof. If D is formally real, the conclusion follows by applying (18.3) to the weak preordering $T(D)$, which consists of all sums of square-products. Now assume D is *not* formally real. Then, vacuously, any $a \in D^*$ is totally positive. Since *char* $D \neq 2$, we can write

$$a = \left(\frac{1+a}{2}\right)^2 - \left(\frac{1-a}{2}\right)^2.$$

By (18.2), -1 is a sum of square-products, so a is also a sum of square-products. QED

(Note that the Corollary fails in general in case *char* $D = 2$. For instance, it clearly does not hold for a nonperfect field of characteristic 2.)

Before we go on, we would like to mention an important class of examples of ordered division rings. Historically, the first example of a *noncommutative* ordered division ring was constructed by Hilbert (in 1903), in connection with his study of the foundations of geometry. Hilbert's example, based on the use of twisted Laurent series, was later generalized by Mal'cev and Neumann in 1948–49. Since we have already covered in detail the Mal'cev–Neumann construction of rings of formal (twisted) Laurent series with well-ordered support, we may as well treat the general case first and return to comment on Hilbert's original example later.

Let R be an ordered ring, with ordering P_0. Let $(G, <)$ be a (multiplicative) ordered group, and let $\omega \colon G \to Aut(R)$ be a homomorphism from G to the group of automorphisms of R. We write $A = R((G, \omega))$ for the Mal'cev–Neumann Laurent series ring constructed in §14. Recall that in A, we multiply Laurent series formally, using the twist law

$$g \cdot r = \omega_g(r)g \quad (g \in G, r \in R).$$

Here, as in §14, ω_g denotes the image of g under ω. The notations introduced in §14 will remain in force in the following discussion.

(18.5) Proposition. *Assume that, for each $g \in G$, ω_g is an order-preserving automorphism of (R, P_0), i.e., $\omega_g(P_0) = P_0$. Let*

$$P = \left\{ \alpha = \sum_{g \in G} \alpha_g g \colon \alpha_{g_0} \in P_0 \textit{ for } g_0 = \textit{least element in } supp(\alpha) \right\}.$$

Then P is an ordering for $A = R((G, \omega))$. If (R, P_0) is an ordered division ring, then so is (A, P).

Proof. Among the axioms for an ordering, (17.1) and (17.3) are both obvious. To check (17.2), let $\alpha = \sum \alpha_g g$ and $\beta = \sum \beta_h h$ be in P. Let g_0 (resp., h_0) be the least element of $supp(\alpha)$ (resp., $supp(\beta)$). Then $g_0 h_0$ is the least element of $supp(\alpha\beta)$, and it appears in $\alpha\beta$ with coefficient $\alpha_{g_0}\omega_{g_0}(\beta_{h_0})$. Since $\alpha_{g_0}, \beta_{h_0} \in P_0$, the hypothesis on ω implies that $\alpha_{g_0}\omega_{g_0}(\beta_{h_0}) \in P_0$, so $\alpha\beta \in P$. Thus, P is an ordering on A. The last conclusion of the Proposition now follows from (14.21). QED

Note that if we regard G as embedded in A (by identifying $g \in G$ with $1 \cdot g$), then, in the notation above, we have $G \subseteq P$. Also, if $g > 1$ in the ordering of G, then $1 - g \in P$, so we end up with $g < 1$ in the ordering of A. Taking $R = \mathbb{Q}$ with its usual ordering P_0, and taking ω to be the trivial homomorphism, we obtain the following nice consequence of (18.5).

(18.6) Corollary. *Any ordered group $(G, <)$ can be embedded, in an order-reversing way, as a subgroup of the multiplicative ordered group of positive elements in an ordered division ring A. If G is commutative, A may be chosen to be an ordered field.*

(Of course, the order-reversing nature of the embedding obtained above should not be a cause of concern. We could have constructed an order-preserving imbedding of $(G, <)$ by first "turning around" the given ordering of G before the Laurent series construction.)

At this point, it is very easy to describe Hilbert's original examples of noncommutative ordered division rings. Let (R, P_0) be an ordered field, and let $G = \langle x \rangle$ be an infinite cyclic group with the ordering cone $\{x^n : n \geq 1\}$. Let σ be any order-preserving automorphism of (R, P_0), and let

$$\omega \colon G \longrightarrow Aut(R)$$

be defined by $\omega_x = \sigma$. The resulting Laurent series division ring $A = R((x, \sigma))$, with multiplication induced by $xr = \sigma(r)x$ $(r \in R)$, has a natural ordering P extending P_0. If σ is not the identity, then (A, P) is a noncommutative ordered division ring. To get an explicit example of such, we may take, for instance, $R = \mathbb{Q}((y))$ with the ordering P_0 obtained from the usual ordering of $\mathbb{Q}$ by the same procedure, i.e.,

$$P_0 = \left\{ \sum_{i=n}^{\infty} a_i y^i : \ n \in \mathbb{Z},\ a_n > 0 \text{ in } \mathbb{Q} \right\}.$$

For σ, we may take, for instance, the (obviously order-preserving) automorphism of R induced by $y \mapsto 2y$. The resulting $(\mathbb{Q}((y))((x, \sigma)), P)$ is Hilbert's original example of a noncommutative ordered division ring.

Let us consider a slight variation of Hilbert's example. Instead of taking $\mathbb{Q}$, let us take a *formally real* field k, and fix an element $c \in k^*$. We take $R = k((y))$, and let σ be the automorphism of R defined by $\sigma(y) = cy$, $\sigma|k = \text{Id}$. For the division ring $A = k((y))((x;\sigma))$, we have the following result.

(18.7) Proposition.

(1) *Whenever* $\sum_i H_i^2 = 0$ *in* A, *we have* $H_1 = H_2 = \cdots = 0$. *In particular,* -1 *is not a sum of squares in* A.

(2) *If* c *has the form* $-(1 + c_1^2 + \cdots + c_r^2)$ *in* k, *then* -1 *is a sum of* $r+1$ *square-products in* A (*so in this case,* A *is not formally real*).

(This Proposition enables us to construct division rings in which -1 is a sum of square-products, but not a sum of squares. The examples obtained here are generally centrally infinite. For some centrally finite examples, see Exercise 1 below.)

Proof. For (2), note that we have $xy = \sigma(y)x = cyx$, so

$$c = xyx^{-1}y^{-1} = x^2(x^{-1}y)^2(y^{-1})^2.$$

If $c = -(1 + c_1^2 + \cdots + c_r^2)$, then

$$-1 = x^2(x^{-1}y)^2(y^{-1})^2 + c_1^2 + \cdots + c_r^2$$

which is a sum of $r+1$ square-products. For (1), assume $\sum_i H_i^2 = 0$ where the H_i's are not all zero. (Here, $c \in k^*$ is arbitrary.) Write $H_i = h_i x^m + \cdots$, where m is chosen such that some $h_i \in k((y))$ is nonzero. Then

$$\textbf{(18.8)} \qquad \begin{aligned} \sum_i H_i^2 &= \sum_i (h_i x^m + \cdots)(h_i x^m + \cdots) \\ &= \left(\sum_i h_i \sigma^m(h_i)\right) x^{2m} + \text{higher terms.} \end{aligned}$$

Now write $h_i = a_i y^n + \cdots$, where n is chosen such that some $a_i \in k$ is nonzero. Then

$$\begin{aligned} \sum_i h_i \sigma^m(h_i) &= \sum_i (a_i y^n + \cdots)(a_i c^{mn} y^n + \cdots) \\ &= \left(\sum_i a_i^2 c^{mn}\right) y^{2n} + \text{higher terms.} \end{aligned}$$

This is a nonzero element in $k((y))$ since $\sum_i a_i^2 \neq 0$ in k. Therefore, by (18.8), $\sum_i H_i^2 \neq 0$ in A. QED

For a non-formally real division ring D, the *level*, $s(D)$, of D is defined to be the smallest integer s such that -1 is a sum of s square-products in D. In case D is a field, A. Pfister proved in 1965 the striking result that $s(D)$ must be a power of 2, and that any power of 2 is the level of some non-formally real field. (For an exposition of Pfister's result, see Lam [73], Ch. 10.) For division rings, however, the situation is quite different. In 1983, Scharlau and Tschimmel proved that the level of a non-formally real division ring can be *any* positive integer. In fact, the following is true.

(18.9) Theorem (Scharlau–Tschimmel). *In* (18.7)(2), *assume that* $-c = 1 + c_1^2 + \cdots + c_r^2 \in k$ *is not a sum of squares of* r *elements in* k. *Then* $-1 = x^2(x^{-1}y)^2(y^{-1})^2 + c_1^2 + \cdots + c_r^2$ *is a shortest representation of* -1 *as a sum of square-products in* A, *i.e.*, $s(A) = r+1$.

The proof of this depends on calculations similar to those in the proof of (18.7), plus a certain nontrivial fact on sums of squares in fields (namely, over k, a product of a sum of p squares and a sum of q squares is always a sum of $p+q-1$ squares). Since it is too much of a digression to prove this fact here, we shall skip the proof of (18.9) altogether. To get a division ring of level $r+1$ from (18.9), it suffices to find a formally real field k in which some element $1 + c_1^2 + \cdots + c_r^2$ is not a sum of r squares. This can be achieved by taking $k = \mathbb{R}(c_1, \ldots, c_r)$ where $c_1, \ldots, c_r$ are independent indeterminates over $\mathbb{R}$, according to a theorem of Cassels (cf. Lam [73], p. 262).

In the rest of this section, we shall study some properties of formally real division rings. The first striking result in this direction was proved by Albert in 1940.

(18.10) Albert's Theorem. *Let D be a formally real division ring. Then the center F of D is algebraically closed in D.*

Proof. Let $d \in D$ be algebraic over F; we need to show that $d \in F$. Assume, for the moment, that $d \notin F$. Then the minimal polynomial, say $t^n + c_1 t^{n-1} + \cdots + c_n$, of d over F has degree $n \geq 2$. Let

$$a = d + \frac{c_1}{n} \notin F.$$

(We use here the fact that *char* $D = 0$.) Then from the equation

$$\begin{aligned} 0 &= \left(a - \frac{c_1}{n}\right)^n + c_1\left(a - \frac{c_1}{n}\right)^{n-1} + \cdots + c_n \\ &= (a^n - c_1 a^{n-1} + \cdots) + c_1 a^{n-1} + \cdots + c_n, \end{aligned}$$

we see that the minimal polynomial of a over F has the form

$$f(t) = t^n + c_2' t^{n-2} + \cdots + c_n' \in F[t].$$

By Wedderburn's Theorem (16.9), we can write

$$f(t) = (t - a_n) \cdots (t - a_1)$$

where $a_1 = a$ and each a_i is a conjugate of a in D. Comparing the coefficients of t^{n-1}, we get $a_1 + \cdots + a_n = 0$. Since D is formally real, by (18.2) there exists an ordering $P \subseteq D^*$. If $a \in P$, then each conjugate a_i of a is also in P, contradicting $a_1 + \cdots + a_n = 0$. If $a \in -P$, a similar contradiction results. Thus, we must have $d \in F$. QED

Remark. Originally, Albert proved (18.10) for *ordered* division rings. However, the ordering is only used as a tool in the proof, and it does not figure in the conclusion of the theorem. Therefore, stating Albert's result in the form (18.10) (assuming (18.2)) gives it a somewhat better perspective.

In the following, we shall record some consequences of Albert's Theorem.

(18.11) Corollary. *Let D be a formally real algebraic division algebra over a field F. Then D is a field. In particular, any formally real centrally finite division ring is a field.*

As an example, consider an ordered field (k, P_0) with a nonidentity automorphism σ such that $\sigma(P_0) \subseteq P_0$. Then, as we saw before, P_0 can be extended to an ordering in the twisted Laurent series ring $D = k((x, \sigma))$, so D is a formally real, noncommutative division ring. By (18.11), D must be centrally infinite. This can also be confirmed as follows. By Exercise 17.13, the automorphism σ has infinite order. Hence, by (14.2), the center of D is a subfield of k, so D is indeed centrally infinite.

The next corollary of (18.10) may be viewed as a self-strengthening of Albert's Theorem. It reveals some rather surprising arithmetic properties of formally real division rings which are not shared by all division rings.

(18.12) Corollary. *Let D be a formally real division ring with center F. Let $a, b \in D$ and $g(t)$ be a nonconstant polynomial in $F[t]$. If $g(a)$ commutes with b, then a already commutes with b.*

Proof. Let D' be the division subring of D generated by F and the two elements a, b, and let $F' \supseteq F$ be the center of D'. Since $c := g(a)$ commutes with a and b, we have $c \in F'$. Then, $g(a) - c = 0$ is a nontrivial polynomial equation satisfied by a over F', so $a \in D'$ is algebraic over the center of D'. Since D' is formally real, Albert's Theorem implies that $a \in F'$, so a commutes with b. QED

Exercises for §18

Ex. 18.1. (This exercise, due to A. Wadsworth, provides an example of a centrally finite division algebra D in which -1 is a square-product, but is *not* a sum of squares.) Let k be a field with at least two orderings, P, P', and let

$e \in P \backslash P'$. Let $F = k((y))$ and let D be the F-algebra with generators i, j and relations

$$i^2 = e, \quad j^2 = y, \quad \text{and} \quad k := ij = -ji.$$

Then D is a 4-dimensional (generalized quaternion) division algebra over F. Show that -1 is a square-product in D (so D has level 1), but $\sum H_\alpha^2 = 0$ in D implies that $H_\alpha = 0$ for all α. (**Hint.** Write $H_\alpha = a_\alpha + b_\alpha i + c_\alpha j + d_\alpha ij$. Then $\sum_\alpha H_\alpha^2 = 0$ gives $\sum a_\alpha^2 + e \sum b_\alpha^2 = y\left(e \sum d_\alpha^2 - \sum c_\alpha^2\right)$. Now argue with the lowest degree terms in y.)

Ex. 18.2. Let (R, P_0) be an ordered division ring, $(G, <)$ be an ordered group, and $A = R((G))$ be the division ring of (untwisted) Laurent series as defined in §14. Let $v \colon A^* \to G$ be the Krull valuation constructed in Exercise 14.10, and let $\mathfrak{m}$ be the Jacobson radical of the valuation ring of v. Show that the ordering $P \subset A$ constructed in (18.5):

$$P = \left\{ \alpha = \sum_{g \in G} \alpha_g g : \ \alpha_{g_0} \in P_0 \text{ for } g_0 = \min(supp(\alpha)) \right\}$$

has the following properties:
(1) $P \cap R = P_0$,
(2) $G \subseteq P$,
(3) P is "compatible" with v in the sense that $1 + \mathfrak{m} \subset P$.
Then show that P is uniquely determined by these three properties.

Ex. 18.3. In the exercise above, show that the valuation ring V of v is given by the convex hull of R in A with respect to the ordering P, i.e.

$$V = \{a \in A : \ -r \le a \le r \text{ for some } r \in P_0 \subseteq R\}.$$

Moreover, show that the residue division ring $V/\mathfrak{m}$ is isomorphic to R.

Ex. 18.4. Keep the notations in Exercise 2. Define two elements $a, b \in A^*$ to be *in the same archimedean class* relative to R if $|a| \le r|b|$, $|b| \le r'|a|$ for suitable $r, r' \in P_0 \subseteq R$. (Here, the absolute values are with respect to P, and are defined in the usual way.) Write $[a]$ for the archimedean class of a (relative to R). We multiply these classes by the rule $[a] \cdot [b] = [ab]$, and order them by the rule: $[a] < [b]$ *iff* $r|a| < |b|$ for all $r \in P_0$. Show that all archimedean classes relative to R form an ordered group which is anti-order-isomorphic to $(G, <)$ itself.

Ex. 18.5. Keep the notations in Exercise 2, and define the valuation topology on A by taking as a system of neighborhoods at 0 the sets

$$\{\alpha \in A : \ \alpha = 0 \text{ or } v(\alpha) > g\} \quad (g \in G).$$

Show that the valuation topology, as a uniform structure, is complete, i.e. every Cauchy net in A has a limit.

Ex. 18.6. (K.H. Leung) An ordered division ring $(D, <)$ is said to be archimedean over a division subring D_0 if, for every $d \in D$, we have $d \le d_0$ for some $d_0 \in D_0$. Construct a noncommutative ordered division ring D which is archimedean over its center. (**Hint.** Let (R, P_0) be an ordered division ring with center $F \subsetneq R$. Extend P_0 to an ordering P in $D = R((x))$ as in (18.5), and show that (D, P) is archimedean over its central subfield $F((x))$.)

Ex. 18.7. A theorem of K.H. Leung states that, *if $(R, <)$ is an ordered ring such that $a^2 + b^2 \geq 2ab$ for all $a, b \in R$, then R must be commutative*. Prove this theorem in the special case when R is an (ordered) division ring. (**Hint.** Replace b by $a + rb$ to get $(rb)^2 \geq arb - rba$. Then

$$|r(ba - ab)| \leq |rba - arb| + |(ar - ra)b| \leq |rb|^2 + |r|^2|b|.$$

In case R is a division ring, this implies that $|ba - ab| \leq c$ for any $c \in P$.) The more ambitious reader might try to prove Leung's result for a general ordered ring.

CHAPTER 7

Local Rings, Semilocal Rings, and Idempotents

In the first two sections of this chapter, we focus our attention on two special classes of rings, namely, local rings and semilocal rings. By definition, a ring R is *local* if $R/rad\ R$ is a division ring, and R is *semilocal* if $R/rad\ R$ is a semisimple ring. Thus, local rings include all division rings, and semilocal rings include all left or right artinian rings. The basic properties of local and semilocal rings are developed, respectively, in §19 and §20. We shall see, for instance, that local rings are connected with the problem of the uniqueness of Krull–Schmidt decompositions, and that semilocal rings are connected with the problem of "cancellation" of modules.

The notion of local rings and semilocal rings has a bearing on the general structure theory of rings through the use of idempotents. By definition, an element e in a ring is called an *idempotent* if $e^2 = e$. The elements 0 and 1 are called the *trivial* idempotents; any idempotent $e \neq 0, 1$ is called a *nontrivial* idempotent. The theory of idempotents turns out to play a much larger role in noncommutative ring theory than in commutative ring theory. We devote §21 and §22 to the theory of idempotents, and prove there the basic facts on primitive idempotents and centrally primitive idempotents. The standard facts about lifting idempotents modulo an ideal are also developed in this section and applied to the study of the Krull–Schmidt decompositions of modules.

This chapter is largely independent of Chapters 3, 4, 5 and 6, and may therefore be read directly after Chapter 2, if the reader is willing to make occasional references to Chapter 3 and Chapter 4.

§19. Local Rings

In commutative algebra, a local ring is defined to be a nonzero ring which has a unique maximal ideal. These local rings form the "local objects" in commutative algebra since for any (commutative) ring R and any prime ideal $\mathfrak{p} \subset R$, one can "localize R at $\mathfrak{p}$" to get a local ring $R_\mathfrak{p}$ with unique maximal ideal $\mathfrak{p}R_\mathfrak{p}$. The process of localization has a natural meaning in algebraic geometry, and is, at the same time, a very powerful algebraic tool with which to study the structure of commutative rings.

In noncommutative algebra, there is a natural generalization of the notion of a local ring. One calls a nonzero ring R *local* if R has a unique maximal left ideal, or, equivalently, if R has a unique maximal right ideal. (The equivalence needs a short proof, which is given below in (19.1).) Many of the properties of commutative local rings can be shown to hold also for general (i.e., not necessarily commutative) local rings. However, in noncommutative algebra, the theory of localization does not work nearly as well as in the commutative case. In fact, even at the outset, it is clear that noncommutative localization is beset with technical difficulties. Due to the lack of a good localization theory, the role of local rings in noncommutative algebra is not nearly as prominent as in the commutative case. Nevertheless, noncommutative local rings do arise naturally, and form an important class for study. In fact, even if one is working only with commutative rings, one needs to consider endomorphism rings of modules, and in many instances, these endomorphism rings are (noncommutative) local rings. The modern formulation of the Krull–Schmidt Decomposition Theorem (the Azumaya version), for example, depends on the notion of noncommutative local rings, and shows clearly the relationship between these local rings and indecomposable modules.

In this section, we study the basic properties of local rings and their applications. In the next section, we shall generalize this class to the class of semilocal rings, which include all left (and right) artinian rings. This paves the way to our study of perfect and semiperfect rings (in Chapter 8) which are special classes of semilocal rings.

Recall that, for any ring R, $U(R)$ denotes the group of units of R, and *rad* R denotes the Jacobson radical of R.

(19.1) Theorem. *For any nonzero ring R, the following statements are equivalent*:

(1) *R has a unique maximal left ideal.*

(2) *R has a unique maximal right ideal.*

(3) *$R/\text{rad } R$ is a division ring.*

(4) *$R \backslash U(R)$ is an ideal of R.*

(5) *$R \backslash U(R)$ is a group under addition.*

(5)$'$ *For any n, $a_1 + \cdots + a_n \in U(R)$ implies that some $a_i \in U(R)$.*

(5)$''$ *$a + b \in U(R)$ implies that $a \in U(R)$ or $b \in U(R)$.*

If any one of these conditions holds, we say that R is a local ring. (To emphasize the role of $\mathfrak{m} = rad\ R$, we shall sometimes say that $(R, \mathfrak{m})$ is a local ring.)

Proof. If we can show that (1) $\Leftrightarrow$ (3), then by symmetry we will also have (2) $\Leftrightarrow$ (3).

(3) $\Rightarrow$ (1) Any maximal left ideal $\mathfrak{m}$ of R contains $rad\ R$. If $R/rad\ R$ is a division ring, then clearly $\mathfrak{m} = rad\ R$, hence (1).

(1) $\Rightarrow$ (3) From (1), it follows that $rad\ R$ is (the unique) maximal left ideal of R. But then $R/rad\ R$ has only two left ideals (namely (0) and itself), so it is a division ring.

(3) $\Rightarrow$ (4) In view of (4.8), (3) implies that any $a \notin rad\ R$ is a unit of R. Thus $R \backslash U(R) = rad\ R$, which is an ideal.

(4) $\Rightarrow$ (5) $\Rightarrow$ (5)$'$ $\Rightarrow$ (5)$''$ are tautologies.

(5)$''$ $\Rightarrow$ (3) Let $a \notin rad\ R$. Take a maximal left ideal $\mathfrak{m}$ such that $a \notin \mathfrak{m}$. Then $\mathfrak{m} + R \cdot a = R$ implies that $1 = m + ba$ for some $m \in \mathfrak{m}$ and some $b \in R$. Since $m \notin U(R)$, (5)$''$ implies that $ba \in U(R)$. In particular, $\bar{a}$ has a left inverse in $\bar{R} = R/rad\ R$. Thus $\bar{R}\backslash\{0\}$ is a group under multiplication and so $\bar{R}$ is a division ring. QED

In the following proposition, we collect a few useful properties of local rings.

(19.2) Proposition. *Let R be any local ring. Then:*

(a) *R has a unique maximal ideal.*

(b) *R is Dedekind-finite (i.e., if $a \in R$ has a left inverse, then $a \in U(R)$).*

(c) *R has no nontrivial idempotents (i.e., any idempotent in R is either* 0 *or* 1).

Proof. (a) A maximal ideal M of R cannot contain any units. Hence

$$M \subseteq R\backslash U(R) = rad\ R,$$

which implies that $M = rad\ R$. (b) follows easily from (4.8). For (c), let $e \in R$ be an idempotent and let $f = 1 - e$. From (19.1)(5$''$), it follows that $e \in U(R)$ or $f \in U(R)$. Since $ef = 0$, this implies that $f = 0$ or $e = 0$.

QED

Note that (a), (b), (c) above are only necessary but not sufficient conditions for R to be a local ring. For instance, (a) is satisfied by any simple ring, but a simple ring need not be local. In the same vein, (b) is satisfied by any commutative ring, but a commutative ring need not be local. Finally, any domain satisfies (c), but a domain need not be local. We invite the reader to give an example of a ring which satisfies all of (a), (b), (c), but which is not local.

Next, let us give some sufficient conditions for local rings.

(19.3) Proposition. (a) *Suppose $R \neq 0$, and every $a \notin U(R)$ is nilpotent, then R is a local ring.* (b) *Suppose R is contained in a division ring D such that for any $d \in D^*$, d or d^{-1} lies in R, then R is a local ring.*

Proof. (a) We shall show that the hypothesis in (a) implies that $R \backslash U(R) \subseteq rad\ R$. Once this is shown, then $R \backslash U(R) = rad\ R$, and (19.1)(4) guarantees that R is a local ring. Let $a \notin U(R)$, and let k be the smallest positive integer such that $a^k = 0$. Then $R \cdot a \subseteq R \backslash U(R)$. For, if some $ra \in U(R)$ $(r \in R)$, then $(ra)a^{k-1} = 0$ implies that $a^{k-1} = 0$, a contradiction. Since $R \backslash U(R)$ consists of nilpotent elements, $R \cdot a$ is a nil left ideal, and so (by (4.11)) $R \cdot a \subseteq rad\ R$, as desired. Next, assume that $R \subseteq D$ has the property in (b), where D is a division ring. It suffices to check that, for nonzero $a, b \in R$, $a + b \in U(R)$ implies that $a^{-1} \in R$ or $b^{-1} \in R$. We may assume that $a + b = 1$. Apply the given hypothesis to the element $c = a^{-1}b \in D$. If $c \in R$, we have

$$a^{-1} = a^{-1}(a + b) = 1 + c \in R.$$

If, instead, $c^{-1} = b^{-1}a \in R$, we have

$$b^{-1} = b^{-1}(a + b) = c^{-1} + 1 \in R. \qquad \text{QED}$$

In the case when D is a field in (b) above, a subring $R \subseteq D$ satisfying the property in (b) is called a *valuation ring* of D. These subrings arise very naturally from Krull valuations. Such valuations can also be studied on division rings. However, the valuation rings R which arise from this study have the additional property that they are *invariant*, i.e., they satisfy also $d^{-1}Rd \subseteq R$ for every $d \in D^*$. (For more details, see Exercises (9), (10).)

We shall now give some examples of local rings. Before we proceed to the noncommutative examples, let us first have a quick review of the possibly more familiar stock of examples of commutative local rings.

(19.4) As mentioned in the opening paragraph of this section, the localization of any commutative ring R at a prime ideal $\mathfrak{p}$ is a local ring $R_{\mathfrak{p}}$ with unique maximal ideal $\mathfrak{p}R_{\mathfrak{p}}$.

(19.5) At any point x of an algebraic variety X, the rational functions on X which are regular at x form a local ring $\mathcal{O}_x$. The functions in $\mathcal{O}_x$ which vanish at x constitute the unique maximal ideal of $\mathcal{O}_x$.

(19.6) At any point x of a Riemann surface X, the ring of germs of functions $X \to \mathbb{C}$ holomorphic at x is a local ring. Its unique maximal ideal consists of germs arising from functions which vanish at x.

(19.7) As we have mentioned before, any valuation ring R of a field is always a (commutative) local ring. For instance, the ring $\hat{\mathbb{Z}}_p$ of p-adic integers for any prime number p is a (discrete, rank 1) valuation ring of the field $\hat{\mathbb{Q}}_p$ of p-adic numbers. The associated valuation measures the divisibility of p-adic numbers by powers of p. The unique maximal ideal of $\hat{\mathbb{Z}}_p$ consists of p-adic integers divisible by p; i.e., $rad(\hat{\mathbb{Z}}_p) = p \cdot \hat{\mathbb{Z}}_p$.

Now let us consider some noncommutative examples. Clearly, any division ring is a local ring. Secondly, if R is any local ring, and $A = R[[x]]$ is the ring of power series in one variable x, then by Exercise 5.6, $rad\ A$ consists of all power series with constant terms in $rad\ R$. Thus, $A/rad\ A \cong R/rad\ R$, which is a division ring. Therefore, A is also a local ring. The same considerations apply if we take A to be a *twisted* power series ring $R[[x;\sigma]]$, where σ is a given automorphism of R. Thus, even if we start with a field R, if there is a nontrivial automorphism σ on R, we get an example of a *noncommutative* local ring $R[[x;\sigma]]$. In the following, we offer a few more examples.

(19.8) Let k be a division ring, and R be the ring of upper triangular $n \times n$ matrices over k. In Example 6 following (4.15), we have shown that $J = rad\ R$ consists of matrices in R with a zero diagonal, and that $J^n = 0$. Let A be the subring of R consisting of matrices in R with a constant diagonal. Clearly $J \subseteq rad\ A$, and, since $A = k \cdot 1 \oplus J$, we have $A/J \cong k$. Thus, A is a local ring with $rad\ A = J$.

(19.9) Let k be a field and V be an n-dimensional vector space over k. Let

$$R = \bigwedge(V) = k \oplus \bigwedge^1(V) \oplus \cdots \oplus \bigwedge^n(V)$$

be the exterior algebra of V. Then

$$\mathfrak{m} = \bigwedge^1(V) \oplus \cdots \oplus \bigwedge^n(V)$$

is a nilpotent ideal in R; in fact, $\mathfrak{m}^{n+1} = 0$. Since $R/\mathfrak{m} \cong k$, it follows that $\mathfrak{m} = rad\ R$, and that R is a local ring.

(19.10) Let k be a field of characteristic $p > 0$, and G be a finite p-group. Then the Jacobson radical of the group algebra $A = kG$ is the augmentation ideal of A, with $(rad\ A)^{|G|} = 0$ (cf. (8.8)). Therefore, $A/rad\ A \cong k$, so A is an (artinian) local ring. This class of local rings can be extended to include group rings of finite p-groups over more general coefficient rings, as follows.

(19.11) Proposition. *Let $(R, \mathfrak{m})$ be a commutative local ring such that $k = R/\mathfrak{m}$ has characteristic $p > 0$. Then, for any finite p-group G, the group algebra $A = RG$ is a local ring with $A/rad\ A \cong k$.*

Proof. Consider any simple left A-module V. This is a cyclic A-module, so it is a finitely generated R-module. By Nakayama's Lemma (4.22), $V \neq 0$

implies that $\mathfrak{m} \cdot V \subsetneq V$. Since $\mathfrak{m} \cdot V$ is clearly an A-submodule of V, we have $\mathfrak{m} \cdot V = 0$. Thus, V may be viewed as a simple left kG-module. By (8.4), G must act trivially on V. Thus, *rad* A contains the ideal I generated by $\mathfrak{m}$ and all $g - 1$ ($g \in G$). Since clearly $A/I \cong k$, we conclude that *rad* $A = I$, and that A is local. (Here, I is the kernel of the reduced augmentation map, obtained by composing the ordinary augmentation map $RG \to R$ with the projection $R \to R/\mathfrak{m} = k$.) QED

We note in passing that (19.11) remains valid when R is a noncommutative local ring. The same proof reduces the consideration to the case when R is a division ring k. In this case, the same ideas used in the proof of (8.8) show that G acts trivially on any simple left kG-module. The result (19.11) is of considerable importance in the theory of integral representations of finite groups.

Another major source of noncommutative local rings lies in the study of indecomposable modules over rings. Recall that, for any ring R, a right R-module $M \neq 0$ is said to be (*directly*) *indecomposable* if M cannot be written as a direct sum of two nonzero R-submodules of M. As is easily seen, the latter condition amounts to the fact that the endomorphism ring $End(M_R)$ has no nontrivial idempotents. This observation leads us to a useful definition.

(19.12) Definition. A nonzero right R-module M is said to be *strongly indecomposable* if $End(M_R)$ is a local ring.

Since a local ring has no nontrivial idempotents by (19.2), it follows that a strongly indecomposable module is always indecomposable. Some examples (and nonexamples) of strongly indecomposable modules are offered below.

(19.13) Any simple module M_R is strongly indecomposable, since, by Schur's Lemma, $End(M_R)$ is a division ring.

(19.14) For $R = \mathbb{Z}$, the right regular module $M_1 = \mathbb{Z}$ is indecomposable, but, as $End(M_1) \cong \mathbb{Z}$ is not a local ring, M_1 is *not* strongly indecomposable. On the other hand, for $M_2 = \mathbb{Z}/p^n\mathbb{Z}$ (where p is any prime), $End(M_2) \cong \mathbb{Z}/p^n\mathbb{Z}$ is a local ring, so M_2 is strongly indecomposable. For M_3, the group of all p^n-th roots of unity, where p is a fixed prime, and n is a varying integer, $End(M_3)$ is the inverse limit $\varprojlim \mathbb{Z}/p^n\mathbb{Z}$, which is isomorphic to the ring of p-adic integers. Since this is a local ring, M_3 is strongly indecomposable.

(19.15) Let k be a field of characteristic $p > 0$, and G be an elementary abelian p-group of order p^2 generated by x, y. Let V be the 3-dimensional right kG-module $k^3 = e_1k \oplus e_2k \oplus e_3k$ with

$$e_1x = e_1, \quad e_2x = e_2, \quad e_3x = e_1 + e_3 \quad \text{and}$$
$$e_1y = e_1, \quad e_2y = e_2, \quad e_3y = e_2 + e_3.$$

Computing in the same basis, it is easy to check that the kG-endomorphisms of V are given by the matrices

$$\left\{ \begin{pmatrix} a & 0 & b \\ 0 & a & c \\ 0 & 0 & a \end{pmatrix} : a, b, c \in k \right\}.$$

Since these matrices form a local ring (see Exercise 2), V is a strongly indecomposable kG-module.

For certain special classes of modules, it may happen that "indecomposability" and "strong indecomposability" become equivalent properties. For instance, this is the case for the class of injective modules, as we shall see in §3 of *Lectures*. There is another, much more classical, class of modules for which indecomposability implies strong indecomposability, namely, the class of modules of finite (composition) length, i.e., modules which satisfy both the ACC and the DCC on submodules. To see this, let us first prove

(19.16) Fitting Decomposition Theorem. *Let R be any ring, and M_R be any right R-module of finite length. For any endomorphism $f \in E := End(M_R)$, we have*

$$M = ker(f^r) \oplus im(f^r)$$

for any sufficiently large integer r.

Proof. Look at the two chains of submodules

$$M \supseteq im(f) \supseteq im(f^2) \supseteq \cdots,$$
$$0 \subseteq ker(f) \subseteq ker(f^2) \subseteq \cdots.$$

Since M has finite length, both chains must stabilize. Now consider any integer r such that

$$im(f^r) = im(f^{r+1}) = \cdots, \quad \text{and}$$
$$ker(f^r) = ker(f^{r+1}) = \cdots.$$

For any such r, we claim that $M = ker(f^r) \oplus im(f^r)$. First, consider any $a \in ker(f^r) \cap im(f^r)$. Writing $a = f^r(b)$ $(b \in M)$, we have $0 = f^r(a) = f^{2r}(b)$, so $b \in ker(f^{2r}) = ker(f^r)$, and so $a = f^r(b) = 0$. Finally, for any $c \in M$, we can write $f^r(c) = f^{2r}(d)$ for some $d \in M$. But then $f^r(c - f^r(d)) = 0$, so we have a decomposition

$$c = (c - f^r(d)) + f^r(d) \in ker(f^r) + im(f^r). \qquad \text{QED}$$

(19.17) Theorem. *Let M_R be an indecomposable R-module of finite length. Then $E := End(M_R)$ is a local ring, and its unique maximal ideal $\mathfrak{m} = rad\ E$ is nil. In particular, M is a strongly indecomposable module.*

Proof. First, we show that any endomorphism $f \in E \backslash U(E)$ is nilpotent. By (19.3), this implies that E is a local ring. To show that f is nilpotent, fix an integer $r > 0$ such that $M = ker(f^r) \oplus im(f^r)$. If $ker(f^r) = 0$, then $im(f^r) = M$. In this case $f^r \in U(E)$ and hence $f \in U(E)$, a contradiction. Therefore, $ker(f^r) \neq 0$. But then the indecomposability of M implies that $im(f^r) = 0$. This means that $f^r = 0$, as desired. QED

(19.18) Remarks.

(A) It can be shown that the maximal ideal $\mathfrak{m} \subset E$ is actually *nilpotent*; more precisely, if the module M_R has composition length n, then $\mathfrak{m}^n = 0$. See Ex. (21.24) below. Local rings $(R, \mathfrak{m})$ with $\mathfrak{m}$ nilpotent are often called *completely primary* rings. For instance, group algebras of finite p-groups over fields of characteristic p are another class of such completely primary rings.

(B) In (19.17), the conclusions do not all hold if we assume only ACC or only DCC on submodules of M, instead of both. In fact, for $R = \mathbb{Z}$, the indecomposable module $M = \mathbb{Z}$ satisfies the ACC on submodules, but $End(M) \cong \mathbb{Z}$ is not local. For the R-module M_3 defined in (19.14), M_3 satisfies the DCC, but not the ACC, on submodules. Here, $End(M_3)$ happens to be local, but it is a domain and its unique maximal ideal is certainly not nil.

The following is a purely ring-theoretic consequence of (19.17).

(19.19) Corollary. *A right artinian ring R $(\neq 0)$ is a local ring iff R has no nontrivial idempotents.*

Proof. ("if" part) Consider the right regular module $M = R_R$. By the Hopkins–Levitzki Theorem (4.15), M has finite length. The endomorphism ring $E = End(M_R)$ (acting on the left on M) is isomorphic to R. If R has no nontrivial idempotents, then M is indecomposable. By (19.17), $E \cong R$ is a local ring. QED

An important application of the notion of a local ring is the formulation of Azumaya's version of the Krull–Schmidt Theorem, nowadays known as the Krull–Schmidt–Azumaya Theorem. We shall now prepare our way for the statement and proof of this result. We begin with the following classical fact.

(19.20) Proposition. *Let R be any ring, and M_R be a right R-module whose submodules satisfy either the ACC or the DCC. Then M can be decomposed into a finite direct sum of indecomposable submodules.* (*We shall say, in short, that M has a Krull–Schmidt decomposition.*)

Proof. To use a more informal language, let us say that a submodule $N \subseteq M$ is "good" if it has a Krull–Schmidt decomposition. Otherwise, we say that N is bad. Note that the zero module is good (being the direct sum of an empty family of indecomposable modules!), any indecomposable submodule $N \subseteq M$ is good, and if $N, N' \subseteq M$ are both good and $N \cap N' = 0$, then $N + N'$ is also good. To prove the Proposition, assume, instead, that M is "bad." Then M cannot be indecomposable, so we have a decomposition $M = M_1 \oplus M_1'$, where $M_1, M_1' \neq 0$. One of the summands must be bad, say M_1. Repeating the argument, we have $M_1 = M_2 \oplus M_2'$, where $M_2, M_2' \neq 0$, and, say, M_2 is bad. This process leads to infinite chains

$$M \supsetneq M_1 \supsetneq M_2 \supsetneq \cdots \quad \text{and}$$
$$(0) \subsetneq M_1' \subsetneq M_1' \oplus M_2' \subsetneq M_1' \oplus M_2' \oplus M_3' \subsetneq \cdots,$$

so M satisfies neither ACC nor DCC on submodules, a contradiction.

QED

Consider the classical case when $R = \mathbb{Z}$, and M is a finitely generated R-module (i.e., a finitely generated abelian group). Then the submodules of M satisfy ACC, and so by (19.20), M is a finite direct sum of copies of $\mathbb{Z}$ and $\mathbb{Z}/p^n\mathbb{Z}$ ($n \geq 1$, p is any prime), these being all the finitely generated indecomposable abelian groups. By the Fundamental Theorem of Abelian Groups, we know that the indecomposable summands in such a decomposition are uniquely determined up to isomorphism. However, it turns out that we are just "lucky" in this particular example. The uniqueness part of this decomposition theorem for finitely generated modules extends to principal ideal domains, but not to Dedekind domains. In fact, for any Dedekind domain R, and nonzero ideals $\mathfrak{A}, \mathfrak{B} \subseteq R$, we have the well-known Steinitz Isomorphism $\mathfrak{A} \oplus \mathfrak{B} \cong R \oplus \mathfrak{A}\mathfrak{B}$, where both sides are considered as R-modules. Thus, if $\mathfrak{A}$ represents an element of order 2 in the class group of R, we will have

$$\mathfrak{A} \oplus \mathfrak{A} \cong R \oplus \mathfrak{A}^2 \cong R \oplus R.$$

Here R and $\mathfrak{A}$ are both indecomposable as R-modules. But, since $\mathfrak{A}$ is not principal, we have $\mathfrak{A} \not\cong R$, so we have two essentially different ways of decomposing R^2 as a direct sum of indecomposables. More explicitly, we can take R to be the ring of algebraic integers $\mathbb{Z}[\theta]$, $\theta = \sqrt{-5}$, which has class number 2. One shows easily that $\mathfrak{A} = (3, 1 + \theta)$ is not principal, and that

$$\mathfrak{A}^2 = (\theta - 2) \cong R.$$

Hence, R^2 decomposes into $R \oplus R$, and also into $N_1 \oplus N_2$ where $N_1, N_2 \cong \mathfrak{A}$ are noncyclic, indecomposable submodules of R^2.

Fortunately, under a suitable hypothesis, we do have a good uniqueness theorem governing the Krull–Schmidt decompositions of a module into (a finite number of) indecomposable summands.

(19.21) Krull–Schmidt–Azumaya Theorem. *Let R be a ring, and suppose that a right R-module M has the following two decompositions into submodules:*

$$M = M_1 \oplus \cdots \oplus M_r = N_1 \oplus \cdots \oplus N_s,$$

where the N_i's are indecomposable, and the M_i's are strongly indecomposable. Then, $r = s$, and, after a reindexing, we have $M_i \cong N_i$ for $1 \leq i \leq r$.

This important result has a rather long and distinguished history. It is the culmination of ideas of Wedderburn (1909), Remak (1911), Schmidt (1913), Krull (1925), and Azumaya (1950). Furthermore, this theorem has been extended to sheaves by Atiyah (1956), and to abstract categories by Gabriel (1962). Before we present the full proof of (19.21), it behooves us to record some of its more classical versions.

(19.22) Corollary (Krull–Schmidt Theorem). *Let M_R be a right R-module of finite length. Then there exists a decomposition*

$$M = M_1 \oplus \cdots \oplus M_r$$

where each M_i is an idecomposable submodule of M. Moreover, r is uniquely determined, and the sequence of isomorphism types of $M_1, \ldots, M_r$ is uniquely determined up to a permutation.

Proof. The existence of the Krull–Schmidt decomposition follows from (19.20). Moreover, by (19.17), each M_i is, in fact, strongly indecomposable. The uniqueness part, therefore, follows from (19.21). QED

(19.23) Corollary. *The two conclusions in the Krull–Schmidt Theorem above apply to any finitely generated right module M_R over a right artinian ring R (in particular, over any finite-dimensional algebra over a field).*

Proof. For a right artinian ring R, we have shown earlier that a finitely generated right R-module M has a composition series (cf. (4.15)). Thus (19.22) applies. QED

We now return to the

Proof of (19.21). Let α_i: $M \to M_i \subseteq M$ and β_j: $M \to N_j \subseteq M$ be the projection maps onto M_i and N_j associated with the two given Krull–Schmidt decompositions. Viewing α_i, β_j as elements of $E := End(M_R)$, we have

$$1 = \alpha_1 + \cdots + \alpha_r = \beta_1 + \cdots + \beta_s,$$

and so $\alpha_1 = \alpha_2\beta_1 + \cdots + \alpha_1\beta_s \in E$. Note that each $\alpha_1\beta_j$ sends M to M_1, so, restricting to M_1, we have

$$1_{M_1} = \sum_{j=1}^{s} \alpha_1\beta_j|_{M_1} \in End(M_1).$$

Since $End(M_1)$ is (by assumption) a local ring, one of the summands above, say $\alpha_1\beta_1|_{M_1}$, is an automorphism of M_1. From this, we see that β_1: $M_1 \to N_1$ is a split monomorphism. Since N_1 is indecomposable, this must be an isomorphism. At this point, we claim that

(19.24) $$M = M_1 \oplus N_2 \oplus \cdots \oplus N_s.$$

If this is the case, then we have

$$N_2 \oplus \cdots \oplus N_s \cong M/M_1 \cong M_2 \oplus \cdots \oplus M_r,$$

and the proof of the theorem proceeds by induction on r. (The case $r = 1$ is, of course, trivial.) To verify (19.24), first note that, since $\beta_1 \colon M_1 \to N_1$ is an isomorphism, M_1 has zero intersection with $ker(\beta_1) = N_2 \oplus \cdots \oplus N_s$. Therefore, we are done if we can show that

$$N_1 \subseteq M_1 + N_2 + \cdots + N_s.$$

Let $a \in N_1$ and write $a = \beta_1(b)$ where $b \in M_1$. Then $\beta_1(a - b) = a - \beta_1(b) = 0$, so

$$a - b \in ker(\beta_1) = N_2 + \cdots + N_s.$$

Adding b, we get $a \in M_1 + N_2 + \cdots + N_s$, as desired. QED

As a simple illustration of the Krull–Schmidt–Azumaya Theorem, consider the case of a finitely generated semisimple module M_R. We know that M has a Krull–Schmidt decomposition into simple modules, say $M = M_1 \oplus \cdots \oplus M_r$ (see Exercise 2.7). Since simple modules are strongly indecomposable (by (19.13)), the theorem above guarantees that the isomorphism types of $M_1, \ldots, M_r$ are determined up to a permutation. Of course, we could have gotten the same conclusion by applying the Jordan–Hölder Theorem to M, since $\{M_1, \ldots, M_r\}$ are the composition factors of M.

As a consequence of (19.23), we shall deduce a theorem of Noether and Deuring on the behavior under scalar extensions of representation modules over a finite-dimensional algebra R over a field k. Let $K \supseteq k$ be any field extension. For any right R-module M, the scalar extension $M^K = M \otimes_k K$ is a right module over $R^K = R \otimes_k K$. It turns out that, in case $dim_k\, M < \infty$, the isomorphism type of M^K determines uniquely the isomorphism type of M. Stated more precisely, we have the following.

(19.25) Noether–Deuring Theorem. *Let R be a finite-dimensional algebra over a field k, and let M, N be right R-modules of finite dimension over k. Let K be any extension field of k. If $M^K \cong N^K$ as R^K-modules, then $M \cong N$ as R-modules.*

Proof. By (7.4), we have a natural isomorphism

(19.26) $$\theta \colon (Hom_R(M, N))^K \to Hom_{R^K}(M^K, N^K).$$

Let $n = dim_k\, M = dim_k\, N$, and think of M, N as the space k^n with two different R-actions. Then $Hom_R(M, N)$ may be identified with a certain k-subspace $\mathcal{S}$ of $\mathbb{M}_n(k)$. By the isomorphism (19.26), $Hom_{R^K}(M^K, N^K)$ may be identified with $\mathcal{S}^K \subseteq \mathbb{M}_n(K)$. Let $S_1, \ldots, S_r$ be a k-basis of $\mathcal{S}$. For commuting indeterminates $x_1, \ldots, x_r$ over K, let

$$f(x_1, \ldots, x_r) = det(x_1 S_1 + \cdots + x_r S_r) \in k[x_1, \ldots, x_r].$$

This is a homogeneous polynomial of degree n. Since $M^K \cong N^K$ as R^K-modules, there exist $a_1, \ldots, a_r \in K$ such that $a_1 S_1 + \cdots + a_r S_r$ is invertible, so $f(a_1, \ldots, a_r) \neq 0$. In particular, we know that f is not the zero polynomial. We now distinguish the following two cases.

Case 1. The field k has more than n elements. By induction on r, it follows easily that, for $f \in k[x_1, \ldots, x_r] \backslash \{0\}$ of total degree $\leq n$, there exist $b_1, \ldots, b_r \in k$ such that $f(b_1, \ldots, b_r) \neq 0$. But then $b_1 S_1 + \cdots + b_r S_r$ gives an R-isomorphism from M to N, as desired.

Case 2. $|k| \leq n$. Take a finite extension $L \supseteq k$ such that $|L| > n$. By Case 1, there exist $b_1, \ldots, b_r \in L$ such that $f(b_1, \ldots, b_r) \neq 0$, so we have $M^L \cong N^L$ as R^L-modules. Let $\alpha_1, \ldots, \alpha_t$ be a k-basis of L. Then, viewed as an R-module,

$$M^L = M \otimes_k L = (M \otimes \alpha_1) \oplus \cdots \oplus (M \otimes \alpha_t)$$

is isomorphic to $t \cdot M$ (direct sum of t copies of M), since each $M \otimes \alpha_i$ is R-isomorphic to M. Therefore, over R, we have $t \cdot M \cong t \cdot N$. Working with the (unique) Krull–Schmidt decompositions of M and N, we conclude easily from (19.23) that $t \cdot M \cong t \cdot N$ implies that $M \cong N$, as desired. QED

We have pointed out before that the uniqueness part of the Krull–Schmidt Theorem does not apply to Krull–Schmidt decompositions of finitely generated modules over Dedekind domains. In the counterexample constructed before, we have a Dedekind ring R with a nonprincipal ideal $\mathfrak{A} \subseteq R$ such that

$$\mathfrak{A} \oplus \mathfrak{A} \cong R \oplus R.$$

At each localization $R_\mathfrak{p}$ ($\mathfrak{p}$ a prime ideal), however, $\mathfrak{A}$ becomes principal, since $R_\mathfrak{p}$ is a discrete valuation ring. Thus, locally, the two decompositions above do not give an example of nonuniqueness of the indecomposable summands. One may ask: *if R is itself a local ring and M_R is an R-module whose submodules satisfy ACC, do we have uniqueness in the Krull–Schmidt decompositions of M?* Unfortunately, the answer is still in the negative. In the following, we shall offer an example of R. Swan which shows the failure of the uniqueness of Krull–Schmidt decompositions of finitely generated modules over, even, commutative noetherian local domains. (We assume, in this presentation, that the reader is familiar with the basic facts of commutative algebra, as contained, for instance in Zariski–Samuel [58].)

Let $(R, \mathfrak{p})$ be a commutative noetherian local domain which is not a field, such that the characteristic of $R/\mathfrak{p}$ is not 2. Let A be the localization of $R[x, y]/(y^2 + x^3 - x^2)$ at the ideal $(\mathfrak{p}, \bar{x}, \bar{y})$ generated by $\mathfrak{p}, \bar{x}$, and $\bar{y}$. Then A is a noetherian local domain with maximal ideal $\mathfrak{m}$ generated by $\mathfrak{p}, \bar{x}$, and $\bar{y}$. Since

$$\mathfrak{m} \supsetneq (\bar{x}, \bar{y}) \supsetneq (0),$$

A has Krull dimension ≥ 2. Let K be the quotient field of A, and let $z = \bar{y}/\bar{x} \in K \backslash A$. We have

$$z^2 = \bar{y}^2/\bar{x}^2 = 1 - \bar{x},$$

so z is integral over A. Let $B = A[z] = A + A \cdot z \subseteq K$. Then

$$\frac{B}{\mathfrak{m} \cdot B} \cong \frac{(A/\mathfrak{m})[t]}{(t^2 - 1)} \cong \frac{A}{\mathfrak{m}} \times \frac{A}{\mathfrak{m}}.$$

From this, it follows that B has exactly two maximal ideals:

$$M_1 = (\mathfrak{m}, z + 1) \quad \text{and} \quad M_2 = (\mathfrak{m}, z - 1).$$

Each of these contract to $\mathfrak{m}$ in A, so by the Cohen–Seidenberg Theorem, $ht(M_i) = ht(\mathfrak{m}) \geq 2$, where "$ht$" denotes the height of a prime ideal. The B-homomorphism $M_1 \oplus M_2 \to B$ induced by the inclusions of M_1, M_2 into B leads to an exact sequence

$$0 \longrightarrow M_1 \cap M_2 \longrightarrow M_1 \oplus M_2 \longrightarrow B \longrightarrow 0.$$

Since the last module is B-free, we have a B-module isomorphism

$$M_1 \oplus M_2 \cong (M_1 \cap M_2) \oplus B.$$

All four modules are finitely generated over B, and therefore finitely generated over A. Also, since they all lie in the quotient field K of A (resp., B), each of them is indecomposable as an A-module (resp., B-module). *We finish by showing that $B \ncong M_i$ ($i = 1, 2$) as A-modules.* (This would provide the desired example of nonuniqueness of Krull–Schmidt decompositions over A.) Indeed, suppose there is an A-isomorphism $f: B \to M_i$ ($i = 1$ or 2). Then f must be given by multiplication by a suitable element of K. Hence f will also be a B-isomorphism, so M_i is a principal ideal of B. Obviously, this contradicts the aforementioned fact that the height of each M_i is at least 2. QED

Another possible complication concerning the Krull–Schmidt uniqueness property is that, in general, if a module M is a direct sum of $r \geq 2$ indecomposable submodules, the number r is *not* uniquely determined by M. Even over very nice commutative, noetherian rings (such as $\mathbb{Z}[x]$), L. Levy has constructed examples of finitely generated modules M that can be written (simultaneously) as a direct sum of r indecomposable submodules for $r = 2, 3, \ldots, n$, where $n \geq 2$ is any given integer! Levy referred to this kind of phenomenon as the "dramatic failure" of the Krull–Schmidt property: it serves to warn us that *nothing* is to be taken for granted as far as isomorphisms of direct sums of indecomposable modules are concerned.

Recently in the mid-90s, two more open problems on the Krull–Schmidt uniqueness property have been resolved, both negatively. Facchini, Herbera, Levy and Vámos have shown that direct sums of indecomposable *artinian* modules do not satisfy Krull–Schmidt uniqueness, and Facchini has shown the same for direct sums of indecomposable *uniserial* modules (modules whose submodules form a chain). An excellent reference for these and other related matters is Facchin's recent book [98].

In spite of all the negative manifestations above, however, it can be shown that, in a *complete*, *local* setting, the uniqueness part of the Krull–Schmidt Theorem survives for the decomposition of finitely generated modules. The proof of this result, which depends on the techniques of lifting idempotents, will be postponed to §21.

To conclude this section, we shall determine the structure of finitely generated projective modules over a local ring. This determination is based upon the following general observation.

(19.27) Lemma. *Let R be a ring and $\bar{R} = R/J$, where J is an ideal of R contained in rad R. Let P, Q be finitely generated projective right R-modules. Then $P \cong Q$ as R-modules iff $P/PJ \cong Q/QJ$ as $\bar{R}$-modules.*

Proof. ("if" part) Consider the following diagram, where $\bar{f}$ is a given isomorphism from P/PJ to Q/QJ:

$$\begin{array}{ccc} P & \longrightarrow & P/PJ \\ \downarrow & & \downarrow \bar{f} \\ Q & \longrightarrow & Q/QJ \end{array} \qquad (19.28)$$

Since P is a projective R-module, there exists an R-homomorphism $f\colon P \to Q$ (marked by the dotted arrow in (19.28)) which makes the diagram commutative. The surjectivity of $\bar{f}$ implies that $im(f) + QJ = Q$. Since Q is finitely generated, we have, by Nakayama's Lemma, $im(f) = Q$, i.e., f is *onto*. But then by the projectivity of Q, there exists a decomposition $P = P' \oplus Q'$ where $P' = ker(f)$ and $f'\colon Q' \to Q$ is an isomorphism. Reducing modulo J, we get

$$P/PJ \cong P'/P'J \oplus Q'/Q'J.$$

The fact that $\bar{f}$ is an isomorphism now implies that $P'/P'J = 0$; i.e., $P' = P'J$. However, being a direct summand of P, P' is also finitely generated as an R-module. Applying Nakayama's Lemma again, we see that $P' = 0$. This means that f is one-to-one; hence $f\colon P \to Q$ is an isomorphism. QED

The lemma leads quickly to the following well-known homological result.

(19.29) Theorem. *Let (R, J) be any local ring. Then any finitely generated projective right R-module P is free.*

Proof. Reducing modulo $J = rad\ R$, P/PJ is a finitely generated projective module over R/J, which is a division ring. Therefore, $P/PJ \cong (R/RJ)^n$ for some integer n. By the Lemma, we conclude that $P \cong R^n$. QED

The theorem above remains true, in fact, for not necessarily finitely generated projective modules over a local ring. This was first proved in 1958 by Kaplansky. Since we do not need this more general result in the sequel, Kaplansky's proof will not be given here.

Theorem (19.29) has many nice applications. For instance, using it, we can deduce a classical result on representations of finite groups in characteristic p. To state this result, consider any right artinian ring R. As a right R-module, R_R has a composition series (by (4.15)). Therefore R_R has a Krull–Schmidt decomposition $U_1 \oplus \cdots \oplus U_n$. Any right R-module isomorphic to some U_i $(1 \leq i \leq n)$ is called a *principal indecomposable R-module*. The following result, due to L.E. Dickson, is one of the earliest known results in modular representation theory.

(19.30) Dickson's Theorem. *Let $R = kG$ where k is a field of characteristic $p > 0$ and G is a finite group. Let U be any principal indecomposable right R-module. Then $dim_k\ U$ is divisible by the order of a p-Sylow subgroup H of G.*

Proof. Let $R = U_1 \oplus \cdots \oplus U_n$ be a Krull–Schmidt decomposition of R_R. Then each U_i is a (cyclic) projective right R-module. By taking coset representatives of G modulo H, we see easily that R is a free right module of rank $[G : H]$ over kH. Therefore, each U_i is a finitely generated projective right kH-module. Since kH is a local ring by (19.10), U_i is a free kH-module by (19.29). Comparing k-dimensions, it follows that $|H|$ divides $dim_k\ U_i$ for every i. QED

In case k is a splitting field for G, the k-dimensions of the principal indecomposable modules of $R = kG$ have another interesting interpretation. In fact, if V is any irreducible right R-module, then the number of times V occurs as a composition factor of R_R is equal to the k-dimension of some principal indecomposable *left* kG-module. (For more details on this, see §25, and especially Exercise 25.2.) Therefore, applying (19.30) to principal indecomposable left modules, we arrive at the following conclusion:

(19.31) Theorem. *Let $R = kG$ be as above, where k is a splitting field of characteristic $p > 0$ for G. Then each irreducible right R-module appears as a composition factor in R_R with a multiplicity divisible by the p-part of $|G|$.*

Exercises for §19

Ex. 19.1. For any local ring R, show that (1) the opposite ring R^{op} is local, and (2) any nonzero factor ring of R is local.

Ex. 19.2. For any field k, show that the 3×3 matrices in (19.15) form a local ring R whose maximal ideal has square zero. Check that $R \cong End_{kG}\ V$ in the example (19.15).

Ex. 19.3. What can you say about a local ring $(R, \mathfrak{m})$ that is von Neumann regular?

Ex. 19.4. (This exercise refines (19.10).) Let $R = kG$ where k is a field and G is a nontrivial finite group. Show that the following statements are equivalent:
(1) R is a local ring.
(2) R_R is an indecomposable R-module.
(3) $R/rad\ R$ is a simple ring.
(4) k has characteristic $p > 0$ and G is a p-group.

Ex. 19.5. Let $\mathfrak{A}$ be an ideal in a ring R such that $\mathfrak{A}$ is maximal as a left ideal. Show that $\bar{R} = R/\mathfrak{A}^n$ is a local ring for every integer $n \geq 1$.

Ex. 19.6. Show that if a ring R has a unique maximal ideal $\mathfrak{m}$, then the center Z of R is a local ring. (In particular, the center of a local ring is a local ring.)

Ex. 19.7. A domain R is called a *right discrete valuation ring* if there is a nonunit $\pi \in R$ such that every nonzero element $a \in R$ can be written in the form $\pi^n u$ where $n \geq 0$ and u is a unit. Show that
(1) R is a local domain;
(2) every nonzero right ideal in R has the form $\pi^i R$ for some $i \geq 0$;
(3) each $\pi^i R$ is an ideal of R; and
(4) $\bigcap_{i \geq 1} (\pi^i R) = 0$.
Give an example of a noncommutative right discrete valuation ring by using the twisted power series construction.

Ex. 19.8. (Brungs) Let R be a nonzero ring such that any collection of right ideals of R has a largest member (i.e. one that contains all the others). Show that (1) R is a local ring, (2) every right ideal of R is principal, and is an ideal. (**Hint.** For (2), let I be a nonzero right ideal, and let I' be the largest right ideal properly contained in I. Show that $I = aR$ for any $a \in I \setminus I'$. If there exists a right ideal which is not an ideal, consider the largest one and get a contradiction.)

Ex. 19.9. For a division ring D and a (not necessarily abelian) ordered group $(G, <)$, a function v: $D^* \to G$ is a called a (Krull) *valuation* if $v(ab) = v(a)v(b)$ for all $a, b \in D^*$ and $v(a + b) \geq \min\{v(a), v(b)\}$ for all $a, b \in D^*$ such that $a + b \neq 0$. Given such a valuation, let

$$R = \{0\} \cup \{r \in D^* : \ v(r) \geq 1\}.$$

(1) Show that R is a local ring.
(2) Show that $aRa^{-1} = R$ for any $a \in D^*$.
(3) Show that for any $a \in D^*$, either $a \in R$ or $a^{-1} \in R$.
(4) Show that any right (resp. left) ideal in R is an ideal.

(5) Show that the (right) ideals in R form a chain with respect to inclusion.
(6) Show that any finitely generated right ideal in R is principal.

Ex. 19.10. Let R be a subring of a division ring D which satisfies the two properties (2), (3) in Exercise 9. Show that there exists an ordered group $(G, <)$ and a valuation v: $D^* \to G$ such that

$$R = \{0\} \cup \{d \in D^* : \; v(d) \geq 1\}.$$

(Such a subring R of a division ring D is called an *invariant valuation ring* of D. If D is a field, the property (2) is automatic; in this case, we get back the usual (commutative) valuation rings in D).

Ex. 19.11. Deduce the fact that a finitely generated projective right module P over a local ring $(R, \mathfrak{m})$ is free from the Krull-Schmidt-Azumaya Theorem (19.21).

Ex. 19.12. Let k be a field with the property that $k(t)$ (the rational function field in one variable over k) is isomorphic to k. (For instance, $\mathbb{Q}(x_1, x_2, \ldots)$ is such a field.) Let θ: $k(t) \to k$ be a fixed isomorphism. Let $p(t)$ be a fixed irreducible polynomial in $k[t]$, and let A be the discrete valuation ring obtained by localizing $k[t]$ at the prime ideal $(p(t))$. On $R = A \oplus A$, define a multiplication by

$$(a, b)(a', b') = (aa', \; ba' + \theta(a)b').$$

The R is a ring with identity $(1, 0)$.
(1) Show that R is a local ring with $rad\ R = A \cdot p(t) \oplus A$.
(2) Show that $\bigcap_{i=1}^{\infty}(rad\ R)^i = (0) \oplus A$, and that this is the prime radical of R.
(3) Show that every right ideal of R is an ideal.
(4) Show that R is right noetherian but not left noetherian.
(This exercise is to be contrasted with Krull's Theorem in commutative algebra which states that, for any commutative noetherian local ring S, $\bigcap_{i=1}^{\infty}(rad\ S)^i = (0)$.)

Ex. 19.13. Show that any finitely generated projective right module P over a right artinian ring R is isomorphic to a finite direct sum of principal indecomposable right modules of R.

Ex. 19.14. Give an example of a local ring whose unique maximal ideal is nil but not nilpotent.

Ex. 19.15. Let (R, J) be a local ring, and M be a finitely generated left R-module. If $Hom_R(M, R/J) = 0$, show that $M = 0$. (**Hint.** Note that $Hom_R(M/JM, R/J) \to Hom_R(M, R/J)$ is injective.)

Ex. 19.16. Let (R, J) be a left noetherian local ring, and M be a finitely generated left R-module. Show that M is a free R-module iff, for any exact sequence $0 \to A \to B \to M \to 0$ of left R-modules, the induced sequence $0 \to A/JA \to B/JB \to M/JM \to 0$ remains exact.

§20. Semilocal Rings

In commutative algebra, local rings are ubiquitous because the localization of any commutative ring at a prime ideal is a local ring. In noncommutative ring theory, it is much harder to define localizations; in particular, it is not clear how to associate a family of local rings to a given ring. Thus, the class of local rings does not play as large a role in the structure theory of rings as their commutative counterparts.

In this section, we shall introduce a new class of rings called *semilocal rings*. This is a fairly big class which includes, for instance, all local rings and all left (resp., right) artinian rings (in particular, all finite-dimensional algebras over fields). Many rings which arise naturally in ring theory are semilocal rings. Also, semilocal rings play a very special role in algebraic K-theory. This section will serve as an introduction to the basic theory of such rings.

In commutative algebra, a semilocal ring is a (commutative) ring which has only a finite number of maximal ideals. The correct generalization of this to arbitrary rings turns out to be the following.

(20.1) Definition. *A ring R is said to be semilocal if $R/rad\ R$ is a left artinian ring, or, equivalently, if $R/rad\ R$ is a semisimple ring.*

Let us first check that this definition is consistent with the one mentioned above in the commutative case.

(20.2) Proposition. *For a ring R, consider the following two conditions:*

(1) *R is semilocal.*

(2) *R has finitely many maximal left ideals.*

We have, in general, (2) $\Rightarrow$ (1). *The converse holds if $R/rad\ R$ is commutative.*

Proof. For both conclusions, we may clearly assume that $rad\ R = 0$. Assume (2) and let $\mathfrak{m}_1, \ldots, \mathfrak{m}_n$ be the maximal left ideals of R. Then $\bigcap_{i=1}^n \mathfrak{m}_i = 0$ and we have an injection of left R-modules

$$R \longrightarrow \bigoplus_{i=1}^{n} R/\mathfrak{m}_i.$$

The latter has a composition series; thus, so does the former. This implies that the ring R is left artinian, so we have (1). Conversely, assume R is commutative and artinian. Since we have assumed that $rad\ R = 0$, R is a direct product of a finite number of fields (for instance, by Wedderburn's Theorem). Then the number of maximal ideals in R is the number of factors in this decomposition. This checks (2). QED

Remark. In (20.2), we do not have (1) $\Rightarrow$ (2) in general. For instance, a matrix algebra over a field is semilocal, but there may be infinitely many maximal left ideals.

In commutative algebra, a major source of semilocal rings is provided by the process of *semilocalization*. If R is any commutative ring and $\mathfrak{m}_1, \ldots, \mathfrak{m}_n$ are distinct prime ideals in R, then for $S = R \backslash (\mathfrak{m}_1, \cup \cdots \cup \mathfrak{m}_n)$, the localization R_S of R at S gives a semilocal ring with exactly n maximal ideals, $(\mathfrak{m}_1)_S, \ldots, (\mathfrak{m}_n)_S$. In the theory of noncommutative rings, however, this process does not work so well, since in general we cannot "localize" a ring at a multiplicative set. Thus, we no longer get quick examples of semilocal rings by semilocalization. But fortuitously, there do exist many interesting sources of semilocal rings in the noncommutative theory. Let us give a list of some of the major sources of noncommutative semilocal rings below.

(20.3) As we mentioned before, *any local ring is semilocal, and any left* (*or right*) *artinian ring is semilocal.* In particular, any finite ring or any finite-dimensional algebra over a field is semilocal.

(20.4) *Let A be any semilocal ring. Then $R = \mathbb{M}_n(A)$ is also a semilocal ring.* In fact, by Example 9 in §4, we have $rad\ R = \mathbb{M}_n(rad\ A)$. Thus,

$$R/rad\ R \cong \mathbb{M}_n(A/rad\ A).$$

Since $A/rad\ A$ is semisimple, $\mathbb{M}_n(A/rad\ A)$ is also semisimple by Exercise 3.1. Therefore, $R = \mathbb{M}_n(A)$ is semilocal. Thus, we can get nice examples of semilocal rings by starting with, say, a *local* ring A and building $R = \mathbb{M}_n(A)$. The semilocal rings R which arise in this way will be characterized in a later section.

(20.5) *A finite direct product of local rings is semilocal.*

The next two Propositions provide many more examples of semilocal rings.

(20.6) Proposition. *Let k be a commutative semilocal ring and R be a k-algebra which is finitely generated as a k-module. Then R is a semilocal ring, and $rad\ R \supseteq (rad\ k)R \supseteq (rad\ R)^n$ for some integer $n \geq 1$.*

Proof. Let $J = rad\ k$. We have shown in (5.9) that $JR \subseteq rad\ R$. View R/JR as a (finitely generated) k/J-module. Since k/J is artinian, R/JR is an artinian module. In particular, R/JR is a (left) artinian ring, so R is semilocal. Also,

$$(rad\ R)/JR = rad(R/JR)$$

is nilpotent, so there exists an integer n such that $(rad\ R)^n \subseteq JR$. QED

(20.7) Proposition. *Let R be a semilocal ring and I be any ideal in R. Then $rad(R/I) = (rad\ R + I)/I$, and R/I is a semilocal ring.*

Proof. Let $J = rad\ R$, and let "bar" denote the quotient map $R \to \bar{R} := R/I$. Clearly

$$\bar{J} = (J + I)/I \subseteq rad\ \bar{R}.$$

Thus, we have

$$(rad\ \bar{R})/\bar{J} = rad(\bar{R}/\bar{J}) = rad(R/(I + J)).$$

Since $R/(I + J)$ is a quotient of the semisimple ring R/J, it is also semisimple, so $rad(R/(I + J)) = 0$. This shows that $rad\ \bar{R} = \bar{J}$. Therefore,

$$\bar{R}/rad\ \bar{R} = \bar{R}/\bar{J} \cong R/(I + J).$$

We have observed that the latter ring is semisimple, so $\bar{R}$ is semilocal.

QED

Many other sources of noncommutative semilocal ring have been found in the recent literature in ring theory. Let us mention here some of these interesting new sources. For any *artinian* module M_R over any ring R, R. Camps and W. Dicks have shown that $End(M_R)$ is always semilocal. For this result, we refer the reader to Camps-Dicks [93]. If M_R is, instead, a *uniserial* module, that is, a module whose submodules form a chain (under inclusion), A. Facchini has shown that $End(M_R)$ is again semilocal (and in fact has at most two maximal right ideals). A proof of this recent result of Facchini is included in an Appendix to this section. The examples of Camps-Dicks and Facchini, together with the examples already given above, show that semilocal rings form a very extensive class of (commutative and) noncommutative rings worthy of our close attention.

Next we shall study some properties of the units of a semilocal ring. We have shown earlier that local rings are Dedekind-finite, so it is not surprising that we have the same property for semilocal rings.

(20.8) Proposition. *A semilocal ring R is Dedekind-finite.*

Proof. By Exercise 3.10, the semisimple ring $R/rad\ R$ is Dedekind-finite. Now apply (4.8). QED

Another important property of semilocal rings (which, in fact, turns out to be a generalization of (20.8)) was discovered by H. Bass in his pioneering work in algebraic K-theory.

(20.9) Bass' Theorem. *Let R be a semilocal ring, $a \in R$, and $\mathfrak{B}$ be a left ideal of R. If $R \cdot a + \mathfrak{B} = R$, then the coset $a + \mathfrak{B}$ contains a unit of R.*

In the case when $\mathfrak{B} = 0$, (20.9) boils down to (20.8). Hence we may view (20.9) as an extension of (20.8). Also, note that we could have stated (20.9) for $\mathfrak{B}$ a principal left ideal, and it would have been an equivalent theorem. However, it turns out to be more convenient to work with arbitrary left ideals. Because of the basic role played by Bass' Theorem in algebraic K-theory, we shall give two different proofs for it below. The reader will see that the methods used in the two proofs are quite different.

First Proof of (20.9). Recalling that $u \in R$ is a unit iff $\bar{u}$ is a unit in $R/rad\ R$, we may replace R by $R/rad\ R$ to assume that R is semisimple. Using the Wedderburn–Artin Theorem, we may further assume that $R = End(V_D)$, where V is a finite-dimensional right vector space over a division ring D. The left ideal $\mathfrak{B} \subseteq R$ gives rise to a subspace $W = \{v \in V\colon \mathfrak{B}v = 0\}$ of V. By Exercise 11.15, $\mathfrak{B}$ equals

$$ann\ W := \{f \in R\colon f(W) = 0\}.$$

Note that the restriction of the action of a on W gives an isomorphism $W \to aW$. To see this, write $1 = ra + b$, where $r \in R$ and $b \in \mathfrak{B}$. If $w \in W$ is such that $a(w) = 0$, then

$$w = (ra + b)w = b(w) = 0,$$

as desired. Now pick a D-automorphism f of V such that $f(w) = a(w)$ for every $w \in W$. Then $f - a \in ann(W) = \mathfrak{B}$, so $a + \mathfrak{B}$ contains the unit f of R.

QED

The proof above was the original one given by Bass. We give next the proof by Swan which is independent of the Wedderburn–Artin Theorem as well as Exercise 11.15.

Second Proof of (20.9). As before, we may assume that R is semisimple. Pick a left ideal $\mathfrak{B}'$ such that $\mathfrak{B} = (Ra \cap \mathfrak{B}) \oplus \mathfrak{B}'$. After replacing $\mathfrak{B}$ by $\mathfrak{B}'$, we may henceforth assume that $R = Ra \oplus \mathfrak{B}$. Consider the exact sequence

$$0 \longrightarrow K \longrightarrow R \stackrel{f}{\longrightarrow} Ra \longrightarrow 0,$$

where f is defined by $f(r) = ra$ $(r \in R)$, and $K = ker\ f$. Let $g\colon R \to K$ be a splitting, so $(f, g)\colon R \to Ra \oplus K$ is an isomorphism. Since $R = Ra \oplus \mathfrak{B}$, there exists an isomorphism $\theta\colon K \to \mathfrak{B}$. Now consider the composition

$$R \stackrel{(f,g)}{\longrightarrow} Ra \oplus K \stackrel{(1,\theta)}{\longrightarrow} Ra \oplus \mathfrak{B} = R$$

which sends $r \in R$ to $ra + \theta(g(r))$. Since this composition is an isomorphism of left R-modules, the image of 1 is a unit $u \in U(R)$. But then

$$u = a + \theta(g(1)) \in a + \mathfrak{B},$$

as desired. QED

The theorem of Bass lends some motivation to the following definition, which arises in the study of algebraic K-theory.

(20.10) Definition. *A ring E is said to have left stable range* 1 *if, whenever $Ea + Eb = E$ $(a, b \in E)$, there exists $e \in E$ such that $a + eb \in U(E)$. (Note that in the special case $b = 0$, this condition amounts to E being Dedekind-finite.)*

Of course, there is a similar notion of *right* stable range 1 for rings too, expressed in terms of comaximal principal right ideals. If we assume the result of Vaserstein in Ex. (1.25), it will follow that *left and right stable range* 1 *are equivalent properties*. However, we do not need this symmetry result for the rest of this section. To show that it is not essential here to assume Vaserstein's result, we shall continue to work with the notion of left stable range 1 through this section, refraining from dropping the adjective "left" which would have been possible as a result of the known symmetry of the stable range property.

Using the terminology of (20.10), Bass' Theorem (20.9) amounts precisely to the fact that *a semilocal ring has left stable range* 1. However, not every ring of left stable range 1 is a semilocal ring: see Exercise 10C.

To show how Definition (20.10) can be used, we shall prove below a Cancellation Theorem for modules. More specific versions of cancellation results will be deduced from it later on.

(20.11) Cancellation Theorem (Evans). *Let R be a ring, and A, B, C be right R-modules. Suppose $E = End(A_R)$ has left stable range* 1 *(e.g., E is semilocal). Then $A \oplus B \cong A \oplus C$ (as R-modules) implies that $B \cong C$.*

Proof. Since $A \oplus B \cong A \oplus C$, there exists a split epimorphism $(f, g)\colon A \oplus B \to A$ with kernel $\cong C$. Let $\begin{pmatrix} f' \\ g' \end{pmatrix}\colon A \to A \oplus B$ be a splitting. Then

$$1_A = (f, g)\begin{pmatrix} f' \\ g' \end{pmatrix} = ff' + gg',$$

so $E \cdot f' + E \cdot gg' = E$. Since E has left stable range 1, there exists $e \in E$ such that $f' + e \cdot (gg') = u \in U(E)$. (Note that gg' belongs to E although g and g' do not.) We have now $(1, eg)\begin{pmatrix} f' \\ g' \end{pmatrix} = u$. From this, we deduce that

$$ker(1, eg) \cong ker(f, g),$$

since each of these kernels is isomorphic to

$$(A \oplus B)/im\begin{pmatrix} f' \\ g' \end{pmatrix}.$$

On the other hand, it is easy to see that $ker(1, eg) \cong B$. Since $ker(f, g) \cong C$, we conclude that $B \cong C$. QED

In practice, (20.11) is applied in the situation of the Corollary below.

(20.12) Corollary. *Let k be a commutative noetherian semilocal ring, and R be a k-algebra which is finitely generated as a k-module. Let A be a finitely generated right R-module, and B, C be arbitrary right R-modules. Then $A \oplus B \cong A \oplus C$ implies that $B \cong C$.*

Proof. It is sufficient to check that $E = End(A_R)$ is a semilocal ring, for then E has left stable range 1 and (20.11) applies. View E as a k-submodule of $End(A_k)$. Since A is finitely generated over R, it is also finitely generated over k. From this, it follows that $End(A_k)$ is finitely generated as a k-module. Since k is noetherian, this implies that E is finitely generated as a k-module. Using (20.6), we conclude that E is a semilocal ring, as desired. QED

Note that the kind of cancellation theorems we proved above, in general, *do not* imply the uniqueness of Krull–Schmidt decompositions for finitely generated modules. For instance, in the context of (20.12), we may still have $A \oplus B \cong C \oplus D$ where A, B, C, D are finitely generated indecomposable R-modules such that $A \not\cong C$, $A \not\cong D$ and $B \not\cong C$, $B \not\cong D$. (Since there is no "common" summand to begin with, (20.12) simply does not apply.) In fact, this was precisely the situation in Swan's example presented in the last section. There, we have $k = R$ which was a commutative noetherian local domain.

To conclude this section, we shall record some results on the cancellation of finitely generated projective modules. The study of the cancellation of such modules is of considerable interest in algebraic K-theory. In general, of course, the cancellation of finitely generated projective modules may not be possible. This leads to the following useful definition: A finitely generated (right) module P over a ring R is said to be *stably free* if there exist two integers m and n such that $P \oplus R^m \cong R^n$ as (right) R-modules. While such a module is always projective, there are many rings (both commutative and noncommutative ones) for which there exist stably free modules which are *not* free. On the positive side, we have the following result.

(20.13) Theorem. *Let R be a ring which has left stable range 1 (e.g., R is a semilocal ring).*

1. *Let A, B, C be right modules, where A_R is finitely generated and projective. Then $A \oplus B \cong A \oplus C$ implies that $B \cong C$.*
2. *R has the invariant basis property, i.e., for natural numbers n and m, $R^n \cong R^m$ (as right R-modules) implies that $n = m$ (unless $R = (0)$).*
3. *Any (finitely generated) stably free module P_R is free.*
4. *$\mathbb{M}_n(R)$ is Dedekind-finite for any integer $n \geq 1$.*

Proof. (1) Choose a right R-module A' so that $A \oplus A' \cong R^n$ for some integer n. Then $A \oplus B \cong A \oplus C$ implies that $R^n \oplus B \cong R^n \oplus C$. It suffices to "cancel" one copy of R at a time, so we may assume that $A = R$. Then the endomorphism ring

$$End(A_R) = End(R_R) \cong R$$

has left stable range 1, so the Cancellation Theorem (20.11) applies. For (2), assume $R^n \cong R^m$ but $n > m$. Canceling R^m, we get $R^{n-m} = 0$, so $R = 0$. For (3), assume that $P \oplus R^r \cong R^s$. If $s < r$, we can cancel R^s to get $R = 0$ and $P = 0$. Thus we may assume that $s \geq r$. Canceling R^r, we get $P \cong R^{s-r}$. For (4), let $\alpha, \beta \in \mathbb{M}_n(R)$ be such that $\alpha\beta = I$. Then α defines a surjective of right R-modules $R^n \to R^n$ which splits by β: $R^n \to R^n$. Thus we have an isomorphism $R^n \cong R^n \oplus ker(\alpha)$. Canceling R^n, we have $ker(\alpha) = 0$, so α: $R^n \to R^n$ is an isomorphism. This says that α is a unit in $\mathbb{M}_n(R)$, so we have $\beta\alpha = I$. QED

Appendix: Endomorphism Rings of Uniserial Modules

We have mentioned in this section that $End(M_R)$ is a semilocal ring if M_R is an artinian module (result of Camps-Dicks) or a uniserial module (result of Facchini) over any ring R. These results provide new interesting classes of semilocal rings, and are therefore highly relevant to this section. The result of Camps-Dicks [93] requires the theory of uniform dimensions, which is not developed in this text; therefore, we will not be able to present it here. On the other hand, Facchini's result is proved by completely elementary means which are easily within our reach. We shall, therefore, cover Facchini's result in this Appendix, with some indications on how this work is related to the further study of the Krull–Schmidt–Azumaya Theorem. Our exposition here is based on Facchini's paper [96].

By definition, a module M_R (over any ring R) is *uniserial* if $M \neq 0$ and the submodules of M form a chain (that is, any two of them are comparable under inclusion). Clearly, any uniserial module is indecomposable. For instance, the group of p^i-th roots of unity ($i = 0, 1, 2, \ldots$), for any prime p, is a uniserial module over $\mathbb{Z}$. (This is the so-called Prüfer p-group.) If R is any discrete valuation ring with uniformizer π, and field of fractions M, then the submodules of M_R are $0, M$, or $\pi^i R$ for some $i \in \mathbb{Z}$. Since these clearly form a chain, M_R gives another example of a uniserial module. In general, nonzero submodules and quotient modules of a uniserial module M are also uniserial.

Throughout the following, *M_R denotes a uniserial right module over any ring R*. It turns out that the endomorphism ring $E := End(M_R)$ is very susceptible to analysis. We start with the following basic observation on R-homomorphisms in and out of M.

(20.14) Lemma. *Let $A \xrightarrow{\alpha} M \xrightarrow{\beta} B$ be R-homomorphisms, where A, B are nonzero right R-modules. Then,*

(1) *$\beta\alpha$ is injective iff α, β are both injective;*

(2) *$\beta\alpha$ is surjective iff α, β are both surjective.*

Proof. The "if" parts in (1) and (2) are true without any assumptions on M and A, B.

(1) ("Only if") Assume $\beta\alpha$ is injective. Clearly, α must be injective, so $\alpha(A) \neq 0$. Since $\alpha(A) \cap ker(\beta) = 0$, the fact that M is uniserial implies that $ker(\beta) = 0$. Therefore, β is also injective.

(2) ("Only if") The argument here is dual to that given above. Assume $\beta\alpha$ is surjective. Clearly, β must be surjective, so $ker(\beta) \neq M$. Since $\alpha(A) + ker(\beta) = M$, the fact that M is uniserial implies that $\alpha(A) = M$. Therefore, α is also surjective. QED

With this lemma, we can study the structure of the endomorphism ring E of M using the following two sets:

$$\mathfrak{m}_1 = \{\alpha \in E : \ \alpha \text{ is not injective}\}, \quad \text{and}$$

$$\mathfrak{m}_2 = \{\alpha \in E : \ \alpha \text{ is not surjective}\}.$$

(20.15) Theorem (Facchini). *$\mathfrak{m}_1$ and $\mathfrak{m}_2$ are completely prime ideals in E, i.e. $E/\mathfrak{m}_i$ $(i = 1, 2)$ are domains. Any proper one-sided ideal of E is contained in $\mathfrak{m}_1$ or $\mathfrak{m}_2$. We have the following two possibilities:*

Case A. *$\mathfrak{m}_1$ and $\mathfrak{m}_2$ are comparable under inclusion. In this case, E is a local ring with unique maximal ideal $\mathfrak{m}_1 \cup \mathfrak{m}_2$.*

Case B. *$\mathfrak{m}_1$ and $\mathfrak{m}_2$ are not comparable. In this case, $E/\mathfrak{m}_i$ $(i = 1, 2)$ are division rings, rad $E = \mathfrak{m}_1 \cap \mathfrak{m}_2$, and $E/\text{rad } E \cong E/\mathfrak{m}_1 \times E/\mathfrak{m}_2$.*

In particular, E is always a semilocal ring.

Proof. Let $\alpha, \beta \in \mathfrak{m}_1$, say with $0 \neq ker(\alpha) \subseteq ker(\beta)$. Then $ker(\alpha) \subseteq ker(\alpha + \beta)$, so $\alpha + \beta \in \mathfrak{m}_1$. This implies that $\mathfrak{m}_1$ is an additive group. If $\alpha \in \mathfrak{m}_1$ and $\beta \in E$, then $\alpha\beta$ and $\beta\alpha$ are both not injective by the lemma, so $\alpha\beta, \beta\alpha \in \mathfrak{m}_1$. This shows that $\mathfrak{m}_1$ is an ideal of E, and the same lemma implies that $E/\mathfrak{m}_1$ is a domain.

The treatment for $\mathfrak{m}_2$ is similar. If $\alpha, \beta \in \mathfrak{m}_2$, we may assume that $\alpha(M) \subseteq \beta(M) \neq M$. Then

$$(\alpha + \beta)(M) \subseteq \beta(M) \neq M \Longrightarrow \alpha + \beta \in \mathfrak{m}_2.$$

By the lemma again, $\mathfrak{m}_2$ is then an ideal, and $E/\mathfrak{m}_2$ is a domain.

Now let $I \subsetneq E$ be any 1-sided ideal. Then I consists of nonunits only, and so $I \subseteq \mathfrak{m}_1 \cup \mathfrak{m}_2$. A standard argument then shows $I \subseteq \mathfrak{m}_1$ or $I \subseteq \mathfrak{m}_2$. In particular, if I is any maximal left ideal of E, we must have $I = \mathfrak{m}_1$ or $I = \mathfrak{m}_2$. By (20.2), this already suffices to show that E is a semilocal ring.

Case A. $\mathfrak{m}_1$ *and* $\mathfrak{m}_2$ *are comparable.* Here, we have a unique maximal left ideal $\mathfrak{m}_1 \cup \mathfrak{m}_2$ (the bigger one of the two), which must be *rad* E. In this case, E is a local ring.

Case B. $\mathfrak{m}_1$ *and* $\mathfrak{m}_2$ *are not comparable.* Here, they are *both* maximal left ideals, and we have *rad* $E = \mathfrak{m}_1 \cap \mathfrak{m}_2$. Since $\mathfrak{m}_1 + \mathfrak{m}_2 = E$, the Chinese Remainder Theorem implies that

(20.16) $$E/rad\ E \cong E/\mathfrak{m}_1 \times E/\mathfrak{m}_2.$$

Each ring $E/\mathfrak{m}_i$ has only two left ideals ((0) and itself), and is therefore a division ring. QED

It can be shown by examples that Case A and Case B can *both* occur, although we shall not dwell on this point here. Given more hypotheses on M (or on R), however, we can often "force" Case A to happen. This is illustrated by the following.

(20.17) Corollary. *Assume that either*

(1) *M is hopfian (every surjective endomorphism is injective), or*

(2) *M is cohopfian (every injective endomorphism is surjective).*

Then, in (20.15), *Case A must occur. In particular, M is strongly indecomposable if it is assumed to be projective, or injective, or noetherian, or artinian.*

Proof. Under (1), we have $\mathfrak{m}_1 \subseteq \mathfrak{m}_2$, and under (2), we have $\mathfrak{m}_2 \subseteq \mathfrak{m}_1$. Therefore, only Case A can occur. If M is either projective or noetherian, it must be hopfian. (The projective case follows from the indecomposability of M, and the noetherian case follows from Exer. (1.12).) Similarly, if M is either injective or artinian, it must be cohopfian. In *any* of these cases, therefore, we are necessarily in Case A so E is *local*; that is, M is strongly indecomposable. QED

Improving upon the above considerations, we can actually say a bit more. Let us first observe the following necessary condition for Case B in (20.15).

(20.18) Lemma. *If Case B holds in* (20.15), *there exist R-submodules $X, Y \subseteq M$ such that*

(20.19) $$0 \neq Y \subseteq X \neq M \quad \text{and} \quad X/Y \cong M.$$

Proof. Assuming we are in Case B, there exist $\alpha, \beta \in E$ such that $\alpha \in \mathfrak{m}_1 \backslash \mathfrak{m}_2$ and $\beta \in \mathfrak{m}_2 \backslash \mathfrak{m}_1$. Thus, α is surjective and not injective, while β is injective and not surjective. Let

$$X = \beta(M) \neq M, \quad \text{and} \quad Y = \beta(ker(\alpha)) \neq 0.$$

Then we have $Y \subseteq X$ in M, with

$$X/Y = \beta(M)/\beta(ker(\alpha)) \cong M/ker(\alpha) \cong M,$$

as desired. QED

Following Facchini and Salce, let us call a module M *shrinkable* if it contains submodules X and Y satisfying (20.19). Thus, M is shrinkable if it is isomorphic to a proper submodule of a proper quotient of itself. Clearly, such a module is rather "special". It can be easily seen, for instance, that such a module is neither noetherian nor artinian. Also, given some hypothesis on R, we can hope to rule out the existence of a shrinkable (uniserial) module M. This idea quickly leads to the following.

(20.20) Proposition. *Assume that R is commutative or right noetherian. Then any uniserial right R-module M is strongly indecomposable.*

Proof. It suffices to show that the new hypothesis on R implies that M is unshrinkable (for then Case A in (20.15) must prevail for M). Assume for the moment that M is shrinkable, with submodules X, Y as in (20.19). We can then come up with a *cyclic* submodule of M that is shrinkable. To do this, fix any element $m \in M \backslash X$. Since M is uniserial, we have $mR \supseteq X$. Restricting an isomorphism φ: $M \to X/Y$ to mR, we have $mR \cong X'/Y$ for some submodule X' between Y and X. Then $0 \neq Y \subseteq X' \neq mR$ shows that mR is shrinkable. Note that $mR \subseteq M$ is still uniserial.

Starting over again, we may assume that M is cyclic. If R is a *right noetherian* ring, then M_R is a noetherian module, and hence M is unshrinkable. Next, assume R is a *commutative* ring. Let us represent the cyclic module M in the form R/I, where I is a right ideal of R. If M is shrinkable, there would exist right ideals X, Y such that

$$I \subsetneq Y \subseteq X \subsetneq R, \quad \text{and} \quad X/Y \cong R/I.$$

Fix any right R-module isomorphism φ: $R/I \to X/Y$. Such φ is induced by *left* multiplication by some element $x \in X$. Invoking now the commutativity of R, we have $xY = Yx \subseteq Y$, and so $\varphi(Y/I) = 0$. This would imply $Y = I$, a contradiction. Therefore, M is again unshrinkable. QED

We can now record a number of consequences of the foregoing results on cancellation problems and Krull–Schmidt uniqueness properties.

(20.21) Corollary. *Any uniserial module is cancellable; that is, if M, B, C are right modules over any ring R and M is uniserial, then*

$$M \oplus B \cong M \oplus C \Longrightarrow B \cong C. \tag{20.22}$$

Proof. This follows from (20.11) since $End(M_R)$ is a semilocal ring according to (20.15). QED

(20.23) Corollary. *Let R be a ring and suppose*

$$M_1 \oplus \cdots \oplus M_r \cong N_1 \oplus \cdots \oplus N_s,$$

where the N_i's are indecomposable, and the M_i's are uniserial. Assume that one of the following holds:

(1) *for any i, M_i is projective, or injective, or noetherian, or artinian;*

(2) *R is either commutative or right noetherian.*

Then $r = s$, and, after a reindexing, we have $M_i \cong N_i$ for $1 \leq i \leq r$.

Proof. By (20.17) and (20.20), each M_i must be strongly indecomposable. Therefore, the Krull–Schmidt–Azumaya Theorem (19.21) applies. QED

Facchini has shown that, if we do not impose any conditions on R or on the uniserial modules M_i, then, even when all N_i's are also uniserial *and* $r = s$, we need not have $M_i \cong N_i$ $(1 \leq i \leq r)$ after reindexing. This shows the failure of the Krull–Schmidt Theorem for direct sums of uniserial modules, answering negatively a question raised by R. Warfield in 1975. On the other hand, Facchini has proved a *weakened* version of the Krull–Schmidt Theorem for direct sums of uniserial modules which introduced a new paradigm in the time-honored investigation of the Krull–Schmidt uniqueness property. For a detailed statement of Facchini's Theorem, we refer the reader to his paper [96], or his book [98].

In closing, let us also make some remarks on *artinian* modules. We have pointed out earlier the result of Camps and Dicks to the effect that the endomorphism ring of any artinian module is semilocal. Therefore, the cancellation result (20.22) also holds for artinian (not necessarily indecomposable) modules M. Moreover, if M is artinian and indecomposable, and R is either commutative or right noetherian, Warfield had shown that M is strongly indecomposable. Therefore, the Krull–Schmidt conclusion in (20.23) will hold if the M_i's are artinian indecomposable, and R is either commutative or right noetherian. However, as we have already pointed out in §19, Facchini, Herbera, Levy and Vámos have shown that the conclusion in (20.23) does not hold in general for artinian indecomposable modules M_i's and N_j's. This answered negatively a question of Krull dating back to 1932.

Exercises for §20

Ex. 20.1. For any ring R, show that the following statements are equivalent:
(1) R is semilocal;
(2) every direct product of simple left R-modules is semisimple;
(3) every direct product of semisimple left R-modules is semisimple;
(4) for any left R-module M, $soc(M) = \{m \in M\colon (rad\ R)m = 0\}$.

(**Hint.** (1) $\Rightarrow$ (4) was Exercise 4.18. (4) $\Rightarrow$ (3) $\Rightarrow$ (2) are easy. Assume (2) and that *rad* $R = 0$. Let $\{\mathfrak{m}_i\}$ be all the maximal left ideals of R. Then we have an injection of left R-modules $R \to \prod R/\mathfrak{m}_i$.)

Ex. 20.2. Let A, B, C be right modules over a ring R. If A has a composition series, show that $A \oplus B \cong A \oplus C$ implies $B \cong C$.

Ex. 20.3. Show that, over any commutative semilocal domain R, any finitely generated projective R-module P is free.

Ex. 20.4. (Jensen-Jøndrup) Construct a commutative semiprimary ring that cannot be embedded in a right noetherian ring.

Ex. 20.5. Show that $R = \begin{pmatrix} \mathbb{Q} & \mathbb{R} \\ 0 & \mathbb{Q} \end{pmatrix}$ is a noncommutative semi-primary ring which is neither right noetherian nor left noetherian.

Ex. 20.5*. Let R be any semiprimary ring, and M be any nonzero left R-module. Show that $(rad\ R)M \neq M$ and $soc(M) \neq 0$. (For a much more general result, see Exercise (24.7) below.)

Ex. 20.6. Let R be a left noetherian semilocal ring such that *rad* R is a nil ideal. Show that R is left artinian. (**Hint.** Use (4.15) and (10.30).)

Ex. 20.7. (cf. Exercise 4.21) Let R be any semilocal ring. For any ideal $I \subseteq R$, show that the natural map $\varphi: GL_n(R) \to GL_n(R/I)$ is onto. (**Hint.** First prove this for $n = 1$.)

Ex. 20.8. For any ring R and any integer $n \geq 1$, show that the following statements are equivalent:
(1) $\mathbb{M}_n(R)$ is Dedekind-finite;
(2) for any right R-module M, $R^n \cong R^n \oplus M$ implies that $M = (0)$;
(3) R^n_R is hopfian, i.e. any right module epimorphism $\alpha: R^n \to R^n$ is an isomorphism.
(A ring R satisfying any of these conditions is called *stably n-finite*. Note that if R is stably n-finite, then R is stably m-finite for any $m \leq n$. For instance, if R has left stable range 1, then R is stably n-finite for all n, by (20.13).)

Ex. 20.9. Let R be either a right noetherian ring or a commutative ring. Show that R is stably n-finite for any n (in the sense of Exercise 8 above).

Ex. 20.10A. (Kaplansky) Show that a ring R has left stable range 1 iff $Ra + Rb = R$ implies that $R \cdot (a + xb) = R$ for some $x \in R$. (**Hint.** For the "if" part, the key is to show that R is Dedekind-finite. Suppose $au = 1$. Then $Ra + R(1 - ua) = R$ implies that some $v := a + x(1 - ua)$ is left-invertible. Right multiply by u and *voilà*!)

Ex. 20.10B. For any ring R with left stable range 1, show that $rad(R)$ is given by the set $\{r \in R\colon\ r + U(R) \subseteq U(R)\}$.

Ex. 20.10C. Show that a ring with left stable range 1 need not be semilocal.

Ex. 20.11. (Camps-Menal) Recall that a subring $A \subseteq R$ is *full* if $A \cap U(R) \subseteq U(A)$ (see Exercise 5.0). If $R = D_1 \times \cdots \times D_n$ where each D_i is a division ring, show that any full subring $A \subseteq R$ is a semilocal ring.

Ex. 20.12. (Cf. Exercise 19.6) For any semilocal ring R, use the last exercise to show that the center $Z(R)$ is semilocal.

§21. The Theory of Idempotents

In the previous sections, we have already had many occasions for using idempotents in rings. Here, we shall try to present a more systematic study of idempotents. In a commutative ring R, whenever we have an idempotent e, the ring R decomposes into a direct product of the two rings $R \cdot e$ and $R \cdot (1 - e)$. For many considerations in commutative ring theory, we can often restrict our attention to rings R which are *indecomposable* (or *connected*); i.e., $R \neq 0$ and R does not decompose into a direct product of two nonzero rings. These are the (commutative) rings which have only the trivial idempotents, 0 and 1. For noncommutative rings, these remarks remain valid if we replace the word "idempotent" everywhere by "central idempotent." Thus, a (nonzero) ring R is indecomposable iff it has no nontrivial *central* idempotents. However, even for these rings, there may be many nontrivial, noncentral idempotents. To understand the structure of these rings, it is often important to study the behavior of their idempotents. Therefore, the theory of idempotents plays a much more prominent role in the study of noncommutative rings than in the study of commutative rings.

The first important facts about idempotents in arbitrary rings were discovered by the American algebraist Benjamin O. Peirce. For any idempotent e in a ring R, we have the following three Peirce decompositions:

(21.1) $R = R \cdot e \oplus R \cdot f$,

(21.2) $R = e \cdot R \oplus f \cdot R$,

(21.3) $R = eRe \oplus eRf \oplus fRe \oplus fRf$,

where $f = 1 - e$ is the "complementary" idempotent to e. We have used the decompositions (21.1), (21.2) before, and (21.3) follows easily from these. Note that (21.1) (resp., (21.2)) is a decomposition of R into left (resp., right) ideals, while (21.3) is a decomposition of R into additive subgroups. Among these subgroups, eRe, fRf are, in fact, rings on their own right, with identities e and f, respectively. As is easily seen, these two rings may be characterized by the equations:

(21.4) $eRe = \{r \in R\colon\ er = r = re\}, \quad fRf = \{r \in R\colon\ fr = r = rf\}$.

In the literature, these are sometimes referred to as the *corner rings* associated with the idempotents e and f.

Also pertinent to the Peirce decomposition is the following observation.

(21.5) Lemma. *e is a central idempotent (i.e., $e \in Z(R)$) iff $eRf = fRe = 0$.*

Proof. For $r \in R$, $erf = 0$ and $fre = 0$ amount to $er = ere = re$. QED

A very good illustration of the Peirce decomposition (21.3) is given by the example of a complete matrix ring $R = \mathbb{M}_n(k)$, where k is some given ring. If r is an integer strictly between 1 and n, and e is the idempotent matrix $diag(1, \ldots, 1, 0, \ldots, 0)$ with r ones, with the complementary idempotent $f = diag(0, \ldots, 0, 1, \ldots, 1)$ with $n - r$ ones, then an easy computation shows that

$$eRe = \left\{\begin{pmatrix} * & 0 \\ 0 & 0 \end{pmatrix}\right\}, \quad eRf = \left\{\begin{pmatrix} 0 & * \\ 0 & 0 \end{pmatrix}\right\},$$

$$fRe = \left\{\begin{pmatrix} 0 & 0 \\ * & 0 \end{pmatrix}\right\}, \quad fRf = \left\{\begin{pmatrix} 0 & 0 \\ 0 & * \end{pmatrix}\right\}$$

where the stars denote, respectively, blocks of sizes $r \times r$, $r \times (n-r)$, $(n-r) \times r$, and $(n-r) \times (n-r)$. In particular, the corner rings eRe and fRf are isomorphic to the matrix rings $\mathbb{M}_r(k)$ and $\mathbb{M}_{n-r}(k)$, respectively. These examples, incidentally, explain where "corner rings" got their name!

Next, we consider two idempotents e, e' in a ring R and compute $Hom_R(eR, e'R)$, the group of R-homomorphisms from eR to $e'R$. Recall that, by our general convention, homomorphisms between right modules are written on the *left* of module elements.

(21.6) Proposition. *Let e, e' be idempotents, and M be a right R-module. There is a natural additive group isomorphism λ: $Hom_R(eR, M) \to Me$. In particular, there is a natural group isomorphism $Hom_R(eR, e'R) \cong e'Re$.*

Proof. Given an R-homomorphism θ: $eR \to M$, consider $m = \theta(e)$. Then

$$me = \theta(e)e = \theta(e^2) = \theta(e) = m.$$

Therefore $m = me \in Me$. We define the desired map λ by setting $\lambda(\theta) = \theta(e)$. Clearly, λ is an injective group homomorphism. To show λ is surjective, consider any $m \in Me$ and define θ: $eR \to M$ by $\theta(er) = mr$, for $r \in R$. Since $er = 0$ implies that $mr \in Mer = 0$, θ is a well-defined R-homomorphism. We have $\lambda(\theta) = \theta(e) = m$, so λ is surjective, as desired. The last conclusion of (21.6) follows by setting $M = e'R$. QED

(21.7) Corollary. *For any idempotent $e \in R$, there is a natural ring isomorphism $End_R(eR) \cong eRe$.*

Proof. Taking $e' = e$ in (21.6), we have a group isomorphism λ: $End_R(eR) \to eRe$. It suffices to show that λ is a ring isomorphism. Let $\theta, \theta' \in End_R(eR)$ and let $m = \theta(e) \in eR$. Then

$$\lambda(\theta'\theta) = \theta'\theta(e) = \theta'(m) = \theta'(em) = \theta'(e)m = \lambda(\theta')\lambda(\theta),$$

as desired. QED

For the next Proposition, recall that two idempotents $\alpha, \beta \in R$ are said to be *orthogonal* if $\alpha\beta = \beta\alpha = 0$.

(21.8) Proposition. *For any nonzero idempotent $e \in R$, the following statements are equivalent:*

(1) *eR is indecomposable as a right R-module.*

(1)′ *Re is indecomposable as a left R-module.*

(2) *The corner ring eRe has no nontrivial idempotents.*

(3) *e has no decomposition into $\alpha + \beta$ where α, β are nonzero orthogonal idempotents in R.*

If the idempotent $e \neq 0$ satisfies any of these conditions, we say that e is a primitive idempotent of R.

Proof. By left-right symmetry, it is enough to show the equivalence of (1), (2), and (3). The equivalence (1) $\Leftrightarrow$ (2) follows from (21.7), since eR is indecomposable iff $End_R(eR)$ has no nontrivial idempotents.

(3) $\Rightarrow$ (2) If eRe has a nontrivial idempotent α, then for $\beta = e - \alpha$ (the complementary idempotent to α in the corner ring eRe), we have the "orthogonal" decomposition $e = \alpha + \beta$, contradicting (3).

(2) $\Rightarrow$ (3) Assume we have a decomposition $e = \alpha + \beta$, where α, β are nonzero orthogonal idempotents in R. Then

$$e\alpha = \alpha^2 + \beta\alpha = \alpha \quad \text{and} \quad \alpha e = \alpha^2 + \alpha\beta = \alpha.$$

By (21.4), $\alpha \in eRe$, which contradicts (2). QED

(21.9) Proposition. *For any idempotent $e \in R$, the following statements are equivalent:*

(1) *eR is strongly indecomposable as a right R-module.*

(1)′ *Re is strongly indecomposable as a left R-module.*

(2) *eRe is a local ring.*

If the idempotent e satisfies any of these conditions, we say that e is a local idempotent. (Clearly, a local idempotent is always a primitive idempotent.)

Proof. The equivalence (1) $\Leftrightarrow$ (2) follows from (21.7), and (1)′ $\Leftrightarrow$ (2) follows from left-right symmetry. QED

Some further characterizations for local idempotents will be given later in this section. These characterizations involve working with the quotient of the ring R modulo its Jacobson radical. To facilitate these considerations, we need to compute first the Jacobson radical of a corner ring.

(21.10) Theorem. *Let e be an idempotent in R, and $J = rad\ R$. Then $rad(eRe) = J \cap (eRe) = eJe$. Moreover, $eRe/rad(eRe) \cong \bar{e}\bar{R}\bar{e}$, where $\bar{e}$ is the image of e in $\bar{R} = R/J$.*

Proof. For the first conclusion, it is enough to prove the following three implications:

(1) $r \in rad(eRe) \Longrightarrow r \in J$.

(2) $r \in J \cap (eRe) \Longrightarrow r \in eJe$.

(3) $r \in eJe \Longrightarrow r \in rad(eRe)$.

For (1), it suffices to show that, for any $y \in R$, $1 - yr$ has a left inverse in R. Working first in eRe, we can find $b \in eRe$ such that $b(e - eye \cdot r) = e$, that is, $b(1 - yr) = e$. Thus,

$$yrb(1 - yr) = yre = yr.$$

Adding $1 - yr$, we get $(1 + yrb)(1 - yr) = 1$, as desired. For (2), we simply note that, for $r \in J \cap eRe$, we have $r = ere \in eJe$. For (3), it suffices to show that, for any $y \in eRe$, $e - yr$ has a left inverse in eRe. Since $r \in eJe \subseteq J$, there exists an $x \in R$ such that $x(1 - yr) = 1$. But then

$$e = ex(1 - yr)e = ex(e - yr) = exe \cdot (e - yr),$$

so $exe \in eRe$ is a left inverse for $e - yr$.

To complete the proof, we have to compute eRe/eJe. Consider the natural map $eRe \to \bar{e}\bar{R}\bar{e}$ which sends ere to $\bar{e}\bar{r}\bar{e}$. This is a well-defined ring homomorphism vanishing on eJe, so it induces a surjection $eRe/eJe \to \bar{e}\bar{R}\bar{e}$. This is an isomorphism, since, if $\bar{e}\bar{r}\bar{e} = 0$, then $ere \in J \cap eRe = eJe$. QED

In the next theorem, we study the relationship between the ideal structure of eRe and that of R.

(21.11) Theorem. *Let e be an idempotent in the ring R.*

(1) *Let $\mathfrak{A}$ be any left ideal of eRe. Then $(R\mathfrak{A}) \cap eRe = \mathfrak{A}$. In particular, $\mathfrak{A} \mapsto R\mathfrak{A}$ defines an injective (inclusion-preserving) map from the left ideals of eRe to those of R.*

(2) *Let $\mathfrak{A}$ be an ideal in eRe. Then $e(R\mathfrak{A}R)e = \mathfrak{A}$. In particular, $\mathfrak{A} \mapsto R\mathfrak{A}R$ defines an injective (inclusion-preserving) map from ideals of eRe to those of R. This map respects multiplication of ideals, and is surjective if e is a full idempotent, in the sense that $ReR = R$.*

Proof. For (1), let $\mathfrak{A}_0 = (R\mathfrak{A}) \cap eRe \supseteq \mathfrak{A}$. Then, since $\mathfrak{A}_0 \subseteq eRe$, we have

$$\mathfrak{A}_0 = e\mathfrak{A}_0 \subseteq e \cdot R\mathfrak{A} = eRe \cdot \mathfrak{A} \subseteq \mathfrak{A}.$$

Therefore, $\mathfrak{A}_0 = \mathfrak{A}$ as claimed. The second conclusion of (1) now follows easily. If $\mathfrak{A} \subseteq eRe$ is, in fact, an ideal, then

$$e(R\mathfrak{A}R)e = eR(e\mathfrak{A}e)Re = (eRe)\mathfrak{A}(eRe) = \mathfrak{A},$$

and, if $\mathfrak{A}'$ is another ideal of eRe, then

$$\begin{aligned}(R\mathfrak{A}R)(R\mathfrak{A}'R) &= R\mathfrak{A}R\mathfrak{A}'R = R(\mathfrak{A}e)R(e\mathfrak{A}')R \\ &= R\mathfrak{A}(eRe)\mathfrak{A}'R = R(\mathfrak{A}\mathfrak{A}')R.\end{aligned}$$

Finally, assume that e is full; i.e., $ReR = R$. For any ideal $\mathfrak{B}$ in R, consider the ideal $\mathfrak{A} = e\mathfrak{B}e$ in eRe. Then

$$\begin{aligned}R(e\mathfrak{B}e)R &= Re(R\mathfrak{B}R)eR = (ReR)\mathfrak{B}(ReR) \\ &= R\mathfrak{B}R = \mathfrak{B}.\end{aligned}$$

This shows the surjectivity of the map in (2). QED

(21.12) Remark. In the case when e is a full idempotent, we also see from the proof above that *rad* R in R corresponds to *rad*(eRe) in eRe under the ideal correspondence in (21.11)(2), since $e(\mathit{rad}\ R)e = \mathit{rad}(eRe)$ by (21.10). However, if e is not full, the ideal map $\mathfrak{A} \mapsto R\mathfrak{A}R$ in (2) above may *not* be surjective. For instance, if R is a commutative ring, we have $R\mathfrak{A}R = \mathfrak{A}$, so the image of the map in (2) is just the set of all ideals in the corner ring eRe.

Using the results above, one can show that many ring-theoretic properties of R are inherited by the ring eRe. The following offers only a small sample of such properties (see also Exercises 11.3 and 21.9).

(21.13) Corollary. *Let $e \neq 0$ be any idempotent in R. If R is Jacobson-semisimple (resp., semisimple, simple, prime, semiprime, left noetherian, left artinian), then the same holds for eRe.*

Proof. The "J-semisimple" case follows from (21.10). The other cases follow easily from (21.11). QED

An important guiding example for the results in (21.10) and (21.11) is the following.

(21.14) Example. Let k be a ring and $R = \mathbb{M}_n(k)$. Let e be the matrix unit E_{11} in R. Then, by an easy computation, $ere = r_{11}e$ for any matrix $r = (r_{ij})$, so as a ring eRe is isomorphic to k. (This is, of course, a special case of the example for Peirce decomposition given after (21.5).) One can also check that e is a full idempotent. Thus, (21.11)(2) recovers the familiar one-one correspondence between the ideals of k and those of $\mathbb{M}_n(k)$, as defined in

(3.1). In particular, Theorem (21.10) is consistent with the earlier result that $rad(\mathbb{M}_n(k)) = \mathbb{M}_n(rad\ k)$.

A more thorough treatment on the relationship between R and eRe for a full idempotent e will be given in *Lectures* in the section on Morita's theory of equivalence of module categories.

In (21.8) and (21.9), we defined the notions of primitive idempotents and local idempotents. Both of these turned out to be left-right symmetric. We shall now introduce the notion of irreducible idempotents: for this, the distinction of left and right becomes necessary, as shown by an example later.

(21.15) Definition. We say that an idempotent e ($\neq 0$) is *right* (resp., *left*) *irreducible* if eR (resp., Re) is a minimal right (resp., left) ideal of R.

Note that, by Brauer's Lemma (10.22), a minimal right ideal $I \subseteq R$ is generated by a right irreducible idempotent iff $I^2 \neq 0$.

(21.16) Proposition. *Let $e \in R$ be an idempotent.*

(1) *If e is right irreducible, then eRe is a division ring.*

(2) *The converse is true if R is a semiprime ring.*

Proof. (1) follows from Schur's Lemma since, by (21.7), $eRe \cong End_R(eR)$. For (2), assume R is semiprime and that eRe is a division ring. Consider any nonzero element $er \in eR$, where $r \in R$. Since R is semiprime, $erRer \neq 0$. Hence $erse \neq 0$ for some $s \in R$. Let ete be the inverse of $erse$ in eRe. Then $(erse)(ete) = e$. Therefore $erR = eR$, so eR is a simple R-module.

QED

We note the following direct consequence of the Proposition.

(21.17) Corollary.

(1) *A right irreducible idempotent is always local.*

(2) *If R is semiprime, then an idempotent is right irreducible iff it is left irreducible.*[1]

(3) *If R is semisimple, then an idempotent is right irreducible iff it is local, iff it is primitive.*

The next result gives a basic relationship between right irreducible idempotents and local idempotents.

---

[1] This fact can also be deduced from (11.9). In fact, the argument for (11.9) is very close to the proof of (21.16).

(21.18) Proposition. *Let e be an idempotent in R, and let $J = rad\ R$, $\bar{R} = R/J$. The following statements are equivalent*:

(1) *e is a local idempotent in R.*

(2) *$\bar{e}$ is a right irreducible idempotent in $\bar{R}$.*

(2)′ *$\bar{e}$ is a left irreducible idempotent in $\bar{R}$.*

(3) *eR/eJ is a simple right R-module.*

(4) *eJ is the unique maximal submodule of eR.*

Proof. We begin by noting that $\bar{R}$ is semiprimitive, hence semiprime. Thus (21.17)(2) gives (2) $\Leftrightarrow$ (2)′. Also, $\bar{e}$ is right irreducible iff $\bar{e}\bar{R}\bar{e}$ is a division ring. But by (21.10),

$$\bar{e}\bar{R}\bar{e} \cong eRe/rad(eRe),$$

so $\bar{e}\bar{R}\bar{e}$ is a division ring iff eRe is a local ring. This gives (2) $\Leftrightarrow$ (1). For the rest of the proof, note that we have an $\bar{R}$-isomorphism λ: $eR/eJ \to \bar{e}\bar{R}$. This gives (2) $\Leftrightarrow$ (3). Finally, assume (3). Then for any right ideal $I \subseteq eR$ not contained in eJ, $\lambda(I + eJ)$ must be $\bar{e}\bar{R}$ since $\bar{e}\bar{R}$ is a simple $\bar{R}$-module. Thus,

$$eR = I + eJ = I + eR \cdot J.$$

By Nakayama's Lemma, $I = eR$. This gives (3) $\Rightarrow$ (4), and (4) $\Rightarrow$ (3) is a tautology. QED

For a local idempotent $e \in R$, there is a useful criterion to decide when a right R-module of finite length has a composition factor isomorphic to the simple module eR/eJ, where $J = rad\ R$.

(21.19) Proposition. *Let $e \in R$ be a local idempotent, and M be a right R-module of finite (composition) length. Then M has a composition factor isomorphic to eR/eJ ($J = rad\ R$) iff $M \cdot e \neq 0$, iff $Hom(eR, M) \neq 0$.*

Proof. Let

$$M = M_0 \supsetneqq M_1 \supsetneqq \cdots \supsetneqq M_r = 0$$

be a composition series of M. First assume $Me \neq (0)$. If $M_ie \subseteq M_{i+1}$ for every i, then $Me = Me^r \subseteq M_r = (0)$, a contradiction. Thus, M has a composition factor V such that $Ve \neq (0)$. Fix an element $v \in V$ such that $ve \neq 0$; then $veR = V$. We have a surjective R-homomorphism λ: $eR \to V$ defined by $\lambda(er) = ver$ for any $r \in R$. The kernel of λ is a maximal submodule of eR, so by (21.18)(4), $ker\ \lambda = eJ$. Therefore $V \cong eR/eJ$, as desired. Conversely, if some M_i/M_{i+1} is isomorphic to eR/eJ, then, since $(eR/eJ) \cdot e \neq 0$, we have $(M_i/M_{i+1}) \cdot e \neq 0$. In particular, $Me \supseteq M_ie \neq 0$. The second "iff" in the Proposition now follows from (21.6). QED

Let us now give an example of a left irreducible (indeed, local) idempotent that is not right irreducible.

Example. Let k be a field, and R be the k-algebra of triangular matrices $\left\{\begin{pmatrix} a & b \\ 0 & c \end{pmatrix}\right\}$ over k. This ring has radical $\left\{\begin{pmatrix} 0 & b \\ 0 & 0 \end{pmatrix}\right\}$ with $(rad\ R)^2 = 0$, so R is not semisimple. For the idempotent $e = \begin{pmatrix} 1 & 0 \\ 0 & 0 \end{pmatrix}$, we have $Re = \left\{\begin{pmatrix} a & 0 \\ 0 & 0 \end{pmatrix}\right\}$ and $eR = \left\{\begin{pmatrix} a & b \\ 0 & 0 \end{pmatrix}\right\}$. Since $dim_k\ Re = 1$, e is clearly left irreducible. However, $eR \supsetneqq rad\ R \supsetneqq (0)$, so e is not right irreducible. (Here, $eRe = \left\{\begin{pmatrix} a & 0 \\ 0 & 0 \end{pmatrix}\right\}$ is isomorphic to k, so e is indeed a local idempotent.) Similarly, we can check that the complementary idempotent $f = 1 - e = \begin{pmatrix} 0 & 0 \\ 0 & 1 \end{pmatrix}$ is right irreducible but not left irreducible.

Next we shall study the notion of isomorphism between idempotents.

(21.20) Proposition. *Let e, f be idempotents in a ring R. Then the following statements are equivalent:*

(1) *$eR \cong fR$ as right R-modules.*

(1)′ *$Re \cong Rf$ as left R-modules.*

(2) *There exist $a \in eRf$ and $b \in fRe$ such that $e = ab$ and $f = ba$.*

(3) *There exist $a, b \in R$ such that $e = ab$ and $f = ba$.*

If e and f satisfy any of these conditions, we say that they are isomorphic idempotents, and write $e \cong f$.

Proof. By left-right symmetry, it is enough to show that (1) $\Rightarrow$ (2) $\Rightarrow$ (3) $\Rightarrow$ (1).

(1) $\Rightarrow$ (2) Fix an R-isomorphism θ: $eR \to fR$. By (21.6), this "corresponds" to the element $b = \theta(e) \in fRe$. Similarly, $\theta^{-1}: fR \to eR$ corresponds to some $a = \theta^{-1}(f) \in eRf$. Under the composition $\theta^{-1}\theta$, e goes to ab. Therefore, $ab = e$ and, similarly, $ba = f$.

(2) $\Rightarrow$ (3) is trivial.

(3) $\Rightarrow$ (1) Given a, b as in (3), we have $be = b(ab) \in fR$ and $af = a(ba) \in eR$. Define θ: $eR \to fR$ and $\theta': fR \to eR$ by $\theta(x) = bx \in fR$ and $\theta'(y) = ay \in eR$. Then

$$\theta'\theta(e) = \theta'(be) = abe = e^2 = e,$$

$$\theta\theta'(f) = \theta(af) = baf = f^2 = f.$$

Hence $\theta'\theta = 1$ and $\theta\theta' = 1$, as desired. QED

Examples.

(A) If R is a commutative ring, then by (3) above, $e \cong f$ means simply that $e = f$. Therefore, the notion of isomorphism between idempotents is of interest only in the noncommutative case.

(B) A typical noncommutative ring is $R = End(M_A)$, where M is a right module over some ring A. For idempotents $e, f \in R$, we claim that *$e \cong f$ in R iff $eM \cong fM$ as A-modules.* To see this, first assume $e \cong f$, so $e = ab$ and $f = ba$ for some $a, b \in R$. Then, left multiplication by b defines an A-homomorphism $\varphi\colon eM \to fM$, and left multiplication by a defines an A-homomorphism $\psi\colon fM \to eM$. It is easy to check that φ and ψ are mutually inverse maps, so we get $eM \cong fM$ as A-modules. Conversely, if $eM \cong fM$ as A-modules, let $\varphi\colon eM \to fM$ be an A-isomorphism with inverse ψ. We can define $b \in R$ by $b|eM = \varphi$ and $b|(1-e)M = 0$; similarly, we can define $a \in R$ by $a|fM = \psi$ and $a|(1-f)M = 0$. A routine calculation shows that $ab = e$ and $ba = f$, so we get $e \cong f$ in R.

(C) Let $M = A_A^n$ where A is any ring; then $R := End(M_A) \cong \mathbb{M}_n(A)$. If $\{E_{ij}\}$ are the matrix units in R, (B) above shows that $E_{ii} \cong E_{jj}$ in R since $E_{ii}M \cong A_A$ for every i. (What are choices of $a, b \in R$ such that $E_{ii} = ab$ and $E_{jj} = ba$?)

(D) *If $\lambda\colon R \to S$ is a ring homomorphism and $e \cong f$ in R, then $\lambda(e) \cong \lambda(f)$ in S.* This follows easily, for instance, from the characterization (21.20)(3) for the isomorphism of idempotents.

Let $e \in R$ be any idempotent, and let $e' = 1 - e$. Then we have a decomposition $R = eR \oplus e'R$, so $P = eR$ is a projective right R-module. For a fixed ideal $I \subseteq R$, write $\bar{R} = R/I$. Then, as is easily verified,

$$P/P \cdot I = eR/eI \cong \bar{e}\bar{R} \quad \text{as } \bar{R}\text{-modules.}$$

From (19.27), we deduce the following result.

(21.21) Proposition. *Let I be an ideal of R inside rad R. Then for idempotents $e, f \in R$, we have $e \cong f$ in R iff $\bar{e} \cong \bar{f}$ in $\bar{R} = R/I$. In particular, if $\bar{e} = \bar{f}$, then $e \cong f$.*

Our next goal is to study the notion of lifting idempotents. If I is an ideal in a ring R, we say that an idempotent $x \in R/I$ can be lifted to R if there exists an idempotent $e \in R$ whose image under the natural map $R \to R/I$ is x. For a general ideal I, we certainly do not expect every idempotent $x \in R/I$ to be liftable. For instance, for $R = \mathbb{Z}$, if we take I to be the ideal generated by $6 = 3^2 - 3$, then $\bar{3}$ is an idempotent in R/I which cannot be lifted to R. We shall soon give some sufficient conditions on $I \subseteq R$ which will guarantee the liftability of idempotents. Before we develop such results, however, let us first explain why it is of interest to study the lifting of idempotents.

(21.22) Proposition. *Let $e \in R$ be an idempotent and $I \subseteq rad\ R$ be an ideal of R. If $\bar{e}$ is primitive in $\bar{R} := R/I$, then e is primitive in R. The converse holds if idempotents of $\bar{R}$ can be lifted to R.*

Proof. We first make the following basic observation about *rad R*:

(21.23) *The only idempotent $\alpha \in rad\ R$ is $\alpha = 0$.*

In fact, consider the complementary idempotent $1 - \alpha$. Since $\alpha \in rad\ R$, $1 - \alpha$ is a unit. But then $1 - \alpha$ must be 1, i.e., $\alpha = 0$. To prove the Proposition, let $e = \alpha + \beta$ be a nontrivial decomposition of e into orthogonal idempotents $\alpha, \beta \in R$. By (21.23), $\alpha \neq 0 \Rightarrow \bar{\alpha} \neq 0$ and $\beta \neq 0 \Rightarrow \bar{\beta} \neq 0$ in $\bar{R}$. Thus, $\bar{e} = \bar{\alpha} + \bar{\beta}$ is a nontrivial decomposition of $\bar{e}$ into orthogonal idempotents $\bar{\alpha}, \bar{\beta} \in \bar{R}$. Conversely, suppose $\bar{e} = x + y$ is a nontrivial decomposition of $\bar{e}$ into orthogonal idempotents $x, y \in \bar{R}$. *Here we assume that these idempotents can be lifted to R.* Let α, β be idempotents of R such that $\bar{\alpha} = x$ and $\bar{\beta} = y$. We have then $\alpha\beta \equiv \beta\alpha \equiv 0 \pmod{I}$. We claim that

(21.24) *There exists an idempotent $\beta' \in R$ orthogonal to α such that $\beta' \equiv \beta \pmod{I}$.*

Assuming this claim, let us first show how to complete the proof. Define $e' = \alpha + \beta'$. This is clearly an idempotent in R, and it is not primitive (since $\alpha, \beta' \neq 0$). However,

$$\overline{e'} = \bar{\alpha} + \overline{\beta'} = \bar{\alpha} + \bar{\beta} = \bar{e} \quad \text{in } \bar{R},$$

so by (21.21), $e' \cong e$ in R. Therefore, e is also not primitive in R, as desired.

To prove the claim (21.24), note that $\beta\alpha \in I \subseteq rad\ R$ implies that $1 - \beta\alpha$ is a unit. Consider the idempotent

$$\beta_0 = (1 - \beta\alpha)^{-1}\beta(1 - \beta\alpha).$$

In $\bar{R}$, we have clearly $\bar{\beta}_0 = \bar{\beta}$. Moreover,

$$\beta_0\alpha = (1 - \beta\alpha)^{-1}\beta(\alpha - \beta\alpha) = 0.$$

However, $\alpha\beta_0$ may not be zero. To remedy this, let $\beta' := (1 - \alpha)\beta_0$. Since $\overline{\alpha\beta_0} = \bar{\alpha}\bar{\beta} = 0$, we have $\overline{\beta'} = \bar{\beta}_0 = \bar{\beta}$. Now not only $\beta'\alpha = (1 - \alpha)\beta_0\alpha = 0$, but also $\alpha\beta' = \alpha(1 - \alpha)\beta_0 = 0$. And β' is an idempotent since

$$\beta'^2 = (1 - \alpha)\beta_0(1 - \alpha)\beta_0 = (1 - \alpha)\beta_0^2 = \beta'. \qquad \text{QED}$$

The idea of the proof above can be refined a little bit to give the following result.

(21.25) Proposition. *Let $I \subseteq rad\ R$ be an ideal of R such that idempotents in $\bar{R} = R/I$ can be lifted to R. Then for any countable (finite or infinite) set of pairwise orthogonal idempotents $\{x_1, x_2, \ldots\}$ in $\bar{R}$, there exists a set of pairwise orthogonal idempotents $\{e_1, e_2, \ldots\}$ in R such that $\bar{e}_i = x_i$ for all i.*

Proof. Suppose we have already found $\{e_1, \ldots, e_n\}$ satisfying the stated conditions. It suffices to show how we can find e_{n+1}. Let α be the idempotent $e_1 + \cdots + e_n$, and let β be an idempotent of R lifting x_{n+1}. Then $\bar{\alpha}$ and $\bar{\beta}$ are orthogonal idempotents in $\bar{R}$. By (21.24), we can find an idempotent e_{n+1} orthogonal to α such that $\bar{e}_{n+1} = \bar{\beta} = x_{n+1}$. Since $e_i = \alpha e_i = e_i\alpha$ for $i \leq n$, e_{n+1} is clearly orthogonal to each of $e_1, \ldots, e_n$. QED

The following example, due to Jacobson, offers an interesting sufficient condition for the existence of a countably infinite set of pairwise orthogonal nonzero idempotents.

(21.26) Example (cf. Exercise 11.9). Let R be any ring which is not Dedekind-finite; i.e., there exist elements $a, b \in R$ such that $ab = 1$ but $e := ba \neq 1$. Then $e^2 = b(ab)a = e$ so e is a (nontrivial) idempotent. For $i, j \geq 0$, let

$$e_{ij} = b^i(1-e)a^j.$$

Then $\{e_{ij}\}$ *is a set of matrix units* in the sense that $e_{ij}e_{k\ell} = \delta_{jk}e_{i\ell}$ (where δ_{jk} are the Kronecker deltas). To see this, note that $a^ib^i = 1$ for all i, and that $a(1-e) = 0 = (1-e)b$. If $j \neq k$, then a^jb^k is either $a^{|j-k|}$ or $b^{|j-k|}$, so

$$e_{ij}e_{k\ell} = b^i(1-e)a^jb^k(1-e)a^\ell = 0.$$

On the other hand, since $1-e$ is an idempotent,

$$e_{ij}e_{j\ell} = b^i(1-e)a^jb^j(1-e)a^\ell = b^i(1-e)a^\ell = e_{i\ell}.$$

Note that each $e_{ij} \neq 0$, for if $b^i(1-e)a^j = 0$, then

$$0 = a^ib^i(1-e)a^jb^j = 1-e,$$

a contradiction. In particular, $\{e_{ii}\colon i \geq 0\}$ is an infinite sequence of nonzero pairwise orthogonal idempotents in R, and R contains an infinite direct sum of nonzero right ideals $\bigoplus_{i\geq 0} e_{ii}R$. This leads to the following observation which generalizes (20.8) (see also Exercise 20.9).

(21.27) Corollary. *Let S be a ring such that $R := S/rad\ S$ does not contain an infinite direct sum of nonzero right ideals (e.g., R is right noetherian). Then S is Dedekind-finite.*

Proof. The work in (21.26) above showed that R is Dedekind-finite. From this, it is easy to deduce that S itself is Dedekind-finite. In fact, assume $ab = 1$ in S. Then we have $ba \in 1 + rad\ S \subseteq U(S)$. Choose $u \in S$ such that $bau = 1$. Left multiplying by a, we get $au = a$, and hence $ba = 1$. QED

Let us now return to the problem of lifting idempotents. We shall try to establish two separate sufficient conditions on an ideal $I \subseteq R$ in order that idempotents in R/I can be lifted to R.

(21.28) Theorem. *Let I be a nil ideal in R (so $I \subseteq \text{rad } R$). Let $a \in R$ be such that $\bar{a} \in \bar{R} := R/I$ is an idempotent. Then there exists an idempotent $e \in aR$ such that $\bar{e} = \bar{a} \in \bar{R}$.*

Proof. For $b = 1 - a$, we have $ab = ba = a - a^2 \in I$, so $(ab)^m = 0$ for some integer $m \geq 1$. By the Binomial Theorem,

$$\begin{aligned} 1 &= (a+b)^{2m} \\ &= a^{2m} + r_1 a^{2m-1} b + \cdots + r_m a^m b^m + r_{m+1} a^{m-1} b^{m+1} + \cdots + b^{2m}, \end{aligned}$$

where the r_i's are integers. Let

$$e = a^{2m} + r_1 a^{2m-1} b + \cdots + r_m a^m b^m \in aR, \quad \text{and}$$

$$f = r_{m+1} a^{m-1} b^{m+1} + \cdots + b^{2m}.$$

Since $a^m b^m = b^m a^m = 0$, we have $ef = 0$ and so $e = e(e+f) = e^2$. Finally, $ab \in I$ implies that $e \equiv a^{2m} \equiv a \pmod{I}$, as desired. QED

As an application of the theorem, we derive the following basic result on idempotents which holds, in particular, for all semiprimary rings and all left (or right) artinian rings.

(21.29) Corollary. *Let R be a semilocal ring such that $I = \text{rad } R$ is a nil ideal.*

(1) *If R has no nontrivial idempotents and $R \neq (0)$, then R is a local ring.*

(2) *A right ideal $\mathfrak{A} \subseteq R$ contains a nonzero idempotent iff $\mathfrak{A}$ is not nil.*

Proof. If R has no nontrivial idempotents, then by (21.28) the same holds for $\bar{R} = R/I$. By the Wedderburn–Artin Theorem, this implies that $\bar{R}$ is a division ring, so R is a local ring. (A somewhat different proof of (1) for left or right artinian rings was given earlier in (19.19).) To prove the nontrivial part of (2), assume $\mathfrak{A}$ is not nil. Since I is nil, the image of $\mathfrak{A}$ in $\bar{R}$ is nonzero, and therefore contains a nonzero idempotent. Let $a \in \mathfrak{A}$ be such that $0 \neq \bar{a} = \bar{a}^2 \in \bar{R}$. By (21.28), there exists an idempotent $e \in aR \subseteq \mathfrak{A}$ such that $\bar{e} = \bar{a} \neq 0$. QED

In order to formulate the second sufficient condition for lifting idempotents, we recall a few facts about the completion of a ring with respect to an ideal. Let I be an ideal in a ring R. We have an inverse system of quotient rings:

$$R/I \twoheadleftarrow R/I^2 \twoheadleftarrow R/I^3 \twoheadleftarrow \cdots.$$

We write $\hat{R} (= \hat{R}_I)$ for the inverse limit $\varprojlim R/I^n$, and call $\hat{R}$ the completion of R with respect to I (or the I-adic completion). We say that R is *I-adically complete* if the natural map $i \colon R \to \hat{R}$ is an isomorphism. This amounts to the following two conditions.

(1) *Injectivity of* i: that is, $\bigcap_{n=1}^{\infty} I^n = (0)$.

(2) *Surjectivity of* i: that is, for any sequence $(a_1, a_2, \ldots)$ such that $a_{n+1} \equiv a_n \pmod{I^n}$ for every n, there exists an element $a \in R$ such that $a \equiv a_n \pmod{I^n}$ for all n.

In dealing with completions, we think of the elements of I^n for large n as being "very small." The sequence $(a_1, a_2, \ldots)$ in (2) above is then a Cauchy sequence, in that $a_m - a_n$ is very small for large m, n. Thus, condition (2) guarantees that the Cauchy sequence $(a_1, a_2, \ldots)$ has a limit a in R. Condition (1) guarantees the uniqueness of this limit, and is essentially a "Hausdorff" condition. If R is not yet I-adically complete, then the element a represented by $(a_1, a_2, \ldots)$ in $\varprojlim R/I^n$ will be a "formal limit" of the Cauchy sequence $(a_1, a_2, \ldots)$. In this case it is suggestive to write $a = \lim_{n \to \infty} a_n$. Thus, if $\bigcap_{n=0}^{\infty} I^n = (0)$, $\hat{R}$ is the topological completion of R in the usual sense if we regard $(R, +)$ as a topological group with $\{I^n : n \geq 0\}$ as a fundamental system of neighborhoods at 0.

(21.30) Remark. *I nilpotent $\Rightarrow$ R is I-adically complete $\Rightarrow$ $I \subseteq$ rad R.* The first implication is clear since, if I is nilpotent, every Cauchy sequence is eventually constant. For the second implication, it suffices to show that, if R is I-adically complete, then $b \in I \Rightarrow 1 - b \in U(R)$. The idea is that an inverse of $1 - b$ is given by $1 + b + b^2 + \cdots$: this series converges since its partial sums form a Cauchy sequence in R. We leave it to the reader to make this heuristic argument mathematically precise.

(21.31) Theorem. *Let I be an ideal in R such that R is I-adically complete. Then idempotents in R/I can be lifted to R.*

Proof. Let $a_1 \in R/I$ be an idempotent. Viewing R/I as $(R/I^2)/(I/I^2)$, we can lift a_1 to an idempotent $a_2 \in R/I^2$, since the ideal $I/I^2 \subseteq R/I^2$ has square zero. Proceeding in this way, we arrive at an element

$$a = (a_1, a_2, \ldots) \in \varprojlim R/I^n = \hat{R} = R,$$

where, for each n, a_n is an idempotent in R/I^n. Clearly then,

$$a^2 = (a_1^2, a_2^2, \ldots) = (a_1, a_2, \ldots) = a,$$

so $a \in R$ is an idempotent lifting $a_1 \in R/I$. QED

From this and (21.22), (21.25), (21.30), we deduce

(21.32) Corollary. *Let (R, I) be as above. Then an idempotent $e \in R$ is primitive in R iff $\bar{e}$ is primitive in R/I, and any countable set of pairwise orthogonal idempotents in R/I can be lifted to a similar set in R.*

How can we get some examples of noncommutative rings R which are I-adically complete with respect to some ideal I? One way to get such examples is through the use of analogous commutative objects. First let us prove the following useful lemma.

(21.33) Lemma. *Let k be a commutative noetherian ring which is I-adically complete with respect to an ideal $I \subseteq k$. Let M be a finitely generated k-module. Then M is I-adically complete in the sense that the natural map $i_M\colon M \to \varprojlim M/I^nM$ is an isomorphism.*

Proof. The kernel of i_M is $N = \bigcap_{n=1}^{\infty} I^nM$. By Krull's Intersection Theorem [Jacobson: 89] (p. 442), we have $I \cdot N = N$. Since N is finitely generated over k, Nakayama's Lemma implies that $N = 0$. It remains to show that i_M is onto. Fix a set of generators $\{m_1, \ldots, m_r\}$ for M and take any element

$$(a_1, a_2, \ldots) \in \varprojlim M/I^nM.$$

(For convenience, we think of the a_i's as elements of M.) Since $I^nM = \sum I^n m_j$, we can write

$$a_{n+1} - a_n = \sum_{j=1}^{r} \beta_{nj} m_j, \quad \text{where } \beta_{nj} \in I^n.$$

Write $a_1 = \sum_{j=1}^{r} \alpha_{1j} m_j$, with $\alpha_{1j} \in k$. Then

$$\begin{aligned} a_n &= a_1 + (a_2 - a_1) + \cdots + (a_n - a_{n-1}) \\ &= \sum_j \alpha_{1j} m_j + \sum_j \beta_{1j} m_j + \cdots + \sum_j \beta_{n-1,j} m_j \\ &= \sum_j \alpha_{nj} m_j, \end{aligned}$$

where $\alpha_{nj} = \alpha_{1j} + \beta_{1j} + \cdots + \beta_{n-1,j}$. Since $\beta_{n-1,j} \in I^{n-1}$, the α_{nj}'s converge to some $\alpha_j \in k$ when $n \to \infty$. Defining $a = \sum_j \alpha_j m_j$, we then have

$$a - a_n = \sum_j (\alpha_j - \alpha_{nj}) m_j \in I^nM \quad \text{for all } n,$$

so $i_M(a) = (a_1, a_2, \ldots)$, as desired. QED

(21.34) Proposition. *Let (k, I) be as in (21.33) and let R be a k-algebra which is finitely generated as a k-module. Then*

(1) *R is IR-adically complete, and idempotents of R/IR can be lifted to R.*

(2) *Assume k is semilocal and $I = \text{rad } k$. If R has no nontrivial idempotents (and $R \neq (0)$), then R is a local ring.*

Proof. (1) follows from (21.33) and (21.31). For (2), note that R/IR is finitely generated as a module over k/I. Under the assumptions in (2), k/I is a (commutative) artinian ring, so R/IR is also a (left and right) artinian ring. If R has no nontrivial idempotents, then by (1), R/IR also has no nontrivial idempotents. By (19.19), R/IR is a local ring. Since $IR \subseteq rad\ R$ by (5.9), this clearly implies that R itself is a local ring. QED

The considerations above lead to the following important result on the direct decompositions of finitely generated modules over algebras in a semi-local, complete setting.

(21.35) Theorem. *Let k be a commutative noetherian semilocal ring which is I-adically complete for $I = rad\ k$. Let R be a k-algebra which is finitely generated as a k-module. Then any finitely generated right R-module M has a Krull–Schmidt decomposition, i.e., $M = M_1 \oplus \cdots \oplus M_r$, where each M_i is an indecomposable R-submodule of M. Moreover, r is uniquely determined, and the sequence of isomorphism types of $M_1, \ldots, M_r$ is uniquely determined up to a permutation.*

Proof. The hypotheses imply that the R-submodules of M satisfy the ACC, so by (19.20), a Krull–Schmidt decomposition $M = M_1 \oplus \cdots \oplus M_r$ exists. Consider the k-algebras $E_i = End_R\ M_i$, which have no nontrivial idempotents. Since $End_k(M_i)$ is finitely generated as a k-module and k is noetherian, E_i is also finitely generated as a k-module. By (21.34)(2), E_i is a local ring, so M_i is *strongly* indecomposable for $1 \leq i \leq r$. The uniqueness part of (21.35) now follows from the Krull–Schmidt–Azumaya Theorem (19.21). QED

The result we just proved above is remarkable because in general, the Krull–Schmidt decompositions of finitely generated modules do not satisfy the uniqueness conclusion of (21.35) over noetherian local rings, as we have shown in an earlier example. Under the completeness assumption on k in (21.35), the uniqueness of Krull–Schmidt decompositions of finitely generated R-modules is restored. This is important for the theory of integral representations of finite groups since (21.35) can be applied to finitely generated modules over kG where G is any finite group and k is the completion of a ring of algebraic integer with respect to any prime ideal.

Exercises for §21

Ex. 21.1. Let e be an idempotent in a ring R. For any right R-module V, we can view Ve as a right eRe-module.
(1) Show that if $0 \to V' \to V \to V'' \to 0$ is an exact sequence of right R-modules, then $0 \to V'e \to Ve \to V''e \to 0$ is an exact sequence of right eRe-modules.
(2) If V_R is irreducible, show that Ve is either zero or is irreducible as an eRe-module.
(3) Show that for any irreducible right eRe-module W, there exists an irreducible right R-module V, unique up to isomorphism, such that $W \cong Ve$.

Ex. 21.2. Define a partial ordering on the set of all idempotents in R by: $e' \leq e$ iff $ee' = e'e = e'$. Call a nonzero idempotent e *minimal* if there is no idempotent strictly between 0 and e. Show that the minimal idempotents in this sense are precisely the primitive idempotents of R.

Ex. 21.2*. Describe the primitive idempotents in R if R is (1) the Boolean ring of all subsets of a nonempty set S, or (2) the ring $End(V_k)$ where V is a nonzero vector space over a division ring k.

Ex. 21.3. Let $e \in R$ be an idempotent, and $f = 1 - e$. Show that for any $r \in R$, $e' = e + erf$ is an idempotent. Writing $f' = 1 - e'$, show that $e = e' + e'sf'$ for some $s \in R$. (**Hint.** Note that $ee' = e$ and $e'e = e$. Try $s = -erf$.)

Ex. 21.4. For idempotents $e, e' \in R$, show that the following statements are equivalent:
(1) $eR = e'R$;
(2) $ee' = e'$ and $e'e = e$;
(3) $e' = e + er(1 - e)$ for some $r \in R$;
(4) $e' = eu$ where $u \in U(R)$;
(5) $R(1 - e) = R(1 - e')$.
If these conditions hold, show that $e' = u^{-1}eu$ for some $u \in U(R)$ (but not conversely). Also, show that these conditions *do not* imply $(1 - e)R = (1 - e')R$.

The next exercise shows that the situation is quite a bit simpler with a class of idempotents called "projections" in rings with involutions.

Ex. 21.4*. Let $(R, {}^*)$ be a ring with an involution *. (This means R is equipped with an additive endomorphism * such that $a^{**} = a$ and $(ab)^* = b^*a^*$ for all $a, b \in R$.) An idempotent $e \in R$ with $e = e^*$ is called a *projection*. For projections e, f in $(R, {}^*)$, show that
(1) $e \leq f$ in the sense of Exercise 2 iff $eR \subseteq fR$, iff $Re \subseteq Rf$.
(2) $e = f$ iff $eR = fR$.
(3) $eR = fR$ iff $Re = Rf$, iff $(1 - e)R = (1 - f)R$, iff $R(1 - e) = R(1 - f)$.

Ex. 21.5. (cf. Exercise 10.9) Let e be an idempotent in a semiprime ring R, and let $S = eRe$. Show that the following are equivalent:
(1) $(eR)_R$ is semisimple;
(2) ${}_S(eR)$ is semisimple;
(3) S is a semisimple ring;
(1)$'$ ${}_R(Re)$ is semisimple;
(2)$'$ $(Re)_S$ is semisimple.

Ex. 21.6. Show that in a von Neumann regular ring R, the intersection of any two principal left ideals $A, B \subseteq R$ is a principal left ideal.

Ex. 21.7. Let R be a von Neumann regular ring.
(1) Show that the center $Z(R)$ is also von Neumann regular.
(2) If R is indecomposable as a ring, then $Z(R)$ is a field.

Ex. 21.8. Show that an idempotent e in a von Neumann regular ring R is primitive iff e is right (resp. left) irreducible, iff eRe is a division ring. (**Hint.** Use (4.23) and (21.16), noting that a von Neumann regular ring is always semiprime.)

Ex. 21.9. Let $e = e^2 \in R$. Show that if R is semilocal (resp. von Neumann regular, unit-regular, strongly regular), so is $S := eRe$.

Ex. 21.10A. (McCoy's Lemma) An element $a \in R$ is said to be *regular* if $a = ara$ for some $r \in R$. Show that a is regular iff there exists $x \in R$ such that $axa - a$ is regular.

Ex. 21.10B. Using Ex. 9 and direct matrix computations (but not using Ex. (6.9) or Ex. (6.10)), show that, for $n \geq 1$, R is von Neumann regular iff $\mathbb{M}_n(R)$ is.

Ex. 21.10C. Let P be any finitely generated projective right module over a von Neumann regular ring A. Show that $End(P_A)$ is a von Neumann regular ring.

Ex. 21.11. For any idempotent $e \in R$, show that $End_R(eR/eJ) \cong eRe/eJe$, where $J = rad\ R$.

Ex. 21.12. Give an example of a nonzero ring in which 1 is not a sum of primitive idempotents. More generally, give an example of a nonzero ring which has *no* primitive idempotents.

Ex. 21.13. Recall that two idempotents $e, f \in R$ are *isomorphic* (written $e \cong f$, or, if necessary, $e \cong_R f$) if $eR \cong fR$ as right R-modules, or equivalently, if $Re \cong Rf$ as left R-modules (see (21.20)). Let S be a subring of R and let e, f be idempotents in S. Does $g \cong_S h$ imply $g \cong_R h$? How about the converse? What happens in the case when $S = gRg$ where $g = g^2$?

Ex. 21.14. Let e, e' be idempotents in R.
(1) If $e \cong e'$ and e is primitive, local, or right irreducible, show that so is e'.
(2) If $e \cong e'$, with $e = ab$, $e' = ba$ where $a, b \in R$, construct an explicit ring isomorphism from eRe to $e'Re'$ using a, b.
(3) Conversely, does $eRe \cong e'Re'$ imply $e \cong e'$?
(4) For any $u \in U(R)$, show that $e \cong u^{-1}eu$.

Ex. 21.15. Let $1 = e_1 + \cdots + e_r = e_1' + \cdots + e_r'$ be two decompositions of 1 into sums of orthogonal idempotents. If $e_i \cong e_i'$ for all i, show that there exists $u \in U(R)$ such that $e_i' = u^{-1}e_iu$ for all i.

Ex. 21.16. Let e, e' be idempotents in R, and $f = 1 - e$, $f' = 1 - e'$ be their complementary idempotents.
(1) Show that e and e' are conjugate in R iff $e \cong e'$ and $f \cong f'$.
(2) If eRe is a semilocal ring, show that e and e' are conjugate in R iff $e \cong e'$.
(3) Is (2) still true if eRe is not assumed to be semilocal?

Ex. 21.16*. (Ehrlich, Handelman) Let R be a von Neumann regular ring. Show that R is unit-regular (in the sense of Exercise 4.14B) iff, for any two idempotents $e, e' \in R$, $e \cong e'$ implies $1 - e \cong 1 - e'$.

Ex. 21.17. Let $1 = e_1 + \cdots + e_r = e'_1 + \cdots + e'_s$ be two decompositions of 1 into sums of orthogonal local idempotents. Show that $r = s$ and that there exists $u \in U(R)$ such that $e'_{\pi(i)} = u^{-1}e_i u$ for all i, where π is a suitable permutation of $\{1, 2, \ldots, r\}$.

Ex. 21.18. Let $e_1, \ldots, e_r$ be idempotents in R which are pairwise orthogonal and isomorphic. For $e = e_1 + \cdots + e_r$, show that $eRe \cong \mathbb{M}_r(e_i R e_i)$ for any i.

Ex. 21.19. Let A be a ring which has no nontrivial idempotents. Let $\{E_{ij}\}$ be the matrix units in $R = \mathbb{M}_n(A)$. True or False: Every idempotent in R is conjugate to $E_{11} + E_{22} + \cdots + E_{ii}$ for some $i \leq n$? (**Hint.** An idempotent in R gives rise to a direct decomposition of $(A^n)_A$ into two projective right A-modules. On the other hand, for $e = E_{11} + \cdots + E_{ii}$, eA^n is isomorphic to $(A^i)_A$ as a right A-module.)

Ex. 21.20. Let J be an ideal in R which contains no nonzero idempotents (e.g. $J \subseteq rad\ R$). Let e, f be commuting idempotents in R.
(1) If $\bar{e} = \bar{f}$ in R/J, show that $e = f$ in R.
(2) If $\bar{e}, \bar{f}$ are orthogonal in R/J, show that e, f are othogonal in R.
Is any of these results true if e, f do not commute? (**Hint.** For (1), note that $e - ef$ and $f - ef$ are idempotents in J.)

Ex. 21.21. Let R be a semilocal ring whose radical is nil. Let I be a right ideal of R. Show that I is indecomposable (as a right R-module) and nonnil iff $I = eR$ where e is a primitive idempotent. (**Hint.** Use (21.29)(2).)

Ex. 21.22. Let R be as in Exercise 21. Show that a nonzero idempotent $e \in R$ is primitive iff every right ideal properly contained in eR is nil. (**Hint.** For the "if" part, assume $eR = \mathfrak{A} \oplus \mathfrak{B}$ where $\mathfrak{A}$, $\mathfrak{B}$ are nonzero right ideals. Then $\mathfrak{A}$, $\mathfrak{B}$ are nil and so $e \in \mathfrak{A} + \mathfrak{B} \subseteq rad\ R$, a contradiction. For the "only if" part, use (21.29)(2) again.)

Ex. 21.23. (Asano) Let R be a ring for which $J = rad\ R$ is nil and $\bar{R} = R/J$ is unit-regular (in the sense of Exercise 4.14B). Show that any nonunit $a \in R$ is a left (resp. right) 0-divisor in R. (In particular, "left 0-divisor" and "right 0-divisor" are both synonymous with "nonunit" in R. In the terminology of Exercise 4.16, ${}_RR$ and R_R are both cohopfian (and hence also hopfian).)

Ex. 21.24. Let M_k be a module of finite length n over a ring k, and let $R = \text{End}(M_k)$, $J = rad\ R$. Show that R is a semilocal ring with $J^n = 0$. (In particular, R is a semiprimary ring.) (**Hint.** Use (19.17) and (21.18) to show that R is semilocal. Then use Fitting's Lemma to show that $f \in J \Longrightarrow f^n = 0$. The nilpotency conclusion $J^n = 0$ is deeper. With the help of Ex. 3.24(2), show first that any nil multiplicative set $S \subseteq R$ is nilpotent.)

Ex. 21.25. Let R be any right artinian ring, and let $C = Z(R)$.
(1) Show that C is a semiprimary ring.
(2) Deduce from (1) that $C \cap rad\ R = rad\ C$.

Ex. 21.26. Let a be a nonsquare element in a field F of characteristic not 2, and let A be a commutative F-algebra with basis $\{1, x, y, xy\}$ such that $x^2 = y^2 = a$. Find the primitive idempotents in A, and show that $A \cong F(\sqrt{a}) \times F(\sqrt{a})$.

Ex. 21.27. Let A be a "look-alike" quaternion algebra over a field F of characteristic 2, i.e. $A = F1 \oplus Fi \oplus Fj \oplus Fk$, where $i^2 = j^2 = -1$ and $k = ij = -ji$. (Of course, $-1 = 1 \in A$, since char $F = 2$.) What are the primitive idempotents in A, and what kind of ring is A?

Ex. 21.28. (Bass) Let G be an abelian group and H be its torsion subgroup. For any commutative ring k, show that any idempotent e of kG belongs to kH. (**Hint.** Reduce to the case when G is free of rank 1, say $\langle t \rangle$. Show that $e \in kG = k[t, t^{-1}]$ is congruent to an idempotent $e_0 \in k$ modulo $N[t, t^{-1}]$ where $N = Nil(k)$. Then $e - ee_0$ and $e_0 - ee_0$ are both nilpotent idempotents, and so $e = ee_0 = e_0$.)

Ex. 21.29. (Bergman) Let A be the real coordinate ring of the 2-sphere S^2, i.e. $A = \mathbb{R}[x, y, z]$ with the relation $x^2 + y^2 + z^2 = 1$. Let σ be the $\mathbb{R}$-automorphism of A defined by

$$\sigma(x) = -x, \quad \sigma(y) = -y, \quad \text{and} \quad \sigma(z) = z.$$

Let $R = A \oplus Ar$, where $r^2 = 1$ and $rh = \sigma(h)r$ for every $h \in A$. Show that for the idempotent $e_0 = (1 - r)/2$ in the ring R, $R/Re_0R \cong \mathbb{R} \times \mathbb{R}$, but the two nontrivial idempotents of $\mathbb{R} \times \mathbb{R}$ cannot be lifted to R.

Ex. 21.30. (Stanley) Let e, f, e', f' be idempotents in a ring R such that $e - e', f - f' \in J$, where J is an ideal in R such that $\bigcap_{n=1}^{\infty} J^n = 0$. Show that $eRf = 0$ iff $e'Rf' = 0$. (**Hint.** It suffices to show that $eRf = 0 \Longrightarrow e'Rf = 0$. With $e' = e + a$ where $a \in J$, $e'^2 = e'$ amounts to $a = ae + ea + a^2$. From this, show inductively that $eRf = 0 \Longrightarrow e'Rf \subseteq a^n Rf \subseteq J^n$ for all $n \geq 1$.)

§22. Central Idempotents and Block Decompositions

In this section, we shall study the block decompositions of rings through the use of central idempotents. For an idempotent e in a ring R, recall that e is central iff $eRf = fRe = 0$, where $f = 1 - e$ is the complementary idempotent of e (see (21.5)). If e is indeed central, then in the Peirce decomposition $R = eR \oplus fR$, both summands are ideals of R. Viewing eR and fR as rings (with identities e and f), we have then a ring isomorphism $R \cong eR \times fR$.

Conversely, if $R = \mathfrak{A} \oplus \mathfrak{B}$ where $\mathfrak{A}, \mathfrak{B}$ are ideals, then, decomposing 1 into $e + f$ where $e \in \mathfrak{A}$ and $f \in \mathfrak{B}$, we see easily that e, f are *central* idempotents, and $\mathfrak{A} = eR$, $\mathfrak{B} = fR$ (see Exer. 1.8). We say that a ring R ($\neq 0$) is *indecomposable* if R is not a direct sum of two nonzero ideals. By the foregoing discussion, this is the case iff R has no nontrivial central idempotents.

Let $c \in R$ be a central idempotent. If $c = \alpha + \beta$ is any decomposition of c into orthogonal central idempotents $\alpha, \beta \in R$, then

$$\alpha = (\alpha + \beta)\alpha = c\alpha \in cR,$$

and similarly $\beta \in cR$. Thus, the decomposition $c = \alpha + \beta$ already occurs in the ring cR. We say that c is *centrally primitive* in R if $c \neq 0$ and c cannot be written as a sum of two nonzero orthogonal central idempotents in R. By the discussion above, this amounts to the condition that cR be indecomposable as a ring (or as an ideal in R).

(22.1) Proposition. *Suppose there exists a decomposition of* $1 \in R$ *into a sum of orthogonal centrally primitive idempotents, say* $1 = c_1 + \cdots + c_r$. *Then*

(1) *any central idempotent* $c \in R$ *is a sum of a subset of* $\{c_1, \ldots, c_r\}$.

(2) $c_1, \ldots, c_r$ *are the only centrally primitive idempotents in* R; *in particular, any two different centrally primitive idempotents in* R *are orthogonal.*

(3) *The decomposition* $1 = c_1 + \cdots + c_r$ *is unique up to a permutation of the summands.*

Proof. (1) If $cc_i \neq 0$, then $cc_i = c_i$ since c_i is the only nonzero central idempotent in c_iR. Therefore, we have

$$c = c(c_1 + \cdots + c_r) = \sum c_i$$

where the summation is over all i such that $cc_i \neq 0$. This proves (1), from which the other conclusions follow immediately. (Note that (3) was also established earlier in (3.8).) QED

Clearly, a ring R can be expressed as a finite direct product of indecomposable rings iff $1 \in R$ can be written as a sum of orthogonal centrally primitive idempotents. In this case, all conclusions of (22.1) apply, and we have

$$R = c_1R \oplus \cdots \oplus c_rR$$

in the notation there. We shall call this the *block decomposition* of R, and call each c_iR a *block* of R. In general, of course, such a block decomposition may not exist. For instance, in the ring $R = \mathbb{Q} \times \mathbb{Q} \times \cdots$, there are infinitely many (centrally) primitive idempotents, so by (22.1), the element 1 *cannot* be written as a sum of (finitely many) orthogonal primitive idempotents in R.

The trouble here is that R fails to satisfy the usual chain conditions. If we impose a suitable condition on the ideals of a ring, then the existence of a block decomposition is guaranteed, as the following result shows.

(22.2) Proposition. *Let R be a ring whose ideals satisfy either the ACC or the DCC (e.g., R is a right or left noetherian ring). Then R has a block decomposition, and all conclusions of* (22.1) *are valid for R.*

Proof. This follows by repeating the argument in the proof of (19.20) for the ideals of R. QED

In order to study the existence of block decompositions in more general circumstances, we shall now introduce certain binary relations on the primitive idempotents of a ring. For a given ring $R \neq 0$, let E be its set of primitive idempotents. For $e, e' \in E$, we define $e \sim e'$ to mean that there exists an $f \in E$ such that $eRf \neq 0 \neq e'Rf$. Recalling from (21.6) that $Hom_R(fR, eR) \cong eRf$, we see that $e \sim e'$ amounts to the fact that there exist nonzero R-homomorphisms

$$fR \longrightarrow eR, \quad fR \longrightarrow e'R$$

for a suitable $f \in E$. In particular, if $e, e' \in E$ are isomorphic (cf. (21.20)), then $e \sim e'$. Also, if $e, f \in E$ and $eRf \neq 0$, then $e \sim f$ (since $fRf \neq 0$).

Since the relation "$\sim$" on E is clearly reflexive and symmetric, it is useful to look at the equivalence relation which it generates. Denoting this equivalence relation by "$\approx$", we have then $e \approx e'$ iff

$$e \sim e_1 \sim e_2 \sim \cdots \sim e_m \sim e'$$

for a sequence of idempotents $e_1, \ldots, e_m \in E$. In the sequel, we shall say that $e, e' \in E$ are *linked* if $e \approx e'$. The following observation is useful in studying the linkage relation on E.

(22.3) Lemma. *Let $e \approx e'$ in E, and let c be a central idempotent in R. Then $e \in cR$ iff $e' \in cR$.*

Proof. First we observe that

(22.4) *For any $e \in E$, we have either $e \in cR$ or $e \in (1-c)R$.*

In fact, from the decomposition $e = ce + (1-c)e$, it follows that either $ce = 0$ or $(1-c)e = 0$, since e is a primitive idempotent (and ce, $(1-c)e$ are orthogonal idempotents). If $ce = 0$, then

$$e = (1-c)e \in (1-c)R,$$

and if $(1-c)e = 0$, then $e = ce \in cR$. This proves (22.4). For the Lemma, it is enough to prove the "only if" part, for which we may assume that $e \sim e'$.

Fix an idempotent $f \in E$ such that $eRf \neq 0 \neq e'Rf$. If $e \in cR$, then

$$0 \neq eRf = ceRf = eR(cf)$$

implies that $cf \neq 0$, so by the foregoing, $f \in cR$. But then

$$0 \neq e'Rf = e'R(cf) = ce'Rf$$

implies that $ce' \neq 0$, so $e' \in cR$. QED

Our next result gives a useful sufficient condition for the existence of block decompositions of a ring, and describes the linkage equivalence classes of E in terms of block decompositions.

(22.5) Theorem. *Assume that $1 \in R$ can be written as $e_1 + \cdots + e_n$ where the e_i's are orthogonal primitive idempotents. Then $1 \in R$ can be written as a sum of orthogonal centrally primitive idempotents (so a block decomposition exists for R). Two primitive idempotents $e, e' \in E$ are linked iff they belong to the same block.*

Proof. The e_i's are distinct elements in E, and "$\approx$" induces an equivalence relation on $\{e_1, \ldots, e_n\}$. Thus, we have a partition of this set into equivalence classes. Let $c_1, \ldots, c_r$ be the various class sums. These are orthogonal idempotents with sum 1. Further, from the definition of "$\approx$", we see that $c_iRc_j = 0$ for $i \neq j$. Thus, for any $a \in R$,

$$c_ia = c_ia(c_1 + \cdots + c_r) = c_iac_i = (c_1 + \cdots + c_r)ac_i = ac_i,$$

so each c_i is central. Next we claim that c_i *is centrally primitive*. For this, it suffices to show that if c is any nonzero central idempotent of c_iR, then $c = c_i$. Say $c_i = e_{i_1} + \cdots + e_{i_m}$, where $\{e_{i_1}, \ldots, e_{i_m}\}$ is an equivalence class. From

$$0 \neq c = cc_i = c(e_{i_1} + \cdots + e_{i_m}),$$

we have, say, $ce_{i_1} \neq 0$. Since c is central in R, this means that $e_{i_1} \in cR$, and so from (22.3), $e_{i_j} \in cR$ for all j. Thus

$$c = c(e_{i_1} + \cdots + e_{i_m}) = e_{i_1} + \cdots + e_{i_m} = c_i,$$

as claimed. Now consider any $e \in E$. By repeated use of (22.4), we see that e belongs to a unique block $R_i := c_iR$. Then

$$0 \neq eR = ec_iR = eR_i = eR_ie_{i_1} + \cdots + eR_ie_{i_m}$$

implies that some $eRe_{i_j} \neq 0$. As observed before, this gives $e \sim e_{i_j}$, and so $e \approx e_{i_1}$. Conversely, by (22.3), any primitive idempotent $e \approx e_{i_1}$ belongs to the block $c_iR = R_i$. This proves the last conclusion in the Theorem. QED

In the classical case of right artinian rings, the decomposition and linkage theory developed above can be described quite a bit more explicitly. For later reference, we record the following consequence of (22.5).

(22.6) Theorem. *Let R be a right artinian ring. Then R has a (unique) block decomposition $R = R_1 \oplus \cdots \oplus R_r$. For primitive idempotents $e, e' \in E$, we have $e \sim e'$ iff eR and $e'R$ have a common composition factor. Thus, two primitive idempotents $e, e' \in E$ belong to the same block iff there exist $e_1, \ldots, e_m \in E$ with $e_1 = e$, $e_m = e'$ such that for any $i < m$, e_iR and $e_{i+1}R$ have a common composition factor.*

Proof. Since R_R has a Krull–Schmidt decomposition, the hypothesis of (22.5) is satisfied. Therefore R has a block decomposition. Now recall (from (19.17)) that any idempotent $f \in E$ is local, so fR/fJ is a simple right R-module for $J = rad\ R$. Conversely, if V is any simple right R-module, then $V \cong fR/fJ$ for some $f \in E$. (To see this, pick an idempotent $x \in \bar{R} = R/J$ such that $V \cong x\bar{R}$. Lifting x to an idempotent $f \in R$, we see from (21.18) that $f \in E$ and $V \cong \bar{f}\bar{R} \cong fR/fJ$.) Now consider $e, e' \in E$. By definition, $e \sim e'$ means that there exists $f \in E$ with

$$eRf \neq 0 \neq e'Rf.$$

By (21.19), this means that eR and $e'R$ both have fR/fJ as a composition factor. This gives the new characterization of " $\sim$ " in (22.6), and the last conclusion of (22.6) follows from (22.5). QED

In the special case when R is a semisimple ring, (22.6) implies, in particular, that $e \sim e'$ iff $e \cong e'$. Thus, $e, e' \in E$ belong to the same block iff $e \cong e'$. The blocks of R are, of course, the simple components of R.

If the identity of a ring R can be decomposed into a sum of orthogonal centrally primitive idempotents, say $1 = c_1 + \cdots + c_r$, then R decomposes into blocks $R_1 \oplus \cdots \oplus R_r$ where $R_i = c_iR$, and likewise the center C of R decomposes into blocks $C_1 \oplus \cdots \oplus C_r$ where $C_i = c_iC =$ center of R_i. The distribution of the primitive idempotents of R into the blocks $R_1, \ldots, R_r$ can often be analyzed by using the properties of C. We shall make this explicit by treating below the classical case of finite-dimensional algebras over an algebraically closed field.

(22.7) Theorem. *Let R be a finite-dimensional algebra over an algebraically closed field k, and let*

$$R = R_1 \oplus \cdots \oplus R_r, \quad C = C_1 \oplus \cdots \oplus C_r$$

be as above. For any primitive idempotent $e \in E$, the action of an element $c \in C$ on the simple module $eR/e \cdot rad\ R$ is that of multiplication by a scalar $\lambda_e(c) \in k$. The map $\lambda_e \colon C \to k$ is a k-algebra homomorphism, and $e, e' \in E$ belong to the same block iff $\lambda_e = \lambda_{e'}$. Moreover, any k-algebra homomorphism $C \to k$ is of the form λ_e for some $e \in E$.

Proof. Since k is algebraically closed, Schur's Lemma implies that

$$End_R(eR/e \cdot rad\ R) = k.$$

Thus, we have the map $\lambda_e\colon C \to k$ defined in the Theorem, and λ_e is clearly a k-algebra homomorphism. Consider a block C_i of C. Since C_i has no nontrivial idempotents, it is a local k-algebra (by (19.19)), with $C_i/\mathit{rad}\ C_i = k$. Thus, C_i has a unique k-algebra homomorphism into k, say λ_i. Now suppose $e \in R_i$. Then each C_j $(j \neq i)$ acts as zero on $eR/e \cdot \mathit{rad}\ R$, so $\lambda_e \mid C_j = 0$, and we must have $\lambda_e \mid C_i = \lambda_i$. The remaining conclusions of the Theorem now follow immediately. QED

Next we shall discuss the problem of lifting central idempotents from a homomorphic image of R to R itself. If I is an ideal of R, a central idempotent of R/I may not lift to a central idempotent in R. For instance, in the ring R of $n \times n$ upper triangular matrices over a field k, the quotient

$$R/\mathit{rad}\ R \cong k \times \cdots \times k \quad (n \text{ copies})$$

has 2^n central idempotents. However, there are no nontrivial central idempotents in R, as one can easily show (Exercise 1). Thus, although idempotents in $R/\mathit{rad}\ R$ can be lifted to R since $\mathit{rad}\ R$ is nilpotent, the nontrivial central idempotents in $R/\mathit{rad}\ R$ cannot be lifted to *central* idempotents in R.

The fact that central idempotents in $R/\mathit{rad}\ R$ may not lift to central idempotents in R is a serious complicating factor in the study of the structure theory of right artinian rings R. By the Wedderburn–Artin Theorem, $R/\mathit{rad}\ R$ splits into a product of (say) n simple artinian rings; however, we may not be able to lift such a splitting. By (22.6), nevertheless, we can decompose R into blocks, say

$$R = R_1 \oplus \cdots \oplus R_r,$$

where clearly $r \leq n$. If n happens to be 1, then of course r is also 1; in this case, the structure of R can be completely determined, as we shall see later (cf. (23.10)) in a more general context. The crucial case is when $r = 1$ and $n > 1$, as in the example of upper triangular matrices. There is no satisfactory classification in this case, but more information will be given later.

We conclude this section by pointing out a few cases in which central idempotents in a quotient ring of R can be lifted to central idempotents in R. First we prove the following curious fact.

(22.8) Lemma. *Let I be an ideal of a ring R such that $\bigcap_{n=1}^{\infty} I^n = 0$. Then an idempotent e is central in R iff its image $\bar{e}$ is central in $\bar{R} := R/I^2$.*

Proof. ("if" part) Assume $\bar{e}$ is central in $\bar{R}$ and let $f = 1 - e$. *We claim that $eRf \subseteq I^n$ for all $n \geq 2$.* If this is true, then we have $eRf = 0$ and similarly $fRe = 0$, so by (21.5), e is central in R. To prove the claim, we proceed by induction on n. Since $\bar{e}$ is central in $\bar{R}$, we have $eI \subseteq Ie + I^2$ and $eRf \subseteq I^2$. Inductively, if $eRf \subseteq I^n$ for some $n \geq 2$, then

$$\begin{aligned} eRf &\subseteq eI^n f \subseteq (Ie + I^2)I^{n-1}f \\ &\subseteq I \cdot eRf + I^{n+1} = I^{n+1}. \qquad \text{QED} \end{aligned}$$

This leads to the following nice result.

(22.9) Theorem. *Let R be a ring and $I \subseteq \operatorname{rad} R$ be a nilpotent ideal. Then the map $e \mapsto \bar{e}$ defines a one-one correspondence between central idempotents of R and those of $\bar{R} = R/I^2$. Further, e is centrally primitive in R iff $\bar{e}$ is centrally primitive in $\bar{R}$. In particular, R is indecomposable iff $\bar{R}$ is.*

Proof. Let x be any central idempotent in $\bar{R}$. Since I^2 is nilpotent, x can be lifted to an idempotent $e \in R$ by (21.28). The nilpotency of I implies that $\bigcap_{n=1}^{\infty} I^n = 0$, so by (22.8), e is central in R. Since I^2 contains no nonzero idempotents of R (see (21.23)), e is the *only* (central) idempotent of R lifting x by Exercise 21.20. This establishes the one-one correspondence in the Theorem. If e has a decomposition into $e_1 + e_2$ where e_1, e_2 are nonzero orthogonal central idempotents in R, then $\bar{e} = \bar{e}_1 + \bar{e}_2$ where $\bar{e}_1, \bar{e}_2$ are nonzero orthogonal central idempotents in $\bar{R}$. Conversely, if $\bar{e} = x_1 + x_2$ where x_1, x_2 are nonzero orthogonal central idempotents in $\bar{R}$, let e_1, e_2 be the (unique) central idempotents in R lifting x_1, x_2. Then $e_1 e_2$ is an idempotent in I^2, so $e_1 e_2 = 0$. Moreover, since e and $e_1 + e_2$ both lift $x = x_1 + x_2$, we have $e = e_1 + e_2$. This shows that e is centrally primitive in R iff $\bar{e}$ is centrally primitive in $\bar{R}$. QED

For right artinian rings R, (22.9) implies in particular that *the blocks of R are in a natural one-one correspondence with those of R/I^2 where $I = \operatorname{rad} R$.* Since $\operatorname{rad}(R/I^2) = I/I^2$, this suggests that the most basic case for considering block decompositions is the case of right artinian rings whose radicals have square zero. Often, this basic case turns out to hold the key in dealing with the general structure theory of right artinian rings.

Another case where we have a positive result on the lifting of central idempotents is based on the following interesting observation.

(22.10) Dade's Lemma. *Let k be a commutative ring and $I \subseteq \operatorname{rad} k$ be an ideal of k. Let R be a k-algebra which is finitely generated as a k-module. Then an idempotent $e \in R$ is central in R iff its image $\bar{e}$ is central in $\bar{R} = R/IR$.*

Proof. ("if" part) Let $f = 1 - e$ and $\bar{f} = 1 - \bar{e}$. Then $\bar{e}\bar{R}\bar{f} = 0$ (since we assume that $\bar{e}$ is central), so $eRf \subseteq IR$. By the Peirce decomposition (21.3), eRf is a k-direct summand of R. Thus $eRf \subseteq IR$ implies that $eRf = I \cdot eRf$. As a k-direct summand of R, eRf is finitely generated as a k-module. Therefore, by Nakayama's Lemma, $eRf = 0$. Similarly, $fRe = 0$. By (21.5), e is central in R. QED

(22.11) Theorem. *Let k be a commutative noetherian ring that is I-adically complete with respect to an ideal $I \subseteq k$. Let R be a k-algebra that is finitely generated as a k-module. Then the map $e \mapsto \bar{e}$ defines a one-one correspondence between central idempotents of R and those of $\bar{R} = R/IR$. Further, e is centrally primitive in R iff $\bar{e}$ is centrally primitive in $\bar{R}$. In particular, R is indecomposable iff $\bar{R}$ is.*

Proof. The proof of (22.9) carries over *verbatim* once we observe the following two facts. First, by (5.9),

$$IR \subseteq (rad\ k)R \subseteq rad\ R,$$

so IR contains no nonzero idempotents of R. Secondly, by (21.34)(1), any central idempotent of R/IR can be lifted to an idempotent of R, and hence to a *central* idempotent of R by (22.10). QED

Note that, under the hypotheses of the Theorem, R above is right (and left) noetherian, so by (22.2), R and $\bar{R} = R/IR$ both admit (unique) block decompositions. By Theorem (22.11), *there is a one-one correspondence between the blocks of R and those of $\bar{R}$.* For instance, if $(k, \mathfrak{m})$ is a commutative noetherian local ring that is $\mathfrak{m}$-adically complete and G is a finite group, then there is a one-one correspondence between the blocks of kG and those of $\bar{k}G$, where $\bar{k}$ is the residue field $k/\mathfrak{m}$ of the local ring k. This is a very useful fact in the integral and modular representation theory of finite groups.

Exercises for §22

Ex. 22.1. Show that the ring R of $n \times n$ upper triangular matrices over any indecomposable ring k is indecomposable.

Ex. 22.2. For two central idempotents e, f in a ring R, show that $e \cong f$ *iff* $e = f$. Using this fact and Exercise 21.16*, show that a strongly regular ring must be unit-regular (a fact proved earlier in Exercise 12.6C). (**Hint.** If $e \cong f$, write $e = ab$, $f = ba$. Then $f = f^2 = b(ab)a = bea = ef$, and similarly $e = fe$.)

Ex. 22.3A. For $e = e^2 \in R$, show that the following are equivalent:
(1) $e \in Z(R)$,
(2) $eR = Re$,
(3) e commutes with all the idempotents of R that are isomorphic to e.
(**Hint.** For (3) $\Rightarrow$ (1), note that, for $r \in R$ and $f = 1 - e$, $e + erf$ and $f + fre$ are both idempotents isomorphic to e: see Ex. 21.4.)

Ex. 22.3B. Let $S = eR$ where $e = e^2 \in R$. Suppose S is an ideal of R not containing any nonzero nilpotent ideal. Show that $e \in Z(R)$, and conclude that R is a direct product of the semiprime ring S and the ring $(1 - e)R$.

Ex. 22.3C. Let S be an ideal in a ring R such that S_R is an artinian R-module and S contains no nonzero nilpotent ideals of R. Show that S is a semisimple ring with an identity e, and that R is the direct product of the ring S with the ring $(1 - e)R$.

Ex. 22.4A. A ring R is said to be *right duo* (resp. *left duo*) if every right (resp. left) ideal in R is an ideal. Show that in a right duo ring R, all idempotents are central. Is every right duo ring also left duo?

Ex. 22.4B. Show that for any ring R, the following are equivalent:
(1) R is strongly regular;
(2) R is von Neumann regular and right duo;
(3) $I \cap J = IJ$ for any left ideal I and any right ideal J;
(4) $I \cap J = IJ$ for any right ideals I, J;
(5) $aR \cap bR = aRbR$ for all $a, b \in R$.

Ex. 22.5. Let S be the set of all central idempotents in a ring R. Define an addition $\oplus$ in S by

$$e \oplus e' = e + e' - 2ee' = (e - e')^2,$$

and define multiplication in S by the multiplication in R. Show that $(S, \oplus, \times)$ is a Boolean ring, i.e. a ring in which all elements are idempotents.

Ex. 22.6. For any ring R, let $S = B(R)$ be the Boolean ring of central idempotents in R, as defined in Exercise 5. For any central idempotent $e \in R$, show that e is centrally primitive in R *iff* e is (centrally) primitive in S.

CHAPTER 8

Perfect and Semiperfect Rings

This chapter will be devoted to the study of two classes of rings, namely, semiperfect rings and left (resp., right) perfect rings. The notion of semiperfect rings is left-right symmetric, while left (resp., right) perfect rings are always semiperfect.

Both of these notions are generalizations of that of one-sided artinian rings. More precisely, they are generalizations of the notion of semiprimary rings defined in (4.15). Recall that a ring R is semiprimary if $R/\mathit{rad}\, R$ is semisimple and $\mathit{rad}\, R$ is nilpotent. The definitions of right perfect and semiperfect rings are the result of an attempt to weaken the nilpotency assumption on *rad R*. As it turns out, right perfect rings can also be characterized by some sort of descending chain condition, namely, the *DCC* on left (not right!) principal ideals. On the other hand, semiperfect rings form a class encompassing both one-sided perfect rings and local rings. All of the above types of rings are semilocal rings, so we have the following inclusion relationships:

$$\begin{array}{c}\{\text{one-sided artinian rings}\}\\ \cap\\ \{\text{semiprimary rings}\}\\ \cap\\ \{\text{right perfect rings}\}\\ \cap\\ \{\text{local rings}\} \subset \{\text{semiperfect rings}\} \subset \{\text{semilocal rings}\}.\end{array}$$

Classically, there was a rich and very well-developed theory of modules over one-sided artinian rings. In the early 1960's, a part of this theory was extended by H. Bass [60] to the wider class of semiperfect rings. However, the passage from 1-sided artinian rings to semiperfect rings is not just a blind generalization. Semiperfect rings turn out to be a significant class of rings from the viewpoint of homological algebra, since they are precisely the rings whose finitely generated (left or right) modules have projective covers. In the same vein, Bass's right perfect rings are precisely the rings all of whose right flat modules are projective. These interesting module-theoretic characterizations led to many more applications of homological methods in ring theory, and helped establish the notions of perfect and semiperfect rings firmly in the literature. For a recent survey on this subject, see Lam [99].

In the first section (§23) of this chapter, we develop the definition and basic properties of perfect and semiperfect rings, using heavily the methods of idempotents in the last chapter. The next section, §24, introduces the homological methods and develops the homological characterizations of perfect and semiperfect rings mentioned above. The last section, §25, gives the decomposition theory of semiperfect rings into blocks, and introduces the notion of basic rings. All of this, together with the classification of finitely generated projective modules, the theory of the Cartan matrix, etc., are a part of the classical theory over artinian rings, but generalize easily to semiperfect rings.

§23. Perfect and Semiperfect Rings

Recalling that a ring R is semilocal if $R/rad\ R$ is semisimple, we introduce the notion of a semiperfect ring as follows.

(23.1) Definition. *A ring R is called semiperfect if R is semilocal, and idempotents of $R/rad\ R$ can be lifted to R.*

If R is left or right artinian, then R is semilocal, and since $rad\ R$ is nilpotent, idempotents of $R/rad\ R$ can be lifted to R. Therefore, R is semiperfect. On the other hand, any local ring R' is also semiperfect, since $R'/rad\ R'$ is a division ring, which has only trivial idempotents. Therefore, semiperfect rings may be viewed as a common generalization of local rings and left/right-artinian rings. Note that semiperfect rings form a proper subclass of the class of semilocal rings. For instance, if R is a commutative semilocal domain with two maximal ideals $\mathfrak{m}_1, \mathfrak{m}_2$, then

$$R/rad\ R \cong R/\mathfrak{m}_1 \times R/\mathfrak{m}_2$$

has two nontrivial idempotents, and these do not lift to R since R has no nontrivial idempotents.

(23.2) Example. *Let k be any local ring. Then $R = \mathbb{M}_n(k)$ is semiperfect.* To see this, first note that

$$R/rad\ R = \mathbb{M}_n(k)/\mathbb{M}_n(rad\ k) \cong \mathbb{M}_n(\bar{k}),$$

where $\bar{k}$ is the division ring $k/rad\ k$. Therefore, $R/rad\ R$ is a simple artinian ring. It remains to show that any idempotent x of $\mathbb{M}_n(\bar{k})$ lifts to $\mathbb{M}_n(k)$. Since x corresponds to a projection of the vector space $\bar{k}^n$ to some subspace, there is an invertible matrix $y \in \mathbb{M}_n(\bar{k})$ such that

$$yxy^{-1} = diag(1, \ldots, 1, 0, \ldots, 0).$$

Let $u \in \mathbb{M}_n(k) = R$ be a lift of y. Then u is automatically a unit of R, and

$$u^{-1}\, diag(1, \ldots, 1, 0, \ldots, 0)\, u$$

is an idempotent of R lifting x. This shows that R is semiperfect. (More generally, we shall show in (23.9) below that, for any semiperfect k, $\mathbb{M}_n(k)$ is semiperfect.)

(23.3) Example. Let k be a commutative noetherian semilocal ring that is I-adically complete for $I = rad\ k$. *Then any k-algebra which is finitely generated as a k-module is semiperfect.* To see this, first note that, by (20.6), R is semilocal and $(rad\ R)/IR$ is a nilpotent ideal in R/IR. By (21.28) any idempotent of $R/rad\ R$ can be lifted to an idempotent in R/IR, and by (21.34), the latter can be lifted to an idempotent in R. This shows that R is semiperfect. As explicit examples, we can take $R = kG$ where G is any finite group, or $R =$ any k-subalgebra of $End_k\ M$, where M is any finitely generated k-module.

(23.4) Example. It is easy to see that a finite direct product of semiperfect rings is semiperfect. Thus, for instance, the direct product of a local ring and a left artinian ring is semiperfect.

One of the most important features of a semiperfect ring is given by the following.

(23.5) Proposition. *In a semiperfect ring R, any primitive idempotent e is local.*

Proof. Since idempotents of $\bar{R} = R/rad\ R$ can be lifted to R, the primitive idempotent e maps to a primitive idempotent $\bar{e}$ of $\bar{R}$. But $\bar{R}$ is semisimple, so $\bar{e}$ is (say left) irreducible. Now (21.18) implies that e is local. QED

This result leads to our first characterization of semiperfect rings.

(23.6) Theorem. *A ring R is semiperfect iff the identity element 1 can be decomposed into $e_1 + \cdots + e_n$, where the e_i's are mutually orthogonal local idempotents.*

Proof. Assume first that R is semiperfect. Then in $\bar{R} = R/rad\ R$, we have a decomposition $\bar{1} = x_1 + \cdots + x_n$ where $x_i \in \bar{R}$ are mutually orthogonal primitive idempotents. (This corresponds to a decomposition of $\bar{R}$ into a direct sum of minimal left ideals.) Let $e_1, \ldots, e_n$ be orthogonal idempotents of R lifting $x_1, \ldots, x_n$. Then (as in the last Proposition) the e_i's are local, and $e := e_1 + \cdots + e_n$ is an idempotent lifting $x_1 + \cdots + x_n = \bar{1}$. But then

$$e = 1 - (1 - e) \in 1 + rad\ R \subseteq U(R)$$

implies that $e = 1$. Conversely, suppose there is given a decomposition $1 = e_1 + \cdots + e_n$, where the e_i's are orthogonal local idempotents. Then we have $\bar{1} = \bar{e}_1 + \cdots + \bar{e}_n$, where the $\bar{e}_i$'s are orthogonal left irreducible idempotents of $\bar{R}$. This gives rise to

$$\bar{R} = \bar{R} \cdot \bar{e}_1 \oplus \cdots \oplus \bar{R} \cdot \bar{e}_n,$$

which shows that $\bar{R}$ is a semisimple ring. To finish the proof, we must show that any idempotent $x \in \bar{R}$ can be lifted to an idempotent of R. Comparing the above decomposition of $\bar{R}$ with $\bar{R} = \bar{R}x \oplus \bar{R}(1 - x)$, we have, after a reindexing,

$$\bar{R}x \cong \bar{R}\bar{e}_1 \oplus \cdots \oplus \bar{R}\bar{e}_i \quad \text{and}$$
$$\bar{R}(1 - x) \cong \bar{R}\bar{e}_{i+1} \oplus \cdots \oplus \bar{R}\bar{e}_n$$

as $\bar{R}$-modules. Then we can find a unit $y \in \bar{R}$ such that

$$yxy^{-1} = \bar{e}_1 + \cdots + \bar{e}_i.$$

(This follows from Exercise 21.16, which is easy to prove here since $\bar{R}$ is a semisimple ring.) Let $y = \bar{u}$ where $u \in R$. Then $u \in U(R)$ and $u^{-1}(e_1 + \cdots + e_i)u$ is clearly an idempotent of R lifting x. QED

(23.7) Remarks. (1) By Exercise 21.17, the decomposition $1 = e_1 + \cdots + e_n$ described in the theorem is unique up to a conjugation by a unit, and a permutation of the idempotents. (2) If R is, in fact, a left artinian ring, the last two results can be seen directly without passing over to $R/rad\ R$. For, if e is a primitive idempotent, then Re is indecomposable as a left R-module, so (19.17) implies that $eRe \cong End(Re)$ is a local ring. Next, take any decomposition of R into a direct sum of indecomposable left ideals. This gives a decomposition of 1 into a sum of orthogonal primitive idempotents, and these are local by what we said above.

(23.8) Theorem. *Let M be a right module over a ring k. Then M is a finite direct sum of strongly indecomposable k-modules iff $R := End(M_k)$ is a semiperfect ring.*

Proof. First assume $M = M_1 \oplus \cdots \oplus M_n$ where each M_i is a strongly indecomposable k-module. Let $e_i \in R$ be the projection of M to M_i associated with this decomposition. Then the e_i's are orthogonal idempotents with sum

1. We check easily that

$$e_iRe_i = \{f \in R : f(M_i) \subseteq M_i \text{ and } f(M_j) = 0 \ \forall j \neq i\}.$$

Therefore, $e_iRe_i \cong End(M_i)_k$. Since these are local rings, the e_i's are local idempotents, as desired. Conversely, suppose R is semiperfect. Then there is a decomposition $1 = e_1 + \cdots + e_n$ as in (23.6). Writing $M_i = e_i(M)$, we have a direct sum decomposition

$$M = M_1 \oplus \cdots \oplus M_n,$$

and, as above, $End(M_i)_k \cong e_iRe_i$. These are local rings by the assumption on the e_i's, so the M_i's are all strongly indecomposable. QED

Note that (23.8) describes quite generally how semiperfect rings can arise. For, given *any* semiperfect ring R, we have a decomposition $1 = e_1 + \cdots + e_n$ as in (23.6). If we take $k = R$ and take M to be the right regular module R_R, then $M = e_1R \oplus \cdots \oplus e_nR$ is a direct sum of strongly indecomposable k-modules, and R is isomorphic to the full endomorphism ring $End(R_R) = End(M_k)$.

(23.9) Corollary. *If k is a semiperfect ring, then so is $\mathbb{M}_m(k)$.*

Proof. We think of $R := \mathbb{M}_m(k)$ as $End(k^m)_k$. Now k_k is a direct sum of strongly indecomposable right k-modules, so the same holds for $(k^m)_k$. We are done by applying the theorem above to $M = (k^m)_k$. QED

The following result determines explicitly the structure of a subclass of semiperfect rings. It may be viewed as an extension of the classical Wedderburn–Artin Theorem.

(23.10) Theorem. *For a ring R, the following are equivalent:*

(1) *R is semiperfect, and $R/rad\ R$ is simple.*

(2) *$R \cong \mathbb{M}_n(k)$ for some local ring k.*

If (1) and (2) hold, then n is uniquely determined, and the local ring k is unique up to an isomorphism. Moreover, R is indecomposable as a ring.

Proof. If (2) holds, then (1) follows from what we did in (23.2), and since

$$R/rad\ R \cong \mathbb{M}_n(k/rad\ k)$$

is indecomposable, so is R. Now assume (1) holds, and decompose 1 into $e_1 + \cdots + e_n$ as in (23.6). The images $\bar{e}_i$ of e_i in $\bar{R} = R/rad\ R$ remain primitive and orthogonal, and we have

$$\bar{R} = \bar{e}_1\bar{R} \oplus \cdots \oplus \bar{e}_n\bar{R}.$$

Since $\bar{R}$ is a simple artinian ring, the simple right $\bar{R}$-modules $\bar{e}_i\bar{R}$ are all isomorphic, and therefore the e_iR's are isomorphic by (21.21). Writing $M = e_1R$, we have $R_R \cong M \oplus \cdots \oplus M$ and hence

$$R \cong End(R_R) \cong End(M_R^n) \cong \mathbb{M}_n(k),$$

where $k := End(M_R) \cong e_1Re_1$ is a local ring. For the uniqueness statement, simply note that if $R \cong \mathbb{M}_n(k)$ for a local ring k, then n is precisely the number of indecomposable summands in a Krull–Schmidt decomposition of R_R, and, since all the summands are isomorphic, k is isomorphic to the endomorphism ring of any of them. QED

If R is a general semiperfect ring, $R/rad\ R$ is a finite direct product of artinian simple rings. Unfortunately, the centrally primitive idempotents of $R/rad\ R$ arising from this decomposition may not lift to central idempotents in R. Thus, the result in (23.10) has no immediate extension to the general case. However, in the commutative case, there will be no difficulties of this sort, and we can prove the following definitive result giving a complete description of the class of commutative semiperfect rings.

(23.11) Theorem. *A commutative ring R is semiperfect iff it is a finite direct product of (commutative) local rings.*

Proof. Any local ring is semiperfect, so a finite direct product of local rings is also semiperfect. Conversely, if R is commutative and semiperfect, decompose 1 into $e_1 + \cdots + e_n$ as in (23.6). Then $R = e_1R \oplus \cdots \oplus e_nR$, and each $e_iR = e_iRe_i$ is a local ring. Therefore, R is a finite direct product of local rings. QED

The next result is well-known in commutative algebra. It fits in very naturally with our discussion of semiperfect rings, so we may as well give a proof for it as a small application of the general theory.

(23.12) Corollary (Akizuki, Cohen). *The following are equivalent for any commutative ring R:*

(1) *R is artinian.*

(2) *R is a finite direct product of artinian local rings;*

(3) *R is noetherian, with Krull dimension 0 (i.e., all prime ideals of R are maximal ideals).*

Proof. (1) $\Leftrightarrow$ (2) follows from the theorem above.* (1) $\Rightarrow$ (3) follows from the Hopkins–Levitzki Theorem (actually, the commutative case was done a few years earlier by Akizuki), and the fact that artinian integral domains are

---

*Alternatively, (2) $\Rightarrow$ (1) is trivial, and (1) $\Rightarrow$ (2) can be deduced from (19.19) and (19.20).

fields. Finally, let us prove (3) $\Rightarrow$ (1). Using the fact that R is noetherian, it is easy to see that, if $R \neq 0$, any ideal of R contains a finite product of prime ideals. In particular, $0 = \mathfrak{m}_1 \cdots \mathfrak{m}_n$ where $\mathfrak{m}_i$ are suitable prime (and hence maximal) ideals. Thus, we have a filtration

$$0 = \mathfrak{m}_1 \cdots \mathfrak{m}_n \subseteq \mathfrak{m}_1 \cdots \mathfrak{m}_{n-1} \subseteq \cdots \subseteq \mathfrak{m}_1 \subseteq R.$$

Each quotient $\mathfrak{m}_1 \cdots \mathfrak{m}_i / \mathfrak{m}_1 \cdots \mathfrak{m}_{i+1}$ is a finitely generated module over $R/\mathfrak{m}_{i+1}$, and therefore has an R-composition series. It follows that R_R has an R-composition series, so R is artinian. QED

Remark. Actually, the equivalence (1) $\Leftrightarrow$ (3) above does have a noncommutative analogue: *A ring R is right artinian iff R is right noetherian and for any prime ideal $\mathfrak{p} \subset R$, $R/\mathfrak{p}$ is simple artinian.* The same proof works in the noncommutative context, if we just recall Exercise 10.4.

Our next goal is to introduce the notion of left and right perfect rings. This depends on a new notion of nilpotency called T-nilpotency, where the letter "T" apparently stands for "transfinite." Its definition is a rather intriguing one.

(23.13) Definition. A subset A of a ring R is called *left* (resp., *right*) *T-nilpotent* if, for any sequence of elements $\{a_1, a_2, a_3, \ldots\} \subseteq A$, there exists an integer $n \geq 1$ such that $a_1 a_2 \cdots a_n = 0$ (resp., $a_n \cdots a_2 a_1 = 0$).

We shall use this notion of T-nilpotency mainly for one-sided ideals J. Note that if J is left or right T-nilpotent, then by applying the above definition to $\{a, a, \ldots\}$, we see that each $a \in J$ is nilpotent, i.e., J is nil. On the other hand, if J is nilpotent, then the condition in (23.13) is fulfilled by picking n to be the index of nilpotency of J, independently of the choice of the sequence $\{a_1, a_2, \ldots\}$ in J. Therefore, we have

(23.14) nilpotent $\Longrightarrow$ left (resp., right) T-nilpotent $\Longrightarrow$ nil

for one-sided ideals. Moreover, we have the following:

(23.15) Proposition. *Let J be a 1-sided ideal of R. If J is right T-nilpotent, then $J \subseteq Nil_* R$ (the lower nilradical of R). In particular, J is locally nilpotent.*

Proof. It suffices to show (after quotienting out $Nil_* R$) that in a semiprime ring R, any 1-sided right T-nilpotent ideal J is zero. Suppose J contains a nonzero element a. Then, since R is semiprime, there exist elements $x_1, x_2, \ldots$ of R such that

$$ax_1a \neq 0, \quad ax_2ax_1a \neq 0, \quad ax_3ax_2ax_1a \neq 0, \quad \ldots, \quad \text{etc.}$$

In the case when J is a right ideal, take $y_0 = a$, $y_1 = ax_1$, $y_2 = ax_2, \ldots$, which all lie in J, and we have

$$y_n \cdots y_1 y_0 \neq 0 \quad \text{for all } n,$$

a contradiction. In the case when J is a left ideal, take $z_1 = x_1 a$, $z_2 = x_2 a, \ldots$, which all lie in J, and we have $az_n \cdots z_2 z_1 \neq 0$ so

$$z_n \cdots z_2 z_1 \neq 0 \quad \text{for all } n,$$

a contradiction.[1] The last statement of the Proposition (for general rings R) follows from the fact that $Nil_* R$ lies in the Levitzki radical of R (cf. (10.32)). QED

Next we shall give two characterizations for the right T-nilpotency of a right ideal $J \subseteq R$. One characterization is in terms of right R-modules, and the other is in terms of left R-modules. The equivalence (1) $\Leftrightarrow$ (2) below is reminiscent of Nakayama's Lemma (4.22). However, there is no finite generation assumption on the modules in the conditions (2) and (2)$'$ below. Therefore, we may view (1) $\Leftrightarrow$ (2) in the following as a general Nakayama Lemma for arbitrary right modules.

(23.16) Theorem. *For any right ideal $J \subseteq R$, the following are equivalent:*

(1) *J is right T-nilpotent.*

(2) *For any right R-module M, $MJ = M \Longrightarrow M = 0$.*

(2)$'$ *For right R-modules $N \subseteq M$, $MJ + N = M \Longrightarrow N = M$.*

(3) *For any left R-module N, $ann_N(J) = 0 \Longrightarrow N = 0$. (Here, $ann_N(J)$ denotes $\{x \in N\colon Jx = 0\}$.)*

Proof. (1) $\Rightarrow$ (3) Assume $ann_N(J) = 0 \neq N$. Fix a nonzero element $x \in N$. Then $a_1 x \neq 0$ for some $a_1 \in J$, and $a_2 a_1 x \neq 0$ for some $a_2 \in J, \ldots$, etc. This gives a sequence $\{a_1, a_2, \ldots\} \subseteq J$ with $a_n \cdots a_2 a_1 \neq 0$ for all n, so J is not right T-nilpotent.

(3) $\Rightarrow$ (2) Assume $M_R \neq 0$; then $A := ann(M)$ is an ideal $\subsetneq R$. Viewing $N := R/A$ as a left R-module, $ann_N(J)$ is given by B/A where

$$B = \{b \in R\colon \ Jb \subseteq A\}.$$

Since $N \neq 0$, (3) gives $B/A \neq 0$, i.e., $A \subsetneq B$. Therefore, $MB \neq 0$. On the other hand, $JB \subseteq A$ implies that $MJB \subseteq MA = 0$, so clearly $MJ \neq M$.

(2) $\Leftrightarrow$ (2)$'$ is clear.

(2) $\Rightarrow$ (1) Consider any $\{a_1, a_2, \ldots\} \subseteq J$. Take $F = \bigoplus_{i=0}^{\infty} e_i R$ with R-basis $\{e_i\colon i \geq 0\}$, and let $M = F/S$, where S is generated as an R-submodule of F by

$$e_0 - e_1 a_1, \quad e_1 - e_2 a_2, \quad e_2 - e_3 a_3, \ \ldots \ , \text{etc.}$$

---

[1] The assumption that J is a 1-sided ideal turns out to be not really essential. A more general conclusion is given in Exercise 1 below.

Then

$$\bar{e}_i = \bar{e}_{i+1}a_{i+1} \in M \Longrightarrow MJ = M \Longrightarrow M = 0$$

by (2). This means that $S = F$, and in particular, there is an expression

$$e_0 = (e_0 - e_1a_1)b_1 + \cdots + (e_{n-1} - e_na_n)b_n$$

for suitable $b_i \in R$. Comparing coefficients of e_i, we have

$$b_1 = 1, \quad b_2 = a_1b_1 = a_1, \quad b_3 = a_2b_2 = a_2a_1, \quad \ldots,$$
$$b_n = a_{n-1}b_{n-1} = a_{n-1}\cdots a_2a_1,$$

and finally

$$0 = a_nb_n = a_na_{n-1}\cdots a_2a_1,$$

proving (1). QED

Using (1) $\Rightarrow$ (2), and repeating the proof of (19.27) without using any finite generation assumptions, we have the following consequence of the Theorem.

(23.17) Corollary. *Let J be any right T-nilpotent right ideal of R. Then, for any two projective right R-modules P and Q,*

$$P/PJ \cong Q/QJ \Longrightarrow P \cong Q.$$

We can now define the notion of left and right perfect rings.

(23.18) Definition. A ring R is called *right* (resp., *left*) *perfect* if $R/rad\ R$ is semisimple and $rad\ R$ is right (resp., left) T-nilpotent. If R is both left and right perfect, we call R a *perfect ring*.

(23.19) Corollary. *A semiprimary ring is always perfect.* (*In particular, any* 1-*sided artinian ring is perfect.*) *On the other hand, any* 1-*sided perfect ring is semiperfect.*

Proof. If $rad\ R$ is nilpotent (as in the case for semiprimary rings), it is certainly left and right T-nilpotent. On the other hand, if $rad\ R$ is left or right T-nilpotent, then it is nil, and idempotents in $R/rad\ R$ can be lifted to R by (21.28). QED

The following result offers various other characterizations for right perfect rings. As it turns out, while 1-sided artinian rings are right perfect, the right perfect rings can in turn be characterized by certain kinds of descending chain conditions. (Note the switch from "right" to "left" in the conditions (2) and (3) below!)

(23.20) Theorem (Bass). *For any ring R, the following are equivalent:*

(1) *R is right perfect.*

(2) *R satisfies DCC on principal left ideals.*

(3) *Any left R-module N satisfies DCC on cyclic submodules.*

(4) *R does not contain an infinite orthogonal set of nonzero idempotents, and any nonzero left R-module N contains a simple submodule.*

Proof. (1) $\Rightarrow$ (2) is rather deep. We shall postpone its proof to the next section. Here we shall prove (2) $\Rightarrow$ (3) $\Rightarrow$ (4) $\Rightarrow$ (1).

(2) $\Rightarrow$ (3). Any descending chain of cyclic submodules in ${}_RN$ can clearly be expressed in the form:

$$(*) \qquad Rx \supseteq Ra_1x \supseteq Ra_2a_1x \supseteq \cdots.$$

Since $R \supseteq Ra_1 \supseteq Ra_2a_1 \supseteq \cdots$ becomes stationary by (2), so does $(*)$.

(3) $\Rightarrow$ (4). By (3), ${}_RN \neq 0$ contains some Rx which is minimal as a nonzero cyclic submodule. Clearly, Rx must be already a simple submodule. Assume, for the moment, that R contains an infinite orthogonal set of nonzero idempotents $\{e_1, e_2, \ldots\}$. Then each $1 - e_1 - e_2 - \cdots - e_n$ is an idempotent, and we have

$$R(1-e_1) \supseteq R(1-e_1-e_2) \supseteq R(1-e_1-e_2-e_3) \supseteq \cdots$$

since $(1-e_{n+1})(1-e_1-\cdots-e_n) = 1-e_1-\cdots-e_{n+1}$. We'll get the desired contradiction (to (3)) if we can show that the inclusions above are all strict. But if

$$1-e_1-\cdots-e_n = r(1-e_1-\cdots-e_{n+1})$$

for some $r \in R$, then right multiplication by e_{n+1} gives

$$e_{n+1} = r(e_{n+1} - e_{n+1}^2) = 0,$$

which is not the case.

(4) $\Rightarrow$ (1). Let $N \neq 0$ be any left R-module. Then there exists a simple submodule $N_0 \subseteq N$, so $ann_N(rad\ R) \neq 0$ since it contains N_0. Applying the criterion (3) in (23.16), we see that $rad\ R$ is right T-nilpotent. Letting $S = R/rad\ R$, it remains only to show that *S is semisimple as a left R-module*. We have the following information on S: (a) *Any nonzero R-submodule of ${}_RS$ contains a simple submodule, and* (b) *any simple R-submodule of ${}_RS$ is a direct summand.* ((b) follows from the fact that S is semiprime; see (10.23).) If ${}_RS$ was not semisimple, then by the argument used in the proof of (4.14), there would exist R-module decompositions

$$S = A_1 \oplus B_1, \quad B_1 = A_2 \oplus B_2, \quad \ldots, \quad \text{etc.},$$

where the A_i's are simple left R-modules. From these decompositions, we can easily construct an infinite orthogonal set of nonzero idempotents in S. Since $rad\ R$ is nil, this would lift to an infinite orthogonal set of nonzero idempotents in R (by (21.25), (21.28)), contradicting the assumption in (4).

QED

At this point, we should mention some further related results of interest. J.E. Björk has proved in general that, for any left module N over any ring, DCC on cyclic submodules of N is equivalent to DCC on finitely generated submodules of N (see Exercise 3). Therefore, two more conditions can be added to (23.20) as characterizations for right perfect rings, namely:

(5) *Any left R-module satisfies DCC on finitely generated submodules.*

(6) *R satisfies DCC on its finitely generated left ideals.*

We should also mention that D. Jonah has added yet another characterization to this list. Remarkably, this last characterization is in terms of a restricted *ascending* chain condition, and also we come back full circle to *right* modules:

(7) *Every right R-module satisfies ACC on its cyclic submodules.*

However, these results are more technical, and we shall not go into the details here. By applying (23.20) to right artinian rings, we do have the following more modest result.

(23.21) Corollary. *If R satisfies DCC on right ideals, then it satisfies DCC on principal left ideals.*

A word on terminology. In spite of the switch from "right" to "left" from (1) in (23.20) to (2) and (3), the name "right perfect ring" for R is nevertheless the correct choice. In fact, as we shall see in the next section, the right perfect condition on a ring is characterized by several other striking homological conditions on the category of *right* R-modules.

(23.22) Example (Bass). *A right perfect ring R need not be left perfect.* In fact, let k be any field, and let J be the set of infinite matrices over k with only finitely many nonzero entries, all occurring *above* the diagonal. Let $R = k \cdot 1 + J$, where 1 here denotes the infinite matrix $diag(1, 1, \ldots)$. We check easily that R is a ring with J as an ideal, and $R/J \cong k$. *We claim that J is right T-nilpotent.* To show this, let us consider any sequence

$$\{a_1, a_2, \ldots\} \subseteq J.$$

We think of J as operating on the left of $e_1k \oplus e_2k \oplus \cdots$. From the definition of J, a_1 acts as zero on $e_{n+1}, e_{n+2}, \ldots$ for some n. Writing $V_i = e_1k \oplus \cdots \oplus e_ik$, we have

$$a_1(V_n) \subseteq V_{n-1}, \quad a_2a_1(V_n) \subseteq a_2(V_{n-1}) \subseteq V_{n-2}, \quad \ldots ,$$

so eventually $a_n \cdots a_2 a_1(V_n) = 0$, and hence $a_n \cdots a_2 a_1 = 0$ in R. This proves our claim, and shows that $J \subseteq rad\ R$. Since $R/J \cong k$, we have $J = rad\ R$ and so R is a local, right perfect ring. However, J is not left T-nilpotent. For, letting a_i be the (infinite) matrix unit $E_{i,i+1} \in J$, we have $a_1 a_2 = E_{1,3}$, $a_1 a_2 a_3 = E_{1,4}$, etc., so no matter how big we choose n, $a_1 a_2 \cdots a_n$ is never zero. Therefore, R is not a left perfect ring!

Next we shall prove the following analogue of our earlier result (23.10) on semiperfect rings.

(23.23) Theorem. *For a ring R, the following are equivalent:*

(1) *R is right perfect, and $R/rad\ R$ is simple.*

(2) *$R \cong \mathbb{M}_n(k)$ for some local ring k whose maximal ideal is right T-nilpotent.*

Proof. (2) $\Rightarrow$ (1). For R as in (2), $R/rad\ R$ is semisimple (in fact simple artinian) as in (23.2). However, it is not clear by using Definition (23.13) that $rad\ R \cong \mathbb{M}_n(rad\ k)$ is right T-nilpotent. Therefore, we shall appeal to the criterion (3) in (23.16) instead. For any left k-module A, the column space A^n is a left $\mathbb{M}_n(k)$-module in a natural way, with the module action given by matrix multiplication. It is not difficult to see that any left $\mathbb{M}_n(k)$-module N is isomorphic to some A^n (namely, take $A = E_{11}N$: see Jacobson's "Basic Algebra II", p. 31). Assume that $N \neq 0$; then $A \neq 0$. Since $rad\ k$ is right T-nilpotent, $rad\ k$ annihilates some nonzero $a \in A$ by (23.16). But then obviously $rad\ R \cong \mathbb{M}_n(rad\ k)$ annihilates the column vector $(a, \ldots, a)^t$. By (23.16) again, $rad\ R$ is right T-nilpotent.

(1) $\Rightarrow$ (2). By (23.10), (1) implies that $R \cong \mathbb{M}_n(k)$ where $n \geq 1$ and k is some local ring. This time, $rad\ R \cong \mathbb{M}_n(rad\ k)$ is right T-nilpotent, and we see easily that $rad\ k$ itself is right T-nilpotent. QED

In the commutative case, of course left T-nilpotency and right T-nilpotency are equivalent. Here we have the following analogue of (23.11) for commutative perfect rings.

(23.24) Theorem. *A commutative ring is perfect iff it is a finite direct product of (commutative) local rings each of which has a T-nilpotent maximal ideal.*

Proof. Since perfect rings are semiperfect, this follows easily from (23.11).
QED

Exercises for §23

Ex. 23.1. Show that any left T-nilpotent set J is locally nilpotent. (**Hint.** The proof involves a "König Tree Lemma" type of argument.)

Ex. 23.2. Let $K \supseteq k$ be a field extension, and let A be a k-algebra such that $rad\ A$ is right T-nilpotent. Show that $A \cap rad(A^K) = rad\ A$. (**Hint.** Use Exercise 1 above to show that $(rad\ A) \otimes_k K$ is a nil ideal.)

Ex. 23.3. (Björk) For any left module M over a ring R, show that if the cyclic submodules of M satisfy DCC, then the f.g. (= finitely generated) submodules of M also satisfy DCC. (**Hint.** Using Zorn's Lemma, show that there exists an $M_0 \subseteq M$ maximal among submodules of M whose finitely generated submodules satisfy DCC. If $M_0 \neq M$, show that M_0 is contained in a submodule M_1 of M such that M_1/M_0 is simple. Now get a contradiction by showing that the finitely generated submodules of M_1 satisfy DCC.)

Ex. 23.4. Using Exercise 3, show that a ring R is right perfect iff the finitely generated submodules of any left R-module satisfy DCC.

Ex. 23.5. (Dischinger)[1] Let R be a ring in which any descending chain $aR \supseteq a^2R \supseteq \cdots$ $(\forall a \in R)$ stabilizes. Show that any descending chain $Ra \supseteq Ra^2 \supseteq \cdots$ $(\forall a \in R)$ also stabilizes. (Such a ring R is known as a *strongly π-regular* ring.)

Ex. 23.6. Recall that an element $b \in R$ is called (von Neumann) *regular* if $b \in bRb$.
(1) (Azumaya) Let R be any strongly π-regular ring, and let $a \in R$. Show that there exists an element $r \in R$ commuting with a such that $a^n = a^{n+1}r$ for some $n \geq 1$. From this, deduce that a^n is regular.
(2) (Kaplansky) Let R be an algebra over a field k, and $a \in R$ be algebraic over k. Show that $aR \supseteq a^2R \supseteq \cdots$ stabilizes, and that a^n is regular for some $n \geq 1$. From this, deduce that any algebraic k-algebra is strongly π-regular.

§24. Homological Characterizations of Perfect and Semiperfect Rings

Perfect and semiperfect rings were introduced by H. Bass around 1960 as "homological generalizations" of semiprimary rings. In this section, we shall develop the homological characterizations of perfect and semiperfect rings discovered by Bass. We begin by studying the notion of small submodules.

(24.1) Definition. Let M be a (right) module over a ring R. A submodule $S \subseteq M$ is said to be *small* (or *superfluous*) if, for any submodule $N \subseteq M$, $S + N = M \Rightarrow N = M$. If S is a small submodule of M, we shall write $S \subseteq_s M$.

(24.2) Examples.

(1) A nonzero direct summand of M is never small. In particular, if M is semisimple, the only small submodule is the zero submodule.

---

[1] Warning: this is an extremely difficult exercise!

(2) Let $M = M_R$ and $J \subseteq rad\ R$ be a right ideal. If either M is finitely generated or J is right T-nilpotent, then $M \cdot J \subseteq_s M$. This follows from Nakayama's Lemma; see (4.22) and (23.16), respectively.

(3) If S is small in M, so is every submodule of S. If S_i $(1 \le i \le n)$ are small in M, so is $\sum S_i$.

(4) If $S \subseteq_s M' \subseteq M$, then $S \subseteq_s M$. In fact, let $S + N = M$ where N is a submodule of M. Then $S + (N \cap M') = M'$, so $N \cap M' = M'$, i.e., $N \supseteq M' \supseteq S$. Then $S + N = M$ leads to $N = M$.

(5) Suppose $S_i \subseteq_s M_i$ $(1 \le i \le n)$. Then $\bigoplus S_i \subseteq_s \bigoplus M_i$. This follows from (3) and (4) above.

(6) If N is a maximal submodule of M, then N contains every $S \subseteq_s M$. (If not, then $S + N = M$ and we have $N = M$, a contradiction.)

Next we introduce a module-theoretic analogue of the radical.

(24.3) Definition. For any (right) R-module M_R, we define *rad* M to be the intersection of all the maximal submodules of M. If there are no maximal submodules in M, we define *rad* M to be M.

The notation *rad* M could lead to confusion if M itself happens to have a ring structure. But in practice we can usually tell easily from the context if *rad* M means the module-theoretic radical or the ring-theoretic radical. Of course, when M is the right regular module R_R, the two concepts coincide.

If $R \neq 0$, then maximal right ideals always exist and we have $rad\ R \neq R$. More generally, if M_R is finitely generated over R, then (by Zorn's Lemma) maximal submodules always exist and we have $rad\ M \neq M$. However, for non-finitely-generated modules M, maximal submodules may not exist, so we may have $rad\ M = M$. The easiest example is the $\mathbb{Z}$-module given by the p-primary component of $\mathbb{Q}/\mathbb{Z}$ for any prime p. Another example is given below in (24.5).

(24.4) Proposition. *Let* $M = M_R$. *Then* (1) *rad* M *is the sum of all small submodules of* M, *and* (2) $MJ \subseteq rad\ M$ *where* $J = rad\ R$. *Equality holds if* R *is a semilocal ring.*

Proof. (1) Let $T = \sum\{S\colon S \subseteq_s M\}$. Then, by Example (6) above, $T \subseteq rad\ M$. For the reverse inclusion, it suffices to show that for any $m \in rad\ M$, $m \cdot R \subseteq_s M$. Let N be any submodule of M such that $N + m \cdot R = M$. Assume, for the moment, that $m \notin N$. Then M/N is a nonzero cyclic module, so by what we said above, M/N has a maximal submodule N'/N. Then N' is a maximal submodule of M, and we must have $m \notin N'$, contradicting $m \in rad\ M$. Therefore, $m \in N$ and $M = N + m \cdot R = N$.

(2) For any maximal submodule $N \subseteq M$, M/N is simple and therefore annihilated by J. This gives $MJ \subseteq N$, and so $MJ \subseteq rad\ M$. Now assume R is semilocal. Then M/MJ is an R/J-module. Since R/J is a semisimple ring,

it is easy to check that $rad(M/MJ) = 0$. But

$$rad(M/MJ) = (rad\ M)/MJ,$$

so we must have $MJ = rad\ M$. QED

(24.5) Example. Let R be a commutative domain with quotient field $K \supsetneq R$. We claim that $rad(K_R) = K$. By (24.4)(1) it suffices to show that any submodule $\frac{a}{b}R \subseteq K$ is small, where $a, b \in R \backslash \{0\}$. After applying the R-automorphism of K_R given by multiplication by b/a, we are reduced to showing that $R \subseteq_s K_R$. Say $R + N = K$, where N is an R-submodule of K. Since $N \neq 0$, we can find a nonzero element $a \in N \cap R$. For any $r \in R \backslash \{0\}$, write $1/ra = r' + b$ where $r' \in R$, $b \in N$. Then $1/r = ar' + ba \in N$. Thus N contains all fractions s/r ($s \in R$), and hence $N = K$.

(24.6) Proposition.

(1) $M' \subseteq M \Rightarrow rad\ M' \subseteq rad\ M$.

(2) $rad\left(\bigoplus_{i \in I} M_i\right) = \bigoplus_{i \in I} rad\ M_i$.

(3) *If F_R is R-free, then $rad\ F = F \cdot rad\ R$.*

Proof. (1) $rad\ M'$ is the sum of all $S \subseteq_s M'$ by (24.4)(1). But by (24.2)(4), each such S is $\subseteq_s M$, so by (24.4)(1) again, $rad\ M' \subseteq rad\ M$.

(2) We shall prove this in the case $I = \{1, 2\}$. The argument in the general case is the same. (Or else we can use induction.) First, by (1),

$$T := rad(M_1 \oplus M_2) \supseteq rad\ M_1 \oplus rad\ M_2.$$

Let $m = (m_1, m_2) \in T$. For any maximal submodule $N \subseteq M_1$, $N \oplus M_2$ is clearly a maximal submodule of $M_1 \oplus M_2$, so $m \in N \oplus M_2$, which implies that $m_1 \in N$. Therefore we have $m_1 \in rad\ M_1$, and similarly $m_2 \in rad\ M_2$. This shows that $T \subseteq rad\ M_1 \oplus rad\ M_2$.

(3) follows easily from (2). QED

The next Theorem shows the remarkable fact that any nonzero projective module (finitely generated or otherwise) always has a maximal submodule.

(24.7) Theorem. *Let P be a nonzero projective right R-module. Then, for $J = rad\ R$, we have $rad\ P = PJ \subsetneq P$.*

Proof. Take a suitable (projective) module Q such that $P \oplus Q$ is a free R-module $F = \bigoplus_{i \in I} e_i R$. By (24.6),

$$rad\ P \oplus rad\ Q = rad\ F = FJ = PJ \oplus QJ.$$

Therefore, $rad\ P = PJ$. The more difficult job is to show that this cannot be equal to P. If P were finitely generated, this would follow from Nakayama's Lemma. However, we must find a different argument in order to get the desired result in the general case. Assume $PJ = P$ and let

$$p = \sum_{i=1}^{n} e_i r_i \in P.$$

(To simplify notation, we shall use integers for elements of the indexing set I.) Let π be the projection of $F = P \oplus Q$ onto P. Since $P = PJ \subseteq FJ$, we can write

$$\pi(e_i) = \sum_{j=1}^{m} e_j a_{ij},$$

where $a_{ij} \in J$ and m is some integer $\geq n$. Then

$$p = \pi(p) = \sum_{i=1}^{n} \pi(e_i) r_i = \sum_{j=1}^{m} e_j \left(\sum_{i=1}^{n} a_{ij} r_i \right). \tag{24.8}$$

Comparing this with $p = \sum_{j=1}^{n} e_j r_j$, we get n equations

$$\sum_{i=1}^{n} (\delta_{ij} - a_{ij}) r_i = 0, \quad \text{for } j = 1, \ldots, n.$$

Since the coefficient matrix of this linear system belongs to

$$I_n + \mathbb{M}_n(rad\ R) = I_n + rad\ \mathbb{M}_n(R) \subseteq U(\mathbb{M}_n(R)),$$

it follows that $r_1 = \cdots = r_n = 0$ and hence $p = 0$. This contradicts the fact that P is a nonzero module. QED

Next we shall define the notion of a projective cover.

(24.9) Definition. For any right R-module M, a *projective cover* of M means an epimorphism θ: $P \to M$ where P_R is a projective module, and $ker\ \theta \subset_s P$. (Sometimes we shall loosely refer to P as a projective cover of M, suppressing the role of θ.)

Note that $ker\ \theta \subseteq_s P$ amounts exactly to the condition that, for any submodule $P' \subseteq P$, $\theta(P') = M \Longrightarrow P' = P$. In general, a projective cover θ: $P \to M$ may not exist. But if it exists, then it is unique in the sense of the following Proposition.

(24.10) Proposition. *If θ: $P \to M$ is a projective cover and θ': $P' \to M$ is an epimorphism where P' is a projective R-module, then there exists a split epimorphism α: $P' \to P$ with $\theta\alpha = \theta'$. If θ': $P' \to M$ is also a projective cover of M, then the α above is an isomorphism.*

Proof. Since P' is projective, there exists an R-homomorphism α: $P' \to P$ with $\theta\alpha = \theta'$. Now $\alpha(P')$ in P maps onto M, so by the remark made after (24.9), $\alpha(P') = P$. Since P is also projective, α splits. If θ' is also a projective cover of M, then $ker\ \theta' \subseteq_s P'$ and hence $ker\ \alpha \subseteq_s P'$ (since $ker\ \alpha \subseteq ker\ \theta'$). But $ker\ \alpha$ is a direct summand of P', so by (24.2)(1), $ker\ \alpha = 0$. Therefore $\alpha: P' \to P$ is an isomorphism. QED

(24.11) Examples and Remarks.

(1) Let P_R be projective, and $J \subseteq rad\ R$ be a right ideal. If either P is finitely generated or J is right T-nilpotent, then the surjection $P \to P/PJ$ is a projective cover of P/PJ, by (24.2)(2).

(2) Let $e \in R$ be an idempotent and $J \subseteq rad\ R$ be a right ideal. Then the surjection $eR \to eR/eJ$ is a projective cover for eR/eJ. This follows by applying (1) to the cyclic projective module eR.

(3) Let θ_i: $P_i \to M_i$ be projective covers for $1 \le i \le n$. Then

$$\bigoplus \theta_i : \bigoplus P_i \longrightarrow \bigoplus M_i$$

is also a projective cover. This follows from (24.2)(5).

(4) Let $M_R \neq 0$, and let θ: $P \to M$ be a projective cover. Then θ gives a one-one correspondence between the maximal submodules of P and those of M. (This is because any maximal submodule of P contains $ker\ \theta \subseteq_s P$.) In particular, $\theta(rad\ P) = rad\ M$, and, since $rad\ P \subsetneq P$ by (24.7), we have also $rad\ M \subsetneq M$. Thus, *any nonzero module M with $rad\ M = M$ cannot admit a projective cover.*

(5) Let R be a J-semisimple ring. Then *a module M_R has a projective cover iff M is already projective.* In fact, suppose θ: $P \to M$ is a projective cover. Then $ker\ \theta \subseteq_s P$ implies that

$$ker\ \theta \subseteq rad\ P = P \cdot rad\ R = 0$$

(see (24.7)). Hence θ: $P \cong M$. It follows, for instance, that over $\mathbb{Z}$, the only modules admitting projective covers are the free abelian groups.

The next result shows why the notion of projective covers is relevant to the study of semiperfect rings.

(24.12) Proposition. *Let R be a semiperfect ring. Then every finitely generated right (resp., left) R-module M has a projective cover. If R is right perfect, then every right R-module M has a projective cover.*

Proof. Let $J = rad\ R$ and $\bar{R} = R/J$. Let $1 = e_1 + \cdots + e_n$ where $\{e_i\}$ are orthogonal local idempotents (see (23.6)). Then

$$\bar{R} = \bar{e}_1\bar{R} \oplus \cdots \oplus \bar{e}_n\bar{R},$$

and every right $\bar{R}$-simple module is isomorphic to some $\bar{e}_i\bar{R}$. Let M be any right R-module, and write

$$M/MJ \cong \bigoplus_{\alpha \in I} \bar{e}_\alpha \bar{R},$$

where I is an indexing set, and for every $\alpha \in I$, $\bar{e}_\alpha$ is one of the e_i's $(1 \le i \le n)$. Let

$$P := \bigoplus_{\alpha \in I} e_\alpha R,$$

a projective R-module. Recalling that $\bar{e}_\alpha \bar{R} \cong e_\alpha R/e_\alpha J$ and using the projectivity of P, we have the following commutative diagram

$$\begin{array}{ccccccc} 0 & \longrightarrow & MJ & \longrightarrow & M & \longrightarrow & M/MJ \cong \bigoplus \bar{e}_\alpha \bar{R} \\ & & & & \uparrow{\scriptstyle \theta} & & \uparrow{\scriptstyle \cong} \\ & & & & P = \bigoplus e_\alpha R & \longrightarrow & \bigoplus e_\alpha R/e_\alpha J \end{array}$$

for a suitable homomorphism θ. Then we have

(A) $\theta(P) + MJ = M$.

(B) $ker\ \theta \subseteq \bigoplus e_\alpha J = PJ$.

Assume that M_R is finitely generated. Then we can take the indexing set I above to be finite. By Nakayama's Lemma (4.22), (A) implies that $\theta(P) = M$. Moreover, $PJ \subseteq_s P$ by (24.2)(2), and hence (B) implies that $ker\ \theta \subseteq_s P$ by (24.2)(3). Therefore, θ: $P \to M$ is a projective cover of M. Next, let M_R be an arbitrary module, but assume R is right perfect. Then by definition J is right T-nilpotent, so far *any* R-module N_R, $NJ \subseteq_s N$ (by (24.2)(2)). Therefore, (A) above implies that $\theta(P) = M$, and since $PJ \subseteq_s P$, (B) implies that $ker\ \theta \subseteq_s P$. So, again, θ: $P \to M$ is a projective cover for M. QED

(24.13) Remark. The construction above takes an especially simple form in the case when R is a local ring. Here we would take $\{m_i\} \subseteq M$ such that $\{\bar{m}_i\}$ form an $R/rad\ R$-basis for the vector space M/MJ. Then we map $\bigoplus_i R \to M$ by sending the ith basis vector to m_i. This is onto and has a small kernel, so it gives the desired projective (in fact, free) cover.

(24.14) Corollary.

(1) *If R is semiperfect, then any finitely generated projective module P_R is isomorphic to a finite direct sum $\bigoplus e_\alpha R$.*

(2) *If R is right perfect, then any projective module P_R is isomorphic to a direct sum $\bigoplus e_\alpha R$.*

Proof. In either case, $P \to P/PJ$ is a projective cover, where $J = rad\ R$. Comparing this with the projective cover for P/PJ constructed in the Proposition, we conclude from the uniqueness of the projective cover (24.10) that $P \cong \bigoplus e_\alpha R$. QED

Before we proceed to the next theorem, we need the following lemma.

(24.15) Lemma. *Let I be an ideal in R and let $\bar{R} = R/I$. Let M be a right $\bar{R}$-module, which is, therefore, also a right R-module. If M_R has a projective cover over R, then $M_{\bar{R}}$ also has a projective cover over $\bar{R}$.*

Proof. Let θ: $P \to M_R$ be a projective cover of M over R. We claim that $\bar{\theta}$: $P/PI \to M_{\bar{R}}$ is a projective cover of M over $\bar{R}$. The kernel of $\bar{\theta}$ is $ker\ \theta/PI$. To show that this is small in P/PI, assume that

$$N/PI + ker\ \theta/PI = P/PI,$$

where N is some R-submodule of P. This implies that $N + ker\ \theta = P$, so $N = P$ since $ker\ \theta \subseteq_s P$. QED

(24.16) Theorem. *For any ring R, the following are equivalent:*

(1) *R is semiperfect.*

(2) *Every finitely generated right R-module has a projective cover.*

(3) *Every cyclic right R-module has a projective cover.*

(*In particular,* (2) *and* (3) *are also equivalent to their analogues for left modules.*)

Proof. Proposition (24.12) gives (1) $\Rightarrow$ (2), and (2) $\Rightarrow$ (3) is trivial. Therefore, we need only prove (3) $\Rightarrow$ (1). *First we try to lift an idempotent $u \in \bar{R}$, where* $\bar{R} = R/J$, $J = rad\ R$. Write

$$\bar{R} = u\bar{R} \oplus v\bar{R}$$

where v is the idempotent $\bar{1} - u \in \bar{R}$. Viewing $u\bar{R}$ and $v\bar{R}$ as (cyclic) R-modules, let θ: $P \to u\bar{R}$ and θ': $Q \to v\bar{R}$ be their respective projective covers over R (guaranteed by (3)). Then

$$\theta \oplus \theta' : \ P \oplus Q \longrightarrow u\bar{R} \oplus v\bar{R}$$

is a projective cover of $\bar{R}_R$. On the other hand, π: $R \to \bar{R}_R$ is also a projective cover, by (24.2)(2). By (24.10), there exists an isomorphism α: $P \oplus Q \to R$ such that $\pi \circ \alpha = \theta \oplus \theta'$. For convenience, let us think of α as an identification, so that we may view P, Q as right ideals of R, with $\pi(P) = u\bar{R}$, $\pi(Q) = v\bar{R}$. Let $1 = e + f$ be the decomposition of 1 into orthogonal idempotents with respect to $R = P \oplus Q$. Then

$$\bar{1} = \pi(1) = \pi(e + f) = \bar{e} + \bar{f}$$

shows that $\bar{e} = u$ and $\bar{f} = v$. *Secondly, we have to show that $\bar{R}$ is semisimple.* For this, it suffices to show that any cyclic right $\bar{R}$-module M is $\bar{R}$-projective (see (2.8)). Viewed as a (cyclic) right R-module, M has a projective cover by (3). Therefore $M_{\bar{R}}$ also has a projective cover over $\bar{R}$ by (24.15). Since $\bar{R}$ is J-semisimple, (24.11)(5) implies that M must be projective as a right $\bar{R}$-module. This completes the proof that R is semiperfect. QED

(24.17) Corollary. *If R is semiperfect, then for any ideal $I \subseteq R$, the quotient ring R/I is also semiperfect.*

Proof. A cyclic right R/I-module M may be viewed as a (cyclic) R-module. Since M_R has a projective cover over R, $M_{R/I}$ also has a projective cover over R/I, by (24.15). Therefore, by the theorem, R/I is semiperfect.

QED

(24.18) Theorem. *R is right perfect iff every right R-module has a projective cover.*

Proof. In view of the second part of (24.12), we need only prove the "if" part. By (24.16), we know already that R is semiperfect. To check that $J = rad\, R$ is right T-nilpotent, we apply the criterion (2) in (23.16). Let $M \neq 0$ be any right R-module. Since M has a projective cover, we have $MJ \subseteq rad\, M \subsetneq M$ (see (24.4)(2) and (24.11)(4)). In particular, for any right R-module M, $MJ = M \Longrightarrow M = 0$, so we are done. QED

From this theorem, it follows, as before, that:

(24.19) Corollary. *Any quotient ring of a right perfect ring is right perfect.*

Next we shall obtain some more characterizations of right perfect rings via the notion of flat modules. So first let us introduce this important homological notion.

(24.20) Definition. A right R-module M_R is called *flat* if the functor $M \otimes_R -$ is exact on the category of left R-modules; in other words, whenever $\mathscr{S}\colon 0 \to A \to B \to C \to 0$ is a short exact sequence of left R-modules, then

$$M \otimes_R \mathscr{S}\colon\ 0 \longrightarrow M \otimes_R A \longrightarrow M \otimes_R B \longrightarrow M \otimes_R C \longrightarrow 0$$

is also short exact (as a sequence of abelian groups). Since it is known that $M \otimes_R A \to M \otimes_R B \to M \otimes_R C \to 0$ above is always exact, the point about flatness is that, whenever $A \to B$ is injective, then so is $M \otimes_R A \to M \otimes_R B$.

It is easy to see that any direct sum of flat modules is flat. In particular, since R_R is flat, any free right R-module is flat. It is also easy to see that a direct summand of a flat module is flat. It follows, therefore, that any projective right R-module is flat. On the other hand, it is not hard to give examples of flat modules which are not projective. For instance, let R be a commutative ring and let $M = (S^{-1}R)_R$, where S is a multiplicatively closed set in R and $S^{-1}R$ denotes the localization of R with respect to S. The functor "$M \otimes_R -$" in this case is the localization functor "S^{-1}", which is well-known to be exact. Therefore, $(S^{-1}R)_R$ is R-flat. However, it need not be R-projective. For example, for $R = \mathbb{Z}$ and $S = \mathbb{Z}\backslash\{0\}$, $S^{-1}R = \mathbb{Q}$ is $\mathbb{Z}$-flat, but certainly not $\mathbb{Z}$-projective.

What is an example of a *nonflat* module? Let $R = \mathbb{Z}$ again and let

$M = \mathbb{Z}/2\mathbb{Z}$. Then "$M \otimes_{\mathbb{Z}} -$" is the "reducing mod 2" functor on the category of abelian groups. This is easily seen to be non-exact. For instance, let $B = \mathbb{Z}/4\mathbb{Z}$, and $A = 2\mathbb{Z}/4\mathbb{Z}$ be its unique subgroup of order 2. Since $A = 2B$, the map induced by the inclusion $A \to B$ is the zero map $A/2A \to B/2B$. However, $A/2A \neq 0$, so $A/2A \to B/2B$ is not injective. This shows that $M = \mathbb{Z}/2\mathbb{Z}$ is not $\mathbb{Z}$-flat. For another example, look at $M = \mathbb{Q}/\mathbb{Z}$ and the injection $A \to B$, where $A = \mathbb{Z}$, $B = \mathbb{Q}$. Here

$$M \otimes A = (\mathbb{Q}/\mathbb{Z}) \otimes \mathbb{Z} \cong \mathbb{Q}/\mathbb{Z} \quad \text{and} \quad M \otimes B = (\mathbb{Q}/\mathbb{Z}) \otimes \mathbb{Q} = 0,$$

so the induced map $M \otimes A \to M \otimes B$ is again not injective, which implies that $\mathbb{Q}/\mathbb{Z}$ is not $\mathbb{Z}$-flat.

As a point of information, we record the following known fact:

(24.21) Proposition. *An abelian group M is $\mathbb{Z}$-flat iff M is torsion-free.*

Since this fact will not be needed here, we shall omit its proof. In view of (24.21), one might think of flat modules as generalizations of torsion-free abelian groups.

In order to give another criterion for checking the flatness of a module, we need the following technical lemma.

(24.22) Lemma. *Let $\mathscr{E}\colon 0 \to K \to F \to M \to 0$ and $\mathscr{E}'\colon 0 \to K' \to F' \to M' \to 0$ be exact sequences of, respectively, right and left modules over a ring R.*

(1) *If F' is flat, then the exactness of $M \otimes_R \mathscr{E}'$ implies that of $\mathscr{E} \otimes_R M'$.*

(2) *If F is flat, then the exactness of $\mathscr{E} \otimes_R M'$ implies that of $M \otimes_R \mathscr{E}'$.*

Proof. By left-right symmetry, it suffices to prove (1). The hypotheses in (1) imply that, in the commutative diagram

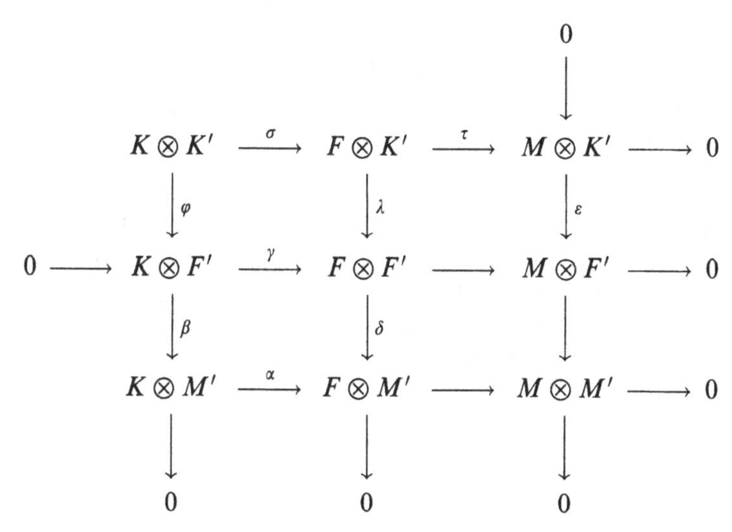

the middle row is exact, and the last column is exact. (Here, $\otimes$ means $\otimes_R$.) An easy diagram chase will show that $K \otimes M' \to F \otimes M'$ is injective (that is, $\mathscr{E} \otimes M'$ is exact). In fact, if $x \in K \otimes M'$ is such that $\alpha(x) = 0$, choose $y \in K \otimes F'$ such that $\beta(y) = x$. Then $\delta(\gamma(y)) = 0 \Longrightarrow \gamma(y) = \lambda(z)$ for some $z \in F \otimes K'$. Since ε is injective, we must have $\tau(z) = 0$ and so $z = \sigma(w)$ for some $w \in K \otimes K'$. Since γ is injective, $\varphi(w)$ must be equal to y and so $x = \beta(y) = \beta(\varphi(w)) = 0$. QED

(24.23) Theorem. *Let $\mathscr{E}: 0 \to K \to F \to M \to 0$ be an exact sequence of right R-modules, where F is R-flat. Then M is flat iff $\mathscr{E} \otimes_R M'$ is exact for every left R-module M'.*

Proof. First assume M is flat. Given any $_RM'$, write down a short exact sequence of left modules

$$\mathscr{E}' : 0 \longrightarrow K' \longrightarrow F' \longrightarrow M' \longrightarrow 0$$

where F' is R-free, in particular R-flat. Since M_R is flat, $M \otimes_R \mathscr{E}'$ is exact by definition, and therefore $\mathscr{E} \otimes_R M'$ is exact by (24.22)(1). Conversely, assume that $\mathscr{E} \otimes_R M'$ is exact for every $_RM'$. Take any exact sequence $\mathscr{E}'$ of left R-modules

$$0 \longrightarrow K' \longrightarrow F' \longrightarrow M' \longrightarrow 0.$$

Since $\mathscr{E} \otimes_R M'$ is exact and F_R is flat, (24.22)(2) implies that $M \otimes_R \mathscr{E}'$ is exact. This checks that M_R is flat. QED

Using this theorem, we can associate a certain flat right R-module M to every sequence $\{a_1, a_2, \ldots\} \subseteq R$, and thus relate the homological notions of flatness and projectivity to the behavior of the descending chain of principal left ideals $Ra_1 \supseteq Ra_2a_1 \supseteq \cdots$ in R.

(24.24) Proposition. *Let $\{a_1, a_2, \ldots\} \subseteq R$ be given. Let F be the free module $\bigoplus_{i=0}^{\infty} e_iR$, and K be its submodule generated (freely) by*

$$\{f_i = e_i - e_{i+1}a_{i+1} : i \geq 0\}.$$

Then the right R-module $M := F/K$ is flat, but M is projective only if the descending chain of principal left ideals $Ra_1 \supseteq Ra_2a_1 \supseteq \cdots$ is eventually stationary.

Proof. To see that M is flat, it suffices (by (24.23)) to check that, for any left R-module M', $K \otimes M' \to F \otimes M'$ is injective (where, again, $\otimes$ means $\otimes_R$). Note that $K = \bigoplus_{i=0}^{\infty} f_iR$, so

$$K \otimes M' = \bigoplus_{i=0}^{\infty} (f_i \otimes M'), \quad \text{and}$$

$$F \otimes M' = \bigoplus_{i=0}^{\infty} (e_i \otimes M').$$

If $\alpha = \sum_{i=0}^{\infty} f_i \otimes x_i \in K \otimes M'$ maps to zero, then

$$\begin{aligned} 0 &= (e_0 - e_1 a_1) \otimes x_0 + \cdots + (e_n - e_{n+1} a_{n+1}) \otimes x_n \\ &= e_0 \otimes x_0 + e_1 \otimes (x_1 - a_1 x_0) + \cdots \\ &\quad + e_n \otimes (x_n - a_n x_{n-1}) - e_{n+1} \otimes a_{n+1} x_n. \end{aligned}$$

Therefore, $x_0 = x_1 = \cdots = x_n = 0$ and so $\alpha = 0$. This proves the flatness of M. Now assume that M is projective. Then

$$0 \longrightarrow K \longrightarrow F \longrightarrow M \longrightarrow 0$$

splits, so there exists an R-homomorphism π: $F \to K$ splitting the inclusion map $K \to F$. Let $\pi(e_i) = \sum_j f_j b_{ij}$ ($b_{ij} \in R$, almost all zero for any given i). Then

$$f_i = \pi(f_i) = \pi(e_i - e_{i+1} a_{i+1}) = \sum_j f_j b_{ij} - \sum_j f_j b_{i+1,j} a_{i+1},$$

so we have $b_{ii} - b_{i+1,i} a_{i+1} = 1$ for all i, and $b_{ij} - b_{i+1,j} a_{i+1} = 0$ for all $i \neq j$. For sufficiently large j, we have then

$$0 = b_{0j} = b_{1j} a_1 = b_{2j} a_2 a_1 = \cdots = b_{jj} a_j \cdots a_2 a_1.$$

As a result, we have

$$\begin{aligned} a_j \cdots a_2 a_1 &= a_j \cdots a_2 a_1 - b_{jj} a_j \cdots a_2 a_1 \\ &= (1 - b_{jj}) a_j \cdots a_2 a_1 \\ &= -b_{j+1,j} a_{j+1} a_j \cdots a_2 a_1 \end{aligned}$$

for sufficiently large j's. This means that the descending chain $Ra_1 \supseteq Ra_2 a_1 \supseteq \cdots$ eventually becomes stationary. QED

We can now prove the following new characterization theorem for right perfect rings in terms of flat modules. The proof of this theorem is completely independent of the earlier characterization theorem (23.20). In particular, this will fill in the missing implication (1) $\Rightarrow$ (2) in the earlier result.

(24.25) Theorem (Bass). (Cf. (23.20)) *For any ring R, the following are equivalent*:

(1) *R is right perfect.*

(2) *R satisfies DCC on principal left ideals.*

(5) *Every flat right R-module M is projective.*

Proof. We shall show that (1) $\Rightarrow$ (5) $\Rightarrow$ (2) $\Rightarrow$ (1).

(1) $\Rightarrow$ (5). By (24.12), M_R has a projective cover θ: $P \to M$ which induces a short exact sequence

$$\mathscr{E}: \ 0 \longrightarrow K \longrightarrow P \longrightarrow M \longrightarrow 0.$$

By (24.23), $\mathscr{E} \otimes_R M'$ is exact for every ${}_RM'$. In particular, $K \otimes_R R/J \to P \otimes_R R/J$ is injective ($J := rad\ R$); that is, $K/KJ \xrightarrow{\alpha} P/PJ$ is injective. But $K \subseteq_s P$ implies that $K \subseteq rad\ P = PJ$ (by (24.4)(1) and (24.7)), so α is the zero map. This forces K/KJ to be zero, which in turn implies that $K = 0$ by (23.16). Therefore, $M \cong P$ is projective.

(5) $\Rightarrow$ (2). Consider any descending chain of principal left ideals, which we can express by $Ra_1 \supseteq Ra_2a_1 \supseteq \cdots$. As in (24.24), we can associate a flat module M to $\{a_1, a_2, \ldots\}$. Since by (5) M is necessarily projective, (24.24) implies that $Ra_1 \supseteq Ra_2a_1 \supseteq \cdots$ eventually becomes stationary.

(2) $\Rightarrow$ (1). Using the numbering of the four statements in (23.20), we have already proved there that (2) $\Rightarrow$ (3) $\Rightarrow$ (4) $\Rightarrow$ (1). Therefore, we do have (2) $\Rightarrow$ (1). However, we have promised to make the present proof independent of that of (23.20). Here is an easy, direct proof of (2) $\Rightarrow$ (1). First, let us show that $J = rad\ R$ is right T-nilpotent. Consider any sequence $\{a_1, a_2, \ldots\} \subseteq J$. Since $Ra_1 \supseteq Ra_2a_1 \supseteq \cdots$ is eventually stationary, we have

$$a_n \cdots a_1 = ra_{n+1}a_n \cdots a_1$$

for some n and some $r \in R$. Then

$$(1 - ra_{n+1})a_n \cdots a_2a_1 = 0 \Longrightarrow a_n \cdots a_2a_1 = 0,$$

since $1 - ra_{n+1}$ is a unit. Finally, we have to show the semisimplicity of $\bar{R} = R/rad\ R$. Since any descending chain of principal left ideals in $\bar{R}$ can be written in the form

$$\bar{R}\bar{x}_1 \supseteq \bar{R}\bar{x}_2\bar{x}_1 \supseteq \bar{R}\bar{x}_3\bar{x}_2\bar{x}_1 \supseteq \cdots,$$

the DCC on principal left ideals of R implies the same for $\bar{R}$. Since $\bar{R}$ is J-semisimple, this guarantees that $\bar{R}$ is semisimple, by (4.14). QED

The work above showing (1) $\Rightarrow$ (5) $\Rightarrow$ (2) $\Rightarrow$ (1) means that our proof of the implication (1) $\Rightarrow$ (2) has to go through the homological condition (5). However, a direct proof for (1) $\Rightarrow$ (2) not using projective or flat modules has been found by R. Rentschler. For a succinct exposition on Rentschler's proof, see Rowen [88], Vol. II, p. 222.

Exercises for §24

Ex. 24.1. For any finitely generated right R-module M, show that $rad\ M \subseteq_s M$.

Ex. 24.2. Let S, M be right R-modules such that $S \subseteq_s M$. Show that M has a projective cover iff M/S does.

Ex. 24.3. Show that a ring R is semiperfect iff every simple right R-module has a projective cover.

Ex. 24.4. For any projective right R-module $P \neq 0$, show that the following are equivalent:
(1) P is a projective cover of a simple R-module;
(2) P has a superfluous and maximal submodule;
(3) *rad* P is a superfluous and maximal submodule;
(4) every maximal submodule of P is superfluous;
(5) $E := End_R(P)$ is a local ring;
(6) $P \cong eR$ for some local idempotent $e \in R$.

Ex. 24.5. Show that a semiprime right perfect ring R is semisimple.

Ex. 24.6. Let R be a ring satisfying one of the following conditions:
(a) Every nonzero right R-module has a maximal submodule;
(b) Every nonzero left R-module has a simple submodule.
Show that $J = rad\ R$ is right T-nilpotent.

Ex. 24.7. For any ring R, show that the following are equivalent:
(1) R is right perfect;
(2) R is semilocal and every right R-module $M \neq 0$ has a maximal submodule;
(3) R is semilocal and every left module $N \neq 0$ has a simple submodule.

Ex. 24.8. Let R be a right perfect ring which satisfies ACC on right annihilators of ideals. Show that $J := rad\ R$ is nilpotent (so R is a semiprimary ring).

Ex. 24.9. (Hamsher, Renault) Let R be a commutative ring. Show that every nonzero R-module has a maximal submodule iff *rad* R is T-nilpotent and $R/rad\ R$ is von Neumann regular.

Ex. 24.10. (Hamsher) Let R be a commutative noetherian ring. Show that every nonzero R-module has a maximal submodule iff R is artinian.

§25. Principal Indecomposables and Basic Rings

In the first half of this section, we study the notion of the basic ring of a semiperfect ring, paving the way to the fuller study of the Morita theory of the equivalence of module categories in *Lectures*. In the second half of this section, we shall specialize to right artinian rings and give explicit examples of principal indecomposable modules and Cartan matrices.

We begin by recalling some relevant facts from §21 and §23. In the following, R shall denote a semiperfect ring, and J shall denote its Jacobson radical. We shall write $\bar{R} = R/rad\ R$, and for any $a \in R$, we write $\bar{a} = a + J$. Also, let E be the set of primitive idempotents in R. Then we have the following.

(25.1) *Any $e \in E$ is a local idempotent* (*cf.* (23.5)).

(25.2) *$\bar{e}\bar{R} \cong eR/eJ$ is a simple right module over $\bar{R}$* (*and hence over R*); *eJ is the unique maximal submodule of eR, so that* $\mathrm{rad}(eR) = eJ$ (*cf.* (21.18)).

The modules eR ($e \in E$) play a very important role in understanding the structure of the semiperfect ring R. We shall call them the *principal indecomposable* modules.[1] Note that these are cyclic, strongly indecomposable projective right modules over R.

(25.3) Theorem.

(1) *There is a natural one-one correspondence between the isomorphism types of principal indecomposable right R-modules and the isomorphism types of simple R-modules. The correspondence is given by*

$$eR \mapsto \bar{e}\bar{R} \cong eR/eJ \quad (\text{for } e \in E).$$

(2) *Let $e_1R, \ldots, e_rR$ represent a complete set of isomorphic types of principal indecomposable right R-modules. Then, for any finitely generated projective right R-module P, there exist unique integers $n_i \geq 0$ such that*

$$P \cong n_1(e_1R) \oplus \cdots \oplus n_r(e_rR).$$

(3) *For the module P in* (2), $\mathrm{End}_R(P)$ *is a semiperfect ring.*

Proof. (1) If $e, e' \in E$, then by (19.27), $\bar{e}\bar{R} \cong \bar{e}'\bar{R}$ implies that $eR \cong e'R$. Since $\bar{R}$ is semisimple, any simple $\bar{R}$-module is isomorphic to $x\bar{R}$ for some right irreducible idempotent $x \in \bar{R}$. Let e be an idempotent of R lifting x. By (21.18), e is a local idempotent; in particular, $e \in E$. Now the principal indecomposable module eR maps to $\bar{e}\bar{R} = x\bar{R}$. This establishes the one-one correspondence asserted in (1). Since the semisimple ring $\bar{R}$ has only finitely many simple right modules (up to isomorphisms), the notation in (2) is justified. For the module P in (2), consider the finitely generated $\bar{R}$-module P/PJ. There exist unique integers $n_1, \ldots, n_r$ such that

$$P/PJ \cong n_1\bar{e}_1\bar{R} \oplus \cdots \oplus n_r\bar{e}_r\bar{R} \cong \frac{n_1(e_1R) \oplus \cdots \oplus n_r(e_rR)}{[n_1(e_1R) \oplus \cdots \oplus n_r(e_rR)]J}.$$

"Lifting" this isomorphism by (19.27), we get

$$P \cong n_1(e_1R) \oplus \cdots \oplus n_r(e_rR).$$

Finally, (3) follows from (2) and (23.8) since each e_iR is a strongly indecomposable R-module. QED

---

[1] We have briefly used this terminology already for right artinian rings near the end of §19.

Now recall that there exists a decomposition $1 = e_1 + \cdots + e_n$ in R, where $\{e_1, \ldots, e_n\}$ is a set of orthogonal primitive idempotents (cf. (23.6)). Since

$$\bar{R} = \bar{e}_1 \bar{R} \oplus \cdots \oplus \bar{e}_n \bar{R},$$

every simple right $\bar{R}$-module is isomorphic to some $\bar{e}_i \bar{R}$. Therefore, we could have assumed that the $e_1, \ldots, e_r$ used in (25.3)(2) are indeed the first r idempotents in the set $\{e_1, \ldots, e_n\}$ above. The existence of the equation $1 = e_1 + \cdots + e_n$ enables us to use the results in §22 on the block decomposition of R. By these results, we have

(25.4) Theorem. *The semiperfect ring R has a unique decomposition*

$$R = c_1 R \oplus \cdots \oplus c_t R$$

where the c_i's are orthogonal centrally primitive idempotents. Two primitive idempotents, $e, e' \in E$ belong to the same block $c_i R$ iff e, e' are linked ($e \approx e'$) in the sense of §22. (Each block $c_i R$ is easily seen to be a semiperfect ring, with identity c_i.)

For instance, if R is semisimple, the blocks of R are just its simple components. And if R is a *commutative* semiperfect ring, then the blocks of R are commutative local rings, by (23.11). In the noncommutative case, however, the indecomposable semiperfect rings R have not been classified. If $R/rad\ R$ is also indecomposable, then (23.10) applies and we know that $R \cong \mathbb{M}_n(k)$ for some $n \geq 1$ and some local ring k. Unfortunately, the indecomposability of R does not imply the indecomposability of $R/rad\ R$ since, in general, central idempotents in $R/rad\ R$ may not lift to central idempotents in R.

Next we shall introduce the construction of a *basic ring* for R.

(25.5) Definition. A *basic idempotent* in a semiperfect ring R is an idempotent of the form $e = e_1 + \cdots + e_r$, where the e_i's are orthogonal primitive idempotents in R such that $e_1 R, \ldots, e_r R$ represent a complete set of isomorphism classes of the principal indecomposables. A *basic ring* of R is a ring of the form eRe, where e is a basic idempotent of R.

Note that a basic idempotent (and hence a basic ring) always exists, since it may be taken as a subsum of a decomposition of 1 into orthogonal primitive idempotents. However, if the basic idempotent e is not 1, then the associated basic ring eRe is *not* a subring of R, since its identity is $e \neq 1$, and $1 \notin eRe$.

(25.6) Proposition. *If e is a basic idempotent, then e is a full idempotent, in the sense that $ReR = R$. Moreover, the isomorphism type of the basic ring eRe is uniquely determined, and it is a semiperfect ring.*

Proof. Write $e = e_1 + \cdots + e_r$ as in (25.5). We may label the simple components $S_1, \ldots, S_r$ of $\bar{R}$ in such a way that $\bar{e}_i\bar{R} \subseteq S_i$. Then $\bar{R}\bar{e}_i\bar{R} = S_i$, and so

$$\bar{R}\bar{e}\bar{R} = \bar{R}(\bar{e}_1\bar{R} \oplus \cdots \oplus \bar{e}_r\bar{R}) = S_1 \oplus \cdots \oplus S_r = \bar{R}.$$

Lifting to R, we have $ReR + J = R$. By Nakayama's Lemma (for the cyclic right R-module R/ReR), we have $ReR = R$. To prove the second statement in the Proposition, let $e' = e_1' + \cdots + e_r'$ be another basic idempotent. Then

$$\bigoplus_{i=1}^{r} e_iR \cong \bigoplus_{i=1}^{r} e_i'R,$$

from which we have

$$eRe \cong End_R(eR) = End_R\left(\bigoplus e_iR\right) \cong End_R\left(\bigoplus e_i'R\right) \cong e'Re'.$$

By (25.3)(3), this is a semiperfect ring. QED

(25.7) Examples.

(1) Let $R = \mathbb{M}_n(k)$, where k is a local ring. Then there is (up to isomorphism) only one principal indecomposable, given, say, by eR where e is the matrix unit E_{11}. For this choice of basic idempotent, the associated basic ring is $eRe = kE_{11} \cong k$.

(2) If R is a commutative semiperfect ring, the only basic idempotent is 1, so the only basic ring for R is R itself.

(3) Let R be the semisimple ring $\prod_{i=1}^{r} \mathbb{M}_{n_i}(D_i)$, where the D_i's are division rings. Using (1), we see easily that any basic ring of R is isomorphic to $D_1 \times \cdots \times D_r$.

The following result shows that a basic ring B of a semiperfect ring R retains much of the structure of R.

(25.8) Theorem. *Let $B = eRe$ be a basic ring of R, as in* (25.5). *Then*

(1) *The map $I \mapsto I \cdot R$ defines an embedding of the lattice of right ideals of B into that of R.*

(2) *The map $I \mapsto RIR$ defines an isomorphism from the lattice of ideals of B onto that of R. This isomorphism respects the multiplication of ideals and takes rad B to rad R.*

(3) *Let E (resp., E_0) be the set of primitive idempotents of R (resp., B). Then $E_0 = E \cap B$, and for $f, f' \in E_0$, $f \cong f'$ in E_0 iff $f \cong f'$ in E, and the same holds for the idempotent relations "$\sim$" and "$\approx$" defined in* §22.

(4) *The map in* (1) *defines a one-one correspondence between isomorphism types of right principal indecomposables of B and those of R. Similarly, the map in* (2) *defines a one-one correspondence between the blocks of B and those of R.*

Proof. (1) holds for any idempotent e, as we have shown in (21.11)(1). Since e is a full idempotent by (25.6), (2) also follows from (21.11)(2).

(3) For any idempotent $f \in B$,

$$fRf = (fe)R(ef) = fBf.$$

Thus, f is primitive (equivalently, local) in B iff f is primitive in R. This shows that $E_0 = E \cap B$. Now consider $f, f' \in E_0$. If $f \cong f'$ in the ring B, we have $f = ab$ and $f' = ba$ for some $a, b \in B$ (cf. (25.20)), so clearly $f \cong f'$ in the ring R. Conversely, if $f \cong f'$ in R, (25.20) implies that $f = ab$, $f' = ba$ for some $a \in fRf'$ and $b \in f'Rf$. But then

$$a \in (fe)R(ef') = fBf' \subseteq B$$

and similarly $b \in B$, so we have $f \cong f'$ in B. Finally, suppose $f, f' \in E_0$ are such that $f \sim f'$ in R. By definition, this means that there exists $g \in E$ such that

$$fRg \neq 0 \neq f'Rg.$$

Choose a $g_0 \in E_0$ such that $g_0R \cong gR$. Then $Rg_0 \cong Rg$, so

$$0 \neq fRg \cong fRg_0 = fBg_0,$$

and similarly $0 \neq f'Bg_0$. Therefore, we have $f \sim f'$ in B. The rest of (3) now follows easily.

(4) If $f, f' \in E_0$, (3) implies that

$$fB \cong f'B \Longleftrightarrow fR \cong f'R,$$

and by the definition of a basic ring, every gR ($g \in E$) is isomorphic to g_0R for some $g_0 \in E_0$. Therefore, the principal indecomposables of B and R are in one-one correspondence via the map in (1). The block decomposition of B is the unique decomposition of B into a finite direct sum of indecomposable ideals. Therefore, under the map in (2), the blocks of B must map to the blocks of R. Alternatively, from the way the centrally primitive idempotents were constructed using the linkage relation "$\approx$", we can check directly that, for each centrally primitive idempotent c of R, ece is a centrally primitive idempotent of B. QED

(25.9) Definition. We say that a semiperfect ring R is *basic* if 1 is a basic idempotent of R, or in other words, if R is its own basic ring.

One should exercise caution so that one does not confuse a basic semiperfect ring with an indecomposable semiperfect ring. For instance, if D_i are division rings, then $D_1 \times \cdots \times D_t$ is basic, but not indecomposable if $t \geq 2$. On the other hand, $\mathbb{M}_n(D_1)$ is indecomposable, but not basic if $n \geq 2$; its basic ring is $\cong D_1$. The only general statement we can make is that R is indecomposable iff one (or all) of its basic rings is (are) indecomposable. As an example, the ring of upper triangular matrices over a division ring is both basic and indecomposable. In any case, we have the following easy criterion for basic semiperfect rings.

(25.10) Proposition. *A semiperfect ring R is basic iff $\bar{R} = R/rad\ R$ is a finite direct product of division rings.*

Proof. This follows easily from the observation that 1 is a basic idempotent in R iff $\bar{1}$ is a basic idempotent in $\bar{R}$. Since $\bar{R}$ is semisimple, the latter happens iff $\bar{R}$ is a finite direct product of division rings. QED

In *Lectures*, the significance of a basic ring B for a semiperfect ring R will become more transparent since it will be seen that B is "Morita equivalent" to R. This means that the category of right modules of R is naturally equivalent to the category of right modules of B. In fact, the basic ring B may be thought of as a "canonical" representative for the Morita equivalence class of the semiperfect ring R. We have arranged our presentation so that the basic ring construction together with the Wedderburn–Artin theory now provide the motivation for the Morita Theorems in *Lectures*. This order of presentation follows closely the sequence of historical events in ring theory, since the construction of the basic ring (for a right artinian ring) by Brauer and Osima preceded Morita's discovery of the theory of module category equivalences by a few years.

To be able to say more about the principal indecomposable modules, we need to impose stronger hypotheses on R. Therefore, we propose to specialize now to the case of right artinian rings. In this case, every finitely generated right R-module has a composition series, with a uniquely determined set of composition factors. In fact, if M_R is finitely generated over R, there is an easy criterion to determine which simple modules can occur as composition factors of M; namely, *for any primitive idempotent $e \in E$, eR/eJ* (where *$J = rad\ R$*) *is a composition factor for M iff $Me \neq 0$* (see (21.19)).

For a right artinian ring R, let $e_1R, \ldots, e_rR$ represent a complete set of isomorphism classes of principal indecomposable right R-modules, and let $V_j = e_jR/e_jJ$, so that $V_1, \ldots, V_r$ represent a complete set of isomorphism classes of simple right R-modules. Let $c_{ij} \geq 0$ be the number of composition factors of e_iR which are isomorphic to V_j. The matrix

$$C = (c_{ij}) \in \mathbb{M}_r(\mathbb{Z})$$

is called the (right) *Cartan matrix* of R. Note that the diagonal elements c_{ii} are ≥ 1, and the sum of the ith row of C is just the (composition) length of e_iR.

Of course, the right Cartan matrix C of R is only determined up to a conjugation by a permutation matrix, since we may permute the principal indecomposables $e_1R, \ldots, e_rR$ in any way. The equivalence relation "$\approx$" partitions $\{e_1, \ldots, e_r\}$ into disjoint blocks, so it is reasonable to assume, after a reindexing, that the $e_1, \ldots, e_r$ are such that the blocks appear one after another. If e_i and e_j occur in different blocks, then V_i (resp., V_j) cannot occur as a composition factor in e_jR (resp., e_iR). Therefore, the Cartan matrix takes on the form

$$diag(C_1, \ldots, C_s),$$

where the C_i's are square matrices (possibly of different sizes). Moreover, each C_i has the following property: For any two rows ρ, ρ' in C_i, there exist a sequence of rows of C_i starting with ρ and ending with ρ', such that any two "adjacent" rows have nonzero entries at some column. This implies that each C_i is "indecomposable," in the sense that no matter how we relabel the idempotents pertaining to C_i, we cannot express C_i in the form $diag(X, Y)$ for nonempty block matrices X and Y. Thus the expression $C = diag(C_1, \ldots, C_s)$ corresponds to the finest possible decomposition of C into diagonal block matrices. Moreover, it is easy to see that if

$$R = B_1 \oplus \cdots \oplus B_t$$

is the block decomposition of R, then $t = s$, and after a reindexing of the blocks, C_i is exactly the Cartan matrix of B_i $(1 \leq i \leq t)$.

For students who have been exposed to some algebraic K-theory, the following remark is in order. The Grothendieck group K_0R for finitely generated projective right R-modules is a free abelian group with $\mathbb{Z}$-basis given by the classes

$$[e_iR] \quad (1 \leq i \leq r),$$

and the Grothendieck group G_0R for finitely generated right R-modules is a free abelian group with $\mathbb{Z}$-basis given by the classes

$$[V_i] \quad (1 \leq i \leq r).$$

If c: $K_0R \to G_0R$ is the homomorphism defined by sending $[P] \in K_0R$ to $[P] \in G_0R$ ("forgetting" that P is projective), then the Cartan matrix C for R is exactly the matrix of c expressed in terms of the bases on K_0R and G_0R described above.

In the special case when $R = kG$ where k is a field and G is a finite group, the Cartan matrix C is of great importance in understanding the representation theory of G in case *char* k divides $|G|$. (If *char* $k \nmid |G|$, kG is semisimple and C is just the $t \times t$ identity matrix.) In this case, a lot more can be said about C. For instance, if the field k is large enough, it is known that C is a *symmetric* matrix, and that the determinant of C is a power of the characteristic of k. (This is a theorem of Brauer, Nesbitt and Nakayama. Note that *det* C measures the size of the cokernel of the Cartan map c: $K_0R \to G_0R$, according to what we said in the last paragraph.) However, these facts belong more properly to the domain of modular representation theory of finite groups, and it would be too much of a digression for us to present them here.

To conclude this section, let us now give some explicit examples of principal indecomposable modules and Cartan matrices. First, let R be the ring of upper triangular $n \times n$ matrices over a division ring k, so $dim_k\, R = n(n+1)/2$. We know from Example 6 of §4 that $J = rad\, R$ consists of matrices in R with a zero diagonal. Let e_i be the matrix unit E_{ii} $(1 \le i \le n)$. Then

$$e_i R e_i = k \cdot E_{ii} \cong k,$$

so the e_i's are (orthogonal) local idempotents, with $1 = e_1 + \cdots + e_n$. Also, $P_i := e_i R$ consists of matrices in R with all rows zero except possibly the ith. Since $dim(P_i)_k = n - i + 1$, and

$$R_R = P_1 \oplus \cdots \oplus P_n,$$

$\{P_1, \ldots, P_n\}$ gives a complete set of right principal indecomposables. This shows that R is basic. Noting that $P_i J = e_i J$ consists of matrices in P_i with a zero diagonal, we can identify $V_i := P_i / P_i J$ with k, with right R-action given by $b \cdot (a_{ij}) = b a_{ii}$. Then $\{V_1, \ldots, V_n\}$ gives a complete set of simple right R-modules. The reader can check easily that $e_i J \cong e_{i+1} R$ as R-modules, so that we have

(25.11)
$$\begin{array}{ccccccccc} P_1 & \supset & P_1 J & \supset & P_1 J^2 & \supset \cdots \supset & P_1 J^{n-1} & \supset & P_1 J^n = 0 \\ \| & & \wr\| & & \wr\| & & \wr\| & & \\ P_1 & & P_2 & & P_3 & \cdots & P_n & & \end{array}$$

This is then a composition series for P_1, whose composition factors are, reading from the top, $V_1, V_2, \ldots, V_n$. Here we have

$$e_1 \sim e_2 \sim \cdots \sim e_n,$$

so there is only one block, i.e., R is indecomposable (in spite of $R/rad\, R \cong k \times \cdots \times k$). The composition factors of P_i are $V_i, V_{i+1}, \ldots, V_n$, each occurring exactly once. Therefore, the (right) Cartan matrix C of R consists of ones on and above the diagonal, and zeros elsewhere. Some of the pertinent facts about the principal indecomposable modules are mentioned below, with proofs largely left to the reader.

(a) *The only submodules of* P_1 *are those appearing in* (25.11). In particular, each P_i has a *unique* composition series. (In the terminology introduced in the Appendix to §20, the P_i's are *uniserial modules.* Modules with unique composition series are, of course, the most classical source of the uniserial modules.)

(b) *Any submodule* M *of a finitely generated projective right module* P *is projective.* In fact, by (25.3)(2), M is a submodule of a finite direct sum of principal indecomposables. Consider an embedding

$$M \subseteq Q_1 \oplus \cdots \oplus Q_m$$

where the Q_i's are principal indecomposables (with possible repetitions), and m is minimal. We shall induct on m. If $m = 1$, M is projective by (a). In general, consider a nonzero projection map, say $f: M \to Q_1$. The image N of f is projective by (a), so $M \cong (\ker f) \oplus N$. Since

$$\ker f \subset Q_2 \oplus \cdots \oplus Q_m,$$

the induction proceeds.

(c) It follows from (b) that all right ideals of R are R-projective. (Any ring R with this property is called a *right hereditary ring*.)

(d) $P_1 = e_1 R$ is an ideal of R. The quotient ring $R/e_1 R$ is isomorphic to the ring of $(n-1) \times (n-1)$ upper triangular matrices. However, as a *left* R-module, $e_1 R$ is no longer indecomposable. In fact, $e_1 R = I_1 \oplus \cdots \oplus I_n$ where $I_i = kE_{1i} = RE_{1i}$, and $I_1, \ldots, I_n$ are all isomorphic as left R-modules, with $I_1 = Re_1$.

(e) Using the left principal indecomposables $P_i' = Re_i$ and the associated simple left R-modules $V_i' = P_i'/JP_i'$, the left Cartan matrix C' of R is easily seen to be the matrix with 1's on and below the diagonal, and zeros elsewhere.

(f) The ring $S = \mathbb{M}_n(k)$ contains R as a subring, so we may view S as both a right R-module and a left R-module. Decomposing S into rows, we see that $S_R \cong n \cdot P_1$ so S_R is a projective R-module. Similarly, decomposing S into columns, we see that ${}_RS \cong nP_n'$, so ${}_RS$ is also a projective R-module.

As a small variation of the above example in the case $n = 2$, consider the ring $R = \begin{pmatrix} \mathbb{Q} & \mathbb{R} \\ 0 & \mathbb{R} \end{pmatrix}$. According to (1.22), R is right artinian but not left artinian. As before, we take $e_1 = E_{11}$ and $e_2 = E_{22}$, with $e_1 R e_1 \cong \mathbb{Q}$ and $e_2 R e_2 \cong \mathbb{R}$. The right principal indecomposables P_1, P_2 are described as before: $P_1 = \begin{pmatrix} \mathbb{Q} & \mathbb{R} \\ 0 & 0 \end{pmatrix}$ and $P_2 = \begin{pmatrix} 0 & 0 \\ 0 & \mathbb{R} \end{pmatrix}$. The corresponding right simple modules are $V_1 = \mathbb{Q}$ with action $b \cdot \begin{pmatrix} a & x \\ 0 & y \end{pmatrix} = ba$, and $V_2 = \mathbb{R}$ with action $z \cdot \begin{pmatrix} a & x \\ 0 & y \end{pmatrix} = zy$, where $a, b \in \mathbb{Q}$, $x, y, z \in \mathbb{R}$. The two left simple modules V_1', V_2' are described similarly, with corresponding left principal indecomposables $P_1' = \begin{pmatrix} \mathbb{Q} & 0 \\ 0 & 0 \end{pmatrix}$ and $P_2' = \begin{pmatrix} 0 & \mathbb{R} \\ 0 & \mathbb{R} \end{pmatrix}$. Note that although R is not left artinian, it is nevertheless semiperfect, so the general theory still applies to its left structure. Here, for $J = rad\ R$, $JP_1' = 0$, so $P_1' \cong V_1'$, but ${}_RP_2'$ *does not* have a composition series, since $JP_2' = \begin{pmatrix} 0 & \mathbb{R} \\ 0 & 0 \end{pmatrix}$ satisfies neither ACC

nor DCC on left R-submodules. In a manner of speaking, JP_2' has "infinitely many" composition factors, each isomorphic to V_1'.

Coming back to artinian rings, let us consider another example:

$$R = \sum_{i=1}^{n} kE_{ii} + \sum_{j=2}^{n} kE_{1j},$$

where k is a division ring. We take $e_i = E_{ii}$ as before, with

$$e_i R e_i = kE_{ii} \cong k;$$
$$e_i R = E_{ii} k \quad \text{if } i \geq 2, \quad \text{and}$$
$$e_1 R = \sum_{j=1}^{n} E_{1j} k.$$

Here, $J = rad\ R = \sum_{j=2}^{n} E_{1j} k$ has square zero, and kills $e_i R$ for all $i \geq 2$. Thus, for $i \geq 2$, $e_i R \cong V_i$, the ith simple right R-module. For $i = 1$, direct calculation shows that

$$e_1 J = J = \bigoplus_{j=2}^{n} E_{1j} k \cong V_2 \oplus \cdots \oplus V_n;$$

this is a semisimple projective R-module. These calculations show that the right Cartan matrix of R has first row $= (1, \ldots, 1)$, and ith row $= i$th unit vector for $i \geq 2$. Here we have $e_i \sim e_1$ for all i, so again there is only one block and R is indecomposable. The principal indecomposables are mutually nonisomorphic, so R is basic as well. It is easy to see that any submodule of any $e_i R$ is projective (for $i = 1$, use the semisimplicity of $e_1 J$). Therefore, as in our earlier example, we can check that any submodule of a finitely generated projective right R-module is projective, and hence R is right hereditary.

As for the left structure of R, we have $Re_1 = kE_{11}$ and for $i \geq 2$:

$$Re_i = kE_{1i} + kE_{ii}, \quad Je_i = kE_{1i},$$

so Re_i has length 2. The left Cartan matrix of R has therefore first column consisting of 1's, and ith column $= i$th unit vector for $i \geq 2$. We leave it to the reader to check that R is also left hereditary.

Exercises for §25

Ex. 25.1. Let R be a finite-dimensional algebra over a field k, and let M be any finite-dimensional right R-module. For any idempotent $e \in R$, show that $dim_k\ Hom_R(eR, M) = dim_k\ Me$.

Ex. 25.2. In Exercise 1, assume k is a splitting field for R. Let $e \in R$ be a primitive idempotent and let $J = rad\ R$. Show that the number of composition factors of M_R isomorphic to eR/eJ is given by $dim_k\ Me$.

Ex. 25.3. Construct a basic idempotent e for the group algebra $R = \mathbb{Q}S_3$, and determine the corresponding basic ring eRe.

Ex. 25.4. Let e be a basic idempotent of a semiperfect ring R, and e' be another element of R. Show that e' is a basic idempotent for R iff $e' = u^{-1}eu$ for some $u \in U(R)$.

References

This short list contains only items explicitly referred to in the text. It is *not* a guide to the literature in ring theory. Readers interested in further reading in the subject should consult the exhaustive list of books on ring theory in the reference section of Rowen's two-volume work listed below. Those interested in the general literature in ring theory will be amply rewarded by consulting L. Small's compilation [Small: 81, 86] of reviews of ring theory papers appearing in the *Mathematical Reviews* before 1984.

S.A. Amitsur [56]: Radicals of polynomial rings, *Canad. J. Math.* **8**(1956), 355–361.

S.A. Amitsur [59]: On the semisimplicity of group algebras, *Mich. J. Math.* **6**(1959), 251–253.

H. Bass [60]: Finitistic dimension and a homological generalization of semiprimary rings, *Trans. Amer. Math. Soc.* **95**(1960), 466–488.

R. Camps and W. Dicks [93]: On semilocal rings, Israel J. Math. **81**(1993), 203–211.

P.M. Cohn [77]: *Skew Field Constructions*, London Math. Soc. Lect. Notes Series, Vol. 27, Cambridge Univ. Press, London/New York, 1977.

P.M. Cohn [95]: *Skew Fields: Theory of General Division Rings*, Encyclopedia of Mathematics, Vol. 57, Cambridge University Press, Cambridge, 1995.

J. Cozzens and C. Faith [75]: *Simple Noetherian Rings*, Cambridge Tracts in Mathematics, Cambridge Univ. Press, London/New York, 1975.

C. Curtis and I. Reiner [62]: *Representation Theory of Finite Groups and Associative Algebras*, J. Wiley-Interscience, New York, 1962.

J. Dauns [82]: *A Concrete Approach to Division Rings*, Heldermann Verlag, Berlin, 1982.

L.E. Dickson [23]: *Algebras and Their Arithmetics*, Univ. of Chicago Press, Chicago, 1923. (Reprinted by Dover Publications, 1960.)

L. Fuchs [63]: *Partially Ordered Algebraic Systems*, Pergamon Press, New York, 1963.

J.A. Eagon [67]: Finitely generated domains over Jacobson semisimple rings/are Jacobson semisimple, *Amer. Math. Monthly* **74**(1967), 1091–1092.

A. Facchini [96]: Krull–Schmidt fails for serial modules, Trans. Amer. Math. Soc. **348**(1996), 4561–4575.

A. Facchini [98]: *Modue Theory: Endomorphism Rings and Direct Decompositions in Some Classes of Modules*, Progress in Math., Vol. 167, Birkhäuser, Boston, 1998.

M. Gerstenhaber and C.T. Yang [60]: Division rings containing a real-closed field, *Duke Math. J.* **27**(1960), 461–465.

I.N. Herstein [68]: *Non-Commutative Rings*, Carus Monographs in Mathematics, Vol. 15, Math. Assoc. of America, 1968.

I.M. Isaacs [76]: *Character Theory of Finite Groups*, Academic Press, New York, 1976.

N. Jacobson [56]: *Structure of Rings*, Coll. Publ., Vol. 37, Amer. Math. Soc., Providence, R.I., 1956.

N. Jacobson [75]: *PI Algebras*, Lecture Notes in Math., Vol. 441, Springer-Verlag, Berlin-Heidelberg-New York, 1975.

N. Jacobson [89]: *Basic Algebra II*, second edition, W.H. Freeman, 1989.

D.A. Jordan [84]: Simple skew Laurent polynomial rings, *Comm. Algebra* **12**(1984), 135–137.

T.Y. Lam [73]: *The Algebraic Theory of Quadratic Forms*, W.A. Benjamin, Addison-Wesley, Reading, Mass., 1973. (Reprinted with Revisions, 1980.)

T.Y. Lam [86]: A general theory of Vandermonde matrices, *Expositiones Math.* **4**(1986), 193–215.

T.Y. Lam [95]: *Exercises in Classical Ring Theory*, Problem Books in Math., Springer-Verlag, Berlin-Heidelberg-New York, 1995.

T.Y. Lam [98]: *Lectures on Modules and Rings*, Graduate Texts in Math., Vol. 189, Springer-Verlag, Berlin-Heidelberg-New York, 1998.

T.Y. Lam [99]: Bass's work in ring theory and projective modules, in "Algebra, K-Theory, Groups, and Education, on the Occasion of Hyman Bass's 65th Birthday", (T.Y. Lam and A. Magid, Eds.), Contemp. Math., Vol. 243, pp. 83–124, Amer. Math. Soc., Providence, R.I., 1999.

T.Y. Lam [01]: Finite groups embeddable in division rings, Proc. Amer. Math. Soc., 2001.

B.H. Neumann [49]: On ordered division rings, *Trans. Amer. Math. Soc.* **66**(1949), 202–252.

M. Nagata [62]: *Local Rings*, J. Wiley-Interscience, New York, 1962.

I. Niven [41]: Equations in quaternions, *Amer. Math. Monthly* **48**(1941), 654–661.

E. Noether [29]: Hyperkomplexe Grössen und Darstellungstheorie, *Math. Zeit.* **30**(1929), 641–692.

D.S. Passman [77]: *The Algebraic Structure of Group Rings*, J. Wiley-Interscience, New York, 1977.

D.S. Passman [91]: *A Course in Ring Theory*, Wadsworth & Brooks/Cole, Pacific Grove, Calif., 1991.

R.S. Pierce [82]: *Associative Algebras*, Graduate Texts in Math., Vol. 88, Springer-Verlag, Berlin-Heidelberg-New York, 1982.

C.E. Rickart [50]: The uniqueness of norm problem in Banach algebras, *Ann. Math.* **51**(1950), 615–628.

L.H. Rowen [88]: *Ring Theory*, Vols. I, II, Academic Press, New York, 1988.

L.W. Small [81, 86]: *Reviews in Ring Theory* (as printed in *Math. Reviews*, 1940–79, 1980–84), Amer. Math. Soc., Providence, R.I., 1981, 1986.

A. Smoktunowicz [00]: Polynomial rings over nil rings need not be nil, Journal of Algebra **233** (2000), 427–436.

O. Zariski and P. Samuel [58]: *Commutative Algebra*, Vols. I, II, Van Nostrand, 1958. (Reprinted by Springer-Verlag as Graduate Texts in Math., Vols. 28, 29.)

Name Index

Subject Index

Graduate Texts in Mathematics

(continued from page ii)

66 WATERHOUSE. Introduction to Affine Group Schemes.
67 SERRE. Local Fields.
68 WEIDMANN. Linear Operators in Hilbert Spaces.
69 LANG. Cyclotomic Fields II.
70 MASSEY. Singular Homology Theory.
71 FARKAS/KRA. Riemann Surfaces. 2nd ed.
72 STILLWELL. Classical Topology and Combinatorial Group Theory. 2nd ed.
73 HUNGERFORD. Algebra.
74 DAVENPORT. Multiplicative Number Theory. 3rd ed.
75 HOCHSCHILD. Basic Theory of Algebraic Groups and Lie Algebras.
76 IITAKA. Algebraic Geometry.
77 HECKE. Lectures on the Theory of Algebraic Numbers.
78 BURRIS/SANKAPPANAVAR. A Course in Universal Algebra.
79 WALTERS. An Introduction to Ergodic Theory.
80 ROBINSON. A Course in the Theory of Groups. 2nd ed.
81 FORSTER. Lectures on Riemann Surfaces.
82 BOTT/TU. Differential Forms in Algebraic Topology.
83 WASHINGTON. Introduction to Cyclotomic Fields. 2nd ed.
84 IRELAND/ROSEN. A Classical Introduction to Modern Number Theory. 2nd ed.
85 EDWARDS. Fourier Series. Vol. II. 2nd ed.
86 VAN LINT. Introduction to Coding Theory. 2nd ed.
87 BROWN. Cohomology of Groups.
88 PIERCE. Associative Algebras.
89 LANG. Introduction to Algebraic and Abelian Functions. 2nd ed.
90 BRØNDSTED. An Introduction to Convex Polytopes.
91 BEARDON. On the Geometry of Discrete Groups.
92 DIESTEL. Sequences and Series in Banach Spaces.
93 DUBROVIN/FOMENKO/NOVIKOV. Modern Geometry—Methods and Applications. Part I. 2nd ed.
94 WARNER. Foundations of Differentiable Manifolds and Lie Groups.
95 SHIRYAEV. Probability. 2nd ed.
96 CONWAY. A Course in Functional Analysis. 2nd ed.
97 KOBLITZ. Introduction to Elliptic Curves and Modular Forms. 2nd ed.
98 BRÖCKER/TOM DIECK. Representations of Compact Lie Groups.
99 GROVE/BENSON. Finite Reflection Groups. 2nd ed.
100 BERG/CHRISTENSEN/RESSEL. Harmonic Analysis on Semigroups: Theory of Positive Definite and Related Functions.
101 EDWARDS. Galois Theory.
102 VARADARAJAN. Lie Groups, Lie Algebras and Their Representations.
103 LANG. Complex Analysis. 3rd ed.
104 DUBROVIN/FOMENKO/NOVIKOV. Modern Geometry—Methods and Applications. Part II.
105 LANG. $SL_2(\mathbf{R})$.
106 SILVERMAN. The Arithmetic of Elliptic Curves.
107 OLVER. Applications of Lie Groups to Differential Equations. 2nd ed.
108 RANGE. Holomorphic Functions and Integral Representations in Several Complex Variables.
109 LEHTO. Univalent Functions and Teichmüller Spaces.
110 LANG. Algebraic Number Theory.
111 HUSEMÖLLER. Elliptic Curves.
112 LANG. Elliptic Functions.
113 KARATZAS/SHREVE. Brownian Motion and Stochastic Calculus. 2nd ed.
114 KOBLITZ. A Course in Number Theory and Cryptography. 2nd ed.
115 BERGER/GOSTIAUX. Differential Geometry: Manifolds, Curves, and Surfaces.
116 KELLEY/SRINIVASAN. Measure and Integral. Vol. I.
117 SERRE. Algebraic Groups and Class Fields.
118 PEDERSEN. Analysis Now.
119 ROTMAN. An Introduction to Algebraic Topology.
120 ZIEMER. Weakly Differentiable Functions: Sobolev Spaces and Functions of Bounded Variation.
121 LANG. Cyclotomic Fields I and II. Combined 2nd ed.
122 REMMERT. Theory of Complex Functions. *Readings in Mathematics*
123 EBBINGHAUS/HERMES et al. Numbers. *Readings in Mathematics*
124 DUBROVIN/FOMENKO/NOVIKOV. Modern Geometry—Methods and Applications. Part III.
125 BERENSTEIN/GAY. Complex Variables: An Introduction.
126 BOREL. Linear Algebraic Groups. 2nd ed.
127 MASSEY. A Basic Course in Algebraic Topology.
128 RAUCH. Partial Differential Equations.
129 FULTON/HARRIS. Representation Theory: A First Course. *Readings in Mathematics*
130 DODSON/POSTON. Tensor Geometry.

131 LAM. A First Course in Noncommutative Rings. 2nd ed.
132 BEARDON. Iteration of Rational Functions.
133 HARRIS. Algebraic Geometry: A First Course.
134 ROMAN. Coding and Information Theory.
135 ROMAN. Advanced Linear Algebra.
136 ADKINS/WEINTRAUB. Algebra: An Approach via Module Theory.
137 AXLER/BOURDON/RAMEY. Harmonic Function Theory. 2nd ed.
138 COHEN. A Course in Computational Algebraic Number Theory.
139 BREDON. Topology and Geometry.
140 AUBIN. Optima and Equilibria. An Introduction to Nonlinear Analysis.
141 BECKER/WEISPFENNING/KREDEL. Gröbner Bases. A Computational Approach to Commutative Algebra.
142 LANG. Real and Functional Analysis. 3rd ed.
143 DOOB. Measure Theory.
144 DENNIS/FARB. Noncommutative Algebra.
145 VICK. Homology Theory. An Introduction to Algebraic Topology. 2nd ed.
146 BRIDGES. Computability: A Mathematical Sketchbook.
147 ROSENBERG. Algebraic K-Theory and Its Applications.
148 ROTMAN. An Introduction to the Theory of Groups. 4th ed.
149 RATCLIFFE. Foundations of Hyperbolic Manifolds.
150 EISENBUD. Commutative Algebra with a View Toward Algebraic Geometry.
151 SILVERMAN. Advanced Topics in the Arithmetic of Elliptic Curves.
152 ZIEGLER. Lectures on Polytopes.
153 FULTON. Algebraic Topology: A First Course.
154 BROWN/PEARCY. An Introduction to Analysis.
155 KASSEL. Quantum Groups.
156 KECHRIS. Classical Descriptive Set Theory.
157 MALLIAVIN. Integration and Probability.
158 ROMAN. Field Theory.
159 CONWAY. Functions of One Complex Variable II.
160 LANG. Differential and Riemannian Manifolds.
161 BORWEIN/ERDÉLYI. Polynomials and Polynomial Inequalities.
162 ALPERIN/BELL. Groups and Representations.
163 DIXON/MORTIMER. Permutation Groups.
164 NATHANSON. Additive Number Theory: The Classical Bases.
165 NATHANSON. Additive Number Theory: Inverse Problems and the Geometry of Sumsets.
166 SHARPE. Differential Geometry: Cartan's Generalization of Klein's Erlangen Program.
167 MORANDI. Field and Galois Theory.
168 EWALD. Combinatorial Convexity and Algebraic Geometry.
169 BHATIA. Matrix Analysis.
170 BREDON. Sheaf Theory. 2nd ed.
171 PETERSEN. Riemannian Geometry.
172 REMMERT. Classical Topics in Complex Function Theory.
173 DIESTEL. Graph Theory. 2nd ed.
174 BRIDGES. Foundations of Real and Abstract Analysis.
175 LICKORISH. An Introduction to Knot Theory.
176 LEE. Riemannian Manifolds.
177 NEWMAN. Analytic Number Theory.
178 CLARKE/LEDYAEV/STERN/WOLENSKI. Nonsmooth Analysis and Control Theory.
179 DOUGLAS. Banach Algebra Techniques in Operator Theory. 2nd ed.
180 SRIVASTAVA. A Course on Borel Sets.
181 KRESS. Numerical Analysis.
182 WALTER. Ordinary Differential Equations.
183 MEGGINSON. An Introduction to Banach Space Theory.
184 BOLLOBAS. Modern Graph Theory.
185 COX/LITTLE/O'SHEA. Using Algebraic Geometry.
186 RAMAKRISHNAN/VALENZA. Fourier Analysis on Number Fields.
187 HARRIS/MORRISON. Moduli of Curves.
188 GOLDBLATT. Lectures on the Hyperreals: An Introduction to Nonstandard Analysis.
189 LAM. Lectures on Modules and Rings.
190 ESMONDE/MURTY. Problems in Algebraic Number Theory.
191 LANG. Fundamentals of Differential Geometry.
192 HIRSCH/LACOMBE. Elements of Functional Analysis.
193 COHEN. Advanced Topics in Computational Number Theory.
194 ENGEL/NAGEL. One-Parameter Semigroups for Linear Evolution Equations.
195 NATHANSON. Elementary Methods in Number Theory.
196 OSBORNE. Basic Homological Algebra.
197 EISENBUD/HARRIS. The Geometry of Schemes.
198 ROBERT. A Course in p-adic Analysis.
199 HEDENMALM/KORENBLUM/ZHU. Theory of Bergman Spaces.
200 BAO/CHERN/SHEN. An Introduction to Riemann–Finsler Geometry.

201 HINDRY/SILVERMAN. Diophantine Geometry: An Introduction.
202 LEE. Introduction to Topological Manifolds.
203 SAGAN. The Symmetric Group: Representations, Combinatorial Algorithms, and Symmetric Functions. 2nd ed.
204 ESCOFIER. Galois Theory.
205 FÉLIX/HALPERIN/THOMAS. Rational Homotopy Theory.
206 MURTY. Problems in Analytic Number Theory.
Readings in Mathematics
207 GODSIL/ROYLE. Algebraic Graph Theory.
208 CHENEY. Analysis for Applied Mathematics.